MODELS OF DISORDER

'The whole world is in a state of chassis!'
Sean O'Casey

MODELS OF DISORDER

*The theoretical physics of homogeneously
disordered systems*

—

J. M. ZIMAN, FRS

H. O. WILLS PROFESSOR OF PHYSICS IN THE
UNIVERSITY OF BRISTOL

CAMBRIDGE UNIVERSITY PRESS

CAMBRIDGE

LONDON   NEW YORK   NEW ROCHELLE
MELBOURNE   SYDNEY

Published by the Press Syndicate of the University of Cambridge
The Pitt Building, Trumpington Street, Cambridge CB2 1RP
32 East 57th Street, New York, NY 10022, USA
296 Beaconsfield Parade, Middle Park, Melbourne 3206, Australia

First published 1979
Reprinted 1982

Printed in Great Britain
at the Alden Press, Oxford

Library of Congress cataloguing in publication data

Ziman, John M. 1925–
Models of disorder

Bibliography: p. 492   Includes index
1. Order–disorder models   I. Title
QC173.39.Z55   530   78-3630
ISBN 0 521 21784 9 hard covers
ISBN 0 521 29280 8 paperback

Contents

—

	Preface	*page ix*
1	**Cellular disorder**	
1.1	Perfect spatial order	1
1.2	Substitutional disorder	5
1.3	Magnetic disorder	6
1.4	Ice disorder	11
1.5	Short-range order	17
1.6	Long-range order	22
1.7	The range of order and ordered domains	25
1.8	Spectral disorder	32
2	**Topological disorder**	
2.1	Atomicity	36
2.2	Disordered linear chains	39
2.3	Physical realizations of one-dimensional systems	43
2.4	Dimensionality and order	47
2.5	Dislocation disorder	51
2.6	Microcrystalline disorder	56
2.7	Atomic distribution functions	58
2.8	Bond network disorder	64
2.9	Amorphous or paracrystalline?	67
2.10	Statistical geometry of bond networks	72
2.11	The Bernal model of a liquid	77
2.12	Analytical theories of the liquid state	87
2.13	Liquid mixtures	96
2.14	Liquid phases of non-spherical molecules	102
2.15	Gas-like disorder	106

3 Continuum disorder

3.1 Continuum models 108
3.2 Homogeneous random fields 110
3.3 Gaussian randomness 113
3.4 Statistical topography 118

4 The observation of disorder

4.1 Diffraction experiments and diffraction theory 122
4.2 Neutron diffraction 127
4.3 Structure determination by X-rays 129
4.4 Small-angle scattering 129
4.5 Diffraction by a mixture 133
4.6 Diffraction effects of substitutional disorder 137
4.7 Diffraction and imaging 140

5 Statistical mechanics of substitutional disorder

5.1 Physical problems and mathematical puzzles 142
5.2 Mean field approximation 144
5.3 Short-range order 147
5.4 Cluster methods 151
5.5 The Ising model in one dimension 161
5.6 The one-dimensional Heisenberg model 165
5.7 The Onsager solution of the two-dimensional Ising problem 171
5.8 Ferroelectric models in two dimensions 178
5.9 The spherical model of ferromagnetism 182
5.10 Graphical expansions 187
5.11 Order as a thermodynamic variable 197
5.12 Scaling and renormalization of critical phenomena 200

6 Thermodynamics of topological disorder

6.1 The linear gas–liquid–crystal 209
6.2 The van der Waals approximation 213
6.3 The Percus–Yevick approximation 218
6.4 Perturbation methods 220
6.5 The virial series 223
6.6 Computer simulation methods 226
6.7 Melting 232
6.8 Entropy and free volume 240

Contents

7   Macromolecular disorder

7.1   Regular solutions   246
7.2   Entropy of macromolecular solutions   248
7.3   Model chains   252
7.4   Random coils   255
7.5   Branching and gel formation   259
7.6   Rubber elasticity   265
7.7   Excluded volume   269
7.8   Random walks on a lattice   272
7.9   Continuum models   272
7.10 Entanglements   282

8   Excitations on a disordered linear chain

8.1   Dynamical, magnetic and electronic excitations   286
8.2   One-dimensional models   291
8.3   Phase-angle representation   295
8.4   Spectral gaps in disordered chains   302
8.5   The spectral density   307
8.6   Local density approximation   312
8.7   Localization of eigenfunctions   315

9   Excitations on a disordered lattice

9.1   The TBA model   321
9.2   The Green function formalism   322
9.3   Propagator and locator expansions   325
9.4   The coherent potential approximation   332
9.5   Local environment corrections to CPA   341
9.6   Spectral bounds and band tails   345
9.7   Spectral moments and continued fractions   348
9.8   Off-diagonal disorder   356
9.9   Anderson localization   358
9.10 Percolation theory   370
9.11 Maze conduction   379

10   Electrons in disordered metals   387
10.1   The NFE model   391
10.2   Screened pseudopotentials   398
10.3   Muffin-tin potentials   404
10.4   The electron spectrum   409
10.5   Many-atom scattering   413
10.6   Scattering operators   416
10.7   Partial-wave representations   421
10.8   The coherent-wave approximation   425
10.9   Cluster scattering   431
10.10 Transport theory

11   Excitations of a topologically disordered network 439
11.1   Dynamics of liquids and glasses   441
11.2   The continuum limit   447
11.3   Ideal tetrahedral coordination   454
11.4   Tree models   456
11.5   The band–gap paradox

12   Dilute and amorphous magnets   460
12.1   The dilute Ising model   464
12.2   Dilute Heisenberg magnets   467
12.3   Amorphous ferromagnets and spin glasses

13   Electrons in 'gases'   472
13.1   Gas-like disorder   475
13.2   The metal–insulator transition   477
13.3   Hopping conduction   481
13.4   Semi-classical electrons in a random potential   487
13.5   Spectral tails in a choppy random potential

       References   492

       Index   511

Preface

—

'My most liquid discoveries, as I thought, of undoubted truths,
have so oft been confuted in recent years'
R. Loveday

Condensed-matter physics has expanded in recent years and shifted its centre of interest to encompass a whole new range of materials and phenomena. Fundamental investigations on the molecular structure of liquids, on amorphous semiconductors, on polymer solutions, on magnetic phase transitions, on the electrical and optical properties of liquid metals, on the glassy state, on metal–ammonia solutions, on disordered alloys, on metallic vapours – and many other interesting systems – now constitute a significant proportion of the activity of innumerable physical and chemical laboratories around the world.

This research is not purely academic: disordered phases of condensed matter – steel and glass, earth and water, if not fire and air – are far more abundant, and of no less technological value, than the idealized single crystals that used to be the sole object of study of 'solid-state physics'. But it is too late, now, to capture the 'physics of disordered systems' within a single book. Indeed, it would be wrong in principle to try to wrench each of these diverse systems from its natural scientific setting: an amorphous semiconductor, for example, behaves much more like a crystalline semiconductor than like a liquid metal; a metal–ammonia solution is more interesting for its exotic chemical properties than as the physical realization of a highly disordered assembly containing free electrons.

But beneath the luxuriance of real materials and observable phenomena, there can be found a common stratum of concepts, hypotheses, models and mathematical deductions that are supposed to belong to a single theory. Apart from some beautiful results concerning the Ising model of magnetism and some bold, if not fully victorious, sallies into the statistical mechanics of the liquid state, this theory scarcely existed before 1960. The quantum theory of electrons in solids, for example, was almost entirely devoted to the properties of perfect or nearly perfect crystalline systems, where lattice-translational and point-group symmetries immensely simplify the mathematical analysis. The theory of disordered systems had not been entirely neglected, but its achievements were fragmentary, phenomenological and lacking in mathematical depth and rigour.

It must be admitted that fififteen years of intense scientific effort to understand disordered systems has still not uncovered a unifying mathematical principle comparable to the Bloch theorem. The theories expounded in this book do not have a doctrinal core from which every truth may be seen to descend. But in the attempt to interpret particular properties of particular materials, a variety of conceptual schemes and mathematical techniques have been formulated, explored, and evaluated. Indeed, the same technique has often been invented several times, in connection with apparently different physical problems. Sometimes, on the other hand, a theoretical discussion appropriate for one class of material has been applied naively to a system with an entirely different underlying structure. Even if the field is by no means intellectually coherent, it appears more confused than it really is because there is no general critical survey of all the relevant published work.

There is confusion, also, concerning the logical status of much of this theory. In this narrow professional role, the theoretical physicist sets up a mathematical representation of some aspect of nature and calculates its potentially observable properties. In formulating a theoretical model he draws upon experimental information; the outcome of his cerebrations and computations must eventually be tested against harsh reality. Although the intermediate steps may be strongly guided by physical intuition they are taken in the realm of theory, where mathematical proof is decisive. It is the responsibility of the theoretical physicist to keep these pathways clear – to combat proliferating phenomenalism, feverish speculation, the tedium of tautological formalism and the hallucinations induced by unrealistic hypotheses.

Our main concern must be with results that are well established and supposedly well understood. To appreciate the significance of a mathematical argument, to see where it comes from, how strong it is and where it leads, it is necessary to grasp the thread for oneself and follow it to its logical conclusion. In applied mathematics and theoretical physics the weak points are seldom manipulative or existential; they occur at the junctions between the physical model and its analytical formulation, or at the stage where, unhappily, some simplifying approximation has to be made if the argument is to arrive at any useful result. In most cases abstract rigour is not called for and standard algebraic transformations or numerical computations can be taken for granted. Nor need we be concerned with mere elaborations, such as the application of an established technique to a system with a more complex crystal structure, that require much labour to execute but little additional insight to conceive.

What I have tried to do, therefore, is to show how certain physically relevant phenomena derive from the defining characteristics of various simple theoretical model systems. To keep in touch with reality the theoretical physicist needs to know something of the actual material systems represented by these models, with some appreciation of the dangers of over-idealizing the initial hypotheses in order to break through the mathematics. In most cases these limitations are easily indicated. On the other hand, problems of interpretation and verification – whether the calculated properties of a particular theoretical model are in fact those observed in a particular experiment – cannot be discussed with brevity, clarity or genuine authority in a work of this kind. The answers to such questions are not to be found in textbooks: one must make up one's own mind in the light of every scrap of observational evidence, every critical comment, every imaginative hypothesis. All reference to the results of *experimental* research on disordered systems is excluded from this book – not because mathematical theory can subsist on its own, but because the facts are too multifarious, too contradictory, too important to be treated so lightly.

Nevertheless, even within these limits, the theory of disordered systems is a very large subject. How could it be reduced to manageable proportions? It would be presumptious, if not physically impossible, to review every topic to the full depth that an expert in that subject might prefer. As a basic selection principle, I assumed that the reader would be a reasonably well educated (i.e. post-graduate or post-doctoral) theoretical physicist wishing to get a broad view of the field, or to inform himself on matters with which he was not acquainted by his own researches. In the first place, he would need the simplest basically sound derivation of each result, set in its mathematical and/or physical context, illustrated diagrammatically, connected with analogous results and hedged in with warnings of possible limitations. From this point of vantage the more sophisticated work on the subject – alternative proofs, more rigorous formulations, elaborations, applications and generalizations – can be grasped in principle, and studied at leisure, in the primary literature.

The first task was to collect together the relevant papers on each topic. I don't pretend that I have actually seen (let alone read!) them all: in the absence of a comprehensive review article, this is where the resident expert has the drop on the rash intruder. Indeed, to avoid all pretence at bibliographic completeness, I have not even cited all the papers that I have actually consulted: reference is made only to those that seemed of permanent worth, either for their originality when published or for what may still be learnt from them. I have also taken a somewhat cautious and sceptical attitude towards many imaginative proposals that have not yet shown themselves to be well founded or fruitful. Nothing is more permanent than a valid theorem; nothing fades more rapidly than an unwarranted speculation. In thus excluding current conjectures I may have made the book less exciting, but I cling to the principle that to ask to be up to date is to admit to being obsolescent. In fact, it is difficult for me to state what 'date' it would be 'up to', since the earlier chapters were first drafted three or four years ago, and it proved beyond my powers to revise them properly in the clearest light of the most recent work! In any case, all references after 1975 are sporadic and consciously fragmentary in coverage.

But where I have really put my head on the block is in making my own selection of themes and trying to expound them in (more or less) my own words. I must surely have made many mistakes of emphasis and glossed over many subtle points of argument. On any single topic, I advise the serious reader to look elsewhere for more reliable knowledge. But although I claim neither originality nor scholarly authority for the work in detail, I think it is interesting to take an overall view of this subject and to see it as something more than a miscellany of models, theorems and techniques. Certain conceptual themes link it together, both physically and mathematically, in ways that are seldom grasped by the specialist on each branch of the subject: that is the message of the book.

The notational conventions of the various specialized fields are not consistent with one another, but I have not tried to impose a superficial unity on the subject by inventing an abstract formalism with uniquely defined symbols. For practical purposes, the dialect of each field is more useful in reading the literature than any artificial universal language. But some changes have been made to avoid local ambiguities, or to emphasize the generality and invariance of certain mathematical quantities that arise at apparently diverse points in the theory. I have also quoted, without proof or bibliographic reference, many basic theorems or foundation principles of quantum mechanics, statistical mechanics, probability theory and 'solid-state theory', assuming that these already would be well known to the reader or accessible in a variety of familiar texts. The theory of disordered systems is not, after all, an elementary scientific discipline, to be entered without long preparation.

This book was written over a period of about five years but often interrupted for weeks or months by absence abroad or other more pressing business. It is an individual work: as always, there was much profit and pleasure in the daily company of scientific colleagues at Bristol and elsewhere, but nobody else need take any responsibility for its contents. But there is more than conventional courtesy in saying that that I am grateful to my personal secretary, Lilian Murphy, for her assistance in many practical aspects of getting a typescript into the hands of the publishers.

JOHN ZIMAN

Bristol, July 1977

1

Cellular disorder

'Substitutes and shadows of things more high in substance and efficacy'
Barrow

1.1 Perfect spatial order

'Disorder' is not mere chaos: it implies defective *order*. To think about a disordered state we must have in mind an ideal of order from which it falls short. It is much easier to characterize disordered systems in terms of their deviations from this ideal than it is to define a perfectly disordered system on which some partial degree of order is to be imposed. The concept of disorder is primitive and intuitive; it belongs with statistical terms such as 'random', 'stochastic', 'unpredictable', which can only be defined within a specific context of what is already known or can be taken for granted.

The highest degree of spatial order (apart from a vacuum!) is found in a *crystal*. Physically we think of an assembly of innumerable identical atoms or molecules, uniformly packed in regular rows and planes to fill all the space available. Mathematically we refer to invariance under the translational operations of a *lattice*; the physical situation at some point $\mathbf{r}$ in space is exactly reproduced at every other point in the set

$$\mathbf{r} + \mathbf{l} \equiv \mathbf{r} + l_1 \mathbf{a}_1 + l_2 \mathbf{a}_2 + l_3 \mathbf{a}_3, \tag{1.1}$$

where $\mathbf{a}_1$, $\mathbf{a}_2$, $\mathbf{a}_3$ are three non-coplanar vectors and l_1, l_2, l_3 are integers. In most actual crystals we also refer to the existence of point group operations, such as rotations about an axis, reflections or inversions, which can transform the system into itself. The fundamental mathematical language of the physics of perfect crystals is, therefore, the theory of finite groups.

But the effect of disorder is always to break some symmetry. In the physics of disordered systems we can no longer rely upon the most powerful mathematical tool in the theory of the condensed state. Theorems and principles that we cheerfully prove and accept for crystalline materials may not be taken for granted in the study of non-crystalline systems, such as liquids or glasses, where lattice translational invariance cannot be assumed.

In some cases, however (§§ 1.2–1.5), something like a crystal lattice may still exist, more loosely defined. In a *substitutional alloy,* for example, we can no longer say that exactly the *same* type of atom is to be found at $\mathbf{r} + \mathbf{l}$ as at $\mathbf{r}$; but we may still assert that there is *an* atom, of one type or another, on each site of the lattice. Or we may still find many of the atoms at or near the sites of a perfect lattice, from which a substantial fraction have been moved or removed. In such cases, the analysis is usually greatly aided by judicious application of suitable approximations drawn from group theory.

In a *topologically disordered* material (chapter 2) such as a liquid no ghost of a crystal lattice remains. But two elementary features of the crystalline state may still be present.

Most crystals are relatively closely packed. In the language of spherical atoms we may say that each atom 'touches' as many of its neighbours as possible, within the restrictions of the basic lattice symmetry, or within the constraints of chemical bonding. One of the standard exercises of elementary metallurgy is to analyse the geometrical ratios allowable for hard spherical atoms of given radii packed together in various ways, and thus to explain the observed crystal structures of alloys and compounds.

But the spherical atoms, even in the face-centred cubic or hexagonal close-packed lattice, do not occupy the whole volume. What should be done with the interstitial regions? The most convenient assumption is to assign to each atom its own polyhedral *Wigner–Seitz cell* (fig. 1.1(*a*)). The vector joining the centre of the atom to the centres of its neighbours is bisected by a plane; the volume enclosed by all such planes is the invariant locality of that atom. In a Bravais lattice this cell is also a unit cell of the crystal lattice and has also the maximum point group symmetry of the crystal. The complete solution of some physical problem, such as self-consistency of electron–electron interaction, inside such a cell is, therefore, a long step towards solution of the problem for the whole crystal. We think of the crystal as a regular assembly of such building blocks, each as complete as possible unto itself, packed together without intervening spaces.

Many disordered systems are also as close-packed as possible, within the same geometrical or chemical constraints. The Wigner–Seitz construction now yields the *Voronoi polyhedra* (fig. 1.1(*b*)) of the system. These are no longer identical regular polyhedra; but because each contains a spherical atom it cannot deviate too markedly from a symmetrical Wigner–Seitz cell

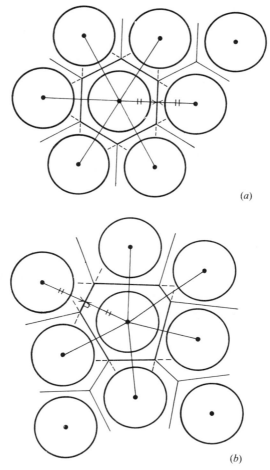

(a)

(b)

Fig. 1.1. (a) Wigner–Seitz cell of a regular lattice. (b) Voronoi polyhedron of disordered system.

of the same volume. The environment of each atom (or molecule) in the
disordered phase may not, after all, be vastly different from what it would
be in a regular crystal of the same average density.

The lattice translational invariance of a crystal is also the condition for
the validity of the Bloch theorem. In abstract language, the finite abelian
group of operations (1.1) has N one-dimensional representations, each
with character

$$\chi^{(\mathbf{k})}(l) = e^{i\mathbf{k} \cdot l}. \tag{1.2}$$

The wave vector **k** is thus a good quantum number for a Bloch state, whether this describes a lattice wave, an electron wave function or a spin wave. The values of **k** consistent with the boundary conditions on the crystal are distributed uniformly in reciprocal space. The unit cell of the direct lattice has, as its reciprocal in **k**-space, the Brillouin zone, into which the 'crystal momentum' may be reduced. All this is standard theory for crystalline materials.

Without long-range order this mathematical edifice falls in ruins. The simplest physical interpretation of the reciprocal lattice – coherent diffraction of plane waves by the Bragg planes of the crystal – no longer makes sense.

Nevertheless, some remnant of the property of translational invariance may still be observed. Close packing, as in a liquid (§ 2.11), implies a high degree of uniformity in the density of the assembly. Any compact volume of the liquid containing, say, a few hundred atoms can be exchanged for a similar volume elsewhere without serious penalties. The material is *statistically homogeneous* over distances of this order and may even be represented approximately as a continuous medium (chapter 3).

If this medium had perfect continuous translational invariance, then momentum would be a good quantum number, just as in free space. We should then derive great advantage from the *momentum representation*

$$\psi_{\mathbf{k}}(\mathbf{r}) = e^{i\mathbf{k}\cdot\mathbf{r}} \tag{1.3}$$

for quasi-independent excitations. But this makes sense only for long wavelengths: it would not be a good approximation for values of k comparable with the reciprocal of the average interatomic spacing.

Nevertheless, in the theory of homogeneously disordered systems we must clutch at straws. The vector **k** may still be a 'fairly good quantum number' whose approximate invariance may be used to improve the convergence of the algebra. To make the most of the uniformity of density of the liquid or glass it is sometimes helpful to introduce a Fourier transformation or *reciprocal space representation* of the structure where, for example (§§ 10.8, 11.2), the translation from the atom at $\mathbf{R}_i$ to the atom at $\mathbf{R}_j$ is associated with a phase factor

$$\chi^{(\mathbf{k})}(\mathbf{R}_i - \mathbf{R}_j) = \exp\{i\mathbf{k}\cdot(\mathbf{R}_i - \mathbf{R}_j)\}. \tag{1.4}$$

This representation, whose analytical properties depend in detail on the actual statistical distribution of the vectors $\mathbf{R}_i$, comes as near as possible to the Bloch representation (1.2) in a crystal without pressing the continuum metaphor too hard. But the power of this analogy, in any particular case, depends upon the degree to which the disordered system resembles an ordered crystal in spatial homogeneity.

1.2 Substitutional disorder

The weakest type of disorder is that of a *substitutional alloy*. It is often found possible to replace an atom of element A (e.g. Ag) in a perfect crystal by an atom of another element B (e.g. Au) with almost no disturbance of the crystal lattice. This phenomenon, which occurs for many different elements in metals, semiconductors and ionic crystals, is of the greatest importance in metallurgy and in other branches of materials science. When the substituted sites do not, of themselves, form a regular lattice we have an example of *substitutional disorder*.

In practice, we apply the theory of disordered systems to idealized models of alloys. Even in the case of a *dilute alloy*, the 'impurity atom' (this is a relative term!) may not be quite the same size as the atom it replaces, so that the lattice in the neighbourhood may be somewhat distorted. There may also be some effect of the substitution on the distribution of electrons in the neighbourhood; for example, an ion of Zn^{++} substituted for Cu^{+} draws a screening charge to itself by its higher valency. The calculation of such effects, even for isolated impurities, is a significant problem in the theory of solids which we ignore here. In other words, we take for granted a procedure by which a specific characteristic of atom A on a given site – its mass, elastic coupling with neighbours, electron bound states, scattering cross-section, etc. – is assigned another value when the site is supposed to be occupied by atom B, without enquiry into the source of such information. Effects due to local lattice distortion or electron screening are assumed to be included in the prescription of the substitution.

This model begs the question whether these characteristics may also depend upon the nature of the atoms on neighbouring sites: an A atom surrounded by other A atoms is not necessarily equivalent to a similar atom within an assembly of Bs. It is well known, for example, that the effective interatomic binding energy is not usually the same for an A–A bond as for an A–B or B–B configuration (cf. § 1.4). Additivity of atomic characteristics must not be assumed for a *concentrated alloy* where the probability of finding many pairs of neighbouring 'impurities' cannot be neglected. But before we can discuss the important physical effects of interactions between the different atomic species in the alloy we usually need to know the solution of the disorder problem itself. In such cases, the substitutional alloy model is only the first step in a self-consistency procedure. Perhaps it is better to talk of *cellular disorder,* emphasizing the variation of properties from cell to cell in a topologically ordered lattice, whilst avoiding the presumption of perfect physical substitution of one component by the other.

Again, the statistical distribution of substituted sites needs to be prescribed. The simplest assumption is that these are distributed at random: the probability that any given site is occupied by a *B* atom is supposed to be c_B, the atomic concentration of atoms of this species relative to the total number of atoms, of whatever type, that might occupy this site. But this assumption of statistical independence of the occupation of neighbouring sites is unrealistic because of the interaction terms in the cohesive energy. We return to this point in §§ 1.3, 1.4 and in chapter 5.

The substitutional impurity could be a point defect of the lattice, such as a vacancy. Although it is not possible, physically, to put a high concentration of vacancies at random into a crystalline solid, this has often been treated as a crude model of a fluid phase. The *hole theory of liquids* (see § 2.11) is based upon this model of a lattice gas (§ 1.5) where, of course, the interatomic forces dominate the distribution of atoms on to the sites of a hypothetical background lattice.

The statistical properties of a system with cellular disorder may often be reduced to those of an *Ising model.* In a binary alloy, for example, we define a variable σ_l which takes the value $+1$ on an *A* site and -1 on a *B* site. We then make the characteristic properties of the constituents depend on the sign of σ_l. For example, suppose that atom *A* has scattering power t_A, whilst atom *B* has the scattering power t_B. Then we ascribe to site *l* the scattering power

$$t_l = \tfrac{1}{2}(1+\sigma_l)t_A + \tfrac{1}{2}(1-\sigma_l)t_B. \tag{1.5}$$

The distribution function for σ_l over the various sites of the lattice defines all the effects due to disorder. The variable σ_l is called the Ising *spin,* for reasons that will be explained in § 1.3.

1.3 Magnetic disorder

In the Heisenberg model of a magnetic material, the *l*th site of a regular crystal carries a localized magnetic moment, proportional to a localized spin variable S_l. If this moment varies randomly from site to site, we have *magnetic disorder.*

Quite apart from its importance as a physical phenomenon in its own right, this type of disorder is of especial interest in the theory of *cooperative behaviour* (see chapter 5). In a magnetic system the energy of interaction between neighbours can seldom be neglected. True *paramagnetic disorder*, without short-range or long-range correlations (see § 1.5), can only be observed at high temperatures, where Curie's law is obeyed. Because of the ease of rotation of the individual spins we cannot freeze this type of magnetic disorder into the assembly, but must be content with the thermal equilibrium distribution at the temperature of observation. To reduce the temperature at which such correlation effects begin to appear we must work with a magnetically dilute crystal – or even with the magnetic moments of the nuclei – where the effects of magnetic disorder on other physical properties would be weak.

At low temperatures, of course, the interactions between the spins give rise to *magnetic order*, as in ferromagnetic, antiferromagnetic and ferrimagnetic crystals. As we shall see, this type of order has analogues in alloy systems. But the fact that S_l is a vector allows great freedom in the pattern of the ordering. Not only do we have simple *ferromagnetic* order, with the moments all lined up in one direction, or *antiferromagnetic order* with moments in opposite directions on two interleaved sub-lattices (fig. 1.2). We may also observe complex *helical* ordering (§ 1.6) with successive moments on sites along a line rotated by uniform angles with one another on a cone (fig. 1.3). Thermal fluctuations in such systems can then give rise to types of magnetic disorder which can only be defined relative to the corresponding ordered phases.

Paramagnetic disorder can also be considered as a type of cellular disorder with effects on other excitation modes (chapter 9). For example, a spin-polarized conduction electron travelling through the crystal will be affected by the variations of spin localized on the atoms. This can be a significant mechanism in the theory of the electrical conductivity of transition metals and may also play a special role in the metal–insulator transition in some transition-metal oxides.

Strictly speaking, the symbol S_l refers to a quantum-mechanical operator, but for many practical purposes little precision is lost by simplifying this to a classical vector in the local direction of magnetization (see § 5.1). This is especially the case where the total magnetic moment on each atom is large. Whenever the interactions between neighbours has to be included the 'quantization' does not apply to each local spin separately, so that the attempt to replace S_l by discrete values of $S_l^{(z)}$, the component of spin along some prescribed direction z, may be quite misconceived.

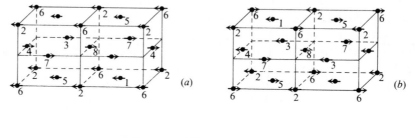

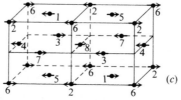

Fig. 1.2. Different types of antiferromagnetic ordering on a face-centred cubic lattice.

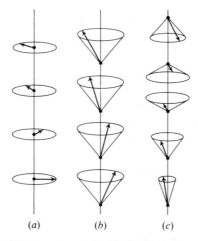

Fig. 1.3. Three different types of helical ordering. In each case, the crystal consists of a stack of hexagonal layers. The magnetic moments of all the atoms in the same layer are parallel. The arrows show the directions of magnetic moments in successive layers.

Nevertheless, the Ising model (§ 1.2), in which σ_l is allowed to have only the values ± 1, is often applied to magnetic systems, thus encouraging direct comparison with the alloy case. But the analogy (§ 1.5) must not be pushed too far. In an alloy, for example, the relative concentrations of the constituents may be given any value within the limits allowed by metallurgical solubility rules, and nearly perfect disorder can be frozen into the system by rapid quenching. In the magnetic Ising model, however, an '*A*' atom turns into a '*B*' atom by mere reversal of spin, so that 'up' spin and 'down' spin sites occur with nearly equal concentration in the paramagnetic region. This equiconcentration can only be altered by very special techniques, as in the partial polarization of nuclear spins at very low temperatures in an immense magnetic field.

At first sight, one may be tempted to regard the thermal fluctuations of local magnetization in, say, a ferromagnetic crystal as a form of cellular disorder, akin to a dilute gas of reversed spins. In this case, however, the Ising model (fig. 1.4(*c*)) is particularly misleading as a guide to the physics. The vector character of the spin variable $\mathbf{S}_l$ asserts itself: instead of complete *reversals* of spin on a few sites, we observed locally correlated *deviations* of spin orientation over substantial regions (fig. 1.4(*b*)). The excitation of nearly independent *spin waves* thus produces quite a different type of disorder which will be discussed in § 1.8. As the temperature is raised this disorder increases, with excitations of shorter and shorter wavelength. The mathematical description of the transition from this phase into the phase of paramagnetic disorder (i.e. fig. 1.4(*a*)) via a regime of *critical fluctuations* (§ 5.11) is the ultimate test of the statistical mechanics of cooperative phenomena.

From this qualitative discussion of well-known physical phenomena we begin to see the deficiencies of the Ising model even in the paramagnetic case. Ignoring quantization we really have three *continuous* variables at each site, $S_l^{(x)}, S_l^{(y)}, S_l^{(z)}$ which may be assigned random values subject to the condition

$$S^2 = S_l^{(x)2} + S_l^{(y)2} + S_l^{(z)2} \tag{1.6}$$

for each *l*. From the point of view, say, of a conduction electron interacting with such objects, this is not at all the same as the sharp bimodal distribution $\sigma_l = \pm 1$ (fig. 1.5). The remarkable fidelity of the Ising model as a description of the statistical mechanics of phase transitions (chapter 5) does not prove that it correctly represents other physical properties of real magnetic systems.

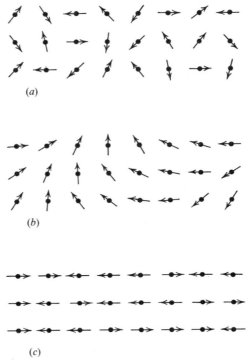

Fig. 1.4. (*a*) Spin disorder in a paramagnet. (*b*) Spin-wave disorder in a ferromagnet. (*c*) Ising spin disorder.

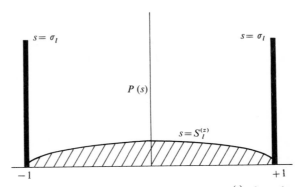

Fig. 1.5. Probability distributions of components $S_l^{(z)}$ of random spin vectors and of random Ising spins σ_l.

Certain solids, such as solid hydrogen and solid deuterium, exhibit order–disorder transitions in which the ordered state is characterized by a regular pattern in the directions of the molecular axes – for example of the longitudinal axes of molecules of H_2 or D_2 – at various sites of the crystal lattice. The physics of the transition to *orientational order* in such systems is complicated by associated transitions from, say, f.c.c. to h.c.p. crystal structure (see e.g. Meyer *et al.* 1972), but is obviously analogous in principle to the magnetic case.

1.4 Ice disorder

Another type of cellular disorder is observed in several of the crystalline phases of water (see e.g. Fletcher 1970). The structure of Ice I – the ordinary phase of atmospheric pressure – is shown in fig. 1.6. The oxygen atoms, which are much larger than the protons, form a regular hexagonal lattice (the wurtzite structure) in which each atom has four neighbours arranged tetrahedrally about it. The bonds between neighbouring oxygens are occupied by the protons. But each proton lies close to one or another of the oxygen atoms that it binds together and every oxygen acquires just two such protons, thus creating a local configuration very close to the arrangement of atoms in the free H_2O molecule.

But the arrangement of protons need not be the same in each unit cell of the crystal. The conditions stated above can be achieved in many different ways, without regular order. In a crystal of $2N$ bonds, there would be 2^{2N} possible ways of arranging the protons. But these would not all satisfy the rules for making H_2O molecules. In fact, out of $2^4 = 16$ ways of arranging the protons on the tetrahedron of bonds about a single oxygen atom, only six ways satisfy the *ice condition* (fig. 1.7). The total number of allowed configurations of the whole crystal must be something like

$$(6/16)^N \, 2^{2N} = (3/2)^N. \tag{1.7}$$

The fact that the proton distribution is actually disordered is shown by the experimental observation that the residual entropy of ice is very close to this theoretical value,

$$S_0 = N\ell \ln (3/2). \tag{1.8}$$

Very similar arguments apply to other phases of ice, where the oxygens may form a diamond structure, or even two interpenetrating tetrahedrally coordinated lattices. From the point of view of an excitation of the crystal, such as lattice wave, *ice disorder* of the protons looks like almost ideal cellular disorder.

Cellular disorder

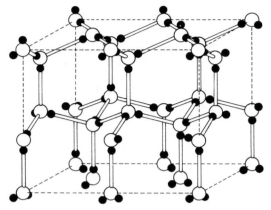

Fig. 1.6. Structure of Ice I. There is one proton to each bond: each oxygen has two close protons.

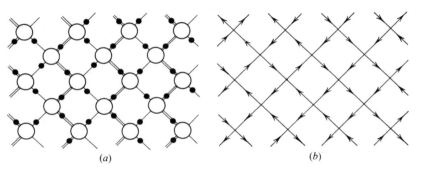

(a)   (b)

Fig. 1.7. (a) Ice disorder in two dimensions. (b) Ice condition is satisfied if each vertex of arrow diagram has two inward and two outward arrows.

Nevertheless, strictly speaking, the cellular disorder in ice is not perfectly random. The Pauling formula (1.8) assumes that the distribution of protons in each cell can be treated as if statistically independent of the state of its neighbours. Consider, however, a closed ring of six bonds. If the positions of the protons on each of the first five oxygens of this ring have been prescribed, the protons in the remaining oxygen cannot be placed at will. This type of disorder is thus subject to *topological* constraints, which slightly alter the statistical properties of the distribution in the neighbourhood of any given site. The combinatorial problem of calculating configurations in this case has not been solved analytically, but successive

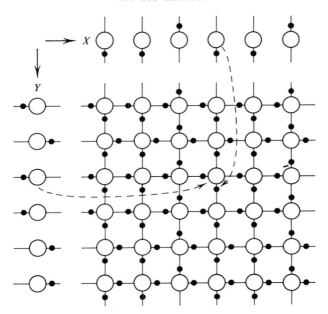

Fig. 1.8. 'Right-angle water.' The configuration of the protons at site (X, Y) is determined by the configurations in column X and row Y.

approximations (§ 5.8) have shown that the true entropy should be about 1 per cent larger than the Pauling value. This is evidently a small effect, but provides us with a hint of the importance of connectivity, dimensionality and other topological properties in the theory of disordered systems.

It is easy to see, for example, that 'ice disorder' cannot occur on a one-dimensional lattice. On a two-dimensional square lattice, the arrangement of molecules of 'right-angle water' may appear to be disordered (fig. 1.8), but the configurational entropy is not an extensive variable. In this model, we insist that the two protons associated with each oxygen atom should not lie on a straight line: the allowed configurations can all be built up by prescribing the phase of alternation of proton shifts along each line of bonds in the X and Y directions. With N atoms, we have $2\sqrt{N}$ such lines, each capable of two alternative phases: the entropy per molecule would be

$$\frac{1}{N} k \ln(2^{2\sqrt{N}}) = \frac{2k}{\sqrt{N}} \ln 2, \qquad (1.9)$$

which tends to zero for large N.

The Pauling model for ice is a special case of a whole class of hydrogen-

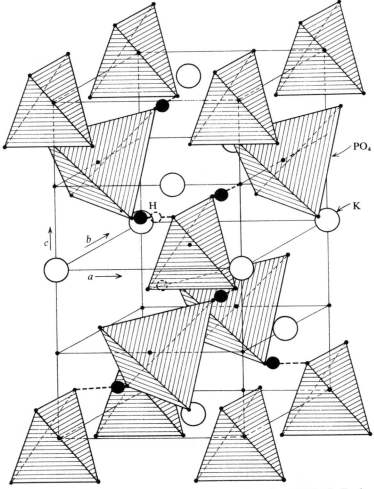

Fig. 1.9. Structure of potassium dihydrogen phosphate ('KDP'). Each
(PO₄) tetrahedron is bonded to each of its four neighbours by a proton
which may be displaced towards either end of its bond.

bonded systems. Ionic compounds homologous with KH_2PO_4 (potassium
dihydrogen phosphate: 'KDP') crystallize in a complicated structure in
which the $(PO_4)^{3-}$ ions occupy the sites of a tetragonal diamond lattice
(fig. 1.9). Each phosphate ion is bonded to its four neighbours by a proton
which may be at one or other end of the bond. Just as in ice, disordered
proton configurations satisfying the ice condition have a large residual

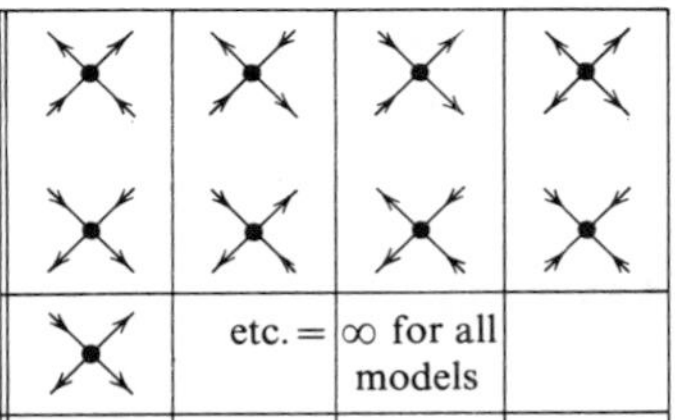

Config-uration				
Model		etc. =	∞ for all models	
'Ice'	0	0	0	∞
'KDP'	0	ϵ	ϵ	∞
'ADP'	ϵ	0	ϵ	∞
'Double Ionized'	$-\tfrac{1}{2}\delta$	$-\epsilon$	$\tfrac{1}{2}\delta$	$-\gamma$

(a)

'KDP'

'ADP' (F-model)

(b)

Fig. 1.10. (a) Energy of local proton configurations in various models. (b) Ferroelectric and antiferroelectric order.

entropy, whose value for the diamond topology is exactly the same as for the hexagonal wurtzite lattice of ice (Nagle 1966).

But, as pointed out by Slater (1941) and by Takahashi (1941), not all the six proton configurations allowed by the ice condition are now equivalent. The c axis of the tetragonal unit cell (formerly a (100) axis of the diamond cube) provides a preferred direction. Configurations in which both protons lie 'above' or 'below' the median plane may have a different energy from configurations in which the protons lie on opposite sides. This energy difference generates an ordered state in which all cells are in the same configuration. Since the combination of an electrostatically $-$ve phosphate ion with two $+$ve protons constitutes an electric dipole, this ordered state may be characterized by a permanent macroscopic electric moment. In KDP, and many of its analogues, the 'up' or 'down' polarization has lower energy than the other configurations, so the material is *ferroelectric*. In the ammonium salts (ammonium dihydrogen phosphate: $NH_4 . H_2PO_4$: 'ADP') the preference is for ordered configurations with antiparallel electric dipoles, thus giving rise to *antiferroelectric* behaviour (see e.g. Känzig 1957; Jona & Shirane 1962).

The Slater model for KDP, and the 'F-model' for ADP (Rys 1963) can thus be characterized very simply by the energy parameter ε separating the two sets of bond configurations (fig. 1.10). An order–disorder transition is to be expected at some critical temperature T_c which must, by dimensional arguments, be some multiple of ε/k.

To set up these models mathematically we may introduce an 'Ising spin' variable σ_i to define the position of the proton on each bond, and then introduce formulae for the energy in terms of the values of these variables at each vertex (see e.g. Suzuki & Fisher 1971). To impose the ice condition, we assert formally that the energy of 'forbidden' configurations is infinite. The eight-vertex model is a generalization in which a large but finite energy is assigned to 'doubly ionized' configurations in which a phosphate group may have either four near protons or none at all (Sutherland 1970). For mathematical reasons most of the work on these models makes the drastic assumption that the three-dimensional, tetrahedrally coordinated lattice has the topology of a square planar lattice, but the mathematical connections between the ferroelectric and antiferroelectric models and the corresponding Ising models for magnetic systems are now well established and will be discussed in chapter 5.

It is by no means certain, however, that these models do justice to the physical complexity of the KDP and ADP materials. The proton is not, for example, constrained to lie precisely at one of only two points along the line of the bond: it is a quantum-mechanical particle, of low mass, moving in a field with two minima between which it can tunnel (Blinc 1959, 1968). To describe this motion quantum-mechanically, a quasi-spin variable (de Gennes 1963a) with components $S_i^{(x)}$, $S_i^{(y)}$, $S_i^{(z)}$ satisfying the usual Pauli conditions is introduced for each proton. The eigenstates of $S_i^{(z)}$ then refer to the two equilibrium positions. With this notation, a Hamiltonian can be written down with a term linear in $S_i^{(x)}$ for the tunnelling frequency and with 2-, 3- and 4-spin coupling terms to match the energy levels of the various configurations of the Slater model. In other words, ferroelectric disorder may demand description in terms of continuous variables analogous to the continuous spin variables of the Heisenberg model of magnetism.

Here again, the mathematical theory has become quite sophisticated (cf. Chock, Résibois, Dewel & Dagonnier 1971) but is really no more than a complicated form of the spin-wave theory of ferromagnetism to be discussed in § 1.8. To make the analysis even more difficult, it seems necessary to include further terms for the coupling of the proton displacements to optical phonons (Blinc & Ribaric 1963). The ferroelectric transition in KDP is thus not merely an order–disorder transition of the proton configuration but is also linked with distortions of the unit cell and relative displacements of the other ions in the crystal. Following this path, we should be led into a discussion of other types of ferroelectric and antiferroelectric material, where the instability of the lattice towards special phonon modes is the dominant mechanism. But these materials no longer exemplify the theory of *disorder* which is our main theme.

1.5 Short-range order

Perfect cellular disorder is seldom achieved: *correlations* between atoms or spins on neighbouring lattice sites cannot be ignored. In a binary alloy, for example, it may be energetically favourable to surround an A atom by B atoms, rather than by atoms of the same type. The system then shows some degree of *short-range order*.

To measure this phenomenon we might look for the total number N_{AB} of $A-B$ type 'bonds' – i.e. neighbouring lattice sites occupied by atoms of different type – and compare this with the number to be expected if the atoms were distributed at random. To define an *order parameter* in this case we must introduce statistical concepts. Thus, the limiting probability of a given bond being of type $A-B$ in a large crystal would be the limit

$$P_{AB} = \underset{N \to \infty}{\text{Lt}} \ (N_{AB}/\tfrac{1}{2}zN) \qquad (1.10)$$

where N sites each with z neighbours are linked by a total of $\tfrac{1}{2}zN$ bonds. If each site were independently occupied with probability c_A or c_B, this quantity would be $2c_A c_B$. The *nearest neighbour correlation parameter* would then be the difference

$$\Gamma_{AB} = \tfrac{1}{2}P_{AB} - c_A c_B. \qquad (1.11)$$

In the literature on alloys (e.g. Muto & Takagi 1955; Guttman 1956; Münster 1962, 1965) this quantity occurs in several other forms. Suppose, for example, that some state of maximum order can be conceived in which P_{AB} takes on some value P_{AB}^M. Dividing Γ_{AB} by $P_{AB}^M - 2c_A c_B$ gives the *Bethe short-range order parameter* which ranges from zero for complete disorder to unity for maximum order. Similarly, the *Cowley order parameter* is obtained by dividing Γ_{AB} by $-c_A c_B$. Since Γ_{AB} is already dimensionless no advantage is gained by these arithmetical manipulations which obscure the connection with more general correlation functions.

Physical intuition suggests that the local ordering effect need not be confined to nearest neighbours. Suppose that we have two sites in the lattice, distant **R** from one another. To describe deviations from randomness we define a probability distribution function $P_{AB}(\mathbf{R})$ which now stands for the fraction of all pairs of sites in this relative position that are occupied by one A atom and one B atom. This appears in a more general *correlation function*

$$\Gamma_{AB}(\mathbf{R}) = \tfrac{1}{2}P_{AB}(\mathbf{R}) - c_A\,c_B \tag{1.12}$$

which may be expected to fall to zero as R increases.

Any analytical theory for the calculation of correlation functions must be based upon the general principles of statistical mechanics. To give meaning to an expression such as (1.12), we should take an *ensemble average*, symbolized by brackets $\langle\ \rangle$, over a quasi-infinite set of copies of our system, relying on various ergodic theorems, etc. to equate this with a time average or a space average over an actual macroscopic sample. In the magnetic case, for example, we might introduce an expression such as

$$\Gamma(\mathbf{R}_{ll'}) = \langle \mathbf{S}_l \cdot \mathbf{S}_{l'}\rangle - \langle \mathbf{S}_l\rangle \cdot \langle \mathbf{S}_{l'}\rangle \tag{1.13}$$

for the correlation of spin directions on pairs of sites such as l and l' related by the vector $\mathbf{R}_{ll'}$. Interpreting these variables as Ising spins, as in (1.5), we have a useful formalism which may be applied to the binary alloy case. Thus, if $\Gamma(\mathbf{R}_{ll'})$ were negative, there would be an excess of atoms of opposite 'spin' (i.e. of opposite type) on these sites; if it were positive, then atoms of the same type are favoured. But the general question of local magnetic order is of importance in its own right.

It is worth noting, as a preliminary to later discussion (§ 5.10), that (1.13) is actually of the form of a *cumulant average* (Kubo 1962),

$$\Gamma(\mathbf{R}_{ll'}) \equiv \langle \mathbf{S}_l \cdot \mathbf{S}_{l'}\rangle_{\mathrm{cum}}, \tag{1.14}$$

which vanishes automatically for any quantity that is the product of independent random variables. Most of the significant physical phenomena in the theory of disordered systems arise from correlations which do *not* vanish in this way.

Short-range order is generated by short-range interactions between the atoms or spins. In the magnetic case, for example, we introduce a *Heisenberg Hamiltonian*, consisting of a sum of terms like

$$\mathcal{H}_{ll'} = -J(\mathbf{R}_{ll'})\mathbf{S}_l \cdot \mathbf{S}_{l'}, \tag{1.15}$$

which would be contributed by the pair of spins at l and l'. By convention, the *exchange parameter* $J(\mathbf{R}_{ll'})$ is given a minus sign, so that *ferromagnetic ordering* (i.e. parallel spins) is favoured when J is *positive*. In the total Hamiltonian of the system,

$$\mathcal{H} = -\tfrac{1}{2}\sum_{l,l'} J(\mathbf{R}_{ll'})\mathbf{S}_l \cdot \mathbf{S}_{l'} - \bar{\mu}\sum_l \mathbf{S}_l \cdot \mathbf{H}, \tag{1.16}$$

the effect of an external magnetic field on the magnetic moment $\bar{\mu}\,\mathbf{S}_l$ of each spin is included.

In general $J(\mathbf{R}_{ll'})$ might be any function of the direction and length of $\mathbf{R}_{ll'}$;

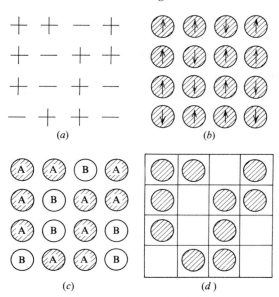

Fig. 1.11. A configuration of the Ising model (*a*) may represent: (*b*) an arrangement of spins; (*c*) an arrangement of atoms in a binary alloy; (*d*) a configuration of a 'lattice gas'.

in practice, we deal mostly with models in which this function is zero beyond nearest or next-nearest neighbours. For certain special types of crystal we can make the interaction anisotropic by treating J as a tensor, i.e.

$$\mathscr{H}_{ll'} = -J_{\parallel} S_l^{(z)} S_{l'}^{(z)} - J_{\perp}\{S_l^{(x)} S_{l'}^{(x)} + S_l^{(y)} S_{l'}^{(y)}\}. \qquad (1.17)$$

The Ising model Hamiltonian is then obtained by quantizing $S_l^{(z)}$ and ignoring $J_{\perp}$, i.e.

$$\mathscr{H} = -\tfrac{1}{2} \sum_{l,l'} J \sigma_l \sigma_{l'} - \bar{\mu} H^{(z)} \sum_l \sigma_l \qquad (1.18)$$

in suitably redefined units.

The Ising model formalism may be used to describe atomic interactions in a binary alloy (fig. 1.11). Suppose, for example, that AA, AB and BB pairs have energy ϕ_{AA}, ϕ_{AB} and ϕ_{BB} respectively. The total energy of the system may then be written

$$\mathscr{E} = N_{AA}\phi_{AA} + N_{BB}\phi_{AB} + N_{AB}\phi_{AB}. \qquad (1.19)$$

But the numbers of pairs of each type are constrained by the relations

$$2N_{AA} + N_{AB} = zN_A \quad \text{and} \quad 2N_{BB} + N_{AB} = zN_B \qquad (1.20)$$

where N_A is the total number of A atoms in the alloy and z is the coordination number, so that

$$\mathscr{E} = \tfrac{1}{2}zN_A\phi_{AA} + \tfrac{1}{2}zN_B\phi_{BB} + N_{AB}(\phi_{AB} - \tfrac{1}{2}\phi_{AA} - \tfrac{1}{2}\phi_{BB}). \quad (1.21)$$

The convention that σ_l should be $+1$ or -1 according as the site is occupied by an A atom or a B atom yields further relations

$$\sum_l \sigma_l = N_A - N_B \quad \text{and} \quad \sum_{l,l'} \sigma_l \sigma_{l'} = 2(N_{AA} + N_{BB} - N_{AB}) \quad (1.22)$$

whence we get

$$\mathscr{E} = -\tfrac{1}{4}(\phi_{AB} - \tfrac{1}{2}\phi_{AA} - \tfrac{1}{2}\phi_{BB}) \sum_{l,l'} \sigma_l \sigma_{l'} - \tfrac{1}{4}z(\phi_{BB} - \phi_{AA}) \sum_l \sigma_l +$$
$$+ \tfrac{1}{4}zN(\phi_{AB} + \tfrac{1}{2}\phi_{AA} + \tfrac{1}{2}\phi_{BB}), \quad (1.23)$$

which is of the same form as (1.18) apart from a constant additive term.

In an alloy, therefore, the short-range order will be similar to that in an Ising magnet with 'exchange parameter'

$$J = \tfrac{1}{2}(\phi_{AB} - \tfrac{1}{2}\phi_{AA} - \tfrac{1}{2}\phi_{BB}) \quad (1.24)$$

in a 'magnetic field' of strength

$$\bar{\mu}H^{(z)} = \tfrac{1}{4}z(\phi_{BB} - \phi_{AA}). \quad (1.25)$$

If J is positive, for example, atoms of like kind tend to cluster together just as parallel spins are favoured in a ferromagnet. If J is negative, there is a tendency to form unlike pairs, as in an antiferromagnet. But if the energy required to substitute an A atom for a B atom were independent of the surroundings, i.e. if we could write

$$\phi_{AA} = \phi_A + \phi_A; \quad \phi_{AB} = \phi_A + \phi_B; \quad \phi_{BB} = \phi_B + \phi_B, \quad (1.26)$$

then J would be zero and our system would be perfectly disordered.

To apply the Ising spin formalism to the ferroelectric and antiferroelectric models of § 1.4, we introduce a convention relating the signs of the variables $\sigma_{l1}, \sigma_{l2}, \sigma_{l3}, \sigma_{l4}$ to the positions of the protons on the four bonds in the lth unit cell of the basic tetragonal diamond lattice. The contribution to the Hamiltonian from the interactions between the protons then takes the form (Suzuki & Fisher 1971)

$$\mathscr{H}_l = -J_0 - \sum_{i<j} J_{ij} \sigma_{li} \sigma_{lj} - J_7 \sigma_{l1} \sigma_{l2} \sigma_{l3} \sigma_{l4}. \quad (1.26a)$$

The parameters J_0, J_{ij} and J_7 can then be written down in terms of the energies assigned to various local configurations. If, for example, the ice condition is rigidly imposed, then all these parameters are infinite in magnitude, but cancel in pairs for allowed values of the σ_{li}. But the 4-spin interaction term J_7 is obviously essential to the analysis, and cannot be dropped without damage to the physical plausibility of the model. Unfortunately this term is an obstacle to an exact theory of the order–disorder transition, even on a planar lattice (§ 5.8).

For a 'lattice gas', the expression for the energy would again be given by (1.23), with ϕ_{AA} for the supposed interaction between the atoms, and $\phi_{AB} = \phi_{BB} = 0$ for the energy associated with 'holes'. Some attention has been given (see Fisher 1967; Stanley 1971) to the properties of *quantal gas* models, whose Hamiltonian may be derived from (1.17) by interpreting the spin variables as operators. We write this out in terms of annihilation and creation operators, a_l, $a_l^\dagger$, for spin deviations from the axis of quantization z and obtain, first of all, an expression like (1.18). This comes by collecting terms in products like $a_l a_l^\dagger$, which is a 'c-number' measuring the 'occupation' of the lth site and can, therefore, be related to the Ising spin variable σ_l. But the non-commuting operators $S_l^{(x)}$, $S_l^{(y)}$ give rise to further *off-diagonal interactions* involving products like $a_l a_{l'}^\dagger$.

There is no strict physical example of an 'alloy' or 'gas' with this type of interaction. The Blinc model of ferroelectricity (§ 1.4) produces a Hamiltonian with terms like (1.17), but because of the short-range interactions implicit in the ice conditions it also contains products of components of three and four quasi-spin operators, just as in (1.26a). The effects of these terms may perhaps be estimated approximately (Blinc & Svetina 1966) but without them the model is not physically realistic.

The mathematical properties of the quantal gas Hamiltonian have been investigated as a model for the *off-diagonal long-range order* (ODLRO) characteristic of superfluidity. In such a system the *order parameter*, Ψ, although a macroscopic variable, behaves like a wave function in having both amplitude and phase. But this type of order implies correlations in the momentum variables, and does not relate to any obvious geometrical structure in real space. This theory, therefore, lies outside the chosen field of the present book.

The essential point about (1.16), (1.18) and (1.23) is that the energy of our system depends directly upon the magnitude of an order parameter such as (1.11) or (1.13). In the Ising model, for example, the statistical homogeneity of our sample allows us to write

$$\sum_{l,l'} \sigma_l \, \sigma_{l'} = 2zN\langle \sigma_l \, \sigma_{l'} \rangle$$
$$= 2zN\{\Gamma_{ll'} + \langle \sigma_l \rangle^2\}, \tag{1.27}$$

where the correlation function is evaluated over an appropriate ensemble. But we cannot calculate the value of $\Gamma_{ll'}$ simply by minimizing the total energy, because there are many different configurations with the same energy. The only possible analytical procedure is to evaluate the *free energy,* in equilibrium, by statistical mechanical arguments. This task is taken up in chapter 5.

1.6 Long-range order

The concept of *long-range order* is intuitively simple. Consider, for example, a classical magnetic system with Hamiltonian (1.16). If $J(\mathbf{R}_{ll'})$ is positive, this system has a minimum energy in the *ferromagnetic ground state,* where each spin vector $\mathbf{S}_l$ is in the direction of the external magnetic field $\mathbf{H}$. In the alloy case, this corresponds to a preference for the clustering of like atoms into large aggregates, leading to *phase separation* of pure A metal from pure B metal – a well-known metallurgical phenomenon.

When J is negative for nearest neighbours we construct the *antiferromagnetic ground state.* Consider, for example, a body-centred cubic lattice, whose sites may be assigned to two interpenetrating simple cubic sub-lattices, α and β, such that every α site is surrounded by β sites, and vice versa. The energy is minimized by giving every α site the spin $\mathbf{S}_\alpha$, say, and every β site the opposite spin $\mathbf{S}_\beta = -\mathbf{S}_\alpha$. The alloy analogue is an equiconcentration ordered phase, with A atoms on the α sub-lattice and B atoms on the β sub-lattice.

The first mathematical problem is to determine the most likely form of ordering for a system with given interactions. This problem is not quite trivial, since it depends upon the nature of the interactions, the concentration of the constituents and the geometry of the underlying crystal lattice. To generate the ordered phase of Cu_3Au, for example, the basic face-centred cubic lattice is subdivided into four interpenetrating simple cubic lattices, of which three are occupied by Cu atoms and the fourth by Au atoms. The possibilities for more and more complex ordered phases in metallic alloys seem almost endless (see e.g. Guttman 1956; Münster 1962 – and an enormous mass of metallurgical literature).

The effects of next-nearest neighbour interactions are not trivial. For the Ising antiferromagnet on an f.c.c. lattice, for example, subdivision into four simple cubic sub-lattices, two with $\sigma_l = +1$ and two with $\sigma_l = -1$, is not the best that can be done. Advantages can accrue from further antiferromagnetic ordering in each sub-lattice (fig. 1.2) – and so on (see e.g. Smart 1966).

In the rare earth metals (see e.g. Martin 1967; Krupicka & Sternberg 1968) even more complicated patterns of order may be observed. Consider a (vector) spin system in which the interaction J_0 between z_0 neighbours in the same layer of sites is positive, whilst the interaction J_1 between spins in neighbouring layers is negative. The spins in each layer will line up parallel, but there will be antiferromagnetic order with each successive layer magnetized in alternate directions. Now add a third exchange coefficient J_2 for interactions between next-nearest layers: what will be the effect? Suppose that the spin systems in successive layers are not completely antiparallel, but that they make an angle ϕ with one another, layer by layer up the crystal (fig. 1.3(a)). It is easy to show that the energy per layer is proportional to

$$\mathscr{E} = -z_0 \, J_0 - 2J_1 \cos \phi - 2J_2 \cos 2\phi, \qquad (1.28)$$

since the contribution of next-nearest layers is proportional to $-J_2 \, \mathbf{S}_i \cdot \mathbf{S}_j$ where these vectors are at angle 2ϕ to one another. Now minimize this with respect to ϕ. In addition to the root at $\phi = \pi$, we find another possible solution when

$$\cos \phi = -J_1/4J_2. \qquad (1.29)$$

Thus, if J_2 is not too small, we may observe *helical* or *spiral order*, with the spin direction rotating round a screw axis as we move normal to the ferromagnetically ordered planes. Observe that the pitch of this screw has nothing to do with the lattice constant of the underlying crystal: the magnetic ordering is a new structure with different group symmetries (cf. Bertaut 1963). Notice, moreover, that the magnetic vector for each layer need not lie in the plane of that layer, provided only that the angle between vectors in successive layers is kept constant. The actual ordered configuration will depend on further energy terms, such as the anisotropy energy for the magnetic vector on each site, but may, for example, correspond to a screw pattern on a cone about the screw axis (fig. 1.3(b) and (c)).

This discussion assumes that the magnetic moments of the atoms are classical vectors. It is worth remarking, however, that the quantum operator properties of each spin cannot be entirely ignored. Suppose, for example, that we attempt to construct the intuitive antiferromagnetic ground state out of the eigenfunctions of $S_l^{(z)}$, chosen with the eigenvalue $+S$ or $-S$ according as l lies on one or the other sub-lattice. It is easy to verify that this is not an eigenfunction of the total Hamiltonian (1.16). The non-commutativity of the off-diagonal operators in (1.17) exchanges spin deviations between neighbouring sites, giving rise to an irreducible 'zero-point motion' disorder (see § 5.6).

The question now arises: how should one characterize deviations from some assumed pattern of long-range order? In the ferromagnetic case, this seems easy enough. We may say, for example, that the vector spin $\mathbf{S}_l$ takes some average value less than its maximum component S. This would be measurable as a reduction of the total magnetic moment $\bar{M}$ of the whole crystal below its maximum value $N\bar{\mu}S$. The *long-range order parameter* in the system might then be written

$$\mathscr{S} = \bar{M}/N\bar{\mu}S = \langle\mathbf{S}_l\rangle/S. \tag{1.30}$$

For a simple antiferromagnet or ferrimagnet, the analogous parameters would be the average magnetizations

$$\bar{\mathbf{M}}_\alpha = N\bar{\mu}_\alpha\langle\mathbf{S}_\alpha\rangle \tag{1.31}$$

on the sub-lattices – or some similar quantities. In the theory of binary alloys, it is conventional to define the *Bragg–Williams order parameter*

$$\mathscr{S} = (r_\alpha - c_A)/c_A = (r_\beta - c_B)/(1 - c_B) \tag{1.32}$$

where r_α is the fraction of α sites 'correctly' occupied by A atoms, and so on. In the language of the Ising model this could be written

$$\mathscr{S} = \frac{\frac{1}{2}\{\langle\sigma_\alpha\rangle - \langle\sigma_\beta\rangle\}}{1 - \frac{1}{2}\{\langle\sigma_\alpha\rangle + \langle\sigma_\beta\rangle\}} \tag{1.33}$$

where $\langle\sigma_\alpha\rangle$ is the average Ising spin on sub-lattice α, etc.

These parameters are not, however, uniquely defined. In the antiferromagnetic case, the sub-lattices must be identified in advance, which implies an unphysical symmetry-breaking operation or observation. The meaning of (1.33) would be destroyed if there were a single *anti-phase domain boundary* crossing the specimen (fig. 1.12). In the case of a binary alloy with J positive the magnitude of $\langle\sigma_l\rangle$ tells us nothing about long-range order, since this quantity is simply the difference of concentration of the two components, and is fixed independently of whether there is clustering and phase separation into pure components. Even a simple ferromagnetic specimen would be expected to have very small net magnetization, since the long-range order on a microscopic scale would be confined to a few large *domains* whose magnetization vectors would largely tend to cancel one another. In the absence of a strong magnetic field defining a physically preferred direction in space, the expectation value of $\langle\mathbf{S}_l\rangle$ over an equilibrium ensemble should be zero.

It is much better, therefore (cf. Domb 1960; Münster 1962, 1965), to define long-range order as the limit of the correlation function at large distances. In general, we study a parameter such as

$$\Gamma_\infty = \operatorname*{Lt}_{R_{ll'}\to\infty}\Gamma(\mathbf{R}_{ll'}) \tag{1.34}$$

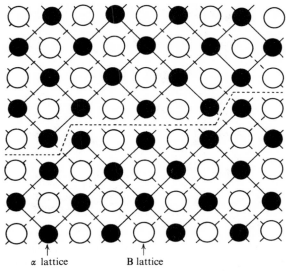

Fig. 1.12. Anti-phase domain boundary.

where $\Gamma(\mathbf{R}_{ll'})$ is defined as in (1.12), (1.13), (1.14) or (1.27). If this limit is not zero, then we have long-range order. In the ferromagnetic case and its analogues, it is easy to prove that this yields the square of the quantity suggested by (1.30), i.e.

$$\Gamma_\infty = \mathcal{S}^2. \qquad (1.35)$$

But in the antiferromagnetic case the function $\Gamma(\mathbf{R}_{ll'})$ reverses in sign as the vector $\mathbf{R}_{ll'}$ steps from sites on the same sub-lattice to sites on different sub-lattices. This alternation plays havoc with the formal definition of the limit (1.34), but we may still measure the degree of long-range order by the magnitude of

$$\Gamma_\infty = \underset{R_{ll'} \to 0}{\mathrm{Lt}} \; |\Gamma(\mathbf{R}_{ll'})| \qquad (1.36)$$

which also satisfies (1.35) when $\mathcal{S}$ is given by (1.32) or (1.33).

1.7 The range of order and ordered domains

From the above discussion it is clear that the state of occupation of a substituted lattice is incompletely defined by the short-range and long-range order parameters (1.11) and (1.36). It is characteristic of a co-operative assembly that a very short-range interaction such as (1.15) can

Cellular disorder

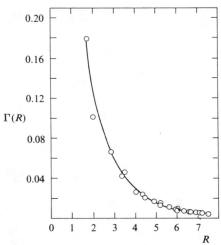

Fig. 1.13. Short-range order, calculated for successive shells of neighbours in a binary alloy, fits closely to the smooth exponential function $R^{-1}\exp(-R/\xi)$.

propagate order over quite large distances, even though it may not be sufficient to produce order at infinite range at the temperature considered. To describe such states of *intermediate-range order*, it is necessary to study the behaviour of the correlation function $\Gamma(\mathbf{R}_{ll'})$ as a function of the distance $\mathbf{R}_{ll'}$ between the sites of the lattice. As we shall see (chapter 5), various theories of critical phenomena predict various forms for this function in its dependence on R, on the temperature, and on the exchange parameter J, but these all behave like

$$\Gamma(R) \sim R^{-n}\exp(-R/\xi) \qquad (1.37)$$

at large distances (fig. 1.13). The exponent n depends upon the dimensionality of the lattice and upon the nature of the interaction, but the dominant factor is the exponential, which falls off very rapidly at distances greater than the *correlation length* ξ. This quantity, or the related length L defined by

$$L^2 = \int R^2\Gamma(R)\mathrm{d}^3R / \int \Gamma(R)\mathrm{d}^3R, \qquad (1.38)$$

thus measures the *range of order* in the material.

The behaviour of ξ as a function of temperature is thus a description of the variation of 'randomness' of a magnetic system or alloy. At very high temperatures, where ξ tends to zero, our assembly is perfectly disordered. As the temperature decreases, short-range order becomes evident over one or two lattice spacings. At lower temperatures ξ becomes very large, giving rise to the phenomena of *critical fluctuations* of spin or concentration. The temperature at which ξ becomes infinite denotes the onset of long-range order, and defines the *critical temperature T_c* for the *order–disorder transition* (in a ferromagnet, the *Curie temperature;* in an antiferromagnet, the *Néel temperature*). Below T_c, the limit Γ_∞ exists as in (1.34), and the system is in an ordered phase.

As we shall see in chapter 4, the range of order can be measured directly by diffraction techniques. The behaviour of ξ as a function of T has been carefully studied in the neighbourhood of T_c, and this general description is well-confirmed experimentally in magnetic and alloy systems (see e.g. Münster 1965). There have also been a number of metallurgical studies of what we might call the *local order* in alloys quenched from temperatures somewhat above T_c, where the correlation function $\Gamma(\mathbf{R}_{ll'})$ (or the equivalent order parameters) can be measured for sites in the first few shells of neighbours (see e.g. Guttman 1956; Münster 1962, 1965; Moss 1965; Moss & Clapp 1968). These observations provide useful information about the nature of the short-range interactions by which order is propagated through the lattice.

The interesting question then arises: to what extent are measurements of the values of $\Gamma(\mathbf{R}_{ll'})$ for half a dozen lattice vectors $\mathbf{R}_{ll'}$ sufficient to determine the distribution of atoms on the lattice sites? We may also ask what constraints there are on these parameters; for example, the correlation between next-nearest neighbours cannot be entirely independent of the two successive nearest neighbour correlations by which this site might be reached.

These general statistical questions do not seem to have been studied in depth. But Gehlen & Cohen (1965) have verified by a Monte Carlo computer simulation that stationary distributions of atoms can be generated with values of the order parameters similar to those observed in real alloys. These distributions make no appeal to any model for the interatomic forces, and come out the same whether one starts from a fully ordered or a perfectly random configuration (fig. 1.14). In other words, the function $\Gamma(\mathbf{R}_{ll'})$ for a few shells of neighbours provides sufficient information to determine many other statistical properties of the atomic distribution in a binary alloy.

In a classical vector spin system, however, the scalar correlation function (1.13) is not sufficient to define the state of local order. The reason is that the components of each vector $\mathbf{S}_l$ are continuous variables. Suppose, for

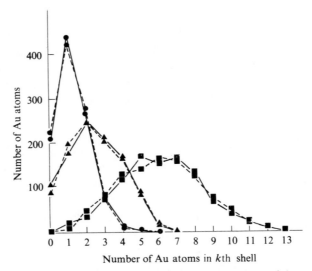

Fig. 1.14. Computer-simulated stationary distributions of Au atoms around Au atom satisfying observed order parameters for Cu_3Au at 450 °C. ●, First shell; ▲, second shell; ■, third shell. Monte Carlo procedures starting from ordered configuration (———) and from random configuration (– – –) converge on the same distributions (Gehlen & Cohen 1965).

example, that $\Gamma(\mathbf{R}_{ll'})$ for nearest neighbours was observed to be only a little less than its maximum possible value $\langle \mathbf{S}_l \cdot \mathbf{S}_l \rangle$. We could not tell whether this was due to there being just a few neighbours with reversed spins or whether the spins on all the neighbouring sites deviated by small amounts from the direction of $\mathbf{S}_l$ (fig. 1.15). The information we really need is contained in the *two-site distribution function* $P(\mathbf{S}_l, \mathbf{S}_{l'})$ which measures the joint probability of finding spins $\mathbf{S}_l$ and $\mathbf{S}_{l'}$ on these two sites for any system in the ensemble. Even in the simplest case, where this probability depends only on the angle θ between the spin directions, the correlation function only measures the mean value of $\cos \theta$

$$\langle \mathbf{S}_l \cdot \mathbf{S}_{l'} \rangle = \iint \mathbf{S}_l \cdot \mathbf{S}_{l'} \, P(\mathbf{S}_l, \mathbf{S}_{l'}) \, d\mathbf{S}_l \, d\mathbf{S}_{l'}$$
$$\sim S^2 \int \cos \theta \, P(\cos \theta) \, d(\cos \theta). \qquad (1.39)$$

To determine $P(\cos \theta)$ fully, we should need more statistical information such as the higher moments of this distribution. The binary alloy/Ising model is a special case because the statistic $\langle \sigma_l \sigma_{l'} \rangle$ is sufficient (as in (1.22)) to define completely the probability distribution for like and unlike atoms

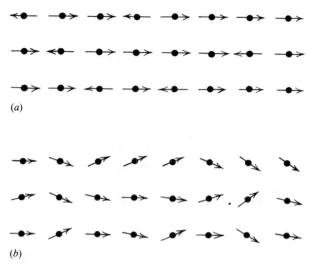

Fig. 1.15. The same value of the local order parameter $\Gamma(\mathbf{R}_{ll'})$ may describe a nearly ordered state with (a) a few *reversed* spins or (b) many *deviated* spins.

on this pair of sites. This mathematically trivial point requires emphasis because of the tendency, in the literature, to refer to 'the order parameter' as if this were an index from which all the statistical properties of the disordered assembly could be uniquely derived.

Near T_c, where the range of order is large, we quite properly think of ξ as a measure of the characteristic size of a cluster of like atoms, or of an ordered domain. But when there is only a modest degree of local order it is misleading to describe the specimen as a collection of 'ordered zones in a disordered matrix' (Greenholz & Kidron 1970). Consider the case of a *random* equiconcentration alloy. It is easy to prove by percolation theory (see § 9.10) that almost every A atom belongs to an *infinite* cluster of similar atoms. Paradoxically, if we are looking for ordered domains, then almost every atom belongs to an *infinite* domain with perfect AB ordering (fig. 1.16). These 'clusters' and 'domains' interpenetrate one another with very complex topology, but within each domain the delineation of smaller groupings, such as $3 \times 3 \times 3$ 'ordered zones', is a matter of arbitrary choice. This argument holds whenever the concentration of the minor constituent exceeds the critical site-percolation probability for the lattice. A typical example of interpenetrating domains is shown in fig. 1.17, which arose in a Monte Carlo calculation for Cu_3Au (Gehlen & Cohen 1965). It is only in a

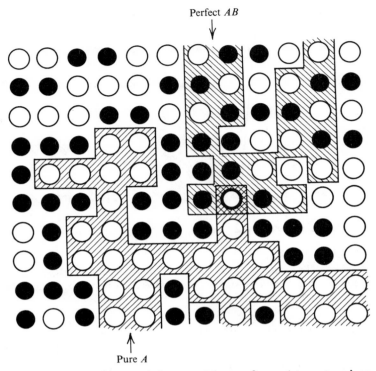

Fig. 1.16. Does this atom belong to a 'cluster of pure *A*-type atoms', or to a 'region of perfect *AB* order'?

Fig. 1.17. Three-dimensional model of ordered domains in Cu_3Au at 450 °C. The shaded regions have their Au atoms on a different sub-lattice from that of the white regions (Gehlen & Cohen 1965).

dilute alloy that high values of the local order parameters may be interpreted in terms of 'clustering' or 'clumping', with domain size comparable with the observed range of order. This again is an elementary mathematical point that is sometimes overlooked in the attempt to give a simple 'physical' interpretation to the observations.

If our system really had a physical tendency to form *compact* blocks of ordered material, what would this imply in statistical language? The intuitive geometrical notion of a 'block' is not clearly defined, but presumably it means that the domain boundaries are smoother than they would be for perfect disorder. To assign a measure to 'smoothness' we need to look simultaneously at four or more sites in the lattice – three to define the tangent plane of the putative boundary and a fourth site that might lie on or near that plane. There would have to be a statistical tendency for sites in this favourable configuration to have more than their fair share of atoms of a particular type. In other words, such a tendency would be quantitatively defined by statistical parameters drawn from a *high order atomic distribution function*

$$P(\sigma_1, \sigma_2, \sigma_3, \sigma_4 \ldots) \tag{1.40}$$

which represents the probability of finding the value σ_1 in site 1, σ_2 on site 2, σ_3 on site 3, etc. As we shall see (chapter 4) it is almost impossible, in practice, to measure this type of distribution function directly, so that a physical tendency towards 'compactness' or 'smoothness' cannot be verified for small clusters. It is only when the range of order is very large that we may begin to think in macroscopic terms, introducing concepts such as *domain-boundary energy, surface tension, Bloch wall thickness*, etc. Such concepts obviously break out of the realm of the theory of homogeneous disorder, since they emphasize the spatial *inhomogeneity* of the transition between one region and another. It is important to recognize, however, that the existence of such inhomogeneities implies strong statistical correlations in the high order distribution functions for atomic positions or spins and cannot be deduced from pair distributions such as (1.39).

Below the critical temperature the statistical properties of a cooperative system are, of course, dominated by the finite long-range order. The parameter Γ_∞ defined in (1.34) rises very rapidly as T falls below T_c, and soon attains nearly its saturation value. This does not mean, however, that the correlation function is independent of range. In general, we may observe behaviour analogous to (1.37), i.e.

$$\Gamma(R) \approx \Gamma_\infty + AR^{-n'} \exp(-R/\xi'). \tag{1.41}$$

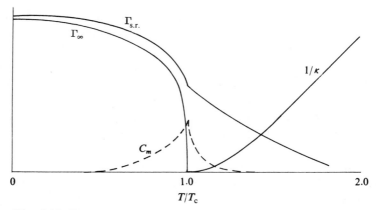

Fig. 1.18. Temperature variation of parameters of a typical ferro-magnet: Γ_∞, long-range order parameter; $\Gamma_{s.r.}$, 'short-range' order parameter; C_m, specific heat; κ, susceptibility.

Here again, the exponent n' depends on the dimensionality of the lattice, and the range of order ξ' and the fluctuation amplitude A will also depend on the temperature. The theory of this dependence will be discussed further in chapter 5. It is interesting to note, however, that the behaviour of ξ' *below* the transition temperature, as a function of $(T_c - T)$, is qualitatively similar to the behaviour of ξ *above* the critical temperature, as a function of $(T - T_c)$. Well below the critical point the range of order is small but it increases rapidly towards infinity as T approaches T_c. For the various models that have been studied carefully (see e.g. Elliott & Marshall 1958; Wu 1966; Kawasaki 1967) the two parameters ξ and ξ' do not behave symmetrically (fig. 1.18) about the critical point (§ 5.11), but the appearance of long-range critical fluctuations within the ordered phase has been experimentally confirmed in magnetic and alloy systems (see Münster 1965). As we go down in temperature, ξ' decreases whilst Γ_∞ is increasing, so that the general picture of a well-ordered system becomes more and more appropriate. Nevertheless, vestiges of additional short-range order can be observed in alloy systems that have been quenched from temperatures substantially below T_c (see e.g. Schwartz 1965).

1.8 Spectral disorder

We have made no use of the Bloch representation of the underlying crystal lattice. Suppose, in general, that the random variable u_l is associated with the site l of the lattice. This could be, for example, a component of the magnetic moment of a localized spin, or the magnitude of a small displacement of an atom from this site. Suppose, further, that the physical recipe for the model is translationally invariant on the lattice. We naturally introduce a Fourier transformation to new variables.

$$U(\mathbf{q}) = \frac{1}{\sqrt{N}} \sum_l u_l \, e^{i\mathbf{q} \cdot l} \qquad (1.42)$$

This transformation has the inverse

$$u_l = \frac{1}{\sqrt{N}} \sum_\mathbf{q} U(\mathbf{q}) e^{-i\mathbf{q} \cdot l} \qquad (1.43)$$

where the summation is over allowed wave vectors $\mathbf{q}$ in a zone of reciprocal space.

Now let us make a simple statistical assumption: let us assume that the *mode amplitudes* $U(\mathbf{q})$ are statistically independent for different allowed values of $\mathbf{q}$. The *static correlation function* for the *site amplitudes* u_l can be calculated immediately. Using (1.42), (1.43) and other standard theorems on lattice transformations, we get

$$\begin{aligned} \Gamma(\mathbf{h}) &\equiv \langle u_l^* \, u_{l+\mathbf{h}} \rangle \\ &= \sum_\mathbf{q} \langle U^*(\mathbf{q}) U(\mathbf{q}) \rangle e^{i\mathbf{q} \cdot \mathbf{h}}. \end{aligned} \qquad (1.44)$$

The correlation function is simply the Fourier transform of the mean square amplitude of excitation over the spectrum of separate modes. This important theorem is discussed further in §§ 3.2, 3.3.

We may describe such a system as being *spectrally disordered,* since the statistical properties are defined for variables in reciprocal space rather than for local site variables. When the circumstances are appropriate, the spectral representation of disorder greatly simplifies the description. Higher moments and correlations can be calculated as required, time variations can easily be included and the connection with the techniques of observation of disorder (chapter 4) is often very direct. Bloch's theorem is being used to the full. Each mode generated by the transformation is a one-dimensional representation of the translational group of the lattice. It is intuitively reasonable that these different representations should be statistically independent.

Nevertheless, the assumption of perfect spectral disorder is seldom justifiable. It would not, for example, make any sense to carry out such a transformation on Ising spin variables, since these can take only the two values ± 1. To impose this condition on (1.43) for each site label l would create a fantastic network of constraints on the complex mode amplitudes $U(\mathbf{q})$, which could not thereafter be treated as statistically independent. To avoid such constraints, which are mathematically quite intractable, the site amplitudes u_l must each have a continuous distribution of values.

The advantages and limitations of the spectral disorder model are exemplified by the standard case of *ferromagnetic spin waves*. We start from a system with the Hamiltonian (1.16), assuming that it is close to the perfectly ordered ferromagnetic state with long-range order parameter (1.30) near to unity. The local variable u_l would then be related to the amplitude of a *spin deviation* from its maximum value $S_l^{(z)} = S_0$. By a series of standard linear transformations we arrive at an expression for the Hamiltonian

$$\mathcal{H} = \sum_q \hbar\omega_q \, a_q^* \, a_q, \tag{1.45}$$

where a_q^*, a_q are spin-wave creation and annihilation operators in the quantum-mechanical case. The transformations leading from (1.16) to (1.45) are merely more complicated versions of (1.43). In this representation the *longitudinal spin correlation function* is given by an expression like

$$\langle S_l^{(z)} \, S_{l+R}^{(z)} \rangle \approx S_0^2 - 2S_0 \frac{1}{N} \sum_q \langle a_q^* \, a_q \rangle \mathrm{e}^{\mathrm{i}q \cdot R}, \tag{1.46}$$

which is obviously analogous to (1.44) (van Hove 1954; de Gennes 1963).

A system with the Hamiltonian (1.45) seems to have very simple properties. Suppose that the exchange interaction extends to z nearest neighbours at some interatomic distance a. Each spin-wave or *magnon* mode oscillates with frequency

$$\hbar\omega_q = \sum_h 2SJ(\mathbf{h})(1 - \mathrm{e}^{-\mathrm{i}q \cdot h}) + \bar{\mu}HS$$

$$\approx 2SJz \, a^2 q^2 + \bar{\mu}HS \tag{1.47}$$

for small values of q. These modes are dynamically independent. At temperature T, each is excited to a mean square amplitude/magnon occupation number given by the Bose–Einstein distribution:

$$\langle a_q^* \, a_q \rangle = \bar{n}_q + \tfrac{1}{2}$$

$$= \{\exp(\hbar\omega_q / \mathit{k}T) - 1\}^{-1} + \tfrac{1}{2}$$

$$\approx \mathit{k}T / \hbar\omega_q \tag{1.48}$$

when $\hbar\omega_q \ll \mathit{k}T$. To calculate the spin correlation functions we merely substitute from (1.47) and (1.48) into (1.46).

But look at the form of this expression for large values of R. The sum becomes an integral whose main contribution comes from the lower limit near $q = 0$. With the approximations (1.47) and (1.48) we get an expression that behaves like

$$\frac{1}{N}\sum_{\mathbf{q}}\langle a_{\mathbf{q}}^* a_{\mathbf{q}}\rangle e^{i\mathbf{q}\cdot\mathbf{R}} \sim \frac{\mathscr{k}TV}{8\pi^3 N}\int\frac{e^{i\mathbf{q}\cdot\mathbf{R}}\,\mathrm{d}^3q}{2SJza^2q^2+\bar{\mu}HS}$$

$$\sim \int\frac{e^{i\mathbf{q}\cdot\mathbf{R}}}{q^2+\kappa^2}\,\mathrm{d}^3q$$

$$\sim \frac{1}{R}\exp(-\kappa R), \tag{1.49}$$

where

$$\kappa^2 = \frac{\bar{\mu}H}{2Jza^2}. \tag{1.50}$$

This is of the form (1.37) with 'range' $1/\kappa$.

But this 'range of order' depends upon the external magnetic field. As H tends to zero, $1/\kappa$ tends to infinity. It is as if the fluctuations of spin away from the ordered ferromagnetic state could themselves acquire long-range order. This nonsense can easily be explained by reference back to the original spin Hamiltonian (1.16). Without a magnetic field the system has complete rotational symmetry for the quantization direction of the spin vectors $\mathbf{S}_l$. The 'long-range order' of the 'spin fluctuations' merely describes the whole assembly of spins swinging round into a different orientation whilst still remaining ferromagnetically aligned spin by spin. To prevent this catastrophe we must either retain the external field or introduce some additional magnetic anisotropy (cf. (1.17)) that prevents the frequency of the mode with $q=0$ going to zero (Kawasaki & Mori 1962; Kawasaki 1967). Serious mathematical difficulties can easily arise in a system with *broken symmetry* if we fail to introduce such symmetry-breaking terms as convergence factors in the analysis.

Even when this difficulty is settled, the magnon representation of magnetic disorder is useful only at low temperatures. In deriving (1.45) and (1.46), higher order terms in the spin deviation operators have been dropped. Even a careful term-by-term expansion in powers of $a_{\mathbf{q}}^*$ and $a_{\mathbf{q}}$ fails to do justice to the simple fact that the spin on a given site cannot be 'deviated' further than the direction of complete reversal. The distribution of the variable u_l, however it may be defined, must be sharply cut off at a value corresponding to this maximum amplitude. But this introduces complex constraints in the Hilbert space of the magnon states, which do not satisfy the assumptions of statistical independence. This point need not be laboured, since it is the fundamental difficulty in approaching the Curie point from the ordered phase (e.g. Martin 1967; Krupicka & Sternberk 1967). We merely emphasize the significance of such considerations in the design of artificial models of disordered systems.

2
Topological disorder

—

2.1 Atomicity

To recognize a case of substitutional disorder we need to identify, by some means, the underlying crystal lattice. In a perfectly *ordered* crystal, all physical attributes are, by definition, strictly periodic. Any observable quantity, such as the electron density, or the one-electron potential, must have the formal mathematical property

$$F(\mathbf{r}) = F(\mathbf{r} + \boldsymbol{l}) \qquad (2.1)$$

for all lattice vectors $\{\boldsymbol{l}\}$ and for any position vector $\mathbf{r}$. When we make random 'substitutions' in such a crystal, we destroy this relationship: some physical parameters are no longer invariant under the translation group (1.1). Nevertheless, some observable quantities must still remain sufficiently periodic for us to identify the basic crystal lattice: there must be some physical procedure by which we can discover the set of lattice vectors $\{\boldsymbol{l}\}$. In a magnetic crystal, for example, the periodicity relation (2.1) might well be true for the spin-averaged electron density, even though the local value of the spin operator $S(\mathbf{r})$ would not be the same from site to site.

In the theory of condensed matter, we usually define such an underlying lattice by the positions of the 'atoms'. To be more precise, we should refer to the positions of the atomic *nuclei*, which are very small and yet carry all the positive charges and nearly all the mass. In dealing with a mixed crystal, for example, the basic lattice structure is implicitly defined by a set of instructions like 'plot the centres of mass of all distinct regions where the density exceeds one mass unit per cubic femtometre' or 'set down the positions of all positive point charges in the specimen, regardless of their magnitude'. Such recipes should produce physical observables that do satisfy the lattice translational symmetry conditions. When we say that a crystal is an ordered array of atoms this is as much as we are entitled to assume. It is then a question for physical investigation whether each nucleus is also the approximate centre for a characteristic region with various other standard properties, such as the charge distribution in closed electron shells, or a magnetic moment from an unpaired spin. For all we know, such properties may be quite sensitive to the randomly substituted environment of the 'atom' we have in mind, and may no longer satisfy the periodicity condition (2.1).

The present chapter deals with systems where the *atomic arrangement* is not an ordered lattice. We introduce into the problem a set of vectors $\{\mathbf{R}_i\}$ where $\mathbf{R}_i$ refers to the position of the nucleus of the ith atom in real space. For simplicity we shall usually assume that the atoms are chemically identical, or occur in identical molecular groupings, without the additional complication of substitutional disorder: but this is not an essential restriction provided that rules are given for the identification of atoms of a particular chemical variety, etc.

Once again, the question of what further physical properties are 'carried' by the nuclei cannot be answered *a priori*. In a liquid metal, for example (§ 10.2), the charge distribution within each ion core is thought to be nearly independent of the environment, but can vary considerably in the interstitial regions. In other words, this function satisfies a weakened form of (2.1), such as

$$F(\mathbf{r} + \mathbf{R}_i) \approx F(\mathbf{r}) \qquad (2.2)$$

for each value of $\mathbf{R}_i$, and for values of $|\mathbf{r}| < r_c$. Theoretical models of *topologically disordered* systems tend to rely upon arbitrary assumptions to cover such points without close attention to the variety of physical circumstances that may in fact occur in each atomic neighbourhood (fig. 2.1).

On the other hand, a constraint such as (2.2) cannot be ignored or averaged away. To say that the map of a 'random' function contains many very similar regions has profound consequences for its statistical properties. A disordered *atomic* system has much more specialized characteristics than, say, the optical density of a storm-wracked cloudy sky. The study of *continuum disorder* will be taken up in chapter 3.

The allowed values of the atomic position vectors $\{\mathbf{R}_i\}$ are, of course, strongly constrained by the physics of the 'atoms' of the specimen. There are only a few systems where it is legitimate, as in an *ideal gas* (§ 2.15), to treat each $\mathbf{R}_i$ as an independent random variable within the total volume of the sample. If the material is made up of well-defined atoms or ions, packed fairly densely, then their mutual impenetrabilities and interactions become

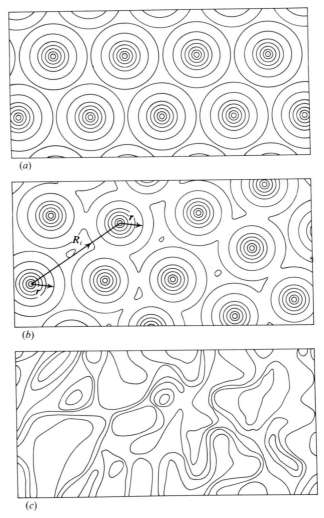

Fig. 2.1. (*a*) Lattice order. (*b*) Topological disorder. (*c*) Continuum disorder.

the dominant influence on the statistical ensemble of arrangements. The main problem, in the present chapter, is to understand the effects of these physical *packing constraints* on the probability distribution of the set $\{\mathbf{R}_i\}$ in typical cases.

2.2 Disordered linear chains

It is natural to start from *one-dimensional* models. If the set of scalar quantities $\{R_i\}$ stands for the arrangement of atoms along a line we have a *linear chain*. An *ordered chain* would be defined as in (1.1) by the set

$$R_{(l)} = la, \tag{2.3}$$

where l is an integer. If the distances R_i are random variables then we say that we have constructed a *one-dimensional liquid* or a *one-dimensional glass*.

Various statistical assumptions may be made concerning the atomic arrangement in such a system. The simplest case is the *linear gas*, where each R_i is an independent variable, distributed with uniform probability along the whole length of the chain. The only statistical parameter that needs to be defined is the *packing density*, or its inverse, the average spacing

$$a = L/N, \tag{2.4}$$

whose limit would remain constant as the length L and number of atoms N become large.

Since the *absolute* coordinate of an atom in the chain is not a significant physical variable it is better to prescribe the statistical properties of the *relative* positions of the atoms. In the linear gas model, for example (§ 6.1), the successive atomic spacings

$$\xi_i = R_{i+1} - R_i \tag{2.5}$$

are independently distributed with the Poisson distribution (cf. Feller 1967)

$$P(\xi)\,\mathrm{d}\xi = a^{-1}\,\mathrm{e}^{-\xi/a}\,\mathrm{d}\xi. \tag{2.6}$$

In some cases it is convenient to use a *Gaussian linear liquid* in which each ξ_i has a normal Gaussian distribution with variance σ_D^2 about the mean value a, even though this is not, apparently, observed in an actual physical system.

To lend an air of verisimilitude to the model we may assert that the atoms are mutually impenetrable and may not approach closer than some minimum diameter D, whilst any gap larger than some length G would be occupied by another atom if it opened up. This is the physical justification for Borland's model (1961), where the interatomic spacing was held to lie in a fixed range

$$D \leqslant \xi_i \leqslant G. \tag{2.7}$$

To establish these limits, we might refer to a Monte Carlo calculation for the packing of equal 'O-spheres' (Smalley 1962), which showed that equal lengths could be distributed at random on a line without overlapping, up to

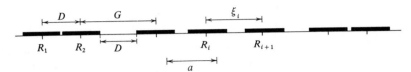

Fig. 2.2. Borland model = packing of O-spheres = car-parking problem.

a density of about 0.75 relative to the close-packed regular density: in other words $D_c = \frac{1}{2}G \approx 0.75a$ would be a reasonable choice for a 'liquid' model. A formal derivation of the value 0.7476 for this limit for the 'car-parking problem' was in fact given by Solomon (1954–5). A similar result may be obtained by playing with a truncated Poisson distribution (2.6) for the gap lengths (fig. 2.2) (Cooper & Aubourg 1972).

Having set up the atomic arrangement we must define the other relevant parameters of the model. To study the *lattice dynamics* of a one-dimensional glass, for example (chapter 8), we postulate that the interatomic forces must vary as a function of the distance between neighbouring atoms. Variation of the overlap integrals between bound states on adjacent atoms provides a *tight-binding model* of electron states in disordered systems (§§ 8.1, 9.1). *Spin-diffusion* models can easily be set up along the same lines by varying the exchange parameters in the Heisenberg Hamiltonian (1.15). For theories of electron propagation in a liquid metal a typical starting point is the disordered *Kronig–Penney model*, in which the electron potential at each atom is a delta function

$$\mathscr{V}(r) = \sum_i \delta(r - R_i). \tag{2.8}$$

This can easily be generalized to an array of *superposed atomic potentials*

$$\mathscr{V}(r) = \sum_i v(r - R_i) \tag{2.9}$$

where each local potential v can be fashioned to satisfy the 'atomicity criterion' (2.2) whilst still showing substantial random variation in the interstitial regions (fig. 2.3).

The physical properties of these various models will be discussed in later chapters. The point to be emphasized here is that a *one-dimensional system cannot be topologically disordered*. In almost every case it is possible to construct a *substitutionally* disordered *regular* array which is mathematically equivalent to the 'linear liquid'.

A general proof of this assertion may not be trivial but the essence of the argument is very simple. Any set of real numbers $\{R_i\}$ can be ordered

(a)

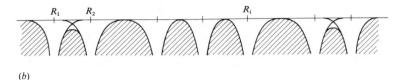

(b)

Fig. 2.3. (a) Disordered Kronig–Penney model. (b) Superposed atomic potentials.

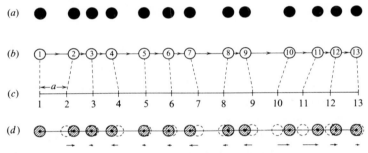

Fig. 2.4. (a) Disordered linear chain. (b) Unique ordering by counting. (c) Equivalent ordered lattice with average spacing a. (d) Chain now topologically ordered, with disordered atomic displacements.

uniquely in magnitude. In other words, we can label the atoms unambiguously by moving from one to the next along the line. The total number of atoms in the length of the specimen is also known. The disordered arrangement is, therefore, equivalent to a unique *ordered* linear chain (2.3) with the same average spacing. The state of disorder may thus be described by simply counting along the chain until we reach the lth atom and writing down the distance $R_{(l)} - la$ that it has moved from its presumed ordered site (fig. 2.4). This transformation is unique but topologically irrelevant, since it can be achieved by a succession of infinitesimal perturbations without discontinuities such as changes of labels.

In cases where only local relative coordinates such as (2.5) appear in the physical prescription, the atomic spacing $\xi_{(l)}$ clearly plays the part of a

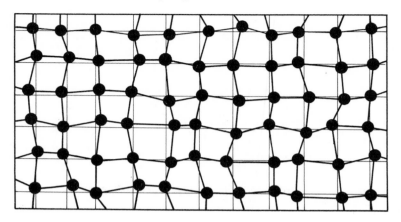

Fig. 2.5. 'Hot solid' is not topologically disordered.

random substitutional variable at the lth site of the equivalent ordered system. This applies, for example, to all problems of lattice dynamics where there is no essential mathematical difference between 'varying the distances between successive atoms' and 'varying the apparent force constants and/or masses in each unit cell'. The same applies to tight-binding and spin-diffusion models. For problems of electron propagation in a potential such as (2.8) or (2.9), the construction of an equivalent substitutionally disordered lattice is more complicated, since it depends upon defining a new potential function in each cell that will simulate the effects of variable spacing between the corresponding atoms of the 'liquid', but the same principles apply (§ 8.1).

In genuine liquids and glasses, irreducible topological disorder in three dimensions certainly occurs. The present argument indicates that the full effects of such disorder cannot be simulated in a linear chain model. Any physical property deduced for a 'one-dimensional liquid' need not necessarily be observed in an actual liquid, glass or gas.

The true analogue of the 'linear liquid' in two or three dimensions is the *hot solid* (fig. 2.5) – an arrangement of atoms having the topology of a regular lattice but with random variations of interatomic spacing. Such a system, however, is best described in the language of substitutional disorder, the random variable being the displacement of each atom from its presumed site in a regular lattice. In practice, this takes us back to the theory of spectral disorder (§ 1.8) in the phonon modes of the system.

2.3 Physical realizations of one-dimensional systems

One-dimensional systems have long been of theoretical interest (see e.g. Lieb & Mattis 1966) as hypothetical models with simple mathematical properties. But this theory (chapter 8) would be more significant if materials with these properties could be found in nature. The search for, or deliberate design of, *quasi-one-dimensional systems* is one of the subsidiary enterprises of solid state physics and chemistry. In recent years, systematic research and chemical ingenuity have provided experimental physicists with a variety of materials that roughly approximate to the theoretical specifications. The theory of the disordered linear chain is thus not entirely academic. The possibilities for the physical realization of *quasi-two-dimensional* or *layer* systems are so manifold, in principle and in practice, that we cannot deal adequately with them in this book.

To realize a *magnetic chain* (see e.g. de Jongh & Miedema 1974; Steiner, Villain & Windsor 1976), one must find a crystalline material where magnetic ions interact with one another along chains which are nearly isolated from one another by large non-magnetic ions, radicals or molecular groupings. In *copper benzoate* $(Cu(C_6H_5COO)_2 \cdot 3H_2O)$, for example, the packing of the large flat benzoate groups allows the magnetic Cu^{++} ions to come within exchange distance along the c axis (fig. 2.6) but keeps these chains well apart in the a and b directions of the monoclinic lattice. This material is said to be a simple linear Heisenberg antiferromagnet (§§ 1.3, 1.5). Very similar behaviour is found in *copper dipyridine dichloride* $(CuCl_2 \cdot 2NC_5H_5)$ where the exchange interaction probably passes through the Cu–Cl–Cu bonds between Cu^{++} ions in chains along the c axis. A whole class of inorganic magnetic chain materials has been found with the general formula $(Alk \cdot Mag \cdot Hal_3)$. The magnetic ion 'Mag' is a transition metal such as Mn^{++}, Ni^{++} or Cu^{++}; 'Alk' stands for an alkali metal or for ammonia; 'Hal' is a halogen such as F or Cl. For example, $CsMnCl_3 \cdot 2D_2O$, where the Mn^{++} ions are linked by superexchange through three shared Cl^- bonds (fig. 2.7) is a linear antiferromagnetic. The magnetic insulation of the $MnCl_3$ chains can be improved by replacing the alkali metal with deuterated *tetramethyl ammonia,* making 'TMMC' $[(CD_3)_4NMnCl_3]$ which shows one-dimensional behaviour down to 1 °K. There is thus quite a diversity of magnetic chain systems for experimental and theoretical analysis.

The first idea one might have for an *electronic chain* would be a long-chain polymer. But these are almost always fully bonded, so that the

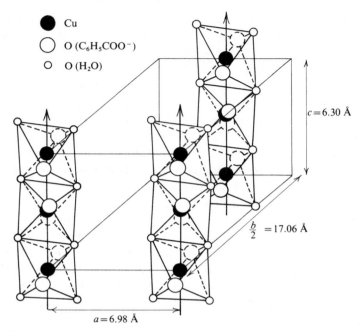

Fig. 2.6. Copper benzoate (Date, Yamazaki, Motokawa & Tazawa 1970) showing magnetic chains. The rest of the lattice is packed with benzoate ions and water molecules.

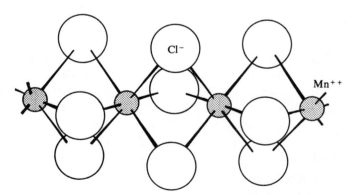

Fig. 2.7. Magnetic chain part of the TMMC structure.

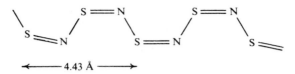

Fig. 2.8. In poly(sulphur nitride), chains linked by unsaturated bonds run along the crystallographic [010] direction.

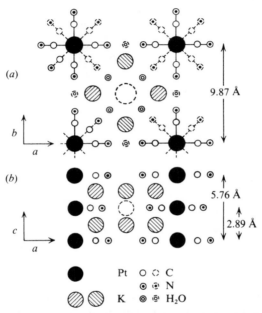

Fig. 2.9. Crystal structure of a tetracyanoplatinate. (a) Projection parallel to, and (b) projection perpendicular to, the tetragonal axis. The large circle in the centre indicates a site for a halogen atom (Krogmann 1969).

physics of the electrons is dominated by correlation effects. The electrons may as well be thought to occupy local bond orbitals as extended Bloch states. However, the 'inorganic' polymer *poly(sulphur nitride)* $(SN)_x$ shows genuine metallic conductivity (Walatka, Labes & Perlstein 1973), due to electron transport along extended chains of chemically unsaturated bonds (fig. 2.8), and thus approximates to a *linear metal*.

Metallic behaviour is also observed in the group of *tetracyanoplatinates* such as $K_2[Pt(CN)_4]Br_{0.3} \cdot 2.3H_2O$ (fig. 2.9) where the Pt d-band is not full

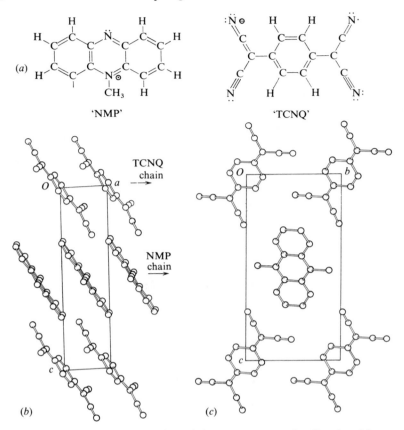

Fig. 2.10. (*a*) *N*-methyl phenazinium tetracyanoquinodimethamide. (*b*) [010] projection. (*c*) [100] projection: the disordered *N*-methyl group is shown on both of its alternative sites.

and where the Pt–Pt distance is only 2.89 Å (Kuse & Zeller 1971). The secret of these structures is that the [Pt(CN)₄] groups, or analogues such as [Pt(C₂O₄)₂], form flat squares which easily pack on one another.

The most interesting linear metals are analogues of a substance called, for short, (NMP)(TCNQ) (fig. 2.10) which have the highest electrical conductivity of any purely organic solids (Melby 1965; Leblanc 1967). In the crystal the NMP groups lose an electron each to the TCNQ. The flat (NMP)⁺ radicals then stack close to one another to form one set of chains, whilst the (TCNQ)⁻ radicals form the other chains (Fritchie 1966). Electrical conduction is thus relatively easy by charge transfer along each chain.

These practical examples have been given in a little detail to show the physical and chemical tricks that are needed to produce systems that conform to a one-dimensional model. The magnetic systems seem to behave more or less according to the theory for ideal linear chains, but it is still not certain whether the electronic transport properties of these systems are due to homogeneous one-dimensional disorder such as might be produced by thermal vibrations or random magnetic spins, or whether they are due to broken chains, inhomogeneities of the bulk material or many-body electron–electron interactions.

2.4 Dimensionality and order

Strangely enough, despite its apparent simplicity, the hypothetical ordered linear chain about which so much is written is physically unrealizable as a genuine one-dimensional system. The cases considered above are really three-dimensional materials in which approximately one-dimensional arrays are embedded.

Suppose, for example, that we had a chain in which the successive atomic spacings (2.5) were supposed to be equal. This could not be achieved precisely: the variable ξ_i would be distributed, at best, with some very small variance σ_D^2. Now consider the distance between atoms separated by N links of the chain. By elementary statistics, this would be distributed about the average value Na with variance $N\sigma_D^2$. In other words, the fluctuations of the absolute length of the chain about its average length would increase in proportion to the number of links.

This suggests that a linear chain lattice would be thermally unstable. In fact, it can be shown under rather general conditions concerning the range of the forces that *spontaneous crystalline order should not exist in one or two dimensions* (Mermin 1968). The complete proof of this theorem is rather lengthy but the special case of a harmonic lattice was originally proved in 1935 by Peierls by the following elementary argument.

From the harmonic model for lattice vibrations the usual transformations (1.43), (1.48), etc. to phonon variables gives a companion result to (1.49), i.e.

$$\langle |u_l - u_{l+R}|^2 \rangle \sim \frac{2\hbar}{MN} \sum_q \frac{1}{\omega_q} \frac{(1 - \cos qR)}{\exp(\hbar\omega_q/kT) - 1}. \tag{2.10}$$

We are interested in the behaviour of this sum for large values of R, so we may replace the denominator by $\hbar\omega_q/kT$ and put $\omega_q = sq$, where s is the velocity of sound. In a d-dimensional lattice, we obtain a d-dimensional integral in reciprocal space, i.e.

$$\langle |u_l - u_{l+\mathbf{R}}|^2 \rangle \sim \frac{kT}{Ms^2} \left(\frac{a}{2\pi}\right)^d \int \frac{(1-\cos \mathbf{q}\cdot\mathbf{R})}{q^2}\, \mathrm{d}^d q. \tag{2.11}$$

For $d=1$ this integral converges to a multiple of R; for $d=2$ it behaves asymptotically like $\ln R$. In each case, therefore, the fluctuations of relative position of atoms on distant sites increase without limit as we move further away. But for $d=3$ the integral converges to a small value that is independent of R so that the assumed lattice order is stable.

A very similar result can easily be proved for the spin-correlation function $\langle |S_l - S_{l+\mathbf{R}}|^2 \rangle$ between distant sites in an ordered ferromagnetic array (Bloch 1930). This function does not, of itself, measure long-range magnetic order and is not, as in (1.49), sensitive to rotations of the whole array. But it can easily be evaluated in the spin-wave representation (1.46): in both ferromagnetic and antiferromagnetic systems it comes out proportional to an integral like (2.11). The fact that this expression increases as R increases, unless $d \geqslant 3$, shows that the assumption of magnetic ordering is not consistent with the magnitude of the fluctuations of relative spin on distant sites. In the absence of a finite magnetic field or magnetic anisotropy to modify the magnon spectrum (1.47), we have apparently demonstrated that *spontaneous ferromagnetic or antiferromagnetic order should not appear in a one- or two-dimensional system.*

This theorem has also been proved exactly for an array of spins with isotropic Heisenberg exchange interactions of finite range (Mermin & Wagner 1966). As with Mermin's theorem, the proof depends upon Bogoliubov's inequality relating ensemble-averaged thermal fluctuations, followed by a succession of further elementary inequalities which lead, in the end, to d-dimensional integrals over a Fourier representation of the dynamical variables, not unlike (2.11).

It does not seem possible to confirm the predictions of the Peierls–Mermin theorem by a direct 'crystallization' experiment. But the absence of magnetic ordering down to the lowest temperatures has been confirmed in several of the magnetic chains referred to in § 2.3. In practice, many such systems are highly anisotropic, so that they would not satisfy the conditions of isotropy assumed for the interactions between spins.

It is easy to prove, however, that *spontaneous order is thermally unstable in a one-dimensional Ising model* (Peierls 1936). Suppose we have such a chain, with all 'spins' +ve (fig. 2.11). This long-range order may be

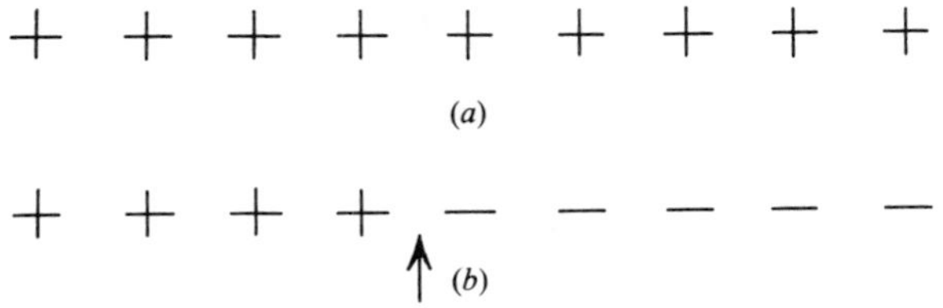

$+ \quad + \quad + \quad + \quad + \quad + \quad + \quad + \quad +$

(a)

$+ \quad + \quad + \quad + \quad - \quad - \quad - \quad - \quad -$

↑ (b)

Fig. 2.11. Long-range order (a) in a linear chain is destroyed (b) by a single break.

destroyed by the introduction of a 'domain boundary' at some arbitrary point of the chain, beyond which all spins are reversed. This can be done in N ways, yielding an entropy gain $k \ln N$. But the increase of exchange energy (1.18) in the reversal is only $2J$, so the total change of free energy is

$$\Delta F = 2J - kT \ln N. \tag{2.12}$$

For large enough N this is always negative at any finite temperature, showing that disorder is favoured thermodynamically.

The two-dimensional Ising lattice does, of course, show spontaneous magnetic order below a transition temperature (see § 5.7). The dimensional argument for this was also given by Peierls. Consider again a domain of reversed spins (fig. 2.12) whose boundary crosses L neighbour–neighbour exchange interactions and has, therefore, cost an energy $2LJ$ to make. How many different domains can be made with just that length of boundary? Roughly speaking, as we traverse the boundary there are 3 choices of direction at each node, so that there might be something like 3^L different shapes of domain. Thus, the free energy change is

$$\Delta F \approx 2LJ - kT \ln 3^L$$
$$= L(2J - kT \ln 3), \tag{2.13}$$

which is positive if

$$kT < 2J/\ln 3. \tag{2.14}$$

This value for the spontaneous magnetization temperature is an underestimate and the true result must depend upon the actual lattice structure but the reliance of the argument on the dimensionality of the system is obvious. More systematic and rigorous analysis of these questions (Griffiths 1971, 1972) has merely confirmed the essential validity of the Peierls argument.

This sort of argument can even explain the Bloch–Mermin–Wagner theorem (Wannier 1959). The Heisenberg exchange energy (1.15) between

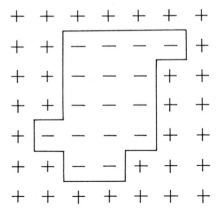

Fig. 2.12. A region of reversed Ising spins.

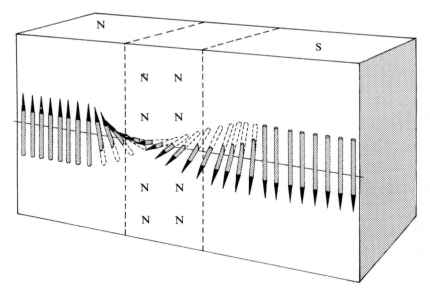

Fig.2.13. Bloch wall between ferromagnetic domains.

two neighbouring spins depends on the angle θ between them. A small deviation from perfect order costs the energy

$$2J(1 - \cos \theta) \sim J\theta^2. \tag{2.15}$$

We can now turn our sharp domain boundary into a *Bloch wall* (fig. 2.13) where the transition from the up spin to the down spin region goes through m steps, at each of which there is a deviation $\theta \sim \pi/m$. Thus, the total energy to make this domain boundary is of the order

$$\Delta E \sim LmJ(\pi/m)^2. \tag{2.16}$$

If now we let m be some small but constant multiple of L, this expression will be independent of L. With the entropy still proportional to L we are back, in effect, at (2.12), showing that the assumed order was not thermally stable. Here again, exchange anisotropy will spoil the proof, and stabilize the magnetization.

In this section we have begun to see the great importance of *dimensionality* in all theories of spatial order and disorder. This fundamental topological characteristic may show its influence either through the convergence of a volume integral of some continuous function as in (2.11), or through some combinatorial factor in the counting of steps along lattice paths as in (2.13). Many examples of each case will be encountered as we proceed.

2.5 Dislocation disorder

The weakest class of topological disorder in a crystal is best described in the language of dislocation theory (see e.g. Nabarro 1967). In a perfect crystal each atom has a uniquely defined set of coordinates: these are simply the integral multiples of the lattice translation vectors (1.1) required to reach this atom from some fixed reference site. The net number of steps along each basis direction – counting forward and backward as positive and negative respectively – is independent of the path chosen and is thus a topological invariant of the lattice. But if the crystal contains a dislocation, one or other of these labels loses its uniqueness. The apparent number of steps between two sites A and B will depend upon the path followed. Each time the path encircles the dislocation line, this number alters by a unit. The dislocated crystal is thus no longer topologically equivalent to a perfect lattice (fig. 2.14).

A crystal containing a random distribution of dislocation lines (fig. 2.15) lacks long-range topological order. It is true that the apparent disorder is confined to the core of each dislocation, and that local crystal order is very

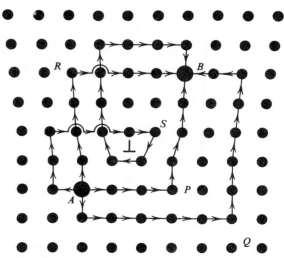

Fig. 2.14. Non-equivalent paths in a dislocated lattice.
$A \rightarrow P \rightarrow B \equiv A \rightarrow Q \rightarrow B \not\equiv A \rightarrow R \rightarrow B \not\equiv A \rightarrow S \rightarrow B.$

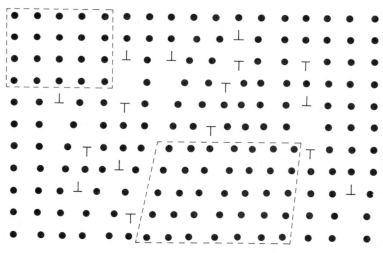

Fig. 2.15. Dislocation disorder with local crystal order.

accurately preserved in any region that is not pierced by a dislocation line. If there happens to be no preponderance of dislocations of a particular sign, then the overall orientational coherence of the original crystal will still be apparent. Yet there can no longer be a unique relationship between the atomic positions in this specimen and the sites of a perfect lattice of the same type.

In practice, this long-range topological disorder is usually ignored. The theory of *dislocation disorder* is dominated by the ability of each dislocation to maintain its independence as a quasi-stable entity, which exists and moves as a whole through the specimen. This is, of course, the key to the modern treatment of the mechanical properties of materials. For the types of physical phenomena discussed in this book the same approximation is also valid. For example, the residual electrical resistivity of a cold-worked metal is usually calculated as if it were the result of many individual scatterings from isolated dislocations (Ziman 1960; Nabarro 1967) with almost no reference to the absence of the long-range order assumed for the Bloch functions of the electrons (cf. Martin & Ziman 1970). Similarly the theory of lattice dynamics in dislocated materials becomes a study of the interaction of phonons with characteristic configurations of one or a few dislocations. In other words, the disorder is treated as mainly localized in the neighbourhood of each dislocation line without significant long-range or cooperative effects.

This approximation is usually justifiable because of the practical difficulty of achieving random dislocation densities exceeding about 10^{13} cm^{-2}, corresponding to average separations of about ten lattice spacings. At higher concentrations, dislocations tend to annihilate one another leaving a residue of point imperfections, or they line up in networks or regular arrays equivalent to grain boundaries. From a macroscopic point of view our specimen has *polygonized* or become *microcrystalline* (fig. 2.16). The regions of true local order have grown, and become more regular in shape, at the expense of the orientational coherence between the local lattice directions in distant grains. The range of topological order has increased but *orientational disorder* has appeared over long distances.

It has sometimes been argued (for references see Nabarro 1967: § 11.1.2) that a *liquid* is essentially a very highly dislocated crystal and that the phenomenon of *melting* is simply the spontaneous production of dislocations at the temperature at which this becomes thermodynamically favourable. Two-dimensional analogue studies of melting seem to follow this pattern (Cotterill & Pedersen 1972). This conjecture is supported by rough

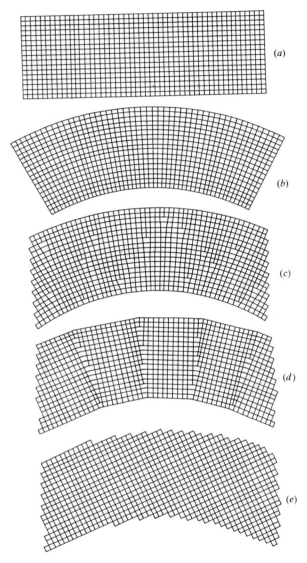

Fig. 2.16. Five states of a single crystal: (*a*) unstrained, (*b*) elastically bent, (*c*) plastically bent, (*d*) polygonized, (*e*) recrystallized.

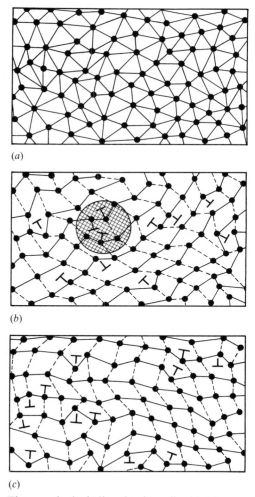

Fig. 2.17. The topological disorder in a liquid (*a*) can always be represented as a crystal containing a high density of strongly interacting dislocations (*b*) but this representation is not unique (*c*).

estimates of the relative balance of potential energy against entropy in the free energy for producing a new dislocation line in a heavily dislocated specimen (e.g. Kosterlitz & Thouless 1972, 1973). It is well known that the elastic strain energy stored in the medium about each dislocation line is proportional to the logarithm of the area per dislocation and thus decreases at high dislocation densities.

But this argument is fallacious at two levels. The estimate of free energy is incorrect, since most of the energy of a dislocation is associated with *atomic* disorder in the *core*, which is independent of the relative spacing of dislocation lines. There is thus no physical foundation for a cooperative catastrophe. More fundamentally, the concept of a *dis*location implies a *location* from which the assembly has been disturbed. The topological characterization of a given dislocation line is uniquely defined only when there remains a great deal of locally perfect lattice against which the discontinuity may be measured. If almost every atom is supposed to lie in the core of a dislocation there is no way of deciding where that dislocation actually lies. The description of topological disorder in the mathematical language of dislocation theory is valid only if the dislocations are sufficiently far apart to be defined unambiguously: otherwise the local disorder produced everywhere by the interaction of core regions is indistinguishable in principle from random close packing, which is best described in simple atomic terms (fig. 2.17).

2.6 Microcrystalline disorder

For a microcrystalline assembly the dislocation description is again inappropriate. It is much more efficient to treat each crystallite separately, as a finite but nearly perfect specimen of a regular crystal, where problems of electron propagation, magnetization, carrier mobility, lattice dynamics, etc. can be solved without regard to the boundary conditions. If each crystallite is larger than the mean free path for microscopic relaxation processes this is a reasonable approximation, since the atomic disorder is confined to the thin surface regions between grains. The scattering of electrons, phonons, magnons, etc. by these grain boundaries can then be estimated as a separate exercise. But, if the macroscopic crystal properties are not perfectly isotropic, the transmission of excitations through these boundaries may depend much more on orientational mismatch of the crystallites than on the actual arrangement of atoms in the contact region. The problem of calculating the bulk properties of such a material is by no means trivial, but the theory of orientational disorder is usually set up in the formalism of a classical macroscopic continuum, without regard to the absence of long-range topological order in the atomic arrangements.

The specimen might be described, for example, in geometrical language. We might assume that each grain has some characteristic diameter or the

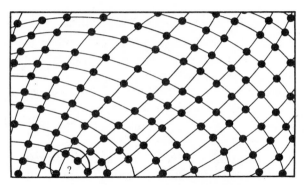

Fig. 2.18. Gubanov model of disorder.

various crystallites have a prescribed range of shapes and sizes. Or we might have to make assumptions about the statistical distribution of the directions of the crystal axes in various grains, including possible correlations between the orientations of neighbouring grains. In other words, we should feed into the computation the sort of information that might be obtained, in principle, by inspection of a photomicrograph of a section through the specimen where grain boundaries and other geometrical features could be clearly distinguished by eye.

It must be emphasized that neither dislocation disorder nor microcrystalline disorder satisfies the conditions of the 'quasi-crystalline' model of Gubanov (1965). In this model the crystal lattice is supposed to retain its topological order, but is so deformed by bending and stretching that orientational coherence is destroyed over large distances (fig. 2.18). This assumption is mathematically convenient because it justifies an adiabatic approximation in which the wave functions are guided continuously to follow the deformation of the lattice, with only weak transitions, calculable by perturbation theory, from the disorder. Unfortunately, this model is physically unrealistic. No condensed phase behaves like this in practice. The plastic deformation of a crystal is most efficiently accommodated by localized topological imperfections such as dislocations and grain boundaries, leaving the maximum amount of unstrained lattice in between. As we have seen, these imperfections act as strong scatterers whose effects cannot be described by residual perturbation terms left over from an adiabatic approximation. On the other hand, glasses and liquids are so greatly disordered beyond one or two atomic spacings that they cannot be represented at all as deformed regular lattices (see §§ 2.8–2.11).

2.7 Atomic distribution functions

A crystal structure is easily described analytically by the general formula
for its lattice sites (1.1). But in the absence of long-range topological or
orientational order a new formal language is required. A mere catalogue of
the coordinates of N atomic centres $\mathbf{R}_1, \mathbf{R}_2, \ldots, \mathbf{R}_N$ is useless since N is an
indefinitely large number. Whether we are dealing with a single large
system or an ensemble of similar systems it is essential to represent the main
features of such a catalogue by *statistical* parameters.

Even for a microcrystalline assembly, the macroscopic parameters of the
grains do not tell us all that we need to know, nor can they be introduced
directly into dynamical or thermodynamical relations involving the microscopic interactions between the atoms. There is no substitute for the
canonical formalism of *atomic distribution functions,* which we shall need in
any case when we study glasses and liquids.

These functions are, of course, the one-body, two-body, three-body . . .,
etc. *probability densities* $n(1), n(1, 2), n(1, 2, 3) \ldots$ The formal definition of
these is that

$$dP(1, 2, \ldots, s) = n(1, 2, \ldots, s)\, d\mathbf{1}\, d\mathbf{2} \ldots ds \qquad (2.17)$$

is the probability of finding an atom centred on the point at $\mathbf{1}$ in the volume
$d\mathbf{1}$, an atom at $\mathbf{2}$ in $d\mathbf{2}$, and so on. For most of the present work we assume
that these functions are time-independent, being deduced by sampling over
a static arrangement, so that we can only count the same atom twice if we
look again at exactly the same spot.

The basic assumption that our specimen is statistically *homogeneous* tells
us that the average density of atoms per unit volume must be constant, i.e.

$$n(1) \equiv n \equiv N/V \qquad (2.18)$$

independently of the position of the point $\mathbf{1}$. Multiple integration of (2.17)
then yields

$$V^{-s} \int_V d\mathbf{1} \int_V d\mathbf{2} \ldots \int_V ds\, n(1, 2, \ldots, s) = n^s. \qquad (2.19)$$

For simplicity of calculation it is then convenient to normalize the probability density to the canonical *distribution function*

$$g(1, 2, \ldots, s) = n(1, 2, \ldots, s)/n^s \qquad (2.20)$$

whose multiple integral per unit volume is unity. In other words,
$g(1, 2, \ldots, s)\, d\mathbf{1} \ldots ds$ measures the probability of finding this sort of
arrangement of s atoms, relative to all hypothetically possible arrangements of this number of atoms in assemblies of the same average density.
These definitions are standard and are given here simply to fix our notation.

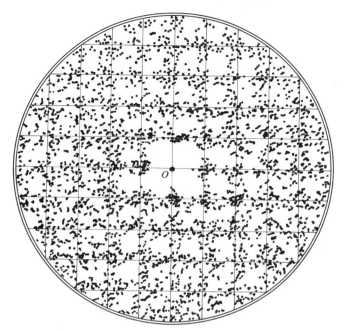

Fig. 2.19. Pair distribution function for model lattice where dislocations destroy long-range order. The vector from each point to the
origin is a value of an interatomic vector $\mathbf{R}_{12}$ from a picture such as
fig. 2.15.

As we shall see in chapter 4, almost all the direct observational evidence
about the atomic arrangements in condensed matter is contained in the *pair
distribution function* $g(1, 2)$. Because of spatial homogeneity this can
depend only on the relative vector separation $\mathbf{R}_{12}$ of points **1** and **2**, i.e.

$$g(1, 2) \equiv g(\mathbf{R}_{12}). \qquad (2.21)$$

For a perfect single crystal this function consists of delta functions at the
lattice sites, i.e.

$$g(\mathbf{R}_{12}) = n^{-1}\, \delta(\mathbf{R}_{12} - \boldsymbol{l}). \qquad (2.22)$$

Absence of long-range topological order, as in a dislocated crystal, would
mainly have the effect of broadening and smearing out the distant peaks of
this function into a uniform continuum (fig. 2.19). We might define the
range of local order empirically as the distance beyond which $g(\mathbf{R}_{12})$
becomes approximately equal to unity.

In a microcrystalline specimen this range would represent a measure of

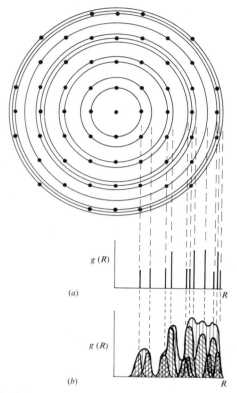

Fig. 2.20. Radial distribution function for (*a*) ideal and (*b*) thermally broadened microcrystalline assembly.

the size of the crystallites. But the orientational disorder, which makes such a specimen macroscopically *isotropic,* must be allowed for in the definition of the statistical distribution functions. However large each grain may be, the hypothetically infinite specimen or ensemble contains similar grains in all possible orientations. The pair distribution function (2.22) must therefore be averaged by rotation over all directions of the lattice vectors $\{l\}$. In other words, we finish up with a set of concentric spherical shells, with radii equal to the lengths of all possible lattice vectors:

$$g(\mathbf{R}_{12}) \equiv g(R) = n^{-1} N(l)\, \delta(R - l) \qquad (2.23)$$

where $N(l)$ is the number of lattice vectors that happen to have the same length l (fig. 2.20).

The structural evidence concerning an isotropic material is often no more than the form of the *radial distribution function* $g(R)$. For the ideal case of a microcrystalline specimen containing only one or two chemical species in simple combinations, it is not impossible to infer the three-dimensional local lattice from the observed peaks in (2.23). But if these peaks are seriously broadened by local disorder, by thermal vibration or by instrumental deficiencies, ambiguities arise and a unique interpretation cannot usually be found. This is a fundamental difficulty in the physics of topologically disordered materials: typical 'local' maps of the atomic arrangements cannot be constructed by formal analytical operations on the radial distribution function, but can only be guessed at and shown to be not inconsistent with this evidence.

A disordered system cannot, therefore, be adequately described analytically unless we go further up the hierarchy of distribution functions. Before looking at 'local order' in glasses and liquids it is instructive to think about the three- and four-body distributions for microcrystalline disorder.

By the principle of homogeneity $g(1, 2, 3)$ is a function only of the relative coordinates $\mathbf{R}_{12}$, $\mathbf{R}_{13}$. In a single crystal these vectors would have to be equal to lattice vectors, i.e.

$$g(1, 2, 3) = n^{-2} \, \delta(\mathbf{R}_{12} - \boldsymbol{l}) \, \delta(\mathbf{R}_{13} - \boldsymbol{l}'). \tag{2.24}$$

To make this expression symmetrical in the triangle of points **1, 2, 3** we might introduce a superfluous delta function for the third side of the lattice triangle, for example

$$g(1, 2, 3) = n^{-3} \, \delta(\mathbf{R}_{12} - \boldsymbol{l}) \, \delta(\mathbf{R}_{23} - \boldsymbol{l}') \, \delta(\mathbf{R}_{31} - \boldsymbol{l}'') \, \delta(\boldsymbol{l} + \boldsymbol{l}' + \boldsymbol{l}''). \tag{2.25}$$

Averaging now over all orientations of this triangle, we have

$$g(1, 2, 3) = n^{-3} \, \delta(R_{12} - l) \, \delta(R_{23} - l') \, \delta(R_{31} - l'') \, \delta(\boldsymbol{l} + \boldsymbol{l}' + \boldsymbol{l}'') \tag{2.26}$$

with, of course, broadening of the peaks into a continuum of unit density beyond the radius of a typical grain (fig. 2.21).

Notice at once that this function does *not* satisfy the *superposition approximation*, which is so often taken as a basis for theories of the liquid state (see § 2.12), i.e. we cannot assume that

$$g(1, 2, 3) \approx g(1, 2) \, g(2, 3) \, g(3, 1). \tag{2.27}$$

This is because not all combinations of lattice lengths l, l', l'' that satisfy the triangle inequalities are actually to be found as sides of a triangle in the local lattice: for example, on a single cubic lattice, if points **2** and **3** are nearest neighbours of point **1** then they are not nearest neighbours of one another (fig. 2.22). Notice, however, that any reasonable model of *one-*

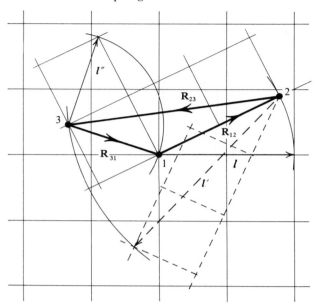

Fig. 2.21. For a microcrystalline assembly, the triplet distribution function $g(1, 2, 3)$ exists only for triangles belonging to the crystal lattice.

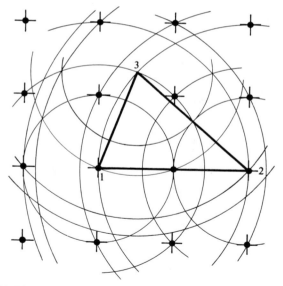

Fig. 2.22. The triangle $(1, 2, 3)$ generated by the superposition approximation does not belong to the lattice.

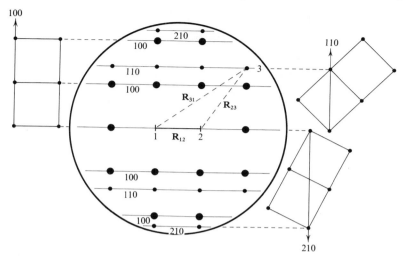

Fig. 2.23. Triplet correlation function for microcrystalline material. When R_{12} has length of a lattice vector, then R_{23} and R_{31} may have the lengths shown to these spots.

dimensional disorder (§ 2.2) would satisfy the superposition principle (Cooper & Aubourg 1972).

The evidence obtainable from the *triplet distribution function* concerning the local crystal structure in a microcrystalline system is still not very direct. The absence from $g(1, 2, 3)$ of certain points that would be generated from $g(1, 2)$ by the superposition approximation merely allows us to map out the relative positions of the atoms in the lattice planes: but these positions would all be superposed on a single diagram without any information concerning the relative orientation of the planes in three dimensions (fig. 2.23). Analytical formulae containing $g(1, 2, 3)$ for a system with long-range orientational disorder cannot, therefore, be expected to tell all about the effects of local order on electronic properties, etc.

To characterize such a material adequately one must go to the *four-body* distribution $g(1, 2, 3, 4)$. For any arbitrary orientation of the crystallite as a whole, the *relative* orientations of three vectors $\mathbf{R}_{12}$, $\mathbf{R}_{13}$, $\mathbf{R}_{14}$ must satisfy the length and angle conditions required of lattice vectors. In other words, we should at last obtain the full description of the local crystalline order that is given for a *single* crystal by $g(1, 2)$. Observe, moreover, that this function *does* satisfy the higher-order superposition principle (Fisher & Kopeliovich 1960)

$$g(1, 2, 3, 4) = \frac{g(1, 2, 3)\, g(1, 2, 4)\, g(1, 3, 4)\, g(2, 3, 4)}{g(1, 2)\, g(1, 3)\, g(1, 4)\, g(2, 3)\, g(2, 4)\, g(3, 4)}, \quad (2.28)$$

which is plausible *a priori* on probabilistic grounds.

The four-body distribution function is not only *necessary* for the analytical description of large-grain microcrystalline assembly; it is also nearly *sufficient*. Until we reach very large numbers of atoms, such as one might find in a whole grain, we can generate the higher order distributions by successive applications of generalizations of (2.28). This actual *closure* of the hierarchy of statistical characteristics of the material is essential if we are to make much use of this formalism in more difficult cases.

2.8 Bond network disorder

Non-crystalline, amorphous, vitreous or *glassy* solids, without discernible crystallinity on the finest microscopic scale, are familiar materials. It is natural to suppose that such a substance appears perfectly homogeneous and isotropic because it is topologically disordered on the atomic scale. General thermodynamic arguments indicate that a vitreous phase is technically only *metastable,* but cannot recrystallize because of local hindrance by the interatomic forces. Such hindrances can easily arise from steric 'jamming' of large irregular molecules, as in most organic glasses, but the most efficient mechanism is complete covalent bonding of the whole specimen, as in the inorganic glasses based on silica.

Stable silicate glasses are formed over wide ranges of composition in innumerable chemical mixtures. It is unlikely that all these materials would conform to a single, simple structure. The ideal type, however, is the *random network model* of Zachariasen (1932). The chemistry of SiO_2 strongly favours an 'infinite molecule' where each silicon atom is surrounded by a tetrahedron of four oxygens, each of which provides a bridge to a further silicon – and so on. These bonding rules can be satisfied by periodic repetition in a number of ways, of which the various crystalline forms of SiO_2 are all examples* (fig. 2.24). In *vitreous silica* the same *local* bonding conditions are supposed to be preserved, but the overall connectivity of the three-dimensional bond network is no longer regular and periodic. This is most simply pictured in two dimensions (fig. 2.25) where

* These rules are the same as for the O and H atoms in ice (§ 1.4). The arrangement of Si atoms in *high tridymite,* an allotrope of quartz, is exactly the same as that of the O atoms in Ice I.

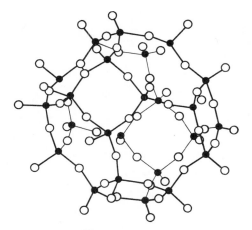

Fig. 2.24. Quartz.

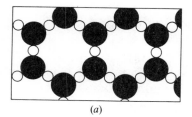

(a)

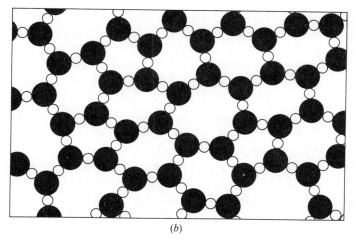

(b)

Fig. 2.25. 'Two-dimensional silica': (a) 'quartz'; (b) 'glass.'

each Si atom is now at the centre of a triangle of bonds. The three-dimensional network is easier to construct because it does less violence to bond lengths and preferred bond angles.

This is obviously a very plausible model for the structure of vitreous silica. It seems to explain most of the physical properties of silica glass extremely well and is not decisively contradicted by any experimental evidence. The question whether it is precisely correct is much more difficult to decide. The best evidence from X-ray diffraction (Bell & Dean 1966; Warren 1972) is consistent with a random network model where the Si–O bond lengths and O–Si–O bond angles are very close to the ideal tetrahedral angle,

$$\theta = 109° \ 28', \tag{2.29}$$

with only the Si–O–Si bond angles varying from 120° to 180°; but this agreement with experiment does not exclude other interpretations.

In more complex glasses the bond rules must be relaxed to accommodate other chemical constituents. In a typical alkali silicate glass, for example, the alkali ions can go into the interstices of the network where they are chemically neutralized by ionized oxygen bond defects. *Borate* glasses probably form *trihedral* networks based on the three covalent bonds of each boron atom with oxygen (cf. fig. 2.25). The actual texture of the specimen also depends upon its history since many inhomogeneities of composition and structure may develop during solidification. All such realities are discussed at length in the vast literature on the physics and chemistry of glasses.

The random tetrahedral network is a common type of topological disorder. Not surprisingly, *vitreous ice,* deposited from the vapour on a surface below −160 °C, is thought to be analogous to vitreous silica (Fletcher 1970; Alben & Boutron 1975). Models of the structure of *liquid water* are usually based on similar networks, with interstitial molecules or non-bonded line-pairs and protons as local defects to provide liquidity (see e.g. Symons 1972). But molecular dynamical computations on an assembly of model molecules with two positive and two negative charges arranged tetrahedrally about each oxygen provide a basis of comparison for these conjectures concerning the structure of water – to the detriment of concepts of 'clusters', 'rings', 'interstitials', etc. (Rahman & Stillinger 1971; Stillinger & Rahman 1972, 1974).

The most elementary examples are *amorphous silicon* and *germanium* which are also prepared by vapour deposition on a cold substrate. There is debate about the reproducibility and homogeneity of these materials, the

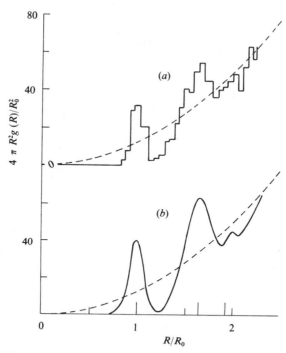

Fig. 2.26. Radial distribution functions: (*a*) for tetrahedral glass model; (*b*) from X-ray diffraction on amorphous Si.

concentration of defects such as 'dangling bonds', etc.; but it is generally agreed that the diamond lattice of the crystalline form of each element has simply been reconnected into a rigid, tetrahedrally bonded network without long-range order (Grigorivici & Manaila 1969). Here again, X-ray diffraction proves nothing positive about the structure but is consistent with theoretical radial distribution functions (fig. 2.26) generated from this model by geometrical analogue or computer simulation (Polk 1971; Henderson & Herman 1972). These materials will occupy our attention quite a lot in this book, not because their physical properties are necessarily equivalent to those of the compound amorphous semiconductors such as the *chalcogenide glasses* (see e.g. Mott & Davis 1971) but because of their theoretical simplicity as elemental materials.

2.9 Amorphous or paracrystalline?

The random network model for the structure of vitreous materials has certainly not gone unchallenged (see e.g. Bartenev 1970 for a review). The problem is to distinguish physically between this model and alternative hypotheses. The best evidence comes from X-ray and neutron diffraction and is never unambiguous. Where several chemical species contribute to the scattering, as in silicate glasses, the task of disentangling these contributions is extremely difficult and uncertain (see § 4.6). But even in the best case of a monatomic material such as amorphous Si or Ge, all that we can measure is the radial distribution function $g(R)$.

What can we learn from $g(R)$? As may be seen in fig. 2.27 this has certain 'features' which can be interpreted physically. It must, for example, be zero for a distance equal to the hard-core diameter of an atom, and then rises to a peak at some typical distance R_0. This is identified with the radius of the *first coordination shell* of atoms. The area under this peak,

$$z = \int_{\text{first peak}} g(R)\, 4\pi R^2 \, \mathrm{d}R, \qquad (2.30)$$

is the *coordination number* of the structure: if the hypothesis of tetrahedral bonding is correct, then the number of *nearest neighbours* should turn out to be $z = 4$. Similarly, the next peak comes from a *second coordination shell* of *next-nearest neighbours* – and so on. But the coordination number is not precisely defined (Pings 1968) and the number of atoms in each shell becomes more and more uncertain as the peaks broaden, merge with one another, and become lost in the continuum background where $g(R) \rightarrow 1$. For many theoretical purposes it is convenient to measure $g(R)$ itself relative to this background: *the total correlation function*

$$h(R) \equiv g(R) - 1 \qquad (2.31)$$

defines the extent of local variation from statistical uniformity about any given atom of the material. The *range of order, L*, would then be defined empirically as a distance such that $h(R) \approx 0$ for $R > L$.

As we have seen, these general features of the radial distribution function do arise quite naturally from a random network model. In the tetrahedral network, for example, the first and second coordination shells are almost identical in spacing and numbers with those in the perfect diamond lattice, but the rotation of tetrahedra about their common bond varies the distance to third neighbours (fig. 2.28); this provides a simple physical explanation of the loss of the third peak in the RDF as one goes from crystalline to amorphous Si (fig. 2.29).

But it is difficult to prove conclusively that such features are inconsistent with *hot solid* or *perturbed microcrystal* disorder. Suppose that the atoms of

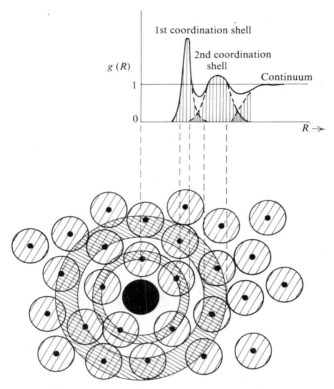

Fig. 2.27. 'Features' of the radial distribution function.

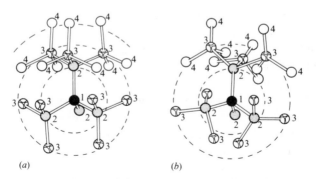

Fig. 2.28. Random network has same nearest and next-nearest neighbour distances.

Topological disorder

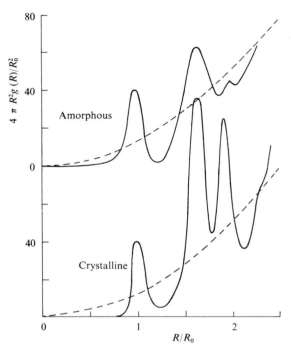

Fig. 2.29. Radial distribution functions of amorphous and crystalline Si.

a perfect crystal have wandered random amounts from the ideal lattice sites – perhaps, as in fig. 2.20, in the course of thermal vibrations. Each sharp peak in the three-dimensional pair distribution function (2.22) is now broadened and reduced in height. In a polycrystalline specimen this distribution must be averaged over all orientations of the local lattice, thus generating a radial function $g(R)$ in which the delta-function peaks (2.23) have been similarly broadened. But for large values of R these peaks become closely spaced; eventually they must overlap, merge into one another, and be lost in the continuum. The parameter L, describing the apparent decay of the correlation function, would thus depend mainly on the amplitude of the deviations from the ideal lattice sites and would have nothing to do with the range of topological order, which is here limited only by the grain size in the material.

This is the interpretation of observed radial distribution functions in glasses and liquids implicit in what is variously called the *crystallite hypothesis* (see Bartenev 1970), the *significant structure theory* (Walter & Eyring

1941; Eyring & Jhon 1969) or the *paracrystalline model* (Hosemann & Bagchi 1962). A slightly different interpretation attributes the apparent broadening of the peaks to diffraction effects from the quasi-crystals themselves, treated as distinguishable objects of finite size L randomly distributed in the medium (Kaplow, Strong & Auerbach 1965; Leadbetter & Wright 1972). These theories differ considerably in their assumptions and in the detail with which they are developed but they all presuppose the existence of fairly large 'clusters' of ordered material within the overall disorder. It is impossible to proceed further with any mathematical theory of an amorphous solid or a liquid without deciding in principle between this sort of picture and a more randomized arrangement of the atoms.

As we have seen, discrimination on the basis of the observable characteristics of the radial distribution function is very uncertain. The statistical characteristics of the perturbations in the paracrystalline model can be adjusted arbitrarily to fit $g(R)$ fairly well; the best that can be done is to try to show that a random network model with simple *a priori* parameters agrees with observation in fine quantitative detail. For a glass with two or more chemical constituents this is almost hopeless.

But an amorphous material must surely have many physical properties that depend upon more subtle statistical characteristics than the pair correlation function. Indeed, one of the aims of this book is to identify such properties. If this aim succeeds then we may have a means of discriminating between the two models.

For example, if each crystallite contains several hundred topologically ordered atoms then this could be more important as a source of coherent diffraction of electrons or phonons than the grain boundaries between them. The assumption that we should be making about the higher-order distribution functions such as $g(1, 2, 3)$ and $g(1, 2, 3, 4)$ would be very different in a paracrystalline model from what we should expect in a random bonded network. Some theories of electron states in disordered systems emphasize the role of the *local connectivity* of the lattice. In the diamond lattice the atoms all belong to 6-membered rings; the relative proportions of 5-, 6- and 7-membered rings in a disordered tetrahedral network might then turn out to be significant statistical parameters in some calculation of an observable physical property.

There are, in fact, good geometrical and mechanical reasons, that cannot be expressed algebraically, for rejecting the paracrystalline hypothesis as a model of the structure of a simple tetrahedral bond network. If a region is topologically ordered then it must conform quite closely to the perfect crystal structure; the stiffness of the tetrahedral bond angles prevents much bending and stretching. But if each crystallite is well-ordered internally, the regions between the grains must be very disordered if they are to fit together without grave defects and strains. For a bonded lattice of low coordination number this is practically impossible unless the boundary region is quite thick: the separate grains can only be held together by substantial quantities of more or less random tetrahedrally bonded material. But then we are assuming observable spatial inhomogeneity in the structure with large grains that ought to be visible by electron microscopy, etc. In other words, the experimental evidence is that the paracrystals, if they exist at all, cannot be more than a dozen or so ångströms in diameter: it is simply not possible to construct a tetrahedral network in which a high proportion of the atoms lie in such regions: if you try to make such a model by bringing small crystals together in arbitrary orientations you will find the grain boundary disorder spreading into the crystallites themselves until they have lost all recognizable existence. Until a protagonist of these models produces an actual three-dimensional structure that conforms to the assumptions he makes we must be doubtful whether it can be done at all, nohow.

There is some temptation to reconcile these models formally by treating the random network as a limiting case of microcrystallinity. But this is mathematically meaningless. A 5-atom tetrahedral group can be extended in a variety of ways, as if it belonged to any one of a number of different lattices, such as diamond, wurtzite, etc. Moreover, the 'boundaries' between such 'crystallites' are indistinguishable from the locally ordered material. The proposed separation into very small crystalline domains is, therefore, entirely arbitrary and its mathematical consequences cannot be supposed exact or unique.

2.10 Statistical geometry of bond networks

It is more profitable to recognize the random tetrahedral network as a special type of space lattice in its own right, with characteristic properties. Crystal lattices are defined, classified and analysed according to their long-range translational order and finite point group symmetries; *non-crystalline lattices* have only statistical homogeneity and short-range order without exact symmetry elements, yet they can still have well-defined statistical properties. The physics of disordered systems leads to the mathematical theory of *statistical geometry,* where models of this kind are studied as ideal cases.

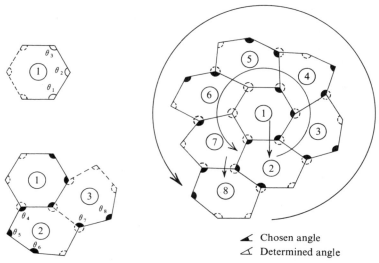

⊾ Chosen angle
⊿ Determined angle

Fig. 2.30. 'Trihedral glass' model of Cooper & Aubourg (1972), showing method of construction.

The axioms and theorems of statistical geometry as an abstract discipline have not, I think, been stated in a coherent fashion, but there are a number of empirical studies of the following type of problem: 'Given an *infinite graph* with prescribed *topological* properties (e.g. that each *vertex* should have a given *valency*), find *statistical* distributions of *metrical* properties (e.g. *lengths* of *edges*, *'bond angles'*, etc.) consistent with the *geometrical* realization of this graph in a stated number of *dimensions*.' Analytical solutions to such problems look almost impossible, but results have been obtained for several networks by mechanical models or computer simulation.

For example, what would be the distribution of bond angles in the two-dimensional *trihedral glass* of fig. 2.25? Cooper & Aubourg (1972) studied this question for a trivalent network, with all bonds of the same length, that was topologically equivalent to a regular hexagonal planar net (fig. 2.30). Each new hexagon added to the net was allowed a choice of any 'free' bond angles from an *a priori* distribution. At the end of the Monte Carlo calculation the overall distribution of angles in the network was measured: surprisingly this distribution was insensitive to the form of the *a priori* disorder function (fig. 2.31). In other words, the hexagonal network may exist as a crystal lattice with all bond angles exactly $2\pi/3$, or as a

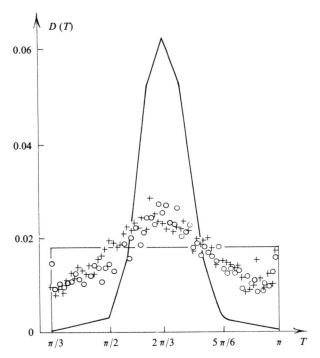

Fig. 2.31. The same final angular distribution is arrived at in the random network, whether one starts from a triangle distribution (o) or a uniform distribution (+).

statistically uniform non-crystalline lattice with this stationary distribution of angles. The pair correlation function for this canonical model is thus of possible physical interest (fig. 2.32).

What happens if we now relax the topological restraints by allowing 5- and 7-membered rings (as in fig. 2.25) whilst trying to optimize the distribution of bond angles to a sharper peak about $2\pi/3$? Do we still get a uniformly disordered lattice, or can an area of glass disorder be contained in a finite region within a perfect crystal? These questions are difficult to pose unambiguously, and have not been answered. Similar questions might be asked about the 'paracrystal' models of Hosemann & Bagchi (1962), which are mostly based upon the topologically regular square planar net. What would be the effect of triangular circuits in this net? Is there a stationary state of uniform disorder for a tetravalent network in the plane?

In three dimensions, attention has been mainly concentrated on tetrava-

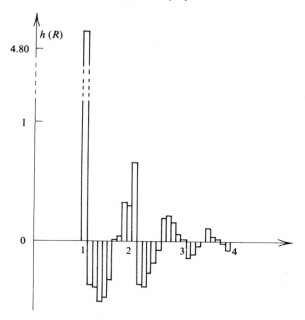

Fig. 2.32. Total correlation function for 'trihedral glass'.

lent nets. As a starting point it is usual to assume that all bonds are of equal length and that adjacent bonds form perfect tetrahedral angles. Disorder is permitted by allowing freedom in the *azimuthal angle* – the relative orientations of the arms of two tetrahedra about their common bond (fig. 2.33). In the perfect diamond lattice this orientation is always 'staggered': merely by permitting a proportion of 'eclipsed' configurations Grigorivici & Manaila (1969) were able to construct large *amorphons* – clusters of topologically disordered material – without other strains on the bonds.

It seems, however (Polk 1971; Paul, Connell & Temkin 1973), that this amorphous phase cannot be continued outwards indefinitely without the introduction of 'imperfections' such as 'dangling bonds' or large vacancies. Complete freedom of the azimuthal angle does not solve the problem. But slight variations of bond length and tetrahedral angle accommodate all the connectivity strains, and permit the construction of a uniformly disordered tetravalent network with well-defined statistical properties (Polk & Boudreaux 1973). Note that carbon bonds would not allow these strains so that ideal 'glassy diamond' does not seem to exist.

As we have seen (fig. 2.26), the radial distribution function for such a

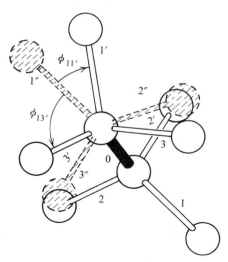

Fig. 2.33. Rotation of bond tetrahedra from the 'staggered' configuration $1''$, $2''$, $3''$ towards an 'eclipsed' configuration.

network is very close to the observed distribution for amorphous Si and Ge. In particular, the third hump in $g(R)$ for the perfect crystal, associated with third neighbours of a central atom, is broadened and lost because of the randomness of the azimuthal angle in the amorphous phase. The mean density of the amorphous structure turns out to be only about 1 per cent less than that of a crystalline array with the same average nearest-neighbour distance – again in reasonable agreement with experiment.

In the Polk model of amorphous structure the ratio of 6-membered to 5-membered rings is approximately 4:1. For a monatomic material such as amorphous Si or Ge, this may not be a significant fact but in a *compound* semiconductor, such as amorphous InSb, this means that we cannot keep identical atoms from occurring as neighbours. But it is possible to build a random structure (Connell & Temkin 1974) with only even-membered rings by allowing slightly larger variations in the tetrahedral angles. This model also has fewer of the (undesirable) 'eclipsed' configurations than the Polk model (fig. 2.34).

The higher-order correlation functions in this idealized model have not been studied but certain qualitative features can be deduced. The triplet distribution function $g(1, 2, 3)$ must be dominated by the tetrahedral configuration for three adjacent atoms, with very little structure beyond these distances. The superposition approximation (2.27) is quite invalid, because

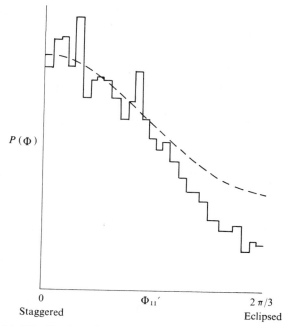

Fig. 2.34. Distribution of azimuthal angles in 'even-membered ring' model (histogram) and in Polk model (dashed curve).

of the low coordination number. But the four-body distribution must always be blurred by rotations about the azimuth angle and is probably well represented by the second-order superposition principle (2.28). It would be interesting to explore these relations quantitatively on the ideal model, to establish canonical statistical functions for calculations of electronic structure, etc.

2.11 The Bernal model of a liquid

For most substances, the atomic volume in the liquid phase is quite close to that of the crystal or glass into which it may be frozen. Liquids are also nearly as incompressible as solids. These elementary facts suggest that the repulsive forces between the atoms or molecules play a dominant role in the liquid state.

The simplest structural model one can imagine is that of an assembly of *hard spheres* – mutually impenetrable spherical atoms – without other interaction. For a rare gas, such as argon, this is a reasonable *a priori* description: the van der Waals attraction is much 'softer' and 'weaker' than the repulsion between closed electron shells when these are forced to overlap. As we shall see, this model is actually quite good for any liquid where the constituents are more or less spherical, as in a liquid metal.

An assembly of hard spheres can be arranged as a close-packed crystal – an ordered *pile* in which each atom touches twelve neighbours. The liquid phase is then to be thought of as a random *heap* of spheres, packed homogeneously as tightly as possible, but without long-range order. The fact that such a system is *fluid* against slowly-varying stresses is, of course, immensely important physically but not relevant to the present discussion. These motions are much slower than the physical phenomena of electron transport, etc. This simple idea, due originally to Bernal (1959) (but see also Rice (1944)), is now seen to be the key to any qualitative or quantitative understanding of the physics of liquids (Rowlinson 1970). The Bernal model for the topological disorder in a simple liquid supersedes various other theoretical approaches based on phenomenological constructs such as 'holes in lattice', 'paracrystals', 'significant structures', 'dislocations', 'glide pencils' (Kotze & Kuhlmann-Wilsdorf 1971), etc. to which occasional reference has been made elsewhere in this work.

What are the topological and geometrical properties of the ideal *random close-packed* (RCP) structure? Despite considerable efforts, very few exact formulae have been proved (see e.g. Finney 1970; Collins 1972). The best information we have is from empirical models, either physically realized with ball-bearings, plasticine, rods, etc. or simulated on a computer. In other words, a list of the coordinates of 'atom centres' is constructed satisfying the geometrical constraint that no two centres shall be closer than *d*, the diameter of an atom.

But this has to be done by trial and error. There is no simple algorithm for computing such a list to produce maximum packing density without long-range order. The various recipes for constructing RCP arrangements do not all converge on the same structure (Finney 1975). For example, if each new sphere is added so as to make contact with three existing spheres, the overall density does not turn out to be quite as high as may be achieved by some readjustments of position at the end (Adams and Matheson 1972; Visscher & Bolsterli 1972; Matheson 1974). Maximization of density is a very complicated geometrical operation involving many different atoms and cannot be reduced to a simple formula. The definition of the RCP structure may not even be well posed mathematically and provides little

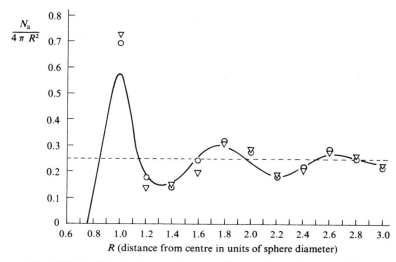

Fig. 2.35. Radial distribution function for random sphere models (▽, Scott; ○, Bernal) compared with observed RDF for liquid argon (——).

information from which other characteristics of the system can be directly deduced.

Instead of trying to build up a static sample of an RCP assembly containing hundreds or thousands of atoms, we may obtain much the same information from *Monte Carlo computations* (see e.g. Wood 1968) where large numbers of allowed configurations of a relatively small number of atoms are sampled according to their statistical weight in a thermodynamic ensemble. Similar information may be obtained by the method of *molecular dynamics* (see e.g. Alder, Hoover & Young 1968; Rahman 1968; Verlet 1968) where the coordinates of the atoms are allowed to move and 'collide' in a physically realistic way. These methods are necessary for the study of liquid models with proper interatomic forces in thermal equilibrium (see § 6.6). Equivalence with the results of the static Bernal model in the limiting case of hard-sphere repulsion has been verified (Finney 1970).

Given a sample of RCP structure, we can readily measure its atomic distribution functions. The radial distribution function $g(R)$ resembles very closely the observed distribution functions for many monatomic liquids (fig. 2.35), showing that the model does not conflict with reality. But thermal motion and genuine interatomic forces do not justify a precise quantitative comparison between this simple theory and experiment. In the

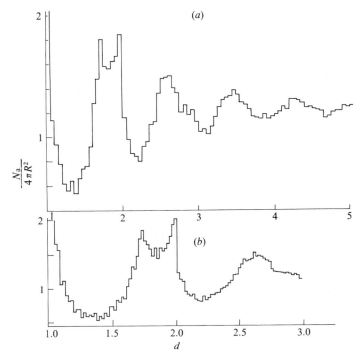

Fig. 2.36. (*a*) Radial distribution function for RCP model. (*b*) The same, to a finer scale.

RCP model, for example, the first coordination shell rises abruptly at $R = d$ because close packing makes almost every sphere touch at least four neighbours (fig. 2.36). The second coordination shell is also well defined but the peak is split. The steep drop at $2d$ is apparently associated with an excess of three-membered *collineations* – three atoms touching one another nearly in a line. Earlier peaks may be due to other special configurations, such as the distances between the apices of two tetrahedra ($1.633d$) or of two coplanar triangles ($1.732d$) with the same base (Finney 1970; Adams & Matheson 1972). In practice, these minor features of the radial distribution function for the idealized hard-sphere model would be smoothed away by thermal fluctuations and by softening the interatomic forces.

Triplet distributions can also be computed directly from the model. The angular distribution of the position of a third atom relative to an axis through a pair of adjacent atoms is shown in fig. 2.37 (Scott & Mader 1964). There is a strong peak at 60° in the first coordination shell, corresponding

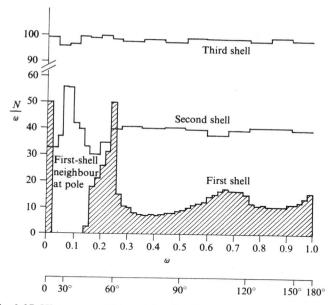

Fig. 2.37. Histogram of the angular distribution of neighbours relative to a first-shell neighbour as a pole. ($\omega = \frac{1}{2}(1 - \cos\theta)$ where θ is polar angle.)

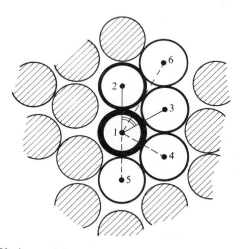

Fig. 2.38. Various neighbours in hard sphere model corresponding to peaks in angular distribution functions.

to atom 3 just touching the fixed pair 1 and 2 (fig. 2.38). Smaller peaks around 110° and 180° can be interpreted as favoured positions of atoms 4 and 5 in the first coordination shell of 1. The sharp peak in the second shell at 30° corresponds to atom 6 fitting between atoms 2 and 3. These peaks could, in fact, be predicted from the radial distribution $g(R)$ of fig. 2.36 using the superposition approximation (2.27) to calculate the triplet distribution $g(1, 2, 3)$: each peak of $g(R)$ then generates one of the special triangles depicted in fig. 2.38. The smearing out of $g(R)$ at larger distances is also consistent with the nearly isotropic angular distribution in the third coordination shell in fig. 2.37.

Detailed numberical investigations (Alder 1964*a*; Rahman 1964*a*; Krumhansl & Wang 1972; Block & Schommers 1975; Tanaka & Fukui 1975) have, in fact, confirmed the closeness of the superposition approximation for RCP models, except when all three atoms are nearly in contact (fig. 2.39). This is very important statistical information concerning liquid disorder, both in the analytical theory of liquids (§ 2.12) and in the theory of electrons in liquid metals (§ 10.5).

It follows that the statistical distribution functions for four or more atoms in a liquid can be calculated with quite sufficient accuracy from the higher-order superposition approximations such as (2.28), using $g(R)$ or $g(1, 2, 3)$ as basic information. For example, the approximate collineations of up to six atoms observed in 'ball and spoke' RCP models (Bernal & King 1968) would be generated by convolution of the favoured 180° angular configuration of triplets, already noted in fig. 2.37. The statistical distribution of the *canonical deltahedra* of the *simplicial graph* of such a model – i.e. the convex polyhedra with triangular faces built up from the lines joining centres of geometrical neighbours (see e.g. Collins 1972) – is then no more than a summary of some special properties of these many-atom distribution functions. The dissection of a given RCP assembly into canonical deltahedra is not a unique process, so that this description is not even topologically invariant, and has no physical significance beyond demonstrating the irregularity of the local arrangements of atoms in the assembly. In this respect, some of the original work of Bernal on this model seems to lead into a blind alley.

But the fact that the triplet superposition approximation is quite good shows that there cannot be any local 'crystalline' order beyond nearest neighbours in the liquid. This follows from (2.27): as shown in fig. 2.22 the superposition approximation cannot hold for an assembly of small crystals. Regular arrays of scores or hundreds of atoms are observed in a Bernal model only when this is built against a flat boundary (Bernal 1964): the

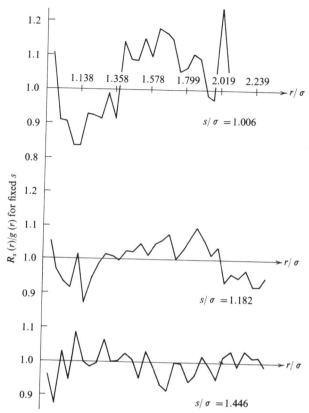

Fig. 2.39. The ratio of the triplet correlation functions $g_3(s, s, r)$ to the superposition approximation for a molecular dynamics computation on a 'Lennard-Jones' model, showing systematic deviations from unity when all three atoms are close together (Krumhansl & Wang 1972).

hexagonal close-packed surface layer then initiates crystallization to a considerable depth into the assembly. It is interesting to note that a typical Monte Carlo realization of a *hard disc fluid* in two dimensions (fig. 2.40) looks rather like an example of microcrystalline disorder (§ 2.6): it is by no means certain that a two-dimensional assembly has a distinct 'liquid' phase (see e.g. Wood 1968; Cotterill & Pedersen 1972; Visscher & Bolsterli 1972). This point is very important for the theory of liquid surfaces and of the nucleation of crystallization in freezing.

Nevertheless, the Bernal model of a liquid has one characteristic property of a solid: it can be thought of as an assembly of well-defined

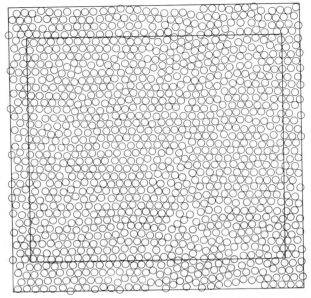

84   *Topological disorder*

Fig. 2.40. Atomic configuration of a 'hard disc fluid' (Wood 1968).

atomic cells. The geometrical dual of the simplicial graph divides space into *Dirichlet regions* or *Voronoi cells* (fig. 1.1) which have many interesting properties. The complete topological classification of the cells to be found in the RCP structure (Bernal 1964; Finney 1970) is not obviously helpful, but the numbers, q, of faces of the cells, corresponding to the numbers of *geometric neighbours* of the atoms, are significant structural indices. In a perfect static, close-packed lattice, all cells have $q = \bar{q} = 12$. Thermal disorder can produce occasional new faces, broadening the distribution of q and raising the average value $\bar{q}$ (fig. 2.41). For a genuine RCP liquid model, q may range from 12 to 17, with an average value $\bar{q} = 14.25$ (Rahman 1966; Finney 1970). In the limiting case of a *gas* of atomic centres – i.e. where the points are distributed in space at random without overlap restraints – this distribution is further broadened and the number of neighbours increases to the exact value $\bar{q} = 15.54$, which can be deduced by an elegant exercise in elementary solid geometry (Meijering 1953). In some cases, we may be counting an atom in the second coordination shell as a neighbour. Integration under the first peak of the RCP radial distribution gives an average *coordination number* $z = 12.3$ (Scott & Mader 1964) which is essentially the same as in the regular close-packed crystal.

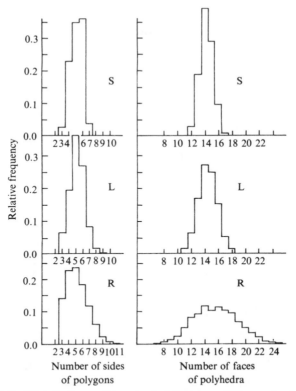

Fig. 2.41. Distributions of topological features of Voronoi cells for Bernal model: S = 'hot solid', L = RCP liquid, R = random gas (Rahman 1966).

The atomic volume is another very important parameter. For an assembly of hard spheres this is best represented as a *packing fraction* η – the proportion of the total volume within the spheres. For regular close packing $\eta = 0.74$. Randomization causes dilatation: studies of RCP assemblies all agree that these settle down to an average density of packing $\bar{\eta} = 0.637$. One of the most significant statements concerning the liquid state was a remark by Bernal (1960) reporting, from experience, on the 'absolute impossibility of forming a homogeneous (irregular) assembly of points of volume intermediate between those of long-range order and closest-packed disorder'.* In other words, a close-packed solid should expand by about 16

---

* This statement requires qualification. For example, for a BCC crystal $\eta = 0.68$. No doubt various crystalline or microcrystalline assemblies of intermediate density could be deliberately constructed if this were required.

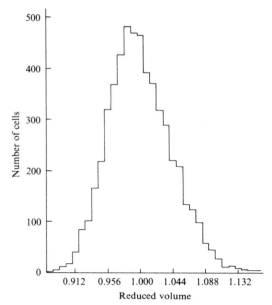

Fig. 2.42. Cell volume distribution for hard sphere RCP model (Finney 1970).

per cent in volume when it melts. This feature of the Bernal theory is confirmed for argon, but is masked in liquid metals by a redistribution of the electron gas.

But this, again, is only an average: as pointed out by Finney (1970), the cell volumes in an RCP assembly vary by as much as 10 per cent either way (fig. 2.42). This demonstrates once more the fundamental difficulty of representing the properties of an assembly of atoms from its pair correlations alone. From the point of view of each atom, the essential difference between the crystalline and liquid structures may not be the disappearance of long-range order but in the appearance of large fluctuations in very local parameters such as the coordination number and atomic volume. These are many-atom statistical features which cannot be deduced analytically from the radial distribution function itself.

At first sight, there seems no obvious connection between the two characteristic types of topological disorder. But a tetrahedral 'glass' structure (§ 2.10) may easily be transformed into quite a good 'liquid' and vice versa (Connell 1975; Chaudhari, Graczyk, Henderson & Steinhard 1975). Suppose, for example, we start from the model of Connell & Temkin (1974)

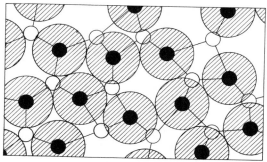

Fig. 2.43. Random close packing generated from random-bonded network with even-membered rings. (The deficiencies of the transformation are exaggerated in the two-dimensional representation.)

which is a random tetrahedral network with only even-membered rings of atoms. The sites in this model can, therefore, be assigned uniquely to two sub-networks in which atoms of type A have only B-type neighbours (fig. 2.43). Suppose now that all the B atoms are removed and then the A atoms expanded until they touch one another. This procedure (which transforms a crystalline diamond lattice into a face-centred cubic lattice) evidently produces a relatively close-packed arrangement without long-range order. But the relationship is not mathematically exact, since the inverse transformation starting from a good RCP structure produces a very imperfectly bonded tetrahedral network, showing once again the subtlety of the most elementary problems of statistical geometry and topological disorder.

2.12. Analytical theories of the liquid state

The Bernal model represents a liquid at its densest, as if it were essentially a disordered solid. An alternative approach is to start from a dilute gas, where the atomic distribution in space is perfectly random, and to calculate the consequences of increasing the density. Going from gas to liquid via the supercritical fluid regime this transformation is continuous, suggesting the possibility of constructing analytical formulae for the distribution functions and thermodynamic properties of liquids by the methods of the statistical theory of dense gases. We know, of course, that good results may be obtained by computer simulation of the liquid state but would appreciate analytic expressions from which useful further properties can be deduced without enormous computational effort. This is a very large topic, with a vast literature (see e.g. Cole 1967; Egelstaff 1967; Rushbrooke 1968; Rowlinson 1969; Croxton 1974) from which we shall draw only the simpler results. In the present chapter we are still concerned only with information about the *structure* of the liquid phase: the more general thermodynamic properties of liquids and especially the theory of structural transformations, such as melting, will be discussed in chapter 6.

It is essential, now, to include temperature effects, and to represent the liquid as a *thermostatic ensemble*. It is also easy to include in the formalism a more realistic interatomic potential, $\phi(R)$, than the hard sphere interaction of the Bernal model. But we neglect *three-body potentials* or *non-central forces* (see e.g. Rowlinson 1969) and write the total potential energy of a configuration of the assembly in the form

$$U(1 \ldots N) = \tfrac{1}{2} \sum_{i,j} \phi(i,j). \qquad (2.32)$$

A useful starting point for classical statistical mechanics is the *partition function*

$$Z = \sum_{\text{configurations}} e^{-\beta \mathcal{H}} \qquad (2.33)$$

where $\beta \equiv 1/kT$. The kinetic energy terms in the Hamiltonian are independent, particle by particle, and may be integrated out, giving

$$Z_{\text{liq}} = Z_{\text{gas}} Q. \qquad (2.34)$$

The *configuration integral*

$$Q = V^{-N} \frac{1}{N!} \int_V \cdots \int_V e^{-\beta U(1 \ldots N)} \, d\mathbf{1} \ldots d\mathbf{N} \qquad (2.35)$$

thus measures the difference between the liquid and an ideal gas with partition function Z_{gas}.

This is equivalent to describing the liquid by a thermostatic canonical ensemble with

$$g(1 \ldots N) = e^{-\beta U(1 \ldots N)} \qquad (2.36)$$

for the complete N-body distribution function (2.20). If we knew this function then all more general properties of the system could be obtained by ensemble averaging. Thus, the two-body distribution is just a multiple integration over all but two atomic positions:

$$g(1, 2) \propto \int_V \cdots \int_V e^{-\beta U(1 \ldots N)} \, d\mathbf{3} \ldots d\mathbf{N}, \qquad (2.37)$$

with appropriate normalization.

Geometrical correlations between the variables impede this integration. But if the assembly were not very dense, so that one could ignore the probability of a third atom being closer to both 1 and 2 than the range of the interatomic potential $\phi(1, 2)$, we could use (2.32) to approximate to (2.37):

$$g(1, 2) \sim e^{-\beta\phi(1, 2)}. \tag{2.38}$$

This we recognize as the simple application of the Boltzmann factor for interatomic correlations in a dilute gas.

It is obviously possible to compute successive correction terms to (2.38) by considering clusters containing more and more atoms. This is the basis of various diagrammatic series expansions for the thermodynamical properties (§ 6.5). But such a series is not properly convergent at liquid densities and can only be evaluated by analytic continuation from an expression for the sum of an infinite sub-set of all its terms. Formulae derived by this method, therefore, are not rigorous and should not be treated with exaggerated respect.

To understand the difference between a gas and a liquid, let us write

$$g(1, 2) \equiv e^{-\beta\psi(1, 2)}. \tag{2.39}$$

The fictitious *potential of average force* $\psi(1, 2)$ includes not only the direct effect $\phi(1, 2)$ of atom 2 on atom 1, but also the average effect of the other atoms around 2: it therefore depends on the 'structure' of the liquid. Indeed, direct differentiation of (2.37), and elementary probability considerations, yield the exact formula:

$$-\frac{1}{\beta g(1, 2)} \nabla_1 g(1, 2) \equiv \nabla_1 \psi(1, 2)$$

$$= \nabla_1 \phi(1, 2) + \frac{n}{g(1, 2)} \int g(1, 2, 3)\nabla_1 \phi(1, 3) \, d3 \tag{2.40}$$

where ∇_i is the gradient operator with respect to the position of atom i. This shows that the effective force acting on an atom can be calculated, provided that we know the three-body distribution function $g(1, 2, 3)$.

To evaluate the two-body distribution we need the three-body distribution: an analogous formula links $g(1, 2, 3)$ with $g(1, 2, 3, 4)$ and so on. To get anywhere, we must step off this staircase and look for another connection between the distribution functions. Various people, at various times – Bogoliubov, Born & Green, Kirkwood, and Yvon – suggested, independently, that one might use the superposition approximation (2.17) to represent $g(1, 2, 3)$ in terms of $g(1, 2)$. The so-called *BBGKY integral equation* that arises from (2.40) can be manipulated, geometrically separated and integrated into a non-linear, one-dimensional integral equation for the radial distribution function $g(R)$ in terms of the inter-

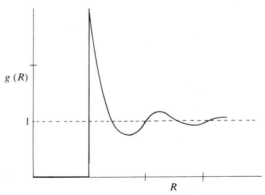

Fig. 2.44. Solution of BBGKY equation for RDF of hard-sphere fluid at typical liquid density ($\eta = 0.45$) (Kirkwood *et al.* 1950).

atomic potential $\phi(R)$, the temperature T, and the particle density n. This equation can be integrated numerically (Kirkwood, Maun & Alder 1950; Levesque 1966) for comparison with experiment.

Actually, direct comparison of such results with the physical properties of a real liquid is not very testing because we do not know the actual form of the interatomic potential $\phi(R)$. It is more instructive (though a significant commentary on the state of the theory of liquids) to solve the BBGKY equation for a standard case, such as a hard-sphere model, and to compare this with the best Monte Carlo computations for the same model (fig. 2.44). The appearance of peaks corresponding to various coordination shells is gratifying but there is obviously something wrong at the first maximum (cf. fig. 2.35). The vertical jump and sharp peak are evidently artefacts of the superposition approximation, which is known to be in error at the closest distances (fig. 2.39).

The BBGKY method is not, therefore, satisfactory as a procedure for calculating the radial distribution function of a liquid from its interatomic potential. To obtain accurate information from first principles about the structure of a liquid we should have to start one step higher with the analogue of (2.40) for $g(1, 2, 3)$ and use the higher-order superposition approximation (2.28) for $g(1, 2, 3, 4)$ (Fisher & Kopeliovich 1960). This equation is said to give very satisfactory results for $g(R)$ and the thermodynamic properties of the liquid, but since it can only be solved by heavy numerical computation it can scarcely be described as an 'analytical' theory.

Having abandoned the mathematical solidity of the staircase of canonical distribution functions we really have no better guide than inspired phenomenology. The most successful line of thought is to concentrate on the *total correlation function* (2.31), i.e.

$$h(1, 2) \equiv g(1, 2) - 1. \tag{2.41}$$

This measures the effects of physical forces and of geometrical constraints upon the statistical independence of the atomic positions. In a famous early paper, Ornstein & Zernike (1914) suggested that $h(1, 2)$ could be considered to consist of a *direct correlation* $c(1, 2)$ between the actual atoms 1 and 2 (analogous to the direct potential $\phi(1, 2)$) and an 'indirect' term by which the correlation is transferred through other neighbouring atoms. This is expressible self-consistently by a convolution formula:

$$h(1, 2) = c(1, 2) + n \int c(1, 3) h(3, 2) \, d3. \tag{2.42}$$

These two equations do not really say anything new: given $g(1, 2)$ then $c(1, 2)$ is determined and vice versa. But we can now argue that $c(1, 2)$ is genuinely of short range and should depend basically on the direct inter-atomic potential $\phi(1, 2)$, which falls off to zero beyond a few atomic diameters. With (2.38) as our model we might try, for example, the *Mayer function*

$$c(1, 2) \approx f(1, 2) \equiv e^{-\beta\phi(1, 2)} - 1, \tag{2.43}$$

which is also reasonably consistent with (2.41) and (2.42); we obtain an integral equation for the unknown function $h(1, 2)$ and hence for the radial distribution function, etc. This equation turns out to be a linearized version of BBGKY, and hence a poor approximation, but the general argument shows promise.

The next step, then, is to modify (2.43) until we get better agreement with the results of model computations. The most successful version is the *Percus–Yevick* (PY) *equation*, in which we write

$$c(1, 2) \approx (e^{-\beta\phi(1, 2)} - 1)e^{\beta\phi(1, 2)} g(1, 2), \tag{2.44}$$

the correction factor being, in effect, just what we should need to put (2.38) right. Plausible justification for this factor can be derived by functional differentiation (Percus 1962; Rushbrooke 1968) or by continuity arguments (Rowlinson 1967), but these are no more convincing than the original derivation by Percus & Yevick (1958) which was based upon somewhat obscure arguments concerning collective coordinates.

It is an interesting commentary on the series method that if we put (2.44) into (2.42) and generate successive terms of a series for g in powers $f(1, 2)$ we do not get all the sub-series of the cluster expansion that can easily be summed to infinity by topological arguments. The 'best' result, in this sense, is obtained from the *hyper-netted chain* approximation

$$c(1, 2) \approx h(1, 2) - \beta\{\phi(1, 2) - \psi(1, 2)\}. \qquad (2.45)$$

But this approximation is unsatisfactory, both for thermodynamic reasons and because the computed solution of the integral equation (2.42) does not agree well with Monte Carlo computations (Rowlinson 1968).

The really strong argument in favour of the PY formula is that it does indeed give results that agree pretty well with Monte Carlo and molecular dynamics computations. It has, moreover, the valuable characteristic that the integral equation for the direct correlation function in the hard-sphere case can be solved exactly. By (2.44), this function must be zero for $R > d$ but within this radius it has the polynomial solution (Thiele 1963; Wertheim 1963, 1964)

$$c(R) = -\frac{(1 + 2\eta)^2}{(1 - \eta)^4} + \frac{6\eta(1 + \frac{1}{2}\eta)^2}{(1 - \eta)^4} \frac{R}{d} - \frac{\eta(1 + 2\eta)^2}{2(1 - \eta)^4} \left(\frac{R}{d}\right)^3, \qquad (2.46)$$

where η is, as before, the packing fraction $\pi d^3 n / 6$. From this function we may derive other correlation functions and distribution functions such as $h(R)$ and $g(R)$ by inversion of the Ornstein–Zernike relation (2.42), but these are not so simple in algebraic form. For comparison with experiment it is often much simpler to deal with $c(R)$ itself, or with its Fourier transform, which can be obtained directly from diffraction observations (see § 4.1). The actual shape of $c(R)$ for a dense, hard-sphere liquid is merely a rounded version of the 'square' function which would arise from (2.43) whose Fourier transform could be written down by inspection. This is, therefore, a useful crude model for the structure of a liquid for 'back-of-envelope' calculations.

The PY hard-sphere formula (2.46) does not contain the temperature: the molecular distribution function seems to depend entirely on the density. This is because there is no energy scale parameter in the interatomic potential. There is no obstacle, however, to computations of solutions of the PY integral equation for various more realistic potentials, such as the *Lennard-Jones potential* (fig. 2.45).

$$\phi_{\mathrm{LJ}}(R) = 4\varepsilon\{(\sigma/R)^{12} - (\sigma/R)^6\}. \qquad (2.47)$$

This is thought to reproduce quite well the overlap repulsion and van der Waals attraction between electronically saturated molecules or atoms such as argon. What is the effect of 'softening' the potential on the structure of the liquid? Solutions of the PY equations tell us little, but Monte Carlo calculations (Finney 1970) show that the distribution of cell volumes (cf.

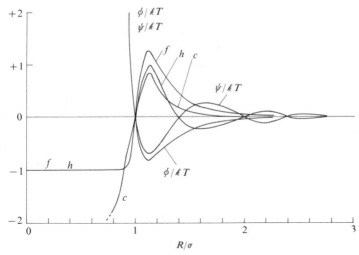

Fig. 2.45. Potential functions and correlation functions for a typical liquid:

$\phi/\mathit{k}T=$ Lennard-Jones potential (2.47),
$f=$ Mayer function (2.43),
$h=$ total correlation function (2.41),
$c=$ direct correlation function (see (2.42)),
$\psi/\mathit{k}T=$ potential of average force (2.39).

fig. 2.44) is further broadened, indicating a higher degree of local disorder in the arrangements. It would be interesting to investigate the structure of a dense assembly of very 'soft' atoms, such as the 'coulomb liquid' of positive point ions immersed in a negative electron gas which was found by Brush, Sahlin & Teller (1966) to have a fluid–solid transition.

The fact is, however, that the PY hard-sphere formula (2.46) is surprisingly successful as a phenomenological parametrization of the two-body distribution function even for liquid metals (Ashcroft & Lekner 1966). Yet there is no basic justification for the fundamental assumption that the direct correlation function must be of short range and the absence of statistical geometry in the model is shown by its failure to forbid solutions when $\eta > 0.74$ – i.e. in the unphysical situation where the packing fraction exceeds that of a close-packed crystal. We must take it, therefore, that the PY relationship (2.44), although convenient in developments of the analytical theory, is more an approximate general consequence of the conditions for random packing of relatively hard spheres than an *a priori* principle. It is not obvious that attempts to construct more accurate formulae based upon more refined heuristic principles of this kind (see e.g. Rushbrooke 1968) could add very much to our understanding of the problem.

In this spirit, for example, the **BBGKY** or **PY** formula can be used with reasonable precision to deduce $\phi(R)$ from an observed radial distribution function $g(R)$. For example, if we have obtained $g(1, 2)$ by a diffraction experiment, then $c(1, 2)$ and $\phi(1, 2)$ follow trivially from (2.41), (2.42) and (2.44). This method seems to disclose significant differences between the interatomic forces in insulators and those in metals (Johnson & March 1963; March 1968; but see Kumaravadivel, Evans & Greenwood 1974).

It is possible, however, by very careful measurements of the density variation of the radial distribution function of an actual liquid, to obtain direct experimental evidence concerning such 'closure assumptions' as the superposition principle or the **PY** equation. The pressure coefficient of $g(R)$ at constant temperature is related analytically to $g(1, 2, 3)$ by an integral that does not explicitly contain the interatomic potential (Raveché & Mountain 1970; Egelstaff, Page & Heard 1971).

The deduction of this formula (Buff & Brout 1955; Mayer 1962; Schofield 1966) is really an analogue of the step from (2.37) to (2.40), but needs to start from the grand canonical ensemble so that we can vary the density. Analogous to (2.35), we construct the grand partition function

$$\Xi = \sum_{N=0}^{\infty} \frac{z^N}{N!} \int_V \cdots \int_V e^{-\beta U(1 \cdots N)} \, d\mathbf{1} \ldots d\mathbf{N}, \qquad (2.48)$$

from which may be deduced the general formula for a probability density (2.17)

$$n(1 \ldots s) = \frac{1}{\Xi} \sum_{N=s}^{\infty} \frac{z^N}{(n-s)!} \int_V \cdots \int_V e^{-\beta U(1 \cdots N)} \, d(\mathbf{s+1}) \ldots d\mathbf{N}. \quad (2.49)$$

This is just a generalized form of (2.37), with the number N of particles in the fixed volume V allowed to fluctuate. The fugacity z looks like an independent variable, but is determined when we know the actual density $n = n(1) = \bar{N}/V$.

Now we differentiate (2.49) with respect to n, using z as an independent variable. By elementary rearrangement of the differentiated series, and systematic use of the same definition we get

$$\frac{\partial n_s(1 \ldots s)}{\partial n} \bigg]_{\beta,V} = \frac{\partial n_s(1 \ldots s)}{\partial z} \bigg]_{\beta,V} \bigg/ \frac{\partial n_1(1)}{\partial z} \bigg]_{\beta,V}$$

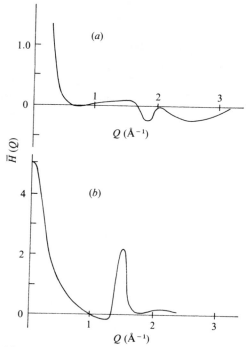

Fig. 2.46. Observed deviations from superposition approximation: (*a*) Rb; (*b*) Argon (Egelstaff *et al.* 1971).

$$= \frac{sn_s(1\ldots s) + \int\{n_{s+1}(1\ldots s+1) - n_s(1\ldots s)n_1(s+1)\}\,d(s+1)}{\int[n_1(1) + \int\{n_2(1,2) - n_1(1)\,n_1(2)\}\,d2]\,d1}. \tag{2.50}$$

For good thermodynamic reasons, the denominator is related to the compressibility, so that the left-hand side can be rewritten rather neatly as a *pressure* derivative. For $s=2$, using (2.50) we get

$$kT\frac{\partial g_2(1,2)}{\partial p}\bigg]_T = \int_V \{g_3(1,2,3) - g_2(1,2)\}\,d3 + \frac{2}{n}g_2(1,2). \tag{2.51}$$

Using this identity, direct diffraction measurements of the pressure variation of the structure factor (§ 4.4) of a liquid such as argon or rubidium can be interpreted as observations of a function such as

$$\bar{H}(Q) = n^2 \int\!\!\int\{g(1,2)\,g(2,3)\,g(3,1) - g(1,2,3)\}\,e^{iQ\cdot R_{12}}\,d2\,d3. \tag{2.52}$$

Since this function is found to deviate considerably from zero (fig. 2.46), the elementary superposition approximation (2.27) cannot be correct (Egel-staff, Page & Heard 1969). Raveché & Mountain (1972) have used this criterion to assess the relative merits of various more complicated closure approximations for $g(1, 2, 3)$, whilst Egelstaff, Page & Heard (1971) have checked the observed behaviour of $\partial g_2/\partial p$ against other heuristic assumptions such as the PY equation, the hyper-netted chain approximation (2.45) and the 'uniform compression model' (Egelstaff 1967) in which $g(R)$ scales with distance as $n^{-\frac{1}{3}}$. Unfortunately the experimental evidence does not decisively favour any single one of these alternative theories for all simple liquids at all densities.

2.13 Liquid mixtures

Liquids containing two or more species of atom or molecule occur in endless variety as *solutions, liquid alloys, molten salts,* etc. We consider only the simplest case of an assembly of two constituents, A and B, in proportions x_A and x_B, of nearly spherical atoms. To calculate the structure and properties of such a system we must know the three interatomic potentials $\phi_{AA}(R)$, $\phi_{BB}(R)$ and $\phi_{AB}(R)$ corresponding to A–A, B–B and A–B interactions. Since these functions are seldom accessible to theory or experiment, quantitative theories of real mixed liquids based on *a priori* atomic parameters cannot be thoroughly tested against the voluminous experimental data. Indeed, the theoretical interpretation of the observed thermodynamic properties of liquid solutions is still very uncertain.

Many of the assumptions made in phenomenological theories of liquid mixtures are unrealistic and self-contradictory (see, for review, Rowlinson 1969, 1970). It is often postulated, for example, that the two constituents are distributed at random and that the mixture behaves as a uniform liquid with average interatomic potential

$$\langle\phi(R)\rangle = \sum_{\alpha=A,B}\sum_{\beta=A,B} x_\alpha\, x_\beta\,\phi_{\alpha\beta}(R). \qquad (2.53)$$

If this function should happen to be of the same type as each $\phi_{\alpha\beta}$, the liquids are said to be *conformal* (Longuet-Higgins 1951). Thus, if ϕ_{AA}, ϕ_{BB} and ϕ_{AB} are of the Lennard-Jones type (2.47), with different depths ε_{AA}, ε_{BB} and ε_{AB} and different diameters σ_{AA}, σ_{BB} and σ_{AB}, then the average parameters $\langle\varepsilon_x\rangle$ and $\langle\sigma_x\rangle$ can be easily calculated for any given relative concentration. The thermodynamic properties of the equivalent uniform liquid can be deduced from those of the constituents by scaling of temperature and volume according to the principle of corresponding states (§ 6.2). Similar results are obtained by a more sophisticated technique where A and B atoms are treated as if each were in its own type of equivalent liquid each with the average potential

$$\langle \phi_\alpha(R) \rangle = \sum_{\beta=A,B} x_\beta \, \phi_{\alpha\beta}(R). \tag{2.54}$$

Phenomenological parameters to indicate deviations from randomness in the distribution of A and B atoms over the 'liquid' lattice (cf. § 1.5) can also be introduced in the hope of improving agreement with experiment, but have not been calculated from first principles.

But the basic assumption (2.53) leads to unphysical results. Consider the elementary case of a mixture of hard spheres of different sizes. The potentials are conformal, but the average potential always has the diameter of the larger constituent: in the equivalent liquid, the smaller atoms always appear to occupy a much larger volume than they really need. Similar paradoxes arise from (2.54). For the softer Lennard-Jones '6–12' potential (2.47) the effects are not so absurd, but one cannot flout the general principle that liquid structure is dominated by the geometrical conditions arising from core repulsion (§§ 2.11, 2.12). Unless both constituents are of the same size and not very different in other interatomic characteristics one cannot base a theory of liquid mixtures on a principle of corresponding states applied to a pure liquid.

To allow for changes in the overall arrangement of atoms in the mixture we must introduce functions describing the statistical distributions of atoms of the same or different types in particular configurations. A natural generalization of (2.21) would be the *partial distribution function* $g_{\alpha\beta}(1, 2)$ telling us the ensemble average probability of finding an atom of type β at point **2**, given that there is an atom of type α at **1**. A good theory of the mixed liquid should tell us the form of $g_{AA}(R)$, $g_{BB}(R)$ and $g_{AB}(R)$, all of which can be determined experimentally in certain cases (§ 4.5).

Analytical equations for the partial distribution functions can be written down fairly easily by the same arguments as for a pure liquid. For example, the identity (2.40) linking $g(1, 2)$ with $g(1, 2, 3)$ can be generalized and 'closed' by the analogue of the superposition approximation (2.27), i.e.

$$g_{\alpha\alpha\beta}(1, 2, 3) \approx g_{\alpha\alpha}(1, 2) \, g_{\alpha\beta}(2, 3) \, g_{\beta\alpha}(3, 1) \tag{2.55}$$

for each choice of atoms. The numerical solutions of these coupled integral equations (Alder 1955) are far from absurd, but suffer from the defects already noted for the BBGKY method. The way is open, however, for further investigations along these lines, using the higher-order super-position principle (2.28), or a generalization of one or other of the *ad hoc* closure relations discussed by Rushbrooke (1968) and studied experimentally by Raveché & Mountain (1972).

Again, an appeal to fluctuation theory (Pearson & Rushbrooke 1957) yields a generalization of the Ornstein–Zernike relation (2.42). To each type of atomic pair we assign a total correlation function $h_{\alpha\beta}$ analogous to (2.41) and a partial direct correlation function $c_{\alpha\beta}$ satisfying

$$h_{\alpha\beta}(1, 2) = c_{\alpha\beta}(1, 2) + \sum_{\gamma = \alpha, \beta} n_\gamma \int c_{\alpha\gamma}(1, 3) \, h_{\gamma\beta}(3, 2) \, d3. \qquad (2.56)$$

The analogue of the PY approximation (2.44) is plausible enough;

$$c_{\alpha\beta}(1, 2) \approx [1 - \exp\{\beta\phi_{\alpha\beta}(1, 2)\}] \, g_{\alpha\beta}(1, 2). \qquad (2.57)$$

Equations (2.56) and (2.57) are merely matrix generalizations of (2.42) and (2.44), which have an exact solution for hard-sphere interactions. Each partial direct correlation function can be written as a finite polynomial in R (Lebowitz 1964; Lebowitz & Rowlinson 1964) whose coefficients are rational functions of the partial packing fractions

$$\eta_A = \tfrac{1}{6}\pi d_A^3 n_A, \quad \eta_B = \tfrac{1}{6}\pi d_B^3 n_B. \qquad (2.58)$$

Fourier transformation of these solutions yields partial structure factors (§ 4.5) which can be compared with experiment in typical cases (Ashcroft & Langreth 1967).

When the constituents differ considerably in size (fig. 2.47), their partial structure factors look quite different, thus demonstrating the fallacy of the 'conformal solution' hypothesis. The thermodynamic accuracy of the PY method for mixtures of hard spheres has been confirmed by comparison with the results of molecular dynamics computation (Alder 1964b). The predicted decrease of volume on mixing (fig. 2.48) is in agreement with empirical model experiments (Mangelsdorf & Washington 1960; Epstein & Young 1962; Visscher & Bolsterli 1972) and follows directly from elementary considerations of random packing. Consider, for example, the extreme case where the B atoms are much smaller than the A atoms. With A atoms alone we make a disordered structure whose packing density, η_A, approaches the maximum value η_0 for the RCP arrangement. The interstices of this assembly, amounting to a relative volume $(1 - \eta_A)$, can now be packed with B-type atoms, until the fraction η_0 of that volume is occupied: the total density of the mixture could thus approach

$$\eta_{\text{mixt}} = \eta_A + \eta_B \leqslant \eta_0 + (1 - \eta_0)\eta_0. \qquad (2.59)$$

In the most favourable case, with $\eta_0 = 0.637$, we could combine a density 0.637 of A atoms with 0.231 of B atoms to achieve a total density $\eta_{\text{mixt}} = 0.868$. Mixture of these two liquids in these proportions is thus

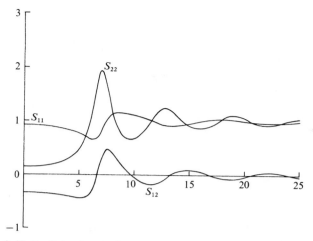

Fig. 2.47. Partial structure factors for mixture of hard spheres of radius ratio 0.7:1.

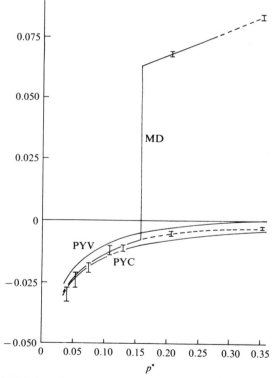

Fig. 2.48. Volume change upon mixing of hard spheres (radius ratio 3:1) as function of pressure, showing good agreement between molecular dynamics computation MD and Percus–Yevick formulae, PYV and PYC, below the point where a 'large sphere solid' separates out (Alder 1964b).

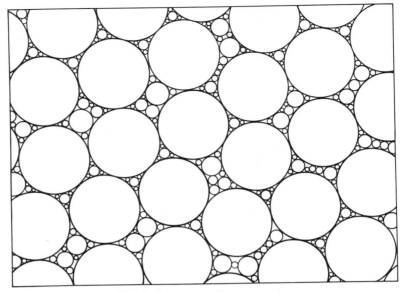

Fig. 2.49. Close packed aggregate.

accompanied by a volume shrinkage of 35 per cent! Indeed, if a third constituent of yet finer grain were available, the remaining interstices could be partially filled – and so on. In other words, an *aggregate* of spheres of a wide range of sizes could, in the limit, be packed to a density approaching unity (fig. 2.49). When the diameters are not very different, a formula such as (2.59) is not valid; but the model experiments of Mangelsdorf & Washington (1960) show clearly that a random mixture of spheres of different sizes can always occupy a smaller volume than the total volume of its separate constituents. In the corresponding two-dimensional models, a good approximation is to assume that three adjacent spheres always touch, so that the relative distributions of various types of interstitial region can be estimated (Dodds 1975).

It is interesting to note that in the extreme case satisfying (2.59) the random mixture occupies a volume 17 per cent less than the total volume of the close-packed *crystals* of its constituents. This may be relevant to the structure of *glassy alloys,* such as those obtained by quenching from the melt (Duwez, Willens & Klement 1960; Duwez 1967; Jones 1973) or by rapid co-evaporation on to a cold surface (Mader *et al.* 1963). Since many of these alloys contain Te or Si, an explanation in terms of bond networks (§ 2.8) is not ruled out. It is known, however (see e.g. Cargill 1975), that

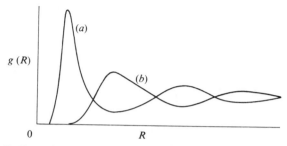

Fig. 2.50. Partial radial distribution function for (*a*) unlike ion pairs and (*b*) like ion pairs, in a model ionic liquid (Woodcock 1972).

relatively stable amorphous phases can be produced in this way for a wide range of alloys of chemically 'metallic' elements – even, at sufficiently low temperatures, of some pure metals. In most such cases it seems that the diffraction information is fully consistent with Bernal-type models of random packing of spheres (§ 2.11), modified for the relative sizes of the constituents. In default of a clear understanding of the detailed electronic mechanisms and structural constraints that prevent such materials from immediately recrystallizing, we must assume that they are most simply described as 'frozen liquids' and attempt to explain their electrical, magnetic and dynamical properties on this basis.

The most extreme type of 'simple' mixed liquid is a *molten salt*. Consider the case of an alkali halide – in particular where the closed-shell repulsive cores of the K^+ and Cl^- ions are almost exactly the same size. The structure of such a liquid is dominated by the electrostatic repulsions and attractions between 'like' and 'unlike' ions. Attempts at analytical formalisms for the partial distribution functions for such systems have made little progress towards realistic results (see e.g. Gillan, Larsen, Tosi & March 1976) but Monte Carlo computations (Woodcock 1972; Lantelme, Turq, Quentrec & Lewis 1974) show clearly what must happen (fig. 2.50). For *unlike* ions the partial distribution function $g_u(R)$ has a fairly sharp peak close in, indicating that each ion is surrounded by a shell of unlike neighbours. The coordination number of this shell is found experimentally (fig. 2.51) to be close to the number of nearest neighbours in the corresponding crystal – i.e. 6 for NaCl (Edwards, Enderby, Howe & Page 1975). The next shell produces a more diffuse peak in $g_l(R)$, with the full complement of about thirteen *like* ions. From here outwards the distribution functions are more or less featureless, corresponding to nearly complete charge cancellation over any larger distance. Some degree of 'molecular' association within the

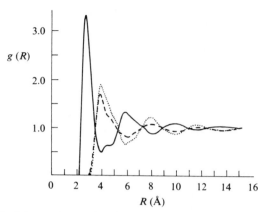

Fig. 2.51. Observed radial distribution functions for molten NaCl: ----, g_{NaNa}; ·····, g_{ClCl}; ——, g_{NaCl} (Edwards *et al.* 1975).

first coordination shell is not ruled out by the experimental evidence (see Powles 1975), but there is no obvious necessity to include such effects in model calculations of the properties of these materials.

2.14 Liquid phases of non-spherical molecules

The *shape* of the molecules in a liquid must have a significant effect on the structural arrangement. But statistical theories of the liquid state almost always treat approximately spherical molecules interacting through central forces. Apart from the molecular dynamics calculations in water (§ 2.8), well-founded mathematical results concerning fluids of non-spherical molecules scarcely exist. The mere fact (Bondi 1968) that the packing density η of most molecular liquids near the melting point is of the order of 0.5–0.6 shows that the space is well filled, but tells us very little about the statistical characteristics of the arrangement. Theories of the thermodynamic properties are essentially phenomenological, and contribute nothing of significance to the mathematical theory of disorder.

Several special cases are, however, of particular interest. When the molecules are very long and very flexible, as in a *long-chain polymer*, it is reasonable to suppose that each chain lies along the path of a simple *self-avoiding random walk*. This assumption is obviously suspect in a molten polymer consisting solely of such molecules, where interactions and entanglements among the chains must surely be important, but is justifiable in a *dilute solution* of the macromolecules in a solvent of low molecular

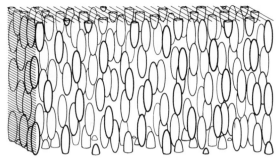

Fig. 2.52. Nematic order.

weight. Since the local microscopic arrangement of the molecules in this mixed liquid scarcely enters into the discussion, the theory of such systems will be reviewed in a later chapter (chapter 7).

Another simplifying assumption is that each molecule is perfectly *rigid* and *axially symmetric,* like an ellipsoid or dumbbell. To give a statistical description of a liquid of such molecules, we should need distribution functions for the relative orientations of the axes of two or more molecules in any given spatial configuration which is, of course, a much more complicated geometrical problem than defining a radial distribution function $g(R)$ for spherical atoms. The best that can be done is to write down a few cluster integrals for a few virial coefficients, and evaluate these assuming typical soft intermolecular forces (Isihara & Hayashida 1951; Kihara 1953; Sweet & Steele 1967) or for a 'hard dumbbell' model (Chen & Steele 1969; Rigby 1970). None of these calculations is applicable to an assembly at liquid densities; but it seems as if the main effect of molecular anisotropy and rotation is to spread and smear out the positional distribution function as if each molecule were simply a large soft sphere.

From experiment, however, we know that very long rigid molecules cannot pack together without inducing correlations in their relative orientations. This is the origin of the numerous fascinating phenomena observed in *liquid crystals* (see e.g. Brown, Doane & Neff 1970; De Gennes 1974; Chandrasekhar 1976 for review and references). In a *nematic* liquid, for example (fig. 2.52), the centres of the molecules do not lie on a regular lattice, but there is long-range order in the alignment of their long axes.

This phenomenon is clearly akin to magnetic ordering (§ 1.3). But the 'quantum chemical' interactions between adjacent molecules of complicated shape and structure cannot be represented by any formula as simple as the Heisenberg spin Hamiltonian (1.16). It must be assumed, for example, that steric constraints due to the repulsion between atoms at close distances are at least as important as the dispersive forces discussed by Maier & Saupe (1959, 1960). Even as the leading term in an empirical formula, the usual $S_i \cdot S_j$ interaction would not be appropriate: the interaction between non-polar molecules must be a minimum for both 'parallel' and 'antiparallel' configurations. For simplicity one might assume something like

$$\mathscr{H}_{ij} = -A(\mathbf{R}_{ij}) \, P_2(\cos \theta_{ij}), \tag{2.60}$$

depending only on the angle θ_{ij} between the axes of the two molecules; but this would neglect important terms containing the orientation of each molecule to the vector $\mathbf{R}_{ij}$ between them. In any case, even this sort of expression cannot be represented by quadratic terms in spin-component, Ising spin or spin-deviation variables (§§ 1.5, 1.8) so that no exact analysis of the thermodynamic behaviour of the system is yet feasible.

There is no doubt, however, that any of the approximate methods to be discussed in chapter 5 can give a satisfactory qualitative description of ordering phenomenon. Until we know the parameters of the intermolecular forces, we can rest content with a *mean field approximation* (Maier & Saupe 1959, 1969) which tells us (§ 5.2) that there should always be some short-range order in the molecular orientations, and that long-range order should set in abruptly below a critical temperature T_c, and should increase further with falling temperature. This is precisely what is observed in many nematic liquids.

Here again, just as in magnetic systems (§ 1.6) there are formal difficulties concerning the definition of long-range order. It is easy enough to assume, phenomenologically, that the molecules are approximately lined up parallel to a *director* vector whose orientation is prescribed in advance, and then to define the order parameter (cf. (1.30)),

$$\mathscr{S} = \langle P_2(\cos \theta_i) \rangle \equiv \langle \tfrac{1}{2}(3 \cos^2\theta_i - 1) \rangle, \tag{2.61}$$

in terms of the angle θ_i made by each molecule relative to this vector. But this definition begs the question whether the director exists: in principle, one should start from an 'internal' procedure, such as taking the long-range limit of the short range order, as in (1.34), i.e.

$$\mathscr{S} = \operatorname*{Lt}_{R_{ij} \to \infty} \langle P_2(\cos \theta_{ij}) \rangle. \tag{2.62}$$

This point is not mere pedantry because, in fact, the director usually varies continuously in direction over substantial distances in the liquid and thus provides a 'local order parameter' in the continuum theory of the nematic phase (Frank 1958). This field variable is obviously very similar to

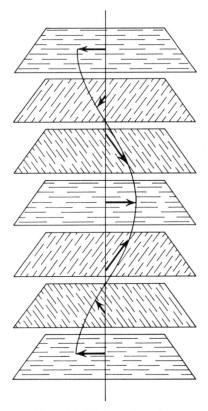

Fig. 2.53. Cholesteric order.

the magnetization vector in a ferromagnetic material or to the sub-lattice magnetization in an antiferromagnet and may be made the basis of a 'Landau theory' of the phase transition (§ 5.11). But there is nothing analogous to the magnetic *anisotropy energy* which tends to align the spins locally along one or other of the symmetry axes of the local crystal lattice and which gives rise to macroscopic domains separated by thin 'walls' (§ 1.7). The statistical isotropy of the arrangement of molecules in a nematic liquid allows the director to bend and twist continuously over macroscopic distances, with only occasional lines of discontinuity (*disclinations*) imposed by the overall topology.*

The study of *order* in liquid crystals is a science in its own right, and will

* Which suggests, incidentally, that the magnetization of an *amorphous* ferromagnet (§ 12.3) may not have the familiar domain structure of crystalline materials.

Topological disorder

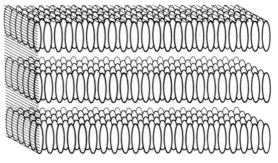

Fig. 2.54. Smectic order.

not be pursued further here. It is sufficient to remark that a *cholesteric* liquid crystal is essentially a nematic with an inbuilt tendency (due to some slight steric asymmetry of the individual molecules) for the director to rotate at a steady rate as one moves normal to it (fig. 2.53). In a *smectic* phase we observe partial ordering in the *position* of the molecules in planes related to the local direction of orientational order (fig. 2.54) and other more complex effects.

2.15 Gas-like disorder

The most extreme degree of topological disorder is to be found in an ideal *gas* where the atoms or molecules are distributed independently, at random, throughout the sample volume. Some geometrical characteristics of such a structure have already been noted (§ 2.11).

Spatial randomness is attained nearly to perfection in an ordinary gas or vapour. But this is done by reducing the density so that the molecules are so far apart that they can be treated as *physically* independent. The interesting phenomena in real gases, whether neutral or ionized, are *dynamical* effects produced by long-range forces acting on the kinetic energy terms in the total Hamiltonian, and these depend very little on any residual correlations in the *positions* of the atoms in space. In *plasma physics,* for example, once we have calculated macroscopic coefficients such as electrical conductivity we move to the level of continuum equations in local density, current, space charge, etc.

It is possible, however, by herioic experimental techniques (see e.g. Ross & Greenwood 1969; Even & Jortner 1974) to follow the behaviour of a metallic fluid from the liquid to the vapour phase through the *supercritical* region. The difficulty in dealing theoretically with these observations is not, as one might expect, due to the thermal motion of the ions but to the

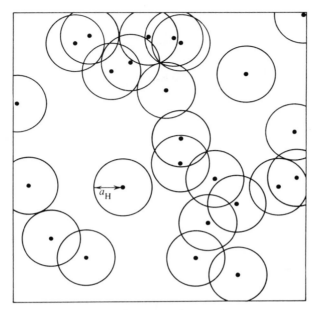

Fig. 2.55. Impurity centres in a semiconductor.

uncertainty in the nature of the interatomic forces. As the density changes, the electron–electron interaction may give rise to atomic or molecular bound states which could drastically alter the effective potential between the ions. This is a problem of considerable delicacy and without convincing evidence concerning, say, any tendency to form molecules or clusters, we can only assume that the structure simply opens up and becomes more random, according, perhaps, to the hard-sphere PY formula (2.46).

The purest example of *dense* gas-like disorder is provided by substitutional impurities such as P in a valence semiconductor such as Si (Mott & Towse 1961; Mott & Davis 1972). Provided that there is no strong chemical tendency towards segregation or clustering as the specimen crystallizes (not necessarily a valid assumption!), the impurities should be distributed at random on the sites of the lattice as in any very dilute alloy (§ 1.2). For the electronic carriers, however, the 'size' of a valence impurity is the effective Bohr radius, a_H, of its lowest impurity level, which may be ten or a hundred times larger than the lattice constant of the underlying crystal. An impurity concentration in the range 10^{-4}–10^{-2} atoms per cent should thus be considered quite 'dense', since the assumption that electronic phenomena occur independently on the individual impurities cannot be justified (fig. 2.55). This system, also, will be discussed at length in chapter 13.

3

Continuum disorder

—

3.1 Continuum models

The atomicity of condensed matter (§ 2.1) is often a source of difficulty in mathematical theories of disordered systems. Important physical quantities such as the electron density and nuclear mass are highly concentrated in very small regions and cannot, therefore, be easily represented by linear combinations of smooth, extended functions. Yet we discover in practice that the 'deep' internal properties of the atoms are not very important; the effects of order, or of substitutional or topological disorder, are produced mainly by small variations of the charge distribution over the general volume of the material rather than by the large, highly localized variations in the atomic cores.

In the construction of phenomenological theories, therefore, it is often convenient to go to an approximate *continuum representation* of the material, where local atomicity is ignored. This is, of course, the natural way of dealing with genuine macroscopic phenomena, such as the propagation of sound through the ocean, or of starlight through the atmosphere, or of radio waves through the ionosphere: a continuous medium is assumed, with a material constitution that determines a local density, elasticity, refractive index, dielectric constant, etc. to be put into the wave equation. This is justified because we are concerned with disturbances of much greater wavelength than the typical interatomic distance. Such an assumption is seldom valid for thermal vibrations or electrons in disordered condensed matter, but the mathematical similarity of these problems to the corresponding macroscopic problems suggests the possible value of setting up models in which fluctuations of density, or variations of local crystalline order, are regarded simply as the physical sources of a variable local potential, density, phonon velocity, etc.

The setting up of such a model is inevitably heuristic and the parameters can seldom be evaluated from first principles. Nevertheless, it is sometimes possible to represent simply the effects of quite complex structural characteristics of the disorder. Consider, for example, the effective potential seen by an electron in a liquid metal. This is a many-electron effect which cannot be represented rigorously as a mere superposition of atomic potentials, but which may depend on many-atom properties of the structure of the liquid, such as the average local atomic density. In § 2.11 (fig. 2.42) we saw that the volumes of the atomic cells are not constant in the liquid state but vary by as much as 10 per cent from the average. We could use a deformation potential formula to relate the electron potential in each cell to the local atomic volume, thus generating a simple continuum model for the electronic properties of liquid metals. Similar arguments might be used to define an effective potential for a carrier near a band edge in an amorphous semiconductor or for local elastic constants in a glass: in each case, the variable quantity would be supposed to depend on local deviations from perfect tetrahedral bonding or from the ideal 'staggered' bond configuration (§ 2.10: fig. 2.33). These particular models are, in fact, too simplified, but they indicate the general line of argument that would be needed to link a continuous random field representation with properties of the underlying atomistic material.

These models are not sufficiently well characterized to be reviewed here, independently of the particular systems they are designed to represent. But some of the mathematical effort in the theory of disordered systems has been directed towards the study of electron or wave propagation in a *random medium,* with analytical properties defined more for mathematical convenience than in relation to any particular structural model. The physical or geometrical significance of such properties is seldom made clear, so that the relevance of deductions concerning electron localization, band gaps, etc. is obscured. In this chapter, therefore, we consider briefly the statistical characterization of a random function $\xi(\mathbf{R})$ in a space $\mathbf{R}$ of one, two or three dimensions and show how some of its geometrical features arise. If $\mathbf{R}$ were a position vector on a plane, then $\xi(\mathbf{R})$ would represent the height of a *random surface*; *statistical topography* leans heavily on a small number of perceptive papers concerned with the mathematical theory of noise (Rice 1944, 1945), of the scattering of waves from rough surfaces (Longuet-Higgins 1957; Beckmann & Spizzichino 1963; Berry 1973) and of homogeneous turbulence (see Batchelor 1953; Yaglom 1962; Lumley 1970).

3.2 Homogeneous random fields

A *random field* $\xi(\mathbf{R})$ can only be defined by its statistical properties. Following the example of spins on a lattice (§§ 1.5, 1.7) or of atomic distributions in space (§ 2.7), we introduce various probability distributions for the values of ξ at points in the field. Thus, $P(\xi, \mathbf{R})$ measures the probability density of the variable ξ in the range $\xi \rightarrow \xi + d\xi$ at the point $\mathbf{R}$. Since our field is assumed to be homogeneous this function must be independent of $\mathbf{R}$. If we had been talking about a quantity varying stochastically in *time,* then we should have described $\xi(t)$, say, as a *stationary* random function: the one-dimensional theory of such functions (Rice 1944, 1945) is only a special case of a homogeneous random field in several dimensions.

For simplicity we assume that any constant component of the field – for example, the average level of a potential – has been removed, so that the mean value of ξ is zero. This mean value may be thought of as an integral over the large but finite volume V in which $\mathbf{R}$ is defined, or over a statistical ensemble of a very large number of identical volumes such that ξ takes on all its possible values at any one point: we rely upon an *ergodic hypothesis* to assert that these are equal, i.e.

$$\langle \xi \rangle = V^{-1} \int \xi(\mathbf{R}) \, d\mathbf{R} = \int \xi P(\xi) d\xi = 0. \tag{3.1}$$

In the first instance we may imagine that we know $P(\xi)$ which may have any positive functional form consistent with (3.1) and with total integral unity. But this would still not provide adequate information about the random field. Just as in (1.39) and (2.17) we need to define the 2-point, 3-point, etc. distribution functions: in general the s-point distribution function $P_s(\xi_1, \mathbf{R}_1; \xi_2, \mathbf{R}_2; \ldots; \xi_s, \mathbf{R}_s)$ measures the probability of finding the value ξ_1 at $\mathbf{R}_1$, ξ_2 at $\mathbf{R}_2$ etc. Only in the non-physical case of a totally, pathologically discontinuous field could we assume that values of ξ at 'neighbouring' points were independently distributed. Mathematically speaking the statistical properties of ξ can only be defined fully by the *probability functional* $P[\xi(\mathbf{R})]$ which is the limit of the s-point distribution as $s \rightarrow \infty$ (Beran 1968). But this formalism is not needed for our present work.

These distribution functions must, of course, satisfy certain identities. Integration over any position variable $\mathbf{R}_1$, say, or ensemble averaging over the corresponding field variable ξ_1, must reduce P_s to P_{s-1} (cf. (2.19)). Again, the values ξ_1 and ξ_2 of the field at two points $\mathbf{R}_1$ and $\mathbf{R}_2$ must tend to statistical independence as the distance $R = |\mathbf{R}_1 - \mathbf{R}_2|$ becomes very large. If our field is *isotropic* as well as homogeneous, the *2-point distribution* becomes very simple: it can only be of the form (cf. (2.33))

$$P_2(\xi_1, \mathbf{R}_1; \xi_2, \mathbf{R}_2) \equiv P_2(\xi_1, \xi_2; R). \tag{3.2}$$

The implicit assumption that $\xi(\mathbf{R})$ is piecewise continuous also tells us that ξ_1 and ξ_2 must tend to the same value as $\mathbf{R}_1$ approaches $\mathbf{R}_2$: thus

$$P_2(\xi_1, \xi_2; \mathbf{R}) \rightarrow \begin{cases} \delta(\xi_1 - \xi_2) P(\xi_1) & \text{as } R \rightarrow 0 \\ P(\xi_1) P(\xi_2) & \text{as } R \rightarrow \infty \end{cases}. \tag{3.3}$$

Within these restraints, however, the *joint probability distribution* P_2 is a function of three variables and, therefore, difficult to represent simply. Once more, the theory of substitutional disorder on a lattice suggests an appropriate representation of its main properties. As in (1.12) and (1.44) we define the *auto-correlation function* of the field:

$$\Gamma(R) = \frac{\langle \xi^*(0)\xi(\mathbf{R})\rangle}{\langle \xi^2 \rangle} = \frac{\iint \xi_1^* \xi_2 \, P(\xi_1, \xi_2; R) \, d\xi_1 \, d\xi_2}{\int |\xi|^2 \, P(\xi) \, d\xi}. \tag{3.4}$$

(In expressions of this kind, it is convenient to allow ξ to be a complex variable whose reality can later be ensured by trivial subsidiary conditions.) By (3.3) this function must be normalized to unity at $R = 0$ and by (3.1) it must tend to zero as R becomes very large. Generally speaking, therefore, we may envisage $\Gamma(R)$ as a monotonically decreasing function of R with the limits

$$\Gamma(R) \rightarrow \begin{cases} 1 & \text{as } R \rightarrow 0 \\ 0 & \text{as } R \rightarrow \infty \end{cases}. \tag{3.5}$$

The analogy with the order parameter of § 1.5 and with the total correlation function (2.41) is very strong. We might use (1.38), for example, to define a *correlation length* or *range of continuity* of the random field; L would represent the typical spatial extent of any topographical feature of $\xi(\mathbf{R})$ such as a 'peak' or 'valley'.

The homogeneity of $\xi(\mathbf{R})$ may be interpreted as translational invariance in a statistical sense (§ 1.1): it is natural to introduce a representation of the field in terms of plane waves $e^{i\mathbf{q}\cdot\mathbf{R}}$ with wave vectors $\mathbf{q}$ chosen to satisfy appropriate boundary conditions in the large volume V. Any member of the ensemble of random fields has a well-defined Fourier representation

$$\xi(\mathbf{R}) = \sum_{\mathbf{q}} \Xi(\mathbf{q}) e^{i\mathbf{q}\cdot\mathbf{R}} \tag{3.6}$$

where the (complex) amplitude $\Xi(\mathbf{q})$ is given explicitly by the inverse integral transform

$$\Xi(\mathbf{q}) = V^{-1} \int \xi(\mathbf{R}) e^{-i\mathbf{q}\cdot\mathbf{R}} \, d\mathbf{R}. \tag{3.7}$$

Continuum disorder

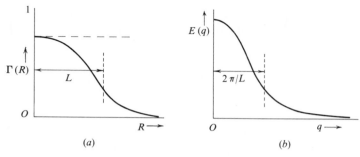

Fig. 3.1. (*a*) Auto-correlation function. (*b*) Spectral density.

In practice (3.6) is always evaluated as an integral, but it is convenient to avoid convergence problems by keeping V finite.

Now let us calculate the auto-correlation function (3.6). Using the ergodic assumption (3.1), we may carry out a familiar manipulation:

$$\Gamma(\mathbf{R}) = (V\langle|\xi|^2\rangle)^{-1} \int \xi^*(\mathbf{R}')\xi(\mathbf{R}'+\mathbf{R}) \, d\mathbf{R}'$$

$$= (V\langle|\xi|^2\rangle)^{-1} \int \sum_{\mathbf{q},\mathbf{q}'} \Xi^*(\mathbf{q}) \, \Xi(\mathbf{q}') e^{i\{-\mathbf{q}\cdot\mathbf{R}'+\mathbf{q}'\cdot(\mathbf{R}'+\mathbf{R})\}} \, d\mathbf{R}'$$

$$= (\langle|\xi|^2\rangle)^{-1} \sum_{\mathbf{q}} |\Xi(\mathbf{q})|^2 \, e^{i\mathbf{q}\cdot\mathbf{R}}. \tag{3.8}$$

This result, which is entirely analogous to (1.44), is true identically for any member of the ensemble. We define

$$E(\mathbf{q}) \equiv \langle|\Xi(\mathbf{q})|^2\rangle \tag{3.9}$$

as the *spectral density function* (or *power spectrum*) of the field. The *Wiener–Kintchine theorem* is the fundamental principle of the whole subject: *for any random field, the spectral density is the Fourier transform of the auto-correlation function.*

From the general properties of $\Gamma(R)$ we can infer the corresponding properties of $E(q)$. This function must be positive definite; for an isotropic field it can depend only on the wave number q and not on direction in reciprocal space. Familiar principles of Fourier analysis applied to (3.5) suggest that $E(q)$ must tend to zero when $q/2\pi$ exceeds L^{-1}, the inverse of the correlation length (fig. 3.1).

The spectral density function is obviously a *necessary* element in the initial prescription of any random field: to what extent is it also *sufficient* as basic information? Writing (3.9) in the form

$$\Xi(\mathbf{q}) = \sqrt{E(\mathbf{q})} \cdot e^{i\phi(\mathbf{q})}, \tag{3.10}$$

we see that the *amplitude* of each (complex) Fourier component of the field is determined: if each value of $\phi(\mathbf{q})$ were also prescribed (3.6) would define $\xi(\mathbf{R})$ uniquely. The randomness of the field may be supposed to lie in the statistical properties of the *phases* of the various components in the Fourier representation (3.6).

3.3 Gaussian randomness

Much the simplest assumption that could be made about the phase variables $\phi(\mathbf{q})$ is that they are statistically independent over the range $-\pi$ to π, for different values of $\mathbf{q}$. This is precisely the assumption of *spectral disorder* on a lattice that we made for the spin-wave and phonon representations in § 1.8. But these are dynamical modes which are supposed not to interact strongly; statistical independence of the phases could be interpreted as a typical *Stosszahlansatz* of statistical thermodynamics. In the present case, the random continuous field is a static representation of the disorder frozen into a system when it was made, so that spectral disorder is not a necessary property. Note, moreover, that the phase angle $\phi(\mathbf{q})$ is not really a measurable local characteristic of the field, since it would depend on the size of the sample volume V and the unspecified boundary conditions by which $\mathbf{q}$ is quantized in that volume. The artificiality of this prescription is emphasized by the fact that if we insist that $\xi(\mathbf{R})$ should be a *real* quantity we must have $\phi(-\mathbf{q}) = -\phi(\mathbf{q})$ for every $\mathbf{q}$.

The advantage is, however, that we generate a canonical form of random field, whose statistical properties are defined to all orders. The distribution of ξ itself is very simple. At the (arbitrary) point $\mathbf{R} = 0$, we have from (3.6) and (3.10) (for real ξ)

$$\xi = \sum_q 2 \cos \phi(\mathbf{q}) \, \Xi(\mathbf{q}) \tag{3.11}$$

with $\phi(\mathbf{q})$ distributed at random over the range $-\pi$ to π. Since the number of terms in this sum is quasi-infinite, this expression satisfies the conditions for the *Central Limit theorem*; as shown in any textbook of probability theory, the variable ξ has a *normal* or *Gaussian* distribution of the standard form

$$P(\xi) = (2\pi)^{-\frac{1}{2}} S^{-1} e^{-\xi^2/2S^2} \tag{3.12}$$

where S^2 is the *variance* of the field, i.e.

$$S^2 \equiv \langle \xi^2 \rangle = \sum_q E(\mathbf{q}). \tag{3.13}$$

Our random field is thus defined by the value of S, and the form of the auto-correlation function $\Gamma(R)$, without recourse to the spectral representation.

The multi-point distributions and higher moments of the field can also be calculated explicitly by standard methods of probability theory. For example, the 2-point distribution (3.2) must be none other than the *joint Gaussian distribution* for variables with correlation $\Gamma(R)$, i.e.

$$P_2(\xi_1, \xi_2; R) = \frac{1}{2\pi S^2[1 - \Gamma^2(R)]^{\frac{1}{2}}} \exp\left\{ -\frac{\xi_1^2 + \xi_2^2 - 2\xi_1 \xi_2 \Gamma(R)}{2S^2[1 - \Gamma^2(R)]} \right\}. \quad (3.14)$$

Because of (3.5) this evidently satisfies (3.3). Since many physical properties of systems represented by random fields depend on this function, this is a particularly useful formula.

But the spectral disorder condition for *Gaussian randomness* is very artificial: under what other general conditions may (3.12) and (3.14) be assumed? According to the Central Limit theorem, this must be the case *whenever the field $\xi(\mathbf{R})$ can be represented as the sum of a large number of independent random variables.* An example of this principle in the case of a stationary random function of time is *Campbell's theorem* (Rice 1944, 1945). The shot-noise current in an electrical circuit may be written as a sum of response functions $F(t)$, in the form

$$I(t) = \sum_{j=-\infty}^{\infty} F(t-t_j) \quad (3.15)$$

where the arrival times of the electrons, t_j, are random and independent. Campbell's theorem states that the distribution of $I(t)$ tends to a Gaussian as the rate of arrival tends to infinity. The analogue for three dimensions would be to represent our field as a superposition of 'potentials' centred on random points $\mathbf{R}_j$ with density N per unit volume, i.e.

$$\xi(\mathbf{R}) = \sum_j v(\mathbf{R} - \mathbf{R}_j). \quad (3.16)$$

From (3.1) we must have

$$\bar{v} \equiv N \int v(\mathbf{R}) \, d\mathbf{R} = 0. \quad (3.17)$$

A trivial extension of the proof of Campbell's theorem, or any equivalent application of the Central Limit theorem, then shows that *as $N \to \infty$ the distribution of ξ tends to the Gaussian form (3.12) with variance*

$$S^2 = \bar{v}^2 = N \int |v(\mathbf{R})|^2 \, d\mathbf{R}. \quad (3.18)$$

The conditions for validity of this result include, of course, the existence of this quantity in this limit – physically unattainable, but not unrealistic as an approximation in some cases.

The correlation function (3.4) in the *superposition representation* turns out to be the auto-convolution of the potential v, normalized to unity:

$$\Gamma(R) = N \int v^*(\mathbf{R}')v(\mathbf{R}' + \mathbf{R}) \, d\mathbf{R}' / \bar{v}^2. \tag{3.19}$$

From the Wiener–Kintchine theorem (3.8) we may evaluate the spectral density, which turns out to be simply the square modulus of the Fourier transform of each potential:

$$E(\mathbf{q}) = |v(\mathbf{q})|^2, \tag{3.20}$$

where

$$v(\mathbf{q}) \equiv N \int v(\mathbf{R}) \, e^{-i\mathbf{q} \cdot \mathbf{R}} \, d\mathbf{R}. \tag{3.21}$$

By Parseval's theorem (3.18) and (3.20) are consistent with (3.13). The formula (3.14) for the 2-point distribution of ξ would again be correct in the high-density limit.

Under what conditions would the Gaussian limit (3.12) be a good approximation? The essential criterion is that the value of ξ at any point in the field should be the sum of a sufficiently large number of independent contributions. To assign a measure to this criterion we may follow the proof of Campbell's theorem given by Rice (1944: equation 1-6-3) to reach the more general expression for the distribution function:

$$P(\xi) \sim \Phi^{(0)} - \frac{1}{3!} \lambda_3 \, \Phi^{(3)} + \left[\frac{1}{4!} \lambda_4 \, \Phi^{(4)} + \frac{1}{72} (\lambda_3)^2 \, \Phi^{(6)} \right] + \dots \tag{3.22}$$

where the derivatives of the Gaussian function are defined by

$$\Phi^{(n)} = \frac{1}{(2\pi)^{\frac{1}{2}}} \frac{1}{S^{n+1}} \left[\frac{d^n}{dx^n} e^{-x^2/2} \right]_{x = \xi/S} \tag{3.23}$$

and the coefficients are the semi-invariants of the potential:

$$\lambda_n = N \int |v(\mathbf{R})|^n \, d\mathbf{R}. \tag{3.24}$$

Let us suppose that $v(\mathbf{R})$ has a limited *range* – for example, a characteristic length L within which $\Gamma(R)$, defined by (3.19), is not zero. It will then be found that the second term of (3.22) is proportional to the *skewness* of the potential, and decreases only as $\{NL^3\}^{-\frac{1}{2}}$ relative to the Gaussian function $\Phi^{(0)}$. This confirms our intuition that the average density of packing of the points $\mathbf{R}_j$ in (3.16) must be considerably larger than one centre for each characteristic volume L^3, so that several potentials must be superposed at each point of the field (fig. 3.2). In a liquid metal, for example, where $v(\mathbf{R})$ might be a screened pseudo-potential for an ion (see § 10.2) these conditions would certainly not apply, so that the assumption of Gaussian randomness or of spectral disorder in the Fourier representation of the total potential is

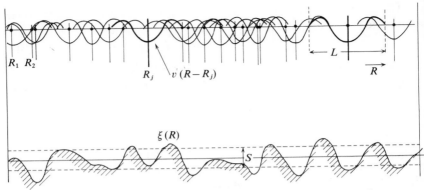

Fig. 3.2. Superposition of randomly placed potentials to produce Gaussian disorder. Note inadequate density at right-hand end.

quite unwarranted. But this criterion might well be satisfied by the carrier potential in an impure semiconductor (§ 2.15), although this could only occur at impurity densities above the limit for the Mott transition to a metallic band (Mott & Twose 1961).

To appreciate the possibilities for *non-Gaussian* fields let us consider a *step surface* (Berry 1973) with the 2-point distribution

$$P_2(\xi_1, \xi_2; R) = P(\xi_1)\delta(\xi_1 - \xi_2)\Gamma(R) + P(\xi_1)P(\xi_2)\{1 - \Gamma(R)\}. \quad (3.25)$$

Whatever the form of $P(\xi)$ or of $\Gamma(R)$, this function satisfies (3.2)–(3.9); but even when $P(\xi)$ is deliberately chosen to be Gaussian, the 2-point distribution is not the same as the joint Gaussian function (3.14). In topographic language, the surface consists of a series of level plateaux (fig. 3.3) with discontinuous jumps to new levels over distances of the order of the correlation length L. By analogy with the telegraph function (Rice 1944), it is convenient to assume that the correlation function is of the form

$$\Gamma(R) \sim e^{-R/L}, \quad (3.26)$$

as if the steps occurred at random intervals. But the topology of the 'cliffs' between adjacent plateaux is not adequately defined by this assumption alone and the model deserves further analytical investigation.

The spectral density is given by (3.8): in the simple case (3.26) this must be *Lorentzian*, of the form

$$E(q) \sim \frac{1}{1 + q^2 L^2}. \quad (3.27)$$

But the discontinuities in $\xi(R)$ produce too many short-wave components so that this spectrum cannot be normalized in three dimensions. Neverthe-

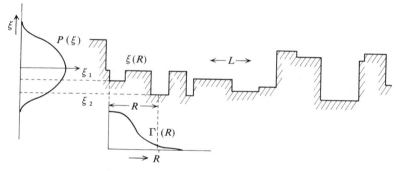

Fig. 3.3. Step surface.

less, (3.10) is still valid in principle: how is it that the conditions for spectral disorder are not satisfied? The answer is that the phases $\phi(q)$ are not uncorrelated: they must be so geared as to produce the level surfaces of the steps. These constraints are of infinite complexity and could never be written down explicitly, yet they cannot be ignored in the statistical prescription of the field. This is a generalization of the impossibility of representing Ising disorder on a lattice in spectral language (§ 1.8).

A step surface is not a satisfactory model for a random field in a real physical system because of the discontinuities. But these could be smoothed out and still leave us with a thoroughly non-Gaussian field. In a liquid metal, for example, we might write

$$\xi(\mathbf{R}) = v(\mathbf{R} - \mathbf{R}_j) \qquad (3.28)$$

when $\mathbf{R}$ lies in the atomic cell of $\mathbf{R}_j$, with a rule for smoothing between cells. The statistical properties of the spectral representation of such a function are quite useless because of phase correlations.

In the extreme case of *gas disorder* (§ 2.15) we can construct a *two-phase Poisson pattern* (Gilbert 1962; Frisch 1965; Beran 1968) by assigning values ξ_1 and ξ_2, at random, in some fixed proportion $\eta/(1-\eta)$, to the Voronoi cells (§ 2.11) of a perfect gas (fig. 3.4). The statistical properties of this model are fully defined by the number density of Poisson points $\mathbf{R}_j$ and by the packing fraction η (§ 2.11) occupied by the phase ξ_1. This is evidently a step surface, with two-point distribution (3.25), where the auto-correlation function can be calculated by elementary probabilistic arguments. Although this model has no obvious realization as a physical system, it has virtues of 'perfect' randomness, in the geometrical sense, which might be exploited mathematically as an ideal case.

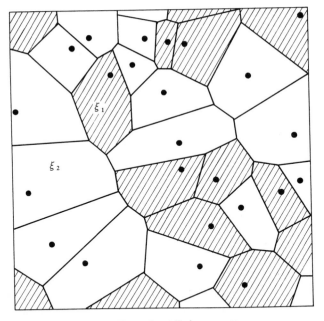

Fig. 3.4. Two-phase Poisson pattern.

3.4 Statistical topography

In many physical situations, the random field variable can be treated as a weak perturbation. But there are cases where the physical behaviour depends in a very non-linear way on $\xi(\mathbf{R})$. Consider, for example, the reflection of very short-wave radiation from a rough surface (Beckmann & Spizzichino 1963): the main non-specular scattering comes from 'geometrical' reflection from the varying tangent planes of local features. We become interested in the distribution of η, the gradient of ξ, as a statistical variable in its own right.

For a Gaussian random field, this is easy to deduce from the spectral representation (3.6), whose gradient is of the form

$$\eta(\mathbf{R}) \equiv \nabla \xi(\mathbf{R}) = \sum_{\mathbf{q}} i\mathbf{q}\, \Xi(\mathbf{q}) e^{i\mathbf{q}\cdot\mathbf{R}}. \tag{3.29}$$

Since the phase angles of $\Xi(\mathbf{q})$ are random, this means that each component of η has a Gaussian distribution, for which the total variance is given by the integrated power spectrum

$$T^2 \equiv \langle |\eta|^2 \rangle = \sum_{\mathbf{q}} q^2 |\Xi(\mathbf{q})|^2. \tag{3.30}$$

From the Wiener–Kintchine theorem (3.8) we may immediately deduce a connection with the auto-correlation function:

$$T^2 = -S^2[\nabla^2 \Gamma(\mathbf{R})]_{\mathbf{R}=0}. \tag{3.31}$$

Thus, the variation of slope on the surface depends essentially on the second derivative of the auto-correlation function of the random field. The sharper the peak of $\Gamma(R)$, the greater the variation in slope.

More difficult statistical problems arise if we are interested in the *topography* of the field. Suppose, for example, that $\xi(\mathbf{R})$ is a potential in which an electron of energy $\mathscr{E}$ is attempting to propagate. On the surface $\xi(\mathbf{R}) = \mathscr{E}$, the Schrödinger equation has a *classical turning point* (Berry & Mount 1972). If such a surface were closed around a minimum of ξ, then there might be a localized bound state below this energy. Such states may occur at the edge of a band in an amorphous or impure semiconductor (§ 11.2). The topology of the level surfaces of the field could thus be a significant feature of the physics of the system, separating free propagation from tunnelling.

The mathematical analysis of such questions is obviously difficult, but not entirely intractable. A simple result, due (once again!) to Rice (1945), is the expected number of zeros per second in a random stationary function of time. This cannot be deduced from $P(\xi)$ alone because it depends on the *continuity* of $\xi(t)$ and hence on the auto-correlation function. But let us suppose that a zero occurs at some point in the range $t_1 < t < t_1 + dt$. If $\xi(t)$ passes through this zero with positive slope η, then at the point t_1 it must have a value in the range $-\eta \, dt < \xi(t_1) < 0$. Suppose that the joint distribution function for the function ξ and its slope η is $P_{\xi,\eta}(\xi, \eta)$. The probability that $\xi(t_1)$ lies in this infinitesimal range must be

$$\int_{-\eta \, dt}^{0} P_{\xi,\eta}(\xi, \eta) \, d\xi = |\eta| P_{\xi,\eta}(0, \eta) \, dt. \tag{3.32}$$

The same result is obtained for negative values of η. The total probability of finding the zero is then the integral of (3.32) over all values of η, i.e.

$$n(0) \, dt = \left\{ \int_{-\infty}^{\infty} |\eta| P_{\xi,\eta}(0, \eta) \, d\eta \right\} dt, \tag{3.33}$$

which is a uniform rate proportional to the length of the interval dt.

To evaluate the integral for Gaussian disorder, we confirm from the spectral representations (3.6) and (3.29) of ξ and η that these variables are uncorrelated. From (3.12) and (3.31), therefore, we have

$$P_{\xi,\eta}(\xi, \eta) = \frac{1}{2\pi ST} \exp\left\{ \frac{-\xi^2}{2S^2} - \frac{\eta^2}{2T^2} \right\}. \tag{3.34}$$

Putting (3.34) into (3.33) and using (3.31)

$$n(0) = \frac{1}{\pi} \frac{T}{S} = \frac{1}{\pi} [-\Gamma''(0)/\Gamma(0)]^{\frac{1}{2}}, \qquad (3.35)$$

where $\Gamma''(0)$ is simply the second derivative of $\Gamma(t)$ at $t=0$. Thus, the average spacing of the zeros of $\xi(t)$ is proportional to the radius of curvature of the auto-correlation function near the origin – a 'length' comparable to the correlation length L defined in (1.38).

Rice obtained several other properties of the function by similar methods. If ξ_1 replaces 0 in (3.32) and (3.33), the integral measures $n(\xi_1)$, the number of times per second that the function passes through a chosen value ξ_1. From (3.34), this substitution simply multiplies (3.35) by the relative probability of finding ξ with this value, i.e.

$$n(\xi_1) = n(0) \exp\{-\xi_1^2/2S^2\}. \qquad (3.36)$$

The average spacing between extremes of ξ can be deduced by applying the argument for the spacing of zeros to the 'slope function' $\eta(t)$ which is itself random, stationary, Gaussian, etc. By analogy with (3.30), the variance of the gradient of η must be the second moment of the spectrum of η and hence, by (3.31), must be proportional to the *fourth* derivative of $\Gamma(t)$. Distinguishing between maxima and minima by the sign of $d\eta/dt$, we deduce from (3.35) that the number of *maxima* per second must be

$$n_{max} = \frac{1}{2\pi} [-\Gamma^{(iv)}(0)/\Gamma''(0)]^{\frac{1}{2}}. \qquad (3.37)$$

The inverse of this rate is yet another characteristic 'range of continuity' of the random function.

The distribution of the *heights* of maxima can also be calculated by a more complicated version of the same general method. For values of ξ_1, somewhat larger than S, the expected number of maxima per second in the range $\xi_1 \rightarrow \xi_1 + d\xi$ is given fairly accurately by

$$n_{max}(\xi_1)\, d\xi \approx \frac{1}{2\pi S^2} [-\Gamma''(0)/\Gamma(0)]^{\frac{1}{2}} \xi_1 \exp\{-\xi_1^2/2S^2\}\, d\xi. \qquad (3.38)$$

Note that this does not contain $\Gamma^{(iv)}(0)$ as in (3.37), but is consistent with (3.36) in that the total number of maxima *above* the value ξ_1 in a given interval now turns out to be the same as the number of times the function goes through ξ_1 with positive slope in this interval.

The application of these techniques to a random field in two or three dimensions obviously leads to further geometrical and analytical complications (Longuet-Higgins 1957, 1960). But the above results still apply to the random function of distance that can be generated by moving along any straight line through the field. The average spacing of zeros and maxima along such a section should be given by (3.35) and (3.37), with appropriate spatial derivatives of $\Gamma(\mathbf{R})$ in place of $\Gamma''(0)$ and $\Gamma^{(iv)}(0)$. It turns out, in fact, by an elaboration of the argument leading to (3.36), that in a two-dimensional field the mean length of the contour $\xi = \xi_1$ per unit area is very close to the inverse of the mean number of crossings of this contour per unit distance along any section; similar approximations must hold in three dimensions. The volume density of maxima can also be calculated, and is consistent with (3.37) and (3.38).

But the *topology* of contours or level surfaces is a much more difficult problem. The stationary points of $\xi(\mathbf{R})$ now consist not only of maxima and minima, but also saddle points, whose relative proportions and arrangement in space are governed by topological constraints. The results obtained so far concerning the pattern and paths of specular points for reflection from random surfaces (Longuet-Higgins 1960) suggest fruitful lines for research, but do not answer explicitly such questions as the depth and dimensions of typical 'basins' in the surface or the statistical parameters for a percolation analysis (§ 13.4) of a contour of given energy. This branch of mathematics seems ripe for further study, in the hope of obtaining results of genuine physical applicability.

4
The observation of disorder

—

'See how he lies at random, carelessly diffused'
Milton

4.1 Diffraction experiments and diffraction theory

We seldom have sufficient information to be quite certain about the arrangement of atoms in a disordered system. If we try to 'see' disorder on the atomic scale by means of a beam of neutrons, X-rays or electrons, we merely observe diffuse scattering from regions of the specimen containing large numbers of atoms. The evidence from a *diffraction* experiment is statistical and is limited, in practice, to 2-body structural characteristics such as the radial distribution function (§ 2.7). Much of the discussion in chapter 2 was concerned with the difficulty of interpreting this sort of information as unique evidence for a particular type of local structure in a liquid or glass. The choice between microcrystalline, random network and random heap models can only be made by inference from knowledge of macroscopic physical properties such as fluidity, or chemical principles such as valence bonding.

At the present time there seems little hope of finding a direct experimental technique that will provide the missing information concerning higher-order distribution functions, etc. This chapter, therefore, does not carry us much further towards the primary goal of setting up well-founded models of real disordered systems. But *diffraction theory* plays such an important part in the experimental study of atomic disorder, and provides a basis for so much of the later theory of the properties of such systems, that a short chapter on the main mathematical principles seems appropriate at this point. It must be made clear, however, that we are not concerned with the practical implementation of the basic theoretical principles discussed here, nor with details of apparatus, correction of experimental data, etc. These are matters of immense technical art and professional skill which can only be expounded by those who know from experience what they entail.

The essence of a diffraction experiment is the measurement of the *intensity* $I(\mathbf{Q}, \mathbf{Q}')$ of radiation scattered into the state $\Psi_{\mathbf{Q}'}(\mathbf{R})$ out of an incident beam of particles in the state $\Psi_{\mathbf{Q}}(\mathbf{R})$. Since the incident and scattered beams are created and collected, respectively, in free space, each of these state functions approximates to a plane wave, i.e.

$$\Psi_{\mathbf{Q}}(\mathbf{R}) \approx e^{i\mathbf{Q} \cdot \mathbf{R}}. \tag{4.1}$$

If the scattering is *elastic* (see §4.2), then the initial and final wave vectors $\mathbf{Q}$ and $\mathbf{Q}'$ must have the same magnitude, as prescribed by the energy or frequency of the incident radiation.

The specimen behaves towards the incident beam as a local 'potential' $\mathscr{U}(\mathbf{R})$; in the case of X-rays, for example, this would be the electron density. The *scattering amplitude* may be calculated from the *Born series* whose leading term is simply the matrix element

$$\langle \mathbf{Q}' | \mathscr{U} | \mathbf{Q} \rangle \equiv \frac{1}{V} \int \Psi_{\mathbf{Q}'}^{*}(\mathbf{R}) \, \mathscr{U}(\mathbf{R}) \, \Psi_{\mathbf{Q}}(\mathbf{R}) \, \mathrm{d}^3\mathbf{R}$$

$$= \frac{1}{V} \int \mathscr{U}(\mathbf{R}) e^{-i\mathbf{q} \cdot \mathbf{R}} \, \mathrm{d}^3\mathbf{R} \tag{4.2}$$

normalized to the volume of the sample. This depends only on the *scattering vector*

$$\mathbf{q} = \mathbf{Q}' - \mathbf{Q}, \tag{4.3}$$

whose magnitude for elastic scattering is easily measured by the *scattering angle* θ between $\mathbf{Q}$ and $\mathbf{Q}'$, i.e.

$$q = 2Q \sin \tfrac{1}{2}\theta. \tag{4.4}$$

The higher terms in the Born series correspond to *virtual multiple-scattering transitions* in which the incident radiation is imagined to have passed into the final state via one or more intermediate plane-wave states of arbitrary energy. But in an actual diffraction experiment we deliberately avoid *real* multiple-scattering processes, which merely smear out the information we are seeking about $\mathscr{U}(\mathbf{R})$ (cf. Pings 1968; McIntyre & Sengers 1968; Enderby 1968): under these conditions the *Born approximation* (4.2) is valid.

The matrix element (4.2) is none other than the Fourier transform of the 'potential' $\mathscr{U}(\mathbf{R})$. If the scattering amplitude itself could be measured directly for all $\mathbf{q}$, then this potential could be reconstructed by Fourier synthesis. But the diffraction apparatus measures *intensity* which does not give information about the *phase* of the scattered radiation. For elastic scattering we may write (dropping geometrical factors)

$$I(\mathbf{q}) = |\mathscr{U}(\mathbf{q})|^2: \tag{4.5}$$

in the language of § 3.2, we may say that the *distribution of the intensity of the diffracted radiation measures the spectral density of the 'potential' in the disordered system.*

If we are talking about a 'structure' then we must be able to identify the atomic sites R_i. In most cases we may assume that $\mathcal{U}(R)$ is a *superposition* of identical 'atomic potentials' centred on these points, i.e.

$$\mathcal{U}(R) = \sum_i u(R - R_i). \tag{4.6}$$

The Fourier transform (4.2) leads by elementary manipulations to

$$I(q) = \left| \frac{1}{V} \int \sum_i u(R - R_i) e^{-iq \cdot R} \, d^3R \right|^2$$

$$= \frac{1}{N^2} \sum_{i,j} e^{-iq \cdot (R_i - R_j)} \left| \frac{N}{V} \int u(R') e^{-iq \cdot R'} \, d^3R' \right|^2$$

$$= N^{-1} S(q) |u(q)|^2, \tag{4.7}$$

where $u(q)$ is the Fourier transform of a single atomic potential, just as in (3.21).

This is, of course, the elementary formula for X-ray or neutron diffraction by any assembly of atoms whether crystalline, amorphous or liquid. The *atomic form factor* $|u(q)|^2$ is supposed to be known independently, being no more than the scattering cross-section of a single atomic potential for the radiation in question. Measurement of $I(\theta)$ is thus interpreted as observation of the *interference function* or *structure factor* of the disordered material, i.e.

$$S(q) = \frac{1}{N} \sum_{i,j} e^{-iq \cdot (R_i - R_j)}. \tag{4.8}$$

In the special case of gas-like disorder (§ 2.15), where the positions of atoms R_i and R_j are statistically independent, we have $S(q) = 1$ for all q: these equations are then equivalent to (3.20). The extent to which the structure factor deviates from unity is thus a measure of the residual order in the atomic arrangement. For a perfect crystal, of course, $S(q)$ consists simply of an array of delta functions at the points of the reciprocal lattice, $q = g$.

The definition (4.8) requires a summation over all sites in a macroscopic specimen of site density $n = W/V$. The basic assumptions of homogeneity and ergodicity allow us to replace this with an ensemble average over site positions measured relative to some standard site at $\mathbf{R}=0$. Separating out the diagonal terms for which $i=j$, we get

$$S(\mathbf{q})=1+\langle e^{-i\mathbf{q}\cdot(\mathbf{R}_i-\mathbf{R}_j)}\rangle$$
$$=1+n\int g(\mathbf{R})e^{-i\mathbf{q}\cdot\mathbf{R}}\,d^3\mathbf{R}, \qquad (4.9)$$

where $g(\mathbf{R})$ is simply the *pair distribution function* (2.21). This is the fundamental result of this chapter. A diffraction experiment can determine the pair distribution of the atoms in our sample, but nothing more.

The inversion of (4.9) to yield $g(\mathbf{R})$ from observed values of $S(\mathbf{q})$ is simple in principle, though not so easy in practice. Normally, our specimen is macroscopically isotropic so that $g(\mathbf{R})$ becomes the *radial distribution function* $g(R)$. The structure factor can depend only on the modulus of $\mathbf{q}$, i.e. on the scattering angle θ defined by (4.4); (4.9) becomes

$$S(q)=1+n\int_0^\infty g(R)\frac{\sin qR}{qR}4\pi R^2\,dR. \qquad (4.10)$$

But, as we saw in § 2.9, the radial distribution function tends to unity for large R. As $q\rightarrow0$, the integral diverges to generate a delta-function singularity $\delta(\mathbf{q})$ which can easily be seen in (4.8). Subtracting this singularity as the Fourier transform of unity on the right, and tacitly ignoring it in the symbol for the structure factor, we get

$$S(q)=1+n\int_0^\infty h(R)\frac{\sin qR}{qR}4\pi R^2\,dR, \qquad (4.11)$$

where $h(R)=g(R)-1$ is the total correlation function (2.31). This integral is well behaved for large R and may, therefore, be inverted by standard methods:

$$g(R)=1+\frac{1}{8\pi^3n}\int_0^\infty \{S(q)-1\}\frac{\sin qR}{qR}4\pi q^2\,dq. \qquad (4.12)$$

It is evident from (4.11) that $S(q)\rightarrow1$ as $q\rightarrow\infty$, so that (4.12) is convergent provided that the singularity $\delta(\mathbf{q})$ has been dropped.

The characteristic features of the structure factor for a liquid or glass (fig. 4.1) can easily be deduced from (4.11). For small values of q we are looking at fluctuations of density over substantial distances so that $S(q)$ is small in magnitude. As q increases, these effects become larger, until the effective wavelength becomes comparable with the interatomic distance. The peaks in $g(R)$ for the various coordination shells then combine by constructive interference to produce a strong peak in $S(q)$. Beyond this point the structure factor falls to a minimum, with further oscillations about unity

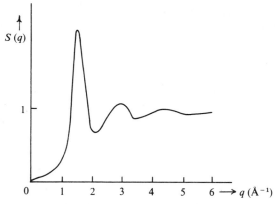

Fig. 4.1. Typical liquid structure factor.

due mainly to the sharp cut-off in $g(R)$ at distances smaller than the closest interatomic spacing. In other words, the *diffraction pattern* consists of one fairly well-defined ring with fainter rings outside it.

If we were dealing with a microcrystalline specimen (§ 2.6) then there would be a succession of peaks in $S(q)$, corresponding to the magnitudes of the various reciprocal lattice vectors of the crystal structure. The diffraction pattern, ideally, would be a set of sharp concentric rings – in fact, the 'powder pattern' of the material. But for very small crystallites these peaks would be broadened and smeared out, so that one could not tell by inspection of the structure factor whether one was dealing with a microcrystalline or an amorphous arrangement. This is simply the Fourier transform of the problems discussed in § 2.9.

Despite the analytical simplicity of (4.12), the Fourier inversion of an observed structure factor to give the radial distribution function is subject to a number of practical difficulties. It is not possible to measure $S(q)$ for all q, so that errors arise from the truncation of the integral at high and low values of q. The art of minimizing such errors is discussed at length in the experimental literature (see e.g. Pings 1968). But many formulae for the properties of disordered systems are written in the Fourier representation, where $S(q)$ itself may appear as a measure of the 'structure' of the medium. It is then more efficient to keep the diffraction data in this form, rather than making transformations to and from the real space function $g(R)$.

It turns out, for example, that the Fourier transform $c(q)$ of the direct correlation function, which plays an important part in the Percus–Yevick theory of liquids (§ 2.12), can be derived directly from the structure factor without any need to integrate (4.12). The defining formula (2.42) for $c(R)$ in terms of $h(R)$ is a convolution integral in **R**-space, whose Fourier transform is a simple product,

$$h(q) = c(q) + c(q)\,h(q): \tag{4.13}$$

hence, by (2.31) and (4.11), we get

$$c(q) = 1 - 1/S(q). \tag{4.14}$$

It is thus very easy, by a Fourier transform of the hard sphere PY solution (2.46), to calculate the structure factor of this model for direct comparison with experiment.

4.2 Neutron diffraction

The advantage of neutron diffraction for structure determination is that the interaction takes place solely with the atomic nuclei. The 'atomic potential' in (4.6) is merely the *Fermi pseudo-potential*

$$u(\mathbf{R}) = b\delta(\mathbf{R}), \tag{4.15}$$

where (in suitable units) the *scattering length* b is a constant characteristic of the particular nucleus and the neutron energy. The Fourier transform of this function is independent of $\mathbf{q}$, so that the diffraction intensity (4.7) contains a constant factor

$$|u(\mathbf{q})|^2 = b^2 \tag{4.16}$$

and measures directly the structure factor of the material.

In practice, this simplicity is compromised by two significant factors. Most chemical elements are mixtures of isotopes with different scattering lengths. In addition to the *coherent diffraction,* for which the value of b in (4.16) should now be $\bar{b}$, the average scattering length of the atoms in the mixture, we observe a background of *incoherent diffraction* proportional to $\{\overline{b^2} - (\bar{b})^2\}$ – the variance of the scattering length about this mean. Fortunately, as may be shown by treating the isotopic mixture as a perfectly random substitutional alloy (§ 4.5), this component is independent of the direction of $\mathbf{q}$ and hence does not spoil accurate determinations of the structure. This important practical point is dealt with at length in all the textbooks (e.g. Egelstaff 1965).

The other effect is the appearance of *inelastic diffraction* due to variation of the structure with time – for example, thermal vibration and diffusion of the atoms. Neutrons of the wavelength most suitable for structure determinations are quite 'slow' and hence are sensitive to such motion. To put this into a formal language we should consider transitions between neutron states of different frequencies (i.e. energies) Ω and Ω'. In place of (4.2) we may write

$$\langle \mathbf{Q}', \Omega' | \mathscr{U} | \mathbf{Q}, \Omega \rangle = \frac{1}{V} \int\int \mathscr{U}(\mathbf{R}, t) e^{i(\omega t - \mathbf{q} \cdot \mathbf{R})} \, d^3 \mathbf{R} \, dt, \qquad (4.17)$$

where $\omega = \Omega' - \Omega$ is the change of energy in the diffraction process. In effect, this matrix element is proportional to the Fourier transform $n(\mathbf{q}, \omega)$ of the *density operator*

$$n(\mathbf{R}, t) = \frac{1}{V} \sum_i \delta\{\mathbf{R} - \mathbf{R}_i(t)\} \qquad (4.18)$$

for the positions $\mathbf{R}_i(t)$ of the atomic nuclei as functions of time. But without knowledge of the phases of the scattered radiation we cannot go back from the inelastic diffraction intensity $I(\mathbf{q}, \omega)$ to this density function. The generalized structure factor $S(\mathbf{q}, \omega)$ is a spectral density function in both space and time: by the Wiener–Kintchine theorem (3.8) it must be the Fourier transform of the *van Hove correlation function* (cf. (3.4))

$$\Gamma(\mathbf{R}, t) = \langle n^*(\mathbf{R}', t') \, n(\mathbf{R}' + \mathbf{R}, t' + t) \rangle. \qquad (4.19)$$

In principle, this is all that we can learn from this sort of experiment.

Inelastic neutron diffraction thus provides valuable evidence concerning the *dynamics* of our system. But it is almost impossible to unscramble the information provided by $S(\mathbf{q}, \omega)$ or $\Gamma(\mathbf{R}, t)$ without a theoretical model for comparison. For a crystal we interpret the observations quite easily (and to very good effect) in the phonon representation; for a glass or liquid the problem of representing the vibrational modes is much more complicated (see § 11.1) and most attempts to describe the *diffusive* motion of the atoms are frankly phenomenological.

But information about the static structure of the system is given by $\Gamma(\mathbf{R}, 0)$ whose Fourier representation contains an integral of $S(\mathbf{q}, \omega)$ over all frequencies. The *static structure factor* that we should use in any formula such as (4.10) is given by

$$S(\mathbf{q}) = \int_{-\infty}^{\infty} S(\mathbf{q}, \omega) \, d\omega. \qquad (4.20)$$

Unfortunately, for elementary geometrical reasons, this integration is not quite the same as the experimental result one would obtain by the simple device of collecting all neutrons diffracted into a given *direction*, regardless of energy. Corrections for this effect are again part of the standard technique of structure determination by neutron diffraction.

Neutrons are sensitive, through their intrinsic spin, to the magnetization of the atoms from which they are scattered. The use of this effect in the study of magnetic disorder will be noted in § 4.6.

4.3 Structure determination by X-rays

X-ray diffraction differs from neutron diffraction in two main respects. On the one hand, the effects of inelasticity of scattering can be ignored. The energy of X-rays of a given wavelength is much higher than that of neutrons. The small change of energy in the diffraction process produces a negligible relative change in the momentum, so that the integration (4.20) for the structure factor is accurately achieved by collecting all the radiation scattered in a given direction.

On the other hand, X-rays are scattered by the *electrons* in an atom or solid, so that the atomic potential $u(\mathbf{R})$ cannot be considered as concentrated at the nucleus. The atomic form factor for X-rays is, in fact, the Fourier transform of the *electron density* $\rho(\mathbf{R})$ in an atom and is not, therefore, independent of the momentum transfer $\mathbf{q}$. To determine the structure factor we must follow (4.7) and divide the observed diffraction intensity by this form factor – usually obtained by independent diffraction measurements on a gas of free atoms. But this begs the question whether the actual electron density in the condensed phase can be represented, as in (4.6), as a superposition of free atom densities. In principle, there should be some redistribution of charge in the interstitial regions. By very careful measurements this effect can be observed in some semiconductor crystals but is, of course, entirely masked by the general disorder in a glass or liquid.

4.4 Small-angle scattering

The diffraction of X-rays with a small scattering vector – i.e. through a small angle from the incident beam – provides information about the state of the specimen over distances larger than a few atomic diameters. Near a phase transition, where the range of order becomes large (§ 1.7), this information can become significant (§ 5.12). The behaviour of $S(q)$ for small q is also important in some theoretical formulae for the electrical properties of liquid metals, etc. (§ 10.1).

The limiting value of the structure factor as $q \rightarrow 0$ simply measures the macroscopic variance of the density of the medium. Starting from the Fourier integral (4.11) we return to the definitions of the distribution functions in § 2.7; the integral over the pair distribution function is then thought of as an ensemble expectation of the *square* of the number of atoms in a volume – remembering that the atom at the origin is not counted again. Thus (Green 1952; Hill 1956)

$$S(0) = 1 + n \int \{g(\mathbf{R}) - 1\} \, d^3\mathbf{R}$$

$$= 1 + \frac{1}{\langle N \rangle} \iint \{n(\mathbf{1}, \mathbf{2}) - n(\mathbf{1}) \, n(\mathbf{2})\} d\mathbf{1} d\mathbf{2}$$

$$= [\langle N^2 \rangle - \{\langle N \rangle\}^2] / \langle N \rangle. \tag{4.21}$$

But the amplitude of fluctuations of macroscopic density is calculable by thermodynamics being simply proportional to the *isothermal compressibility* κ_T. In general, therefore, for a thermodynamic system such as liquid or crystal, we have

$$S(0) = n \text{\emph{k}} T \kappa_T. \tag{4.22}$$

Under ordinary circumstances this is all that we need to know: as q decreases, $S(q)$ falls quickly from its main peak to a value near $S(0)$, where it stays nearly constant as $q \to 0$. In other words, the density fluctuations reach their thermodynamic amplitude for regions of only a few dozen atoms and are then more or less uncorrelated. For most liquids, in fact, this limiting value of $S(0)$ is a few per cent. It is interesting to note that the variance of the cell volume of an RCP model of hard spheres (Finney 1970: see fig. 2.42) is only 0.0016, which is much smaller than the thermodynamic variance. In other words, ideal hard-sphere packing is much more uniform in density than is required by the thermal fluctuations in an actual liquid. Notice, however, that a *glass* is not in thermodynamic equilibrium so that there is nothing to prevent large fluctuations of density having been frozen in without having to satisfy (4.22) for the compressibility in the rigid phase.

The reference to density fluctuations in (4.21) strongly suggests that we should think about small-angle X-ray scattering (SAXS) as a *continuum* effect, without regard to the contributions of the individual atoms. Following (4.19), for example, and smoothing the sum in (4.18) into a continuous density variable $n(\mathbf{R}, t)$, we see that $S(q)$ for small q is nothing more than the spectral function (3.8) for variations of 'local' density, about the mean value n. The variable

$$\Delta n(\mathbf{R}, t) = n(\mathbf{R}, t) - n \tag{4.23}$$

is a random field of continuum disorder, of the kind discussed in chapter 3.

This provides a very simple model for the analogous phenomenon of *optical scattering* (see e.g. McIntyre & Sengers 1968) by fluids and glasses.

We simply assume that the local refractive index, $\varepsilon(\mathbf{R})$, varies from place to place by an amount proportional to the local density, i.e.

$$\Delta\varepsilon(\mathbf{R}) = \gamma\Delta n(\mathbf{R}) \tag{4.24}$$

(with time variation as well, if necessary). This variation will appear in the wave equation for light propagation in this medium as a small perturbation term, which gives rise to transitions between the incident and scattered beams (cf. § 10.1). The argument is no deeper than the Born approximation: it is very easy to show that the intensity of the scattered radiation for a change of wave vector $\mathbf{q}$ is proportional to γ^2 and to the spectral function $S(q)$ of the density field $n(\mathbf{R})$. Inelastic diffraction due to time variation of the density is also accounted for, exactly as in (4.17). Optical scattering may have technical advantages, especially for measurements of small momentum transfer and small changes of energy, but it responds to precisely the same structural features as small-angle X-ray scattering.

But (4.22) only tells us the limiting value of the structure factor: it does not tell us the form of the spectrum for non-zero values of q. In the neighbourhood of a *critical point* the compressibility of a fluid can become very large, so that the behaviour of $S(q)$ between $S(0)$ and the main peak comes into question. This point was discussed long ago in the famous paper by Ornstein & Zernike (1914). Notice the very simple relation (4.14) between the Fourier representation of the direct correlation function $c(q)$ and the structure factor. Evidently, if $S(q)$ becomes large near $q=0$ then $c(q)$ tends to a value near unity:

$$c(0) = 1 - 1/S(0). \tag{4.25}$$

On the other hand, when $S(q)$ rises rapidly and exceeds unity near the main peak, $c(q)$ behaves quite normally: this is the basic principle of the PY equation (2.44). The assumption is, therefore, that $c(q)$ is well behaved throughout this range, and may be expanded as a power series in q^2 (remember that q is the modulus of a vector quantity $\mathbf{q}$) near the limiting value (4.25). We thus write

$$c(q) \approx c(0) - \xi_0^2\, q^2, \tag{4.26}$$

with ξ_0 to be determined. Putting this into (4.14) we have

$$S(q) = \frac{1}{\{1-c(0)\} + \xi_0^2\, q^2}$$
$$= \frac{S(0)}{1 + S(0)\,\xi_0^2\, q^2} \tag{4.27}$$

The *Ornstein–Zernike spectrum* of the density fluctuations is evidently

Lorentzian (as in (3.27)) in this approximation. Applying the Wiener–Kintchine formula (2.8) we obtain the density–density correlation function in the medium, in the form

$$\Gamma(R) \sim \frac{1}{R} e^{-R/\xi_1} \qquad (4.28)$$

for large values of R. This is of the form (1.37), with the correlation length

$$\xi_1 = \{S(0)\}^{\frac{1}{2}} \xi_0. \qquad (4.29)$$

The basic *physical* assumption of the Ornstein–Zernike theory is that the parameter ξ_0 is governed by the local structural order in the fluid and is not, therefore, strongly influenced by the proximity of the critical point. Approaching this point, however, $S(0)$ can vary rapidly with temperature and become very large; the high compressibility of the medium is associated with long-wave *critical fluctuations* of density. The *critical opalescence* observed optically yields sensitive evidence concerning, for example, the exponent of the temperature variation of the long-range order parameter under these conditions (cf. Stanley 1971). The range of order in a liquid crystal above the critical temperature can also be studied by this method (Stinson & Litser 1973). It is evident, however, from the form of (4.28) that the disorder observed over distances greater than ξ_1 is best considered as a macroscopic *inhomogeneity* present in a large specimen whose local fluid structure is given, as a function of local density, by the usual statistical arguments.

In view of the heuristic assumptions in the Ornstein–Zernike formula (4.27), it is interesting to note that the exponent of q in the observed spectrum is actually quite close to 2. But model calculations for two-dimensional systems give results that are quite different from (4.27).

In the present section, our main aim is to draw attention to the information concerning large-scale disorder that can be obtained by small-angle X-ray and optical scattering. From the general argument of § 3.3, however, it is clear that an observed structure factor cannot be interpreted unambiguously as evidence for a particular type of density variation in the medium. The dynamical theory of critical fluctuations proves that these must satisfy the conditions for Gaussian disorder (cf. §§ 1.8, 3.3): but a Lorentzian spectrum such as (4.27) could equally well be characteristic of a 'step surface' model as in (3.25). There is, in fact, a considerable literature (e.g. Fournet 1957; Beeman, Kaesberg, Anderegg & Webb 1957) concerning small-angle X-ray diffraction from dispersions of small 'particles', such as colloidal suspensions. In such cases, $S(q)$ is taken to be unity for all values of q, whilst the scattered intensity is governed by the factor $|u(q)|^2$, now redefined as the diffraction factor of a whole 'particle'. If this object happens to be more or less spherical, then the formula for *Rayleigh scattering* would be appropriate, i.e.

$$|u(q)|^2 \propto \left[3 \frac{\sin qD - qD \cos qD}{q^3 D^3} \right]^2. \qquad (4.30)$$

Such observations are often analysed in terms of the *Guinier approximation* whereby the dependence of the scattering on q for small values of q is written

$$|u(q)|^2 \propto \exp\{-\tfrac{1}{3}q^2 D^2\}, \qquad (4.31)$$

where D may be defined more explicitly as the radius of gyration of each particle. But the fact that an observed diffraction pattern happens to fit such a formula does not prove that the system actually consists of sharply defined, approximately spherical objects distributed at random in a statistically uniform matrix. This point is important in the interpretation of small-angle diffraction from, say, a glass; density or concentration fluctuations with a spectrum like (4.31) might equally well be suspected (cf. Seward & Uhlmann 1972).

4.5 Diffraction by a mixture

The formula for diffraction by a mixture of various atomic species (e.g. a liquid alloy) is an elementary extension of (4.7). Let us assume, as in (4.6), that the scattering potential can be written as a superposition

$$\mathcal{U}(\mathbf{R}) = \sum_\alpha \sum_{i(\alpha)} u_\alpha(\mathbf{R} - \mathbf{R}_{i(\alpha)}) \qquad (4.32)$$

where the atom at $\mathbf{R}_{i(\alpha)}$ has potential u_α. The Fourier transform (4.2) can be manipulated as in (4.7) into the general formula

$$I(\mathbf{q}) = N^{-2} \sum_{\alpha,\beta} \sum_{i(\alpha), j(\beta)} \exp\{-i\mathbf{q} \cdot (\mathbf{R}_{i(\alpha)} - \mathbf{R}_{j(\beta)})\} u_\alpha^*(\mathbf{q}) \, u_\beta(\mathbf{q}). \qquad (4.33)$$

In this expression we segregate the terms for which $\alpha = \beta$ and $i(\alpha) = j(\beta)$: these contribute a structure-independent term, weighted according to the relative concentrations c_α of the components, i.e.

$$I_0(\mathbf{q}) = N^{-1} \sum_\alpha c_\alpha |u_\alpha(\mathbf{q})|^2$$

$$= N^{-1} \left\{ |\bar{u}(\mathbf{q})|^2 + \sum_\alpha c_\alpha |\Delta u_\alpha(\mathbf{q})|^2 \right\} \qquad (4.34)$$

where $\Delta u_\alpha(\mathbf{q})$ is the deviation of $u_\alpha(\mathbf{q})$ from the average scattering potential

$$\bar{u}(\mathbf{q}) = \sum_\alpha c_\alpha u_\alpha(\mathbf{q}). \tag{4.35}$$

The remaining terms in (4.33) now refer to pairs of distinct atoms (although often of the same species) whose relative vector positions $(\mathbf{R}_{i(\alpha)} - \mathbf{R}_{j(\beta)})$ are distributed statistically over the ensemble with partial distribution functions $g_{\alpha\beta}(1, 2)$ and $h_{\alpha\beta}(1, 2)$ as defined in § 2.13. Following (4.11) we define the corresponding *partial structure factors* (without the singularity at $q = 0$)

$$S_{\alpha\beta}(\mathbf{q}) = 1 + n \int h_{\alpha\beta}(\mathbf{R}) e^{i\mathbf{q}\cdot\mathbf{R}} \, d^3\mathbf{R}. \tag{4.36}$$

Rewriting the sums in (4.33) as ensemble averages we obtain, at once, a contribution

$$I_s(\mathbf{q}) = N^{-1} \sum_{\alpha,\,\beta} c_\alpha c_\beta \{S_{\alpha\beta}(\mathbf{q}) - 1\} u_\alpha^*(\mathbf{q}) u_\beta(\mathbf{q}). \tag{4.37}$$

Combining (4.34) and (4.37) and cancelling a term in $|\bar{u}(\mathbf{q})|^2$, we get a general formula for the total scattered intensity:

$$I(\mathbf{q}) = N^{-1} \left\{ \sum_{\alpha,\,\beta} c_\alpha c_\beta \, S_{\alpha\beta}(\mathbf{q}) u_\alpha^*(\mathbf{q}) u_\beta(\mathbf{q}) + \sum_\alpha c_\alpha |\Delta u_\alpha(\mathbf{q})|^2 \right\}. \tag{4.38}$$

It is obvious from the form of (4.38) that a single observation of an X-ray or neutron diffraction pattern does not tell us the separate partial structure factors and hence gives only limited information about the liquid, glass or solid alloy. But if we could assume (as in the theory of conformal solutions discussed in § 2.13) that we had a genuine *random substitutional mixture* then we could set all the partial structure factors equal and factorize the first term. Thus if

$$S_{\alpha\beta}(\mathbf{q}) \approx S(\mathbf{q}) \quad \text{for all } \alpha, \beta, \tag{4.39}$$

then

$$I(\mathbf{q}) \approx N^{-1} \left\{ S(\mathbf{q}) |\bar{u}(\mathbf{q})|^2 + \sum_\alpha c_\alpha |\Delta u_\alpha(\mathbf{q})|^2 \right\}. \tag{4.40}$$

The first term represents *coherent diffraction,* as if from an assembly of identical atoms each with the average potential (4.35) and the common structure factor (4.39). But the variation of the actual atomic potentials about this mean value gives rise to additional *incoherent diffraction.* In the case of neutron diffraction from an isotopic mixture (§ 4.2), this contribution is simply proportional to the variance of the scattering lengths, i.e.

$$\overline{(\Delta b)^2} = \overline{b^2} - (\bar{b})^2. \tag{4.41}$$

But, as we saw in § 2.13, the assumption (4.39) is seldom justifiable in a liquid mixture. We know, for example, that the partial structure factors are considerably affected by small differences in hard sphere radius (cf. fig. 2.47). The next level of approximation is to assume that each partial distribution function or structure factor $S_{\alpha\beta}$ depends only on the nature of the atomic species α and β but not on their relative concentrations c_α, c_β in the mixture. In a series of liquid alloys of Au and Sn, for example (Wagner, Halder & North 1967), we assume that the probability of finding an atom of Au within a certain distance of a given atom of Sn is given by (cf. 2.20)

$$n_{Sn(Au)}(\mathbf{R}) = n_{Sn}\, g_{Sn(Au)}(\mathbf{R}) \qquad (4.42)$$

where $g_{Sn(Au)}$ is the same function of $\mathbf{R}$, whatever the number density n_{Sn} of Sn atoms in the fluid. The advantage of this assumption is that, given direct measurements of $I(q)$, for X-rays or neutrons, at various different concentrations of the constituents, we can solve (4.38) for the various partial structure factors (Halder & Wagner 1967; North & Wagner 1970; Isherwood & Orton 1969, Buhner & Steeb 1970).

Unfortunately, even this approximation is not sound. If we look, for example, at the partial structure factors calculated for hard-sphere, binary liquid mixtures by the PY method (Ashcroft & Langreth 1967) we find considerable variation with concentration. A better method in principle is to exploit the different neutron-scattering lengths of different isotopes of the same chemical species. As pointed out by Vineyard (1958) and by Keating (1963), the diffraction observed from three alloy samples, each of the same chemical composition but with varying relative concentrations of the isotopes of one of the constituents, yields sufficient information to determine the three separate partial structure factors. In fig. 4.2, for example, the results for the alloy composition Cu_6Sn_5 were obtained from diffraction by specimens containing pure ^{63}Cu, pure ^{65}Cu, and the natural $^{63}Cu/^{65}Cu$ mixture respectively (Enderby, North & Egelstaff 1966). This technique is obviously much more elaborate and expensive than X-ray diffraction (but see Ramesh & Ramaseshan 1971) by alloys of varying chemical composition, but seems the only way to get unambiguous information concerning the atomic arrangements in liquid mixtures without the intervention of theoretical models and questionable interpretative assumptions. For example, the results obtained by Page & Mika (1971) on liquid CuCl and by Edwards, Enderby, Howe & Page (1975) on NaCl (fig. 4.3) do not fit into any of the model theories discussed in § 2.13.

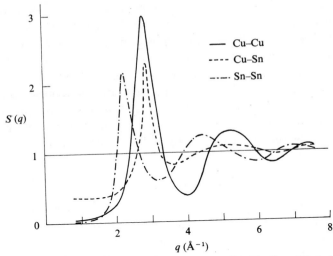

Fig. 4.2. Partial structure factors in Cu–Sn liquid alloy (Enderby, North & Egelstaff 1966).

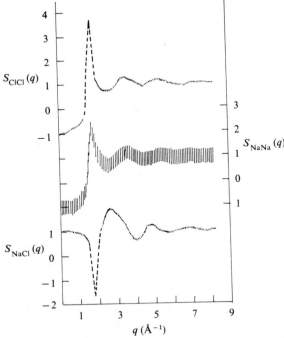

Fig. 4.3. Partial structure factors for molten NaCl from which radial distribution functions of fig. 2.50 were derived (Edwards *et al.* 1975).

4.6 Diffraction effects of substitutional disorder

The concept of substitutional or cellular disorder (§ 1.2) implies that we can specify a basic arrangement of atomic sites in space – typically, but not necessarily, an ordered lattice. This arrangement could be defined statistically by a set of *site distribution functions* such as a pair distribution function $g(1, 2)$. If the atoms were placed at random on to these sites, as assumed in (4.40), then each partial distribution function $g_{\alpha\beta}$ would be the same as the site distribution. Correlations in the atomic distributions on these sites, represented by an order parameter such as the correlation function $\Gamma_{\alpha\beta}(\mathbf{R})$ discussed in (1.12), would appear statistically as deviations of the actual partial distributions from this equality, i.e.

$$\Gamma_{\alpha\beta}(\mathbf{R})\, g(\mathbf{R}) \equiv g_{\alpha\beta}(\mathbf{R}) - g(\mathbf{R}). \tag{4.43}$$

The various partial distributions cannot, however, be independent functions. The *total* number of *alternative* atom types that may be put on site 1, in relation to a *particular* atom on site 2, must be the same, whatever the type of atom at 2; as in (1.20), we have the identity

$$g(1, 2) \equiv \sum_{\alpha} c_{\alpha}\, g_{\alpha\beta}(1, 2) \tag{4.44}$$

for each type β. Fourier transformation yields similar identities for the partial structure factors (4.36), which must add up to the site structure factor: for each β,

$$S(\mathbf{q}) \equiv \sum_{\alpha} c_{\alpha}\, S_{\alpha\beta}(\mathbf{q}). \tag{4.45}$$

These identities allow us to rewrite the total diffracted intensity (4.38) in a form analogous to (4.40), i.e.

$$I(\mathbf{q}) = N^{-1} S(\mathbf{q})|\bar{u}(\mathbf{q})|^2 +$$
$$N^{-1} \sum_{\alpha} \sum_{\beta > \alpha} c_{\alpha}\, c_{\beta}\{1 + S(\mathbf{q}) - S_{\alpha\beta}(\mathbf{q})\}|u_{\alpha}(\mathbf{q}) - u_{\beta}(\mathbf{q})|^2. \tag{4.46}$$

The coherent diffraction term may be described, as before, as the effect of putting an average atomic potential on each site. The incoherent diffraction is now somewhat more complicated than it was in the purely random mixture, since it contains effects due to any partial ordering of the atomic substitutions. In fact, measurement of this term, by X-ray or neutron diffraction from a binary alloy, together with knowledge of c_{α}, c_{β}, $u_{\alpha}(\mathbf{q})$, $u_{\beta}(\mathbf{q})$, provides us with values of

$$S(\mathbf{q}) - S_{\alpha\beta}(\mathbf{q}) = -h \int \Gamma_{\alpha\beta}(\mathbf{R})\, g(\mathbf{R}) e^{-i\mathbf{q}\cdot\mathbf{R}} d^3\mathbf{R} \tag{4.47}$$

from which the order parameter $\Gamma_{\alpha\beta}(\mathbf{R})$ can be deduced by Fourier inversion. This is the method used to investigate long-range and short-range order (§§ 7.6, 7.7) near the order–disorder critical point in such materials (Münster 1965).

In practice, we may not be interested in the precise geometrical arrangement of the atomic sites, expressed through the factor $g(\mathbf{R})$ in (4.47). It is often convenient to follow the example of § 4.4, and set up a continuum model where the field variables are the concentration functions $c_\alpha(\mathbf{R})$ for each component. We naturally measure each of these as deviations from the mean concentration c_α, and in a binary mixture we may use the identity that the concentrations of the two components must add up to 1. The variable

$$\Delta c(\mathbf{R}) \equiv [c_A(\mathbf{R}) - c_A] - [c_B(\mathbf{R}) - c_B] \qquad (4.48)$$

is a good measure of these deviations, since the average local *potential* will deviate from the ensemble average by an amount

$$\Delta \mathscr{U}(\mathbf{R}) = \tfrac{1}{2}(u_A - u_B)\Delta c(\mathbf{R}) \qquad (4.49)$$

where u_α is the potential seen by the diffracted beam in the neighbourhood of an atom of type α. This implies, of course, that we are looking at long-wave or small-angle diffraction, just as we were in the optical scattering model (4.24). It is easy to show that diffraction by the perturbation (4.49) is governed by a *concentration–concentration* structure factor

$$S_{cc}(\mathbf{q}) = \int \langle \Delta c(0)\, \Delta c(\mathbf{R}) \rangle e^{-i\mathbf{q}\cdot\mathbf{R}}\, d^3\mathbf{R}, \qquad (4.50)$$

together with a factor proportional to $|u_A - u_B|^2$. In accordance with the general principles of chapter 3, we observe that this structure factor is simply the spectral density of the *concentration–concentration correlation function,* which can easily be rewritten in terms of partial distribution functions, i.e.

$$\langle \Delta c(0)\, \Delta c(\mathbf{R}) \rangle = c_A^2\, g_{AA}(\mathbf{R}) + c_B^2\, g_{BB}(\mathbf{R}) - 2c_A\, c_B\, g_{AB}(\mathbf{R}). \qquad (4.51)$$

Thus, we may write the structure factor $S_{cc}(\mathbf{q})$ as a sum of the partial structure factors $S_{AA}(\mathbf{q})$, etc. This model, therefore, may be linked with (4.46) to interpret the diffraction as the effect of fluctuations of relative atomic concentration in the continuous material of the alloy.

The parallel with the theory of density fluctuations in liquids (§ 4.4) can be carried further. Exactly analogous to (4.22), there is a general thermodynamic formula (Münster 1965; Bhatia & Thornton 1970) for the amplitude of macroscopic fluctuations of concentration on an assembly of N sites with total Gibbs free energy G:

$$S_{cc}(0) = \langle (\Delta c)^2 \rangle = Nk\,T/(\partial^2 G/\partial c^2)_{T,\,P,\,N}. \qquad (4.52)$$

This tells us the limiting value of the structure factor, as $q \to 0$, in terms of measurable macroscopic variables. It is further argued that the spectrum of fluctuations for long wavelengths should be well represented by the Ornstein–Zernike formula (4.27). The corresponding expression (4.28) for the correlation function $\Gamma(R)$ is a convenient starting point for the detailed study of the range of order in binary alloys and analogous systems.

The implicit assumption of these formulae is that the total site density is quite uniform. But in a *liquid alloy* we may expect fluctuations of local atomic *density,* described as in (4.23) by a variable such as $\Delta n(\mathbf{R})$, in addition to fluctuations of local relative *concentration*. The continuum description implies knowledge of three correlation functions – *density–density, concentration–concentration,* and *density–concentration* – analogous to (4.51). But the corresponding three structure factors – S_{dd}, S_{cc}, S_{cd} – must be expressible algebraically in terms of the three conventional partial structure factors $S_{\alpha\beta}$ that appear, for example, in (4.38). Moreover, in the limit of long wavelengths, fluctuations of the continuum variables can be evaluated from the macroscopic thermodynamic properties of the system as in (4.22) and (4.52). In other words (Bhatia & Thornton 1970; McAllister & Turner 1972) we can use measurements of compressibility, partial vapour pressures, etc. to evaluate all three partial structure factors $S_{AA}(0)$, $S_{BB}(0)$, $S_{AB}(0)$ at the limit $q \to 0$. It is interesting to note (fig. 4.4) that in a liquid alloy such as Na–K the partial distributions are found to be more or less independent of concentration over a wide range, as assumed in (4.42), but this is evidently a bad approximation in the liquid K–Hg alloy. It can be shown, for example (Bhatia, Hargrove & March 1973), that the differences in the partial structure factors are related to the energy required to substitute one atom for another, i.e.

$$S_{AA}(0) + S_{BB}(0) - 2S_{AB}(0) \approx 4J_{AB}/kT, \qquad (4.53)$$

where J_{AB} might be the parameter defined in (1.24) for an Ising model. This quantity thus measures the deviation of the mixture from the ideal of a conformal solution discussed in § 2.13.

It is clear from the above analysis that diffraction by other types of cellular disorder such as ice disorder (§ 1.4), ferroelectric and antiferroelectric disorder (§ 1.4) and magnetic disorder (§ 1.3), can provide information about correlation functions analogous to (4.19) or (4.51). For example, the theory of neutron diffraction by a magnetic crystal (Elliott & Marshall 1958; Marshall & Low 1968) is centred on the study of the correlation function

$$\Gamma^{\alpha,\beta}(\mathbf{R}, t) = \langle S^{\alpha}(0, 0) \, S^{\beta}(\mathbf{R}, t) \rangle \qquad (4.54)$$

140   *The observation of disorder*

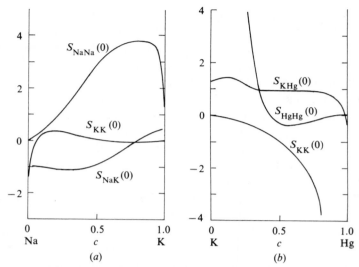

Fig. 4.4. Partial structure factors at infinite wavelength in liquid alloys:
(a) Na–K; (b) K–Hg (McAlister & Turner 1972).

which is a time-dependent generalization of (1.13). As in (4.19), this time
dependence gives rise to inelastic diffraction which can be interpreted (for
$T < T_c$) as the excitation of spin-wave modes (§ 1.8) whose dispersion
function can thus be investigated. The observation of critical scattering due
to strong fluctuations of magnetization near the transition temperature is
also closely analogous to the corresponding phenomena for liquids (§ 4.4)
and for alloys and can similarly be analysed in terms of the Ornstein–
Zernike formula (4.27) which is, indeed, found to be quite well satisfied
when the effects of inelasticity have been allowed for (Münster 1965).

4.7 Diffraction and imaging

The limitation of all pure diffraction methods for the study of disordered
systems is that only the structure factor $S(\mathbf{q})$ can be determined. This
follows from the elementary formula (4.7); all phase information in the
scattered radiation is lost because the observable intensity is the square
modulus of the diffracted wave.

Nevertheless, according to (4.2), the scattered amplitude $\Psi(\mathbf{q})$ from the
potential $\mathcal{U}(\mathbf{R})$ is proportional to the Fourier transform $\mathcal{U}(\mathbf{q})$ of that
potential; if we knew $\Psi(\mathbf{q})$ itself, then we could readily reconstruct $\mathcal{U}(\mathbf{R})$.

This is, of course, the basic principle of a *microscope* (Born & Wolf 1970), where the arrangement of lenses and apertures produces a function $\Psi_1(\mathbf{R})$ in the image plane that is a filtered Fourier transform of $\Psi(\mathbf{q})$ i.e.

$$\Psi_1(\mathbf{R}) = \int F(\mathbf{q})\, \Psi(\mathbf{q})\, e^{i\mathbf{q}\cdot\mathbf{R}}\, d^3\mathbf{q}. \tag{4.55}$$

If the *transfer function* $F(\mathbf{q})$ were constant for all values of $\mathbf{q}$ then the observable quantity $|\Psi_1(\mathbf{R})|^2$ would faithfully reproduce $|\mathscr{U}(\mathbf{R}|^2$, the square of the potential at each point in the specimen.

In practice, of course, $F(\mathbf{q})$ is cut off at large values of $\mathbf{q}$, both by geometrical factors and by the wave number of the incident radiation. This limits the detail that may be observed in the image. For our present purposes, optical radiation is far too coarse and X-rays cannot be focussed. In principle, however, an *electron microscope* at sufficiently high voltage should be able to show up details of disordered structure, down perhaps to the atomic level.

The observation of dislocation (§ 2.5) and microcrystalline (§ 2.6) disorder in thin metallic films is a familiar use of this technique. It remains to be seen, however, whether the method can provide useful information concerning, for example, the local structure in an amorphous semiconductor or glass. Rudee & Howie (1972) have reported diffraction fringes in images of thin films of amorphous Ge and Si which they attribute to interference between diffracted waves from local crystallites which happen to be in the Bragg orientation. But Berry & Doyle (1973) argue that this may well be an artefact of the transfer function when applied to a Gaussian amplitude $\Psi(\mathbf{R})$ (§ 3.3), which must itself be the sum of contributions from many local regions in a film that is actually much thicker than any supposed crystallite. The interpretation of such images depends in very subtle and complex ways on instrumental conditions (see McFarlane 1975).

5

Statistical mechanics of substitutional disorder

—

5.1 Physical problems and mathematical puzzles

The fundamental problem in the physics of disordered systems is the
mechanism of the disorder itself. In almost every case, the ultimate source
of 'randomness' in the distribution of atoms, spins, etc. is thermal: the
specimen is defined as a member of a thermodynamic ensemble in equilibrium at the temperature of observation or as being in a metastable state
frozen from equilibrium at some higher temperature. It is, of course, always
possible to produce disorder by radiation damage, plastic deformation,
magnetic or dielectric cycling, etc., but the actual effects of each form of
treatment would depend on the details of the material and the apparatus.
From a theoretical point of view, the merit of *thermodynamic disorder* is
that it depends only on the energy spectrum of the specimen and on the
temperature and is, therefore, mathematically calculable and physically
reproducible.

In previous chapters the main features of thermodynamic disorder on a
lattice (chapter 1) and in space (chapter 2) were sketched out, with particular attention to the characteristic order parameters, correlation functions
and distribution functions that may be observed experimentally or that
play an important part in the physical properties of the system. But the
mathematical theory of the variation of the disorder with temperature,
with the interactions between atoms or spins and with the dimensionality of
the assembly was not developed in detail. Conventionally this theory
belongs to the realm of *statistical mechanics* which must be considered an
integral part of the theory of disordered systems. One cannot, for example,
separate the discussion of the equilibrium positions of the atoms in a
thermally disordered liquid from the dynamical analysis of the motion of
the atoms about such equilibrium, or the theory of the electronic states

(which actually determine the interatomic forces) in the same assembly. We shall find, moreover, that many of the mathematical techniques that are used in statistical mechanics have close analogues in other parts of the theory. Apart from one or two exact solutions for special statistical mechanical models we are forced to make approximate assumptions of statistical independence (random phase, coherent potential, mean field, etc.) which are essentially the same for a wide variety of physical phenomena.

The real difficulty for us both, gentle reader, is that the literature on the statistical mechanics of order–disorder transitions and critical phenomena has now grown to enormous proportions. By what principle can we select material relevant to the present book?

In this chapter we return to the study or disorder on a regular lattice. In chapter 1, we found that there are really only four distinct elementary models of physical significance:

(1) the *Ising model*, with Hamiltonian (1.18) which can represent a binary alloy or a special type of magnetic system;

(2) the *classical Heisenberg model*, with Hamiltonian (1.16) which is a fairly good approximation for many real ferromagnetic and antiferromagnetic systems;

(3) the *quantal Heisenberg model*, where the Hamiltonian (1.16) is interpreted more exactly, especially for $S = \frac{1}{2}$;

(4) the *ice model* of § 1.4, whose Hamiltonian (1.26a) applies to ice and to various ferroelectric materials.

The mathematical analysis of these *physical models* has prompted the study of a number of artificial 'models' whose equations have no counterpart in reality – the 'modified KDP model', the 'eight-vertex model', the 'spherical model', the 'hard square gas' model, the 'quantal gas model' and many others. The solution of these equations often throws light on the mathematical properties of the physical models, but shades off into solving very difficult mathematical problems without external relevance. Admirable reviews of these developments are to be found in the volumes edited by Domb & Green (1972). Experimental properties of various model magnetic systems are reviewed at length by de Jongh & Miedema (1976).

The physical models themselves may be elaborated endlessly with more and more bizarre lattices, various types of interaction between the sites, etc. Experience has shown, however, that these complications do not introduce any essentially new mathematical features in the problem. The order–disorder phenomena are dominated by the *strength* of the short-range interaction represented by a single *nearest-neighbour* 'exchange' parameter J in a model Hamiltonian such as*

$$\mathcal{H} = -\tfrac{1}{2} \sum_{l,l'} J \, \mathbf{S}_l \cdot \mathbf{S}_{l'} - \mathbf{H} \cdot \sum_l \mathbf{S}_l \tag{5.1}$$

and by the *dimensionality* of the lattice l, which we take to be *linear, square planar* or *simple cubic* as the case may be. A great part of the primary literature of the subject is thus conveniently excluded from further discussion; numerous references will be found in Domb & Green (1972) and in many works on magnetism, alloys, ferroelectricity, etc.

Finally, we shall make the connection with macroscopic thermodynamics by evaluating the statistical mechanical *partition function*

$$Z = \mathrm{Tr}\{\exp(-\beta\mathcal{H})\}, \tag{5.2}$$

from which the free energy or other thermodynamic potentials may be deduced by standard identities. We shall not distinguish between 'alloy' and 'magnetic' systems (§ 1.5), nor shall we attempt to discuss all the different thermodynamic properties such as specific heat, susceptibility, etc. which can then be calculated, but will concentrate mainly on the behaviour of the order parameters as we pass from one phase to another. Here again, we are on the borders of the very large topic of the general theory of phase transitions and critical phenomena which is dealt with at length by Brout (1965), Fisher (1967a, 1972, 1974), Domb & Green (1972, 1974, 1976) and many others.

5.2 Mean field approximation

The simplest phenomenological approach to all these problems is to assume, tentatively, that the expectation value $\langle \mathbf{S} \rangle$ of each 'spin' is not zero and then to show that there is a self-consistent solution for this quantity. Taking (5.1) as the Hamiltonian, with J +ve, we argue that the *effective field* $\mathbf{H}_{\mathrm{eff}}$ acting on the lth spin is given

$$\mathbf{H}_{\mathrm{eff}} = zJ\langle \mathbf{S} \rangle + \mathbf{H}, \tag{5.3}$$

where z is the coordination number of the lattice. But this spin itself must have just this expectation value under conditions of thermal equilibrium, i.e.

$$\langle \mathbf{S} \rangle = \langle \mathbf{S}_l \rangle \equiv \mathrm{Tr}\{\mathbf{S}_l \exp(\beta\mathbf{H}_{\mathrm{eff}} \cdot \mathbf{S}_l)\}/\mathrm{Tr}\{\exp(\beta\mathbf{H}_{\mathrm{eff}} \cdot \mathbf{S}_l)\}. \tag{5.4}$$

Solving these two equations simultaneously we may find non-zero values of

* For simplicity, the 'magnetic moment' $\bar{\mu}$ in (1.16) is incorporated in the 'magnetic field' $\mathbf{H}$.

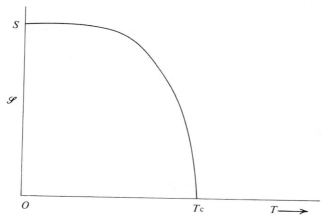

Fig. 5.1. Mean field approximation for order parameter of Ising model as function of temperature.

$\langle \mathbf{S} \rangle$. For example, if we reinterpret $\mathbf{S}_l$ as an Ising spin variable σ, and we let the external field $\mathbf{H} \to 0$, these reduce to an implicit equation for the LRO parameter

$$\langle \sigma \rangle \equiv \mathscr{S} = \tanh\left(\frac{zJ}{\not{k}T} \mathscr{S} \right), \tag{5.5}$$

which has a non-zero solution if

$$\not{k}T < \not{k}T_c = zJ. \tag{5.6}$$

The behaviour of $\mathscr{S}$ as a function of T is shown in fig. 5.1. The results for the classical and quantal Heisenberg models are essentially the same: below the *critical temperature* T_c, the average spin $\langle \mathbf{S} \rangle$ rises rapidly from zero and closely approaches its maximum ordered value S, which it attains at $T = 0$.

This, of course, is the good old *Curie–Weiss theory* of ferromagnetism. In the theory of alloys, it is called the *Bragg–Williams method*. In general, it is referred to as the *internal field* or *mean field approximation* (MFA) although a more exact name would be the *coherent field approximation*, since $\mathbf{H}_{\text{eff}}$ represents the coherent part of the statistically fluctuating fields exerted on each spin, through the exchange interaction from its neighbours. In other words, the emphasis is entirely on the *long-range* order (§ 1.6), without attention to possible short-range spin correlations.

For all its weaknesses, which will be amply emphasized in later sections, the method is sound in principle and extremely useful in practice. A sharp *order–disorder transition* is predicted at a *critical temperature* T_c which is given approximately by (5.6), in terms of the exchange parameter J. To deal with more complicated systems, such as the antiferromagnetic and helical spin orderings of figs. 1.9 and 1.10, or the various ferroelectric and antiferroelectric structures of § 1.4, it is merely a matter of coupling the various sub-lattices, each with its own average spin, by the appropriate exchange interactions.

To make correct thermodynamic use of the MFA method we must derive the phenomenological equation (5.4) from the partition function (5.2). The basic assumption (see e.g. Domb 1960; Brout 1965) is that there is no correlation between spins on neighbouring sites, apart from the condition that the ensemble average of each spin need not be zero. For an Ising model this means that the total numbers of $+$ and $-$ spins are fixed; by (1.22)

$$N^{\pm} = \tfrac{1}{2}(1 \pm \mathscr{S})N. \tag{5.7}$$

But these spins can be distributed over the N sites of the lattice in

$$W(\mathscr{S}) = \frac{N!}{N^+! \, N^-!} \tag{5.8}$$

ways. By the Boltzmann principle, $\ell \ln W(\mathscr{S})$ would, in fact, be our estimate of *entropy* of this assembly.

The *energy* of each configuration of spins in (5.8) depends in detail on the arrangements of like and unlike neighbours. Somewhat arbitrarily, we assign to every configuration the same energy, as if each spin had exactly its mean value, i.e.

$$\mathscr{E} = -zNJ\mathscr{S}^2. \tag{5.9}$$

The free energy is thus estimated to be

$$F = -\ell T \ln Z = zNJ\mathscr{S}^2 - \ell T \ln \left\{ \frac{N!}{\{\tfrac{1}{2}N(1+\mathscr{S})\}! \, \{\tfrac{1}{2}N(1-\mathscr{S})\}!} \right\}, \tag{5.10}$$

which depends only on the LRO parameter $\mathscr{S}$. Following the usual procedures of statistical mechanics, we evaluate the factorials by Stirling's formula: the condition

$$\partial F/\partial \mathscr{S} = 0 \tag{5.11}$$

for minimum free energy leads by a few elementary algebraic steps to the MFA condition (5.5). The corresponding conditions for long-range order in more complex systems may be deduced by similar arguments. Formulae for specific heat, magnetic susceptibility, etc. follow from (5.10) and (5.11) by the usual thermodynamic identities.

5.3 Short-range order

The fundamental defect of the MFA method is that correlations between spins on neighbouring sites are totally ignored. This is especially misleading above the critical point, since it fails to take note of the increasing range of order (§ 1.7) that signals the approach of the phase transition. It is not true, for example, that the specific heat vanishes for $T > T_c$, as would be predicted by (5.10).

It is easy to estimate the strength of such correlations. Consider, for example, an *isolated* pair of Ising spins, σ_1, σ_2, which interact directly through an exchange integral of strength J_{12}. The Boltzmann factors for the four configurations of this pair give

$$\langle \sigma_1 \sigma_2 \rangle \approx \tanh(\beta J_{12}) \tag{5.12}$$

which is of the order $1/z$ when $T \sim T_c$.

For spins (or components of an alloy) in equilibrium with one another in a lattice we would expect to satisfy the *quasi-chemical condition* (Guggenheim 1935) for the relative concentrations of 'diatomic molecules' in the reaction

$$AA + BB \rightleftharpoons 2AB, \tag{5.13}$$

i.e.

$$\frac{\langle N_{AB} \rangle^2}{\langle N_{AA} \rangle \langle N_{BB} \rangle} = \frac{4 \exp(-\beta \phi_{AB})}{\exp(-\beta \phi_{AA}) \exp(-\beta \phi_{BB})}$$

$$= 4 \exp \beta J) \tag{5.14}$$

if the molecular binding energies are as in (1.24). In the strongly interacting Ising system this formula is not, of course, exact. But by manipulation of the identities (1.20) we can produce a closed approximation for the nearest-neighbour correlation in the presence of LRO, i.e.

$$\langle \sigma_1 \sigma_2 \rangle \approx [1 + \zeta^2 - 2\zeta\{\zeta^2 + (\zeta^2 - 1)(1 - \mathscr{S}^2)\}^{\frac{1}{2}}]/(1 - \zeta^2), \tag{5.15}$$

where $\zeta \equiv \exp(-2\beta J)$. When $\mathscr{S} = 0$ this does not reduce to (5.12) because we are not dealing now with an isolated pair.

This formula, in fact, can be used to evaluate all the thermodynamic properties of the system in the *quasi-chemical approximation* (QCA) (see e.g. Domb 1960). By (1.27) and a standard thermodynamic identity relating free energy F to internal energy E we write

$$-T^2 \frac{\partial}{\partial T}\left(\frac{F}{T}\right) = E$$

$$= zNJ\langle \sigma_1 \sigma_2 \rangle, \tag{5.16}$$

which can be integrated to a formula for F analogous to (5.10). The condition (5.10) for minimum free energy predicts stable long-range order below a critical temperature

$$\frac{\cancel{k} T_c}{zJ} = \frac{-2/z}{\ln(1 - 2/z)}.$$ (5.17)

This expression lies somewhat below the MFA formula (5.6), to which it tends at 'infinite coordination number'. The fluctuations implicit in the concept of short-range order thus reduce T_c. Indeed, the phase transition for a one-dimensional Ising model, with $z = 2$, is at $T_c = 0$, in general agreement with the exact solution (§ 5.5). The QCA method is thus somewhat more realistic than the MFA method and has the same advantages of physical simplicity and closed analytical form.

But of course the SRO described by (5.15) should really extend beyond the nearest neighbours even when $T > T_c$. To demonstrate this effect (Brout 1965), we may attempt to generalize (5.3). Let us calculate the 'effective exchange' $J_{\text{eff}}(1, 2)$ at the site 1, where the spin is known to be σ_1, with the spin on the neighbouring site which is known to have value σ_2. This field is made up of a direct interaction with σ_2, together with any effects due to other spins σ_l where $l \neq 1, 2$, which may be to some extent 'polarized' by σ_1 and σ_2. We may write this

$$J_{\text{eff}}(1, 2)\sigma_2 = J_{12}\sigma_2 + \sum_{l \neq 1, 2} J_{1l} \bar{\sigma}_l$$ (5.18)

where, of course, J_{1l} is zero unless sites 1 and l are neighbours.

To give precise meaning to $\bar{\sigma}_l$, we must introduce spin probability distributions $g_2(\sigma_1, \sigma_2)$, $g_3(\sigma_1, \sigma_2, \sigma_3)$ etc., analogous to the atomic distribution functions (2.20). Allowing for the probability $g_2(\sigma_1, \sigma_2)$ of having chosen these values of σ_1 and σ_2 in the first place, we define this average by

$$\bar{\sigma}_l \equiv \sum_{\sigma_l = \pm 1} \sigma_l g_3(\sigma_1, \sigma_2, \sigma_l)/g_2(\sigma_1, \sigma_2).$$ (5.19)

If now we multiply (5.18) by σ_2, we determine the value of $J_{\text{eff}}(1, 2)$. But now we have, so to speak, decoupled the spins σ_1 and σ_2 from the rest of the lattice: this must be the exchange parameter that should be used in (5.12) to calculate the actual correlation between σ_1 and σ_2 in the complete assembly. From these equations we get

$$\tanh^{-1}\{\beta^{-1}\langle\sigma_1\sigma_2\rangle\} = J_{12} + \sum_{l \neq 1, 2} \sum_{\sigma_l = \pm 1} J_{1l}\sigma_l\sigma_2 g_3(\sigma_1, \sigma_2, \sigma_l)/g_2(\sigma_1, \sigma_2).$$ (5.20)

By definition

$$\langle\sigma_1\sigma_2\rangle \equiv \sum_{\sigma_1, \sigma_2 = \pm 1} \sigma_1\sigma_2 g_2(\sigma_1, \sigma_2).$$ (5.21)

The formula (5.18) is thus an identity involving the successive distribution functions g_2, g_3 etc., analogous to the identity (2.40) which was the starting point of the BBGKY theory of the structure of liquids. This is, in fact, only the first in a hierarchy of such identities that can be derived by algebraic manipulation from the Ising partition function.

$$Z = \sum_{\sigma_1 \ldots \sigma_N = \pm 1} \exp\left\{\tfrac{1}{2}\beta \sum_{l,l'} J_{ll'} \sigma_l \sigma_{l'}\right\}. \tag{5.22}$$

One might attempt to solve (5.20) by introducing the superposition approximation (2.27). A grosser approximation is to write

$$g_3(\sigma_1, \sigma_2, \sigma_l)/g_2(\sigma_1, \sigma_2) \approx g_2(\sigma_2, \sigma_l), \tag{5.23}$$

which ignores the effect that the value of σ_1 may have on the correlation between spins σ_2 and σ_l. In the absence of LRO we may linearize the left-hand side of (5.20) to give

$$\langle \sigma_1 \sigma_2 \rangle \approx \beta J_{12} + \sum_{l \neq 1, 2} \beta J_{1l} \langle \sigma_l \sigma_2 \rangle, \tag{5.24}$$

which may be treated as a set of linear equations for the propagation of the SRO parameter (1.14), i.e.

$$\Gamma(l) \equiv \langle \sigma_0 \sigma_l \rangle. \tag{5.25}$$

Indeed (5.24) is the lattice version of the Ornstein–Zernike convolution formula (2.42). Transforming $\Gamma(l)$ to its reciprocal space representation as in (1.42) and dropping the condition $l \neq 2$, we get the exact analogue of the Fourier transform (4.13), i.e.

$$\Gamma(\mathbf{q}) \approx \frac{\beta J(\mathbf{q})}{1 - \beta J(\mathbf{q})}. \tag{5.26}$$

By the fundamental principles of § 3.2, $\Gamma(\mathbf{q})$ is the spectral density function for the spin fluctuations that are associated with SRO correlations of spin. This approximate relationship with the Fourier transform, $J(\mathbf{q})$, of the exchange interaction (cf. (1.47)) is not accidental and may be deduced directly by application of the *random phase approximation* to the elementary excitations of the Ising Hamiltonian (Brout 1965).

For small values of $\mathbf{q}$, the function $J(\mathbf{q})$ can be expanded in the form

$$J(\mathbf{q}) \equiv \sum_{\mathbf{h}} J_{\mathbf{h}} \, e^{i\mathbf{q} \cdot \mathbf{h}}$$
$$\approx zJ(1 - \tfrac{1}{3}q^2 a^2 \ldots) \tag{5.27}$$

where the vector $\mathbf{h}$ runs over z nearest neighbours each at distance a from the central site. Thus (5.26) becomes

$$\Gamma(\mathbf{q}) \approx \frac{\beta z J}{(1 - \beta z J) + (\tfrac{1}{3}\beta z J a^2) q^2}, \tag{5.28}$$

which is of the Ornstein–Zernike form (4.27). As in (1.37) or (4.28), we verify that the spin correlation falls off exponentially with a range

$$\xi_1 = \left(\frac{\frac{1}{3}\beta z J a^2}{1 - \beta z J}\right)^{\frac{1}{2}}$$

$$= \left(\frac{\frac{1}{3}T_c}{T - T_c}\right)^{\frac{1}{2}} a. \qquad (5.29)$$

This approximate theory thus describes quite well the increasing range of order amongst the spins as we approach a critical temperature T_c which turns out to be the point at which, according to the MFA formula (5.6), LRO suddenly appears (presumably the approximations in going from (5.20) to (5.26) are too crude to reproduce the quasi-chemical value (5.17) of T_c). This discussion is consistent with the macroscopic thermodynamic treatment in § 4.4 which predicts a connection between the spin fluctuations and the magnetic susceptibility and also proves the validity of the Ornstein–Zernike spectral formula for fluctuations of concentration in an alloy, already mentioned in § 4.6.

This method can be extended (Brout 1965) to the calculation of the fluctuation spectrum *below* the critical temperature. In this case the linearized equation (5.24) must be generalized to allow $\langle \sigma_1 \rangle$ and $\langle \sigma_2 \rangle$ to have the non-zero value $\mathscr{S}$, which is given as a function of temperature by (5.5). The range or order in (1.41) behaves like

$$\xi' = \left\{\frac{1}{3}T_c \Big/ \left(T_c - \frac{T}{1 - \mathscr{S}^2}\right)\right\}^{\frac{1}{2}} a, \qquad (5.30)$$

which again tends to infinity at the MFA critical temperature T_c. But as $T \to 0$, $(1 - \mathscr{S}^2)$ goes exponentially to zero and fluctuations about the ordered state disappear: in an Ising model we cannot excite the long wavelength spin waves that are responsible for the fluctuation catastrophe (1.49).

The approximations introduced in going from (5.20) to (5.26) introduce serious errors, especially in the neighbourhood of the transition. Nevertheless, if we treat T_c as the observed critical temperature, it is surprising how well such simple formulae as (5.29) and (5.30) describe the order–disorder phenomena in many physical systems. Theoretical improvements to such features as the *critical exponent* $\frac{1}{2}$ in the factor $(T - T_c)^{-\frac{1}{2}}$ in (5.29) cost almost as much in additional mathematical effort as the experimental research required to verify them!

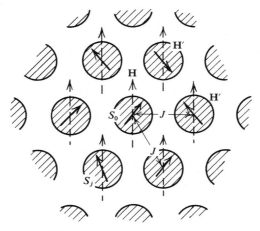

Fig. 5.2. Bethe–Peierls cluster.

5.4 Cluster methods

It demands little mathematical originality to suggest that the MFA results might be improved by applying the same arguments to a *cluster* of spins as to a single spin. This is a characteristic strategy in the theory of disordered systems; we take account of local interactions and correlations by treating in detail a compact group of spins or atoms embedded in a medium whose properties represent the average macroscopic behaviour of the bulk material. This strategy is often successful but it has its dangers. Unlike a single spin on a well-defined site, such a cluster is an artificial construct. In a homogeneous material there is no unique procedure for separating one such group from another. The boundary that we introduce between the cluster and the surrounding medium has no meaning in reality.

For this reason, there is no canonical sequence of approximations leading from the MFA equations to the exact equations of the complete system. Various schemes have been proposed, based upon generalizations of the various ways in which the MFA method itself may be set up.

The most elementary approach (Bethe 1935; Peierls 1936) was originally developed as a generalization of the effective field argument (5.3). In the *Bethe method* we consider a central spin S_0, together with the first *shell* of z neighbours $S_1 \ldots S_z$ (fig. 5.2), with which it interacts directly through the exchange parameter J. These spins, in turn, interact with the remainder of the specimen in a complicated manner which is represented approximately by an unknown effective field $\mathbf{H}'$. In other words, the cluster behaves as if it had a Hamiltonian

$$\mathscr{H}_{\text{cluster}} = -\sum_{i=1}^{z} J\,\mathbf{S}_0 \cdot \mathbf{S}_i - \mathbf{H}' \cdot \sum_{i=1}^{z} \mathbf{S}_i - \mathbf{H} \cdot \mathbf{S}_0, \qquad (5.31)$$

where any external field $\mathbf{H}$ is counted as acting separately on the central spin but is included in $\mathbf{H}'$ for the shell of neighbours.

For an Ising model, it is simply a matter of enumerating the various configurations to evaluate the partition function

$$Z_{\text{cluster}} = \text{Tr}\{\exp(-\beta\mathscr{H}_{\text{cluster}})\} \qquad (5.32)$$

and the ensemble average of any spin as in (5.4), i.e.

$$\langle \mathbf{S}_j \rangle = \text{Tr}\{\mathbf{S}_j \exp(-\beta\mathscr{H}_{\text{cluster}})\}/Z_{\text{cluster}}. \qquad (5.33)$$

But in the actual lattice the central spin of the cluster is not especially distinguished, and must have the same average value as any spin in the first shell. The analogue of the MFA consistency condition (5.5) is now

$$\langle \mathbf{S}_0 \rangle = \langle \mathbf{S}_i \rangle, \qquad (5.34)$$

from which $\mathbf{H}'$ can be determined and eliminated from (5.33) (see e.g. Domb 1960; Burley 1972).

For an Ising model the Bethe method gives analytical formulae for $\langle \sigma \rangle$ as a function of T which behave very similarly to the MFA equation (5.5). In the limit of vanishing external field, the effective field $\mathbf{H}'$ is zero above a critical temperature T_c: below this temperature there are solutions for which the LRO parameter $\mathscr{S} = \langle \sigma \rangle$ tends rapidly to unity as $T \to 0$. It turns out, in fact, that the critical temperature predicted by the Bethe method is exactly the same as the value (5.17) given by the quasi-chemical approximation. These two methods, starting apparently from different assumptions, are entirely equivalent.

The derivation (5.31)–(5.34) does, however, show what is lacking from this approximation. The critical behaviour depends only on the coordination number z, and not on the connectivity of the lattice. The cluster Hamiltonian (5.31) neglects the interactions between spins in the first shell, which would certainly modify the spectrum of the group. As in (5.23) we are throwing away correlations involving three or more spins, such as would be induced if these were linked in a circuit by the exchange parameter J.

In other words, the Bethe/QCA method is an exact solution for the Ising model on a *Bethe Lattice* (cf. Runnels 1972). In the language of graph theory (Harary 1967; Wilson 1972) this is an infinite regular (Cayley) *tree*: i.e., a *connected graph without circuits* (fig. 5.3). The Bethe lattice is important in the theory of disordered systems as an artificial mathematical model

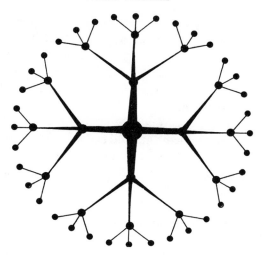

Fig. 5.3. Bethe lattice, or regular tree, of coordination number 4.

for which some theoretical techniques give exact solutions. But, although such a lattice is easy to define axiomatically, it cannot be realized as a physical system: to preserve the geometrical homogeneity of the structure and the equivalence of the z branches from each vertex, one would have to think of each branch as a step along a new dimension in a regular lattice of an *infinite* number of spatial dimensions. On the other hand, topologically speaking, a tree has many 'one-dimensional' characteristics; it is simply connected and there is only one path between any two vertices. But it is not so restricted 'locally' as a linear chain (§ 2.4) and has some of the combinatorial properties of a lattice of higher dimensionality. It is obvious, for example, that the number of chains of length L that may be traced from a given vertex is exactly

$$n(L) = z(z-1)^{L-1}, \qquad (5.35)$$

since there are $(z-1)$ choices at each new vertex arrived at. This is a convenient model for the number of *self-avoiding walks* in a regular lattice which is fundamental to many theoretical problems (§ 7.7). From this it is also easy to deduce, by the Peierls argument of § 2.4, an approximation for the critical temperature below which an ordered phase on a Bethe lattice would be stable against chains of reversed spins.

Many characteristic problems of the theory of disordered systems are much more easily solved on a Bethe Lattice (e.g. §§ 7.5, 9.10, 9.11) than on any realistic crystal lattice. But the dimensional ambivalence of this model can give rise to disturbing paradoxes. Thus (von Heimburg & Thomas

1974) the appearance of spontaneous order below the QCA critical temperature (5.17) applies only to the 'interior' of the Ising model. The number of sites on the 'surface' of a tree tends to the constant proportion $(z-2)/(z-1)$ of the total number in the system and should not, strictly speaking, be neglected. But, when surface spins are included, the assembly behaves like a linear chain (§§ 2.4, 5.5), with zero *total* magnetization for $T > 0$.

On the face of it, the Bethe method is not limited to Ising models. If we interpret **S** in (5.31)–(5.34) as a quantum spin operator, we are calculating the order–disorder properties of a Heisenberg ferromagnet (1.16). Numerical evaluation of the various matrix expressions gave promising results for the critical behaviour (Weiss 1948) until it was shown (Anderson 1950) that these equations have an 'anti-Curie point' (at $kT = 0.269J$ in a simple cubic lattice) below which the ferromagnetic order disappears. The fundamental inconsistencies inherent in this and several similar methods are discussed by Strieb, Callen and Horwitz (1963). It seems that the search for a 'closed', 'compact' representation of the behaviour of a Heisenberg ferromagnet in more than one dimension has not got beyond the simple MFA formula, which entirely fails to do justice to such important phenomena as the spin-wave excitations at low temperatures (§ 1.8).

Returning to the Ising model, it is natural to generalize the Bethe method by building larger clusters containing several interacting 'shells' of neighbours. Self-consistency conditions like (5.34) then determine the 'internal field' in each shell (Fosdick & James 1953). An analogous approach by Cowley (1950, 1965) assumes that each shell has some degree of average polarization (i.e. 'order') relative to the spin on the central atom and thus contributes a modified combinatorial factor (cf. (5.10)) to the entropy of the cluster. As in (5.11), variation of the free energy provides a set of equations for the local order parameters. This method may well prove a convenient approximation when dealing with the transition to a rather complicated pattern of ordered sub-lattices, as in many alloys. It is interesting to note, however, that the transition temperature is just the MFA value (5.6). Cowley's equations can in fact be derived (Clapp 1964) by judicious approximation from the first of the hierarchy of *Green function identities* (Doman & ter Haar 1962; Callen 1963), which are apparently more elaborate versions of the identities like (5.20) involving pair, triplet, etc. spin correlations. The Cowley method is thus quite close to Brout's linearization (5.24) of the first of these identities, from which was derived the Ornstein–Zernike formula (5.28) for the propagation of order.

A more systematic way of introducing larger clusters is to make successive improvements to the combinatorial factor (5.8) in the expression (5.10) for the free energy. To derive the Bethe/QCA equations, for example, we must take account (at least approximately) of the correlations between neighbouring spins implicit in the quasi-chemical condition (5.14). This is easy enough (see e.g. Domb 1960) and it is then fairly simple (Guggenheim & McGlashan 1951) to count various configurations of triplets or tetrahedra of spins, thus taking better account of their contributions to the internal energy and entropy. It turns out, in fact (Hijmans & de Boer 1955, 1956), that the equations arrived at by these and other methods appear at various levels of approximation in the *cluster variation procedure* (Kikuchi 1951; Kurata, Kikuchi & Watari 1952) which is probably the best technique for approximating self-consistently to the combinatorial factors for statistical distributions on a lattice.

To illustrate the *Kikuchi pseudo-assembly method* let us consider a simple lattice with coordination number z – for example, the planar square lattice of fig. 5.4. We have a stock of A and B atoms in the fixed proportion $x_1 : x_2$ where, of course, $x_1 + x_2 = 1$. As in (5.8), the number of ways of putting these atoms on the N sites of the lattice is just

$$W_1 = Q_1^{(1)} = N!/\{ \bullet \},$$ (5.36)

where we define the symbol

$$\{ \bullet \} \equiv (x_1 N)! \, (x_2 N)!.$$ (5.37)

Combining this number with the average energy gave us the MFA formula (5.10) for the free energy.

Now suppose, instead, that we had to draw from a stock of the different *pairs* of atoms tabulated in fig. 5.4(e). Allowing for the degeneracy ($\beta_2 = 2$) of the (AB) type pair, we suppose that these various types occur in the proportions $y_1 : y_2 : y_3$, which must, of course, satisfy the normalization condition

$$\sum_i \beta_i y_i = 1.$$ (5.38)

For consistency, moreover, the individual atoms of the pairs must also occur in the proportions $x_1 : x_2$. As in (1.20), this subjects the y_i to one further independent condition

$$x_1 = y_1 + y_2,$$ (5.39)

leaving just one of these parameters to be determined.

In how many ways can we distribute pairs from this population on to the

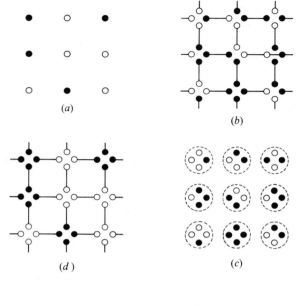

Fig. 5.4. Kikuchi pseudo-assembly method: (*a*) distribution of 'atoms' on sites; (*b*) distribution of pairs on bonds; (*c*) pseudo-assembly of atoms on sites; (*d*) restricted pair distribution equivalent to (*a*); (*e*) pair concentrations and energies ε_i.

$\frac{1}{2}zN$ 'bonds' of the lattice. The number of arrangements such as the one shown in fig. 5.4(*b*) is

$$Q^{(2)}_{\frac{1}{2}z} = \frac{(\frac{1}{2}zN)!}{(\frac{1}{2}zy_1N)! \, [(\frac{1}{2}zy_2N)!]^2 (\frac{1}{2}zy_3N)!}$$

$$= (N!/\{\bullet\!\!-\!\!\bullet\})^{\frac{1}{2}z} \tag{5.40}$$

where, as in (5.37), we define the factorial symbol

$$\{\bullet\!\!-\!\!\bullet\} \equiv \prod_{i=1}^{3} [(y_i N)!]^{\beta_i}. \tag{5.41}$$

Stirling's formula has been used to replace $(\frac{1}{2}zN)!$ by $(N!)^{\frac{1}{2}z}$, etc. since N is very large.

But $Q^{(2)}_{\frac{1}{2}z}$ is really much too large as an estimate of W. As we see in fig. 5.4(*b*) it includes a high proportion of 'incorrect' arrangements, in which both A and B atoms appear on the same site. To estimate the correction factor

$$\Gamma_{(2)} = W/Q^{(2)}_{\frac{1}{2}z}, \tag{5.42}$$

let us think of the total number of ways in which such distributions of atoms might be arrived at directly, by repeated drawings from the 'single atom' stock. For the *pseudo-assembly* of fig. 5.4(*c*), where there are now z atoms assigned to each lattice site, the number of arrangements in the proportions $x_1 : x_2$ is

$$Q^{(1)}_z = \frac{(zN)!}{(x_1zN)!\,(x_2zN)!}$$
$$= (N!/\{ \bullet \})^z. \tag{5.43}$$

Of these configurations, however, only the number $Q^{(1)}_1$, given by (5.36), are 'correct' in that all the atoms assigned to a given site are of the same type: fig. 5.4(*d*). Thus, the proportion of 'correct' distributions in the pseudo-assembly is only

$$\Gamma'_{(2)} = Q^{(1)}_1/Q^{(1)}_z$$
$$= (\{ \bullet \}/N!)^{z-1}. \tag{5.44}$$

The basic approximation of the Kikuchi method is to assume that the proportion of 'correct' arrangements in the random pair distribution (fig. 5.4(*b*)) is the same as in the corresponding pseudo-assembly (fig. 5.4(*c*)), i.e.

$$\Gamma_{(2)} \approx \Gamma'_{(2)}. \tag{5.45}$$

Our estimate of the combinatorial factor for the entropy of the assembly comes from (5.40)–(5.45), i.e.

$$W_{(2)} = \Gamma'_{(2)}\, Q^{(2)}_{\frac{1}{2}z}$$
$$= \{ \bullet \}^{z-1}\{ \bullet\!\!-\!\!\bullet \}^{-\frac{1}{2}z}(N!)^{1-\frac{1}{2}z}. \tag{5.46}$$

The separation into pairs is particularly convenient for the energy (cf. (1.27)), which is given exactly by summing contributions from each type, i.e.

$$\mathscr{E} = \frac{1}{2}zN\{-y_1 + 2y_2 - y_3\}J. \tag{5.47}$$

We now follow (5.11) and vary the free energy with respect to the independent variable in the set y_i. The resulting equations have well-defined solutions which can be identified with regimes of order and disorder

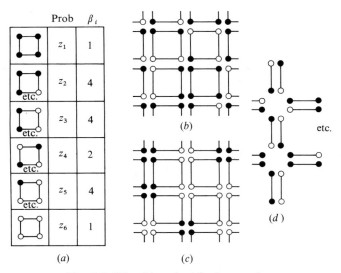

Fig. 5.5. Kikuchi method for 'squares'.

separated by a critical temperature: indeed, this approximation for the combinatorial factor exactly reproduces the Bethe/QCA formula (5.17).

For a tree, in fact, the approximation (5.45) becomes exact: errors only arise when we try to apply this relation to a cycle of lattice sites where n independent choices of atomic type are equivalent to only $(n-1)$ choices of pair type, since the nature of the closing bond is then determined. To make a significant improvement, therefore, it is necessary to build up our distribution from larger 'clusters' containing such cycles.

To take account of the multiple connectivity of the planar square lattice, for example, we draw from a stock of characteristic 'squares' of atoms, in the relative proportions z_i tabulated in fig. 5.5(a). These proportions must, of course, be consistent with the overall proportions of A and B atoms, and with the proportion of the various bonds, i.e.

$$y_1 = z_1 + 2z_2 + z_3; \quad y_2 = z_2 + z_3 + z_4 + z_5; \quad y_3 = z_3 + 2z_5 + z_6, \quad (5.48)$$

which still leaves us with two further independent parameters that can later be varied to minimize the free energy.

The calculation of the combinatorial factor follows the same general lines. We first construct all arrangements of squares, as in fig. 5.5(b), consistent with the proportions z_i in stock. This can be done in

$$Q_1^{(4)} = N! / \{ \square \} \quad (5.49)$$

ways, where $\{\ \square\ \}$ is defined as in (5.41), with z_i in place of y_i. But this number is far too large and must be reduced by a factor $\Gamma_{(4)}$ to count only the 'correct' distributions of fig. 5.5(c). We approximate to this by $\Gamma'_{(4)}$ which is the proportion of 'correct' arrangements in a pseudo-assembly, fig. 5.5(d), made up of all arrangements of 'double pairs' on the lattice. Since these objects are distributed independently the factor $\Gamma'_{(4)}$ can be evaluated exactly by comparison between, say, the combinatorial factors for the pair distributions in fig. 5.4 and the factor $Q_8^{(1)}$ for the distribution of eight atoms on each site. This yields

$$\Gamma'_{(4)} = \left[Q_2^{(2)} \frac{Q_1^{(1)}}{Q_4^{(1)}} \right] \Big/ \left[Q_4^{(2)} \frac{Q_4^{(1)}}{Q_8^{(1)}} \right]. \tag{5.50}$$

In this approximation, therefore, the combinatorial factor in the free energy is

$$W_{(4)} = \Gamma'_{(4)} \, Q_1^{(4)} = \frac{\{\,\bullet\!\!-\!\!\bullet\,\}^2}{\{\,\bullet\,\}\{\,\square\,\}}. \tag{5.51}$$

Varying the free energy with respect to the independent parameters in the set z_i, we find an order–disorder transition at the temperature

$$\frac{kT_c}{zJ} = \frac{1}{2 \ln\{(5 + \sqrt{17})/4\}} = 0.606 \tag{5.52}$$

which is somewhat below the Bethe/QCA result (5.17) for this lattice. This approximation to T_c was, in fact, obtained in quite a different manner by Kramers & Wannier (1941) (§ 5.10). The corresponding calculation for squares on a three-dimensional simple cubic lattice gives results that are very close to the best estimates by series expansions (§ 5.10): these are only slightly improved by taking a larger cluster such as a cube or tetrahedron.

One would imagine that the only way to obtain more accurate results by the Kikuchi method would be to build up the lattice out of much more complicated multiply-connected clusters, extended in all dimensions – for example a grid of four squares on the planar lattice (fig. 5.6(a)). This obviously demands immense algebraic and computational labour. It seems, however (Kikuchi & Brush 1967), that clusters extended in one dimension (fig. 5.6(b)), do just as well. This is because a two-dimensional square lattice may be constructed out of 'ladders' (fig. 5.6(c)), by 'one-dimensional' steps, for which the combinatorial factor can be written down exactly.

Even more rapid convergence is achieved by building the square lattice

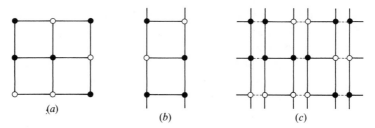

Fig. 5.6. (*a*) Larger cluster. (*b*) 'Ladder' cluster. (*c*) Ladder pseudo-assembly.

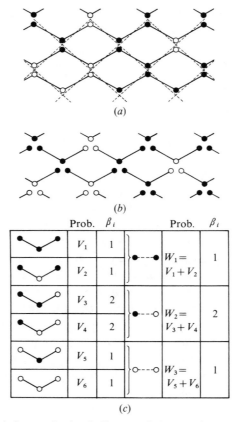

	Prob.	β_i		Prob.	β_i
	V_1	1	$\bullet\text{---}\bullet$ $W_1 =$ $V_1 + V_2$		1
	V_2	1			
	V_3	2	$\bullet\text{---}\circ$ $W_2 =$ $V_3 + V_4$		2
	V_4	2			
	V_5	1	$\circ\text{---}\circ$ $W_3 =$ $V_5 + V_6$		1
	V_6	1			

(*c*)

Fig. 5.7. (*a*) Square lattice built out of zig-zag chains. (*b*) Triplets. (*c*) Probability factors.

diagonally (fig. 5.7(*a*)), by the addition of zig-zag chains (Temperley 1961). The combinatorial factor for a single chain is given quite accurately by building it up out of triplets of atoms in the V configuration (fig. 5.7(*b*)). If we write $\{\mathbf{V}\}$ for the analogue of (5.41), and for the corresponding function for the distribution of the diagonal pairs of atoms on such clusters, we get by very simple arguments (Kikuchi & Brush 1967; Burley 1972)

$$W_{\cdot \mathrm{V}\cdot} = \frac{\{\bullet \cdots \bullet\}\{\bullet\!-\!\bullet\}^2}{\{\bullet\}\{\mathbf{V}\}^2} \tag{5.53}$$

which turns out to be exactly equivalent to the square approximation (5.51).

The advantage of a cluster method is that it gives a *compact representation* of the properties of the model. Instead of trying to sum several different series whose convergence is marginal we discover the different regimes of order and disorder in the different roots of a *closed* set of algebraic equations. These equations are often quite sufficiently accurate when theory is to be compared with experiment. For example, in any actual three-dimensional Ising system, we seldom know the exchange parameter J to anything like the precision (1–2 per cent) that would be needed to distinguish the transition temperature calculated by the Kikuchi method, as in (5.53), from the exact value.

From a more formal point of view, however, these methods are unsatisfactory, especially for the theoretical behaviour of one- and two-dimensional models. However large we make the clusters we cannot predict correctly the analytical behaviour of the thermodynamic variables as they pass through the critical region. To get at such characteristics as the critical exponents (cf. (5.29)) we must look for *exact* analytical solutions of the statistical mechanical problem, without the arbitrary assumptions of superposition, statistical independence, etc. that are perforce made in the cluster methods in order to achieve closure of the equations. The range of such exact solutions is really very limited, but they deserve careful study.

5.5 The Ising model in one dimension

The Ising problem on a linear chain has a very simple exact solution. The simple connectivity of the chain and the discreteness of the states of each component 'spin' turn the evaluation of the partition function into a purely algebraic question. For a closed ring the partition function (5.2) can be written in the form

$$Z_1 = \sum_{\sigma_1 = \pm 1} \sum_{\sigma_2 = \pm 1} \cdots \sum_{\sigma_N = \pm 1} \exp\{\beta J(\sigma_1\sigma_2 + \sigma_2\sigma_3 + \ldots + \sigma_N\sigma_1)$$

$$+ \beta H(\sigma_1 + \sigma_2 + \ldots + \sigma_N)\}$$

$$= \sum_{\sigma_1 = \pm 1} \sum_{\sigma_2 = \pm 1} \cdots \sum_{\sigma_N = \pm 1} V_{\sigma_1\sigma_2} V_{\sigma_2\sigma_3} \ldots V_{\sigma_N\sigma_1} \qquad (5.54)$$

where we introduce the symbol

$$V_{\sigma_l\sigma_{l+1}} \equiv \exp\{\beta J\sigma_l\,\sigma_{l+1} + \tfrac{1}{2}\beta H(\sigma_l + \sigma_{l+1})\}. \qquad (5.55)$$

We can think of (5.55) as defining a 2×2 *transfer matrix* V, whose rows and columns are labelled by the values $+1$ and -1 of the variables σ_l and σ_{l+1}. In this language, the partition function (5.54) simply consists of the trace of the matrix product of V with itself, N times: i.e.

$$Z = \mathrm{Tr}[V^N]. \qquad (5.56)$$

Now suppose we find a representation in which V is diagonal, i.e.

$$V_{\alpha\beta} = \lambda_\alpha\,\delta_{\alpha\beta}. \qquad (5.57)$$

The product and trace operations in (5.56) are trivial in this representation: thus

$$Z = \sum_\alpha (\lambda_\alpha)^N. \qquad (5.58)$$

In fact, since N is a very large number, we need consider only the largest of these eigenvalues, which will depend of course on the temperature and magnetic field. From the free energy

$$F = -\mathit{k}T \ln Z = -N\mathit{k}T \ln \lambda_{\mathrm{M}}(H,\, T) \qquad (5.59)$$

we can immediately derive the thermodynamic properties of the system.

In the present case, we may write (5.55) in the form

$$V_{\sigma\sigma'} = \begin{pmatrix} \exp(K+B) & \exp(-K) \\ \exp(-K) & \exp(K-B) \end{pmatrix} \qquad (5.60)$$

where

$$K = \beta J \quad \text{and} \quad B = \beta H. \qquad (5.61)$$

The eigenvalues of this matrix are

$$\lambda_\pm(H,\, T) = e^K \cosh B \pm \{e^{2K} \sinh^2 B + e^{-2K}\}^{\frac{1}{2}}, \qquad (5.62)$$

of which the larger is λ_+. Putting (5.62) into (5.59) and using various thermodynamic relations, we can calculate the magnetization, specific heat, magnetic susceptibility, etc. of the chain, as a function of the temperature and external magnetic field.

The most significant feature of (5.62) is that λ_{M} has no poles, branch points or zeros for all positive values of H and T. We thus confirm the general proof (§ 2.4) that the one-dimensional Ising model does not have a stable, ordered phase. It is easy to show, for example, that the magnetization always tends to zero as $H \rightarrow 0$.

It is worth noting, however, that the zero-field *susceptibility* tends to infinity as $T \rightarrow 0$. This suggests that the phase transition of the Ising chain exists in principle but is physically inaccessible at $T_c = 0$. For confirmation of this interpretation let us calculate (e.g. Thompson 1972) the correlation between the spins at sites l and l' along the chain. In the notation of (5.54) and (5.56) this may be written

$$\langle \sigma_l \sigma_{l'} \rangle = Z^{-1} \sum_{\{\sigma\}} V_{\sigma_1 \sigma_2} \ldots V_{\sigma_{l-1} \sigma_l} \sigma_l V_{\sigma_l \sigma_{l+1}} \ldots V_{\sigma_{l'-1} \sigma_{l'}} \sigma_{l'} V_{\sigma_{l'} \sigma_{l'+1}} \ldots V_{\sigma_N \sigma_1}$$

$$= Z^{-1} \mathrm{Tr}[V^l S V^{l'-l} S V^{N-l'}]. \tag{5.63}$$

In the representation (5.55) the operation of measuring σ at a given site takes the form

$$S_{\sigma \sigma'} = \begin{pmatrix} 1 & 0 \\ 0 & -1 \end{pmatrix}. \tag{5.64}$$

But in the representation (5.57) in which V is diagonal, this matrix reads

$$S_{\alpha \beta} = \begin{pmatrix} 0 & 1 \\ 1 & 0 \end{pmatrix}. \tag{5.65}$$

The correlation function (5.63) can be evaluated in this representation by direct matrix multiplication:

$$\langle \sigma_l \sigma_{l'} \rangle = Z^{-1} \mathrm{Tr} \left[\begin{pmatrix} \lambda_+^l & \cdot \\ \cdot & \lambda_-^l \end{pmatrix} \begin{pmatrix} \cdot & 1 \\ 1 & \cdot \end{pmatrix} \begin{pmatrix} \lambda_+^{l'-l} & \cdot \\ \cdot & \lambda_-^{l'-l} \end{pmatrix} \begin{pmatrix} \cdot & 1 \\ 1 & \cdot \end{pmatrix} \begin{pmatrix} \lambda_+^{N-l'} & \cdot \\ \cdot & \lambda_-^{N-l'} \end{pmatrix} \right]$$

$$= (\lambda_+^{N-(l'-l)} \lambda_-^{l'-l} + \lambda_-^{N-(l'-l)} \lambda_+^{l'-l}) / (\lambda_+^N + \lambda_-^N)$$

$$\rightarrow (\lambda_- / \lambda_+)^{l'-l} \tag{5.66}$$

for very large N. From (5.62) we find that the ratio of the eigenvalues when $H = 0$ is just $\tanh K$. In other words, we always have exponentially decaying short-range order (1.37) along the chain, but the range of order tends to infinity at the point $T = 0$, where the two roots of (5.62) tend to equality. This is a special case of a general theorem according to which *long-range order exists if and only if the maximum eigenvalue of the transfer matrix is asymptotically degenerate* (Kac 1968).

These properties of the Ising chain with nearest neighbour interactions are also to be found in any one-dimensional system where the interactions are of finite range. Suppose, for example, that the exchange parameter $J(l' - l)$ does not vanish until $|l - l'|$ exceeds some fixed length L. We divide the chain into blocks of L spins. Each block can now interact only with the two neighbouring blocks, and may be considered a complicated entity with 2^L different states. These states are used to label the rows and columns of a generalized transfer matrix analogous to V. The algebraic steps from (5.56) to (5.59) then follow and we deduce the thermodynamic properties of the system from the largest eigenvalue of this matrix. A theorem of Frobenius concerning the eigenvalues of a finite matrix of positive elements tells us that λ_M is not degenerate, so that there is no phase transition.

This follows, in fact, from the argument of Peierls (1936). As we saw in § 2.4, the stability of the ordered phase against a spin reversal of part of the chain depends upon the change of free energy (2.12). In general this would take the form

$$\Delta F = 2 \sum_{l=1}^{N} l J(l) - \mathit{k} T \ln N, \qquad (5.67)$$

since there would be l pairs of reversed spins, distant l apart, interacting with strength $J(l)$, bridging the break in order. If this sum is finite, then the entropy term always wins and the order is thermodynamically unstable.

What happens if the range of the interaction becomes infinite? According to (5.67) we may expect a transition to an ordered phase if

$$\sum_{l=1}^{N} l J(l) \sim \int_0^N l J(l) \, dl \qquad (5.68)$$

increases more rapidly than $\ln N$ for large values of N. This is the interpretation of the theorems of Ruelle (1969) and of Dyson (1969) which show that if

$$J(l) \sim 1/l^{1+\alpha} \qquad (5.69)$$

then a transition occurs only when $0 < \alpha < 1$. For $\alpha > 1$ the system behaves like the linear chain with finite range; but the Ising problem has not been solved for the intermediate case where the interaction falls off as $1/l^2$ along the chain (cf. Thouless 1969; Griffiths 1971).

Another way of going to an interaction of infinite range is to write

$$J(l) \sim \gamma J \exp(-\gamma l), \qquad (5.70)$$

for which, of course, the expression (5.68) always converges. By exploiting the analogy between the partition function of this model and a Markoff process (Kac 1968) it is possible to represent the transfer matrix as the kernel of an integral equation and hence to find the largest eigenvalues to put into (5.59). The interesting result is that these eigenvalues become degenerate in the limit $\gamma \to N^{-1} \to 0$, giving a phase transition at a temperature $2J/k$. What really happens in this limit is that every spin is interacting very weakly with every other spin so that the whole chain is a single homogeneous 'cluster'. We have transformed the model into a lattice of infinite coordination number – i.e. of infinite dimensionality – for which the MFA solution (5.6) is known to be exact.

5.6 The one-dimensional Heisenberg model

To evaluate the partition function for a linear chain of *classical* spin vectors, with isotropic or anisotropic *Heisenberg* interactions (1.15) or (1.17), we must introduce a transfer matrix with an infinite number of rows and columns corresponding to the values of the continuous variables for the orientation of each spin (cf. Thompson 1972). But the determination of the eigenvalues of such an integral operator is not always a trivial problem with a direct solution. All we really need to know is that the thermodynamic behaviour of any such system is essentially the same as for the corresponding Ising chain. This follows very simply by a direct calculation of the partition function of an open, isotropic, Heisenberg chain in zero magnetic field; in the notation of (5.54)–(5.61) this may be written

$$Z = \int \frac{d\Omega_1}{4\pi} \int \frac{d\Omega_2}{4\pi} \cdots \int \frac{d\Omega_N}{4\pi} \exp\{K(\mathbf{S}_1 \cdot \mathbf{S}_2 + \mathbf{S}_2 \cdot \mathbf{S}_3 \ldots + \ldots \mathbf{S}_{N-1} \cdot \mathbf{S}_N)\}$$

(5.71)

where $d\Omega_l$ is the element of solid angle over which we evaluate the vector $\mathbf{S}_l$ (here assumed to be of unit length). Now we integrate from the end of the chain, using the direction of $\mathbf{S}_{N-1}$ as the polar axis for the integral over $d\Omega_N$, and so on. This gives us, immediately

$$Z \approx \left[\int_0^\pi \tfrac{1}{2} \exp(K\cos\theta) \sin\theta \, d\theta \right]^N$$

$$= (\sinh K/K)^N$$

(5.72)

from which we determine the free energy, etc. This is almost the same result as for the Ising model, where (5.59) and (5.62) make this function $(\cosh K)^N$. Similarly, the correlation function looks almost exactly like (5.66), with $(\coth K - K^{-1})$ replacing $\tanh K$ as the ratio of eigenvalues.

The partition function for the *quantum-mechanical* Heisenberg Hamiltonian has not been evaluated exactly, even in one dimension. We did find, however (§ 2.4), that the ground state of a *ferromagnetic* linear chain, although itself ordered, is thermally unstable against the excitation of spin waves (§ 1.8). This confirms a reasonable conjecture that this quantum system, like the Ising and classical models, has no phase transition at a non-zero temperature.

In the present context, the most interesting feature of the Heisenberg model is that the *ground state of an antiferromagnetic linear chain is not spatially ordered.* The exact proof of this theorem for $S=\frac{1}{2}$, by Hulthén (1938), using the method of Bethe (1931), is one of the few rigorous results in this whole subject and has turned out to be the key to several other exactly soluble models (§ 5.8).

We note, first of all, that the 'ferromagnetically' ordered state in which all the spins are pointing in the same direction, is an eigenstate of the Hamiltonian

$$\mathcal{H} = -J \sum_l \mathbf{s}_l \cdot \mathbf{s}_{l+1} \tag{5.73}$$

where $\mathbf{s}_l$ is a vector made up of Pauli spin matrices normalized to unity, i.e.

$$s_l^{(z)} = \begin{pmatrix} 1 & 0 \\ 0 & -1 \end{pmatrix}, \quad \text{etc.} \tag{5.74}$$

For this system, the total component of spin along the axis of quantization z, i.e.

$$S^{(z)} = \sum_l s_l^{(z)}, \tag{5.75}$$

commutes with the Hamiltonian and is, therefore, a constant of the motion. Using standard properties of the Pauli matrices, we verify that $|0\rangle$ is an eigenstate of $\mathcal{H}$ and of $S^{(z)}$, i.e.

$$\mathcal{H}|0\rangle = -NJ|0\rangle \tag{5.76}$$

and

$$S^{(z)}|0\rangle = N|0\rangle. \tag{5.77}$$

But, when the exchange parameter J is negative, the state $|0\rangle$ is at the top of the spectrum of 2^N eigenstates of $\mathcal{H}$. What is the lowest state? We think at once of the antiferromagnetic ordered state in which up and down spins alternate along the chain. This is an eigenstate of $S^{(z)}$, with zero eigenvalue, but it is easy to verify that it is *not* an eigenstate of $\mathcal{H}$. The effect of the off-diagonal parts of $\mathbf{s}_l \cdot \mathbf{s}_{l+1}$ is to interchange the signs of the spins on sites l and $l+1$, so that the simple 'alternating' state is not reproduced when we multiply by this term in $\mathcal{H}$. In fact, if we try to build up an eigenstate of $\mathcal{H}$ starting from this ordered arrangement, we shall find ourselves eventually with a mixture of all possible arrangements of equal numbers of up and down spins along the chain. This is the essential nature of the antiferromagnetic ground state.

To calculate the coefficients in this mixture, we introduce a symbolism for the states that can be obtained from $|0\rangle$ by reversing one or more spins. The symbol $|l_1, l_2, \ldots, l_r\rangle$ stands for a state in which there are r 'down' spins, at sites $l_1 < l_2 < \ldots < l_r$ along the chain. By arranging these site labels in ascending order we specify each state uniquely. Taking all such distinct choices, for each value of r between 0 and N, we generate 2^N orthogonal basis functions for the representation of $\mathcal{H}$. But because $S^{(z)}$ is a constant of the motion, each eigenfunction of $\mathcal{H}$ must lie entirely in a sub-manifold corresponding to a fixed value of $S^{(z)}$ – i.e. to a fixed value of r. In general, therefore, we can express such an eigenfunction in the form

$$\Psi = \sum_{l_1 < l_2 < \ldots < l_r} a(l_1, l_2, \ldots, l_r)|l_1, l_2, \ldots, l_r\rangle \tag{5.78}$$

where the sum runs over all different ways of assigning r labels to N sites of the chain.

The unknown coefficients $a(l_1, \ldots, l_r)$ are obtained by applying the Hamiltonian (5.73) to this function. The effect of the diagonal part of $\mathcal{H}$ acting on $|l_1, \ldots, l_r\rangle$ is simply to subtract from (5.76) a quantity proportional to N_a, the number of antiparallel neighbours in this configuration. But the off-diagonal part of $\mathcal{H}$ exchanges the spins of each antiparallel pair in turn. We thus get

$$\mathcal{H}|l_1, \ldots, l_r\rangle = -J(N - 2N_a)|l_1, \ldots, l_r\rangle - 2J \sum_{l_1', l_2', \ldots, l_r'} |l_1', \ldots, l_r'\rangle \tag{5.79}$$

where $l_1', \ldots, l_r'$ is any set of labels which differs from $l_1, \ldots, l_r$ by the interchange of spins on any one pair of neighbours. The condition for Ψ to be an eigenstate of $\mathcal{H}$ can be written

$$2\mathcal{E}a(l_1, \ldots, l_r) = \sum_{l_1', \ldots, l_r'} [a(l_1, \ldots, l_r) - a(l_1', \ldots, l_r')] \tag{5.80}$$

where $2\mathcal{E}J = NJ + E$ measures the distance in energy E between this state and the 'ferromagnetic' state $|0\rangle$.

To solve this set of difference equations we apply the Bloch theorem. The translational symmetry of the lattice suggests the introduction of wave-like modes. For example, the solution of (5.80) for $r = 1$ is given by a single spin-wave excitation,

$$a(l) = \exp(ikl), \tag{5.81}$$

with energy

$$\mathcal{E}(k) = (1 - \cos k). \tag{5.82}$$

We naturally assume that our ring is closed: to satisfy the general periodic boundary condition

$$a(l_1, l_2, \ldots, l_j, \ldots, l_r) \equiv a(l_1, l_2, \ldots, l_j + N, \ldots, l_r) \qquad (5.83)$$

we must have

$$k = 2\pi n/N \quad \text{for } n = 0, 1, \ldots, N-1. \qquad (5.84)$$

For two reversed spins we expect to introduce two independent spin-wave modes, i.e. a function of the form

$$a(l_1, l_2) = \exp(ik_1 l_1 + ik_2 l_2). \qquad (5.85)$$

It is easy to verify that this function satisfies (5.80) when written out in the form

$$2\mathscr{E}a(l_1, l_2) = 4a(l_1, l_2) - a(l_1 - 1, l_2) - a(l_1 + 1, l_2) - a(l_1, l_2 - 1) - a(l_1, l_2 + 1) \qquad (5.86)$$

with

$$\mathscr{E}(k_1, k_2) = (1 - \cos k_1) + (1 - \cos k_2). \qquad (5.87)$$

But the form of (5.86) is incorrect when the two reversed spins are on adjacent sites, i.e. when $l_2 = l_1 + 1$. To reproduce (5.80) in this case we must by some means ensure that some of the terms in (5.86) automatically cancel, i.e.

$$2a(l_1, l_1 + 1) = a(l_1, l_1) + a(l_1 + 1, l_1 + 1). \qquad (5.88)$$

This condition can only be achieved if we construct $a(l_1, l_2)$ out of a linear combination of the two degenerate solutions obtained by interchanging k_1 and k_2 in (5.85). The neatest way of combining such solutions is by a relative phase factor ϕ_{12}, i.e.

$$a(l_1, l_2) = \exp i(k_1 l_1 + k_2 l_2 + \phi_{12}) + \exp i(k_2 l_1 + k_1 l_2 - \phi_{12}). \qquad (5.89)$$

Substituting (5.89) in (5.88) we get

$$\cot \tfrac{1}{2}\phi_{12} = \cot \tfrac{1}{2}k_1 - \cot \tfrac{1}{2}k_2. \qquad (5.90)$$

This is, indeed, precisely the sort of condition on the relative phases of the spin-wave modes, induced by the necessity of correctly describing the local spin reversal operators, to which we referred in § 1.8. The antiferromagnetic ground state that we are trying to construct would not, therefore, necessarily have the Gaussian randomness (3.14) characteristic of pure spectral disorder.

The beauty of the *Bethe Ansatz* (5.89) is that it does not become more complicated as we increase the number of reversed spins. In general, we write

$$a(l_1, l_2, \ldots, l_r) = \sum_P \exp\left\{ i \sum_i k_{Pj} l_j + \tfrac{1}{2}i \sum_{j,m} \phi_{PjPm} \right\} \qquad (5.91)$$

where P is any permutation of the integers $1, 2, \ldots, r$ that carries j into Pj. As in (5.87) each term of the sum has the same energy

$$\mathscr{E}(k_1, k_2, \ldots, k_r) = \sum_j (1 - \cos k_j), \tag{5.92}$$

and we can arrange each of the phase factors ϕ_{PjPm}, just as in (5.88) and (5.90), to take care of the case where l_j and l_m happen to be adjacent sites. This condition, indeed, is exactly as in (5.90):

$$\cot \tfrac{1}{2}\phi_{jm} = \cot \tfrac{1}{2}k_j - \cot \tfrac{1}{2}k_m. \tag{5.93}$$

Finally, our solution (5.91) must satisfy all versions of the cyclic boundary conditions (5.83). In place of the simple Bloch distribution of wave numbers (5.83) we must include the effects of the phase relations between degenerate modes. The values of $k_1 \ldots k_r$ are to be chosen from the set satisfying

$$Nk_j = 2\pi n_j + \sum_{m \neq j} \phi_{jm}, \quad n_j = 0, 1, \ldots, N-1. \tag{5.94}$$

We now look for the lowest eigenstate out of this set. Symmetry suggests that this must be a state with $S^{(z)} = 0$, i.e. for which $r = \tfrac{1}{2}N$. To find the distribution of wave vectors $k_1, \ldots, k_r$ that makes $\mathscr{E}(k_1, \ldots, k_r)$ as large as possible, we study the conditions (5.93) and (5.94) carefully for mutual consistency. Looking, for example, at the effects of the boundary conditions on possible choices of sign for the wave numbers k_1 and k_2 in the pair excitation (5.89) we find that these can be real and give maximum contribution to the energy (5.87), if $n_2 \geqslant n_1 + 2$. Applying this condition to the allowed values of all the integers n_j, we find that these must be evenly spaced, 2 units apart, across the whole range of values from 1 to N. The state of lowest energy is actually achieved by the choice

$$n_j = 2j - 1; \quad j = 1, 2, \ldots, \tfrac{1}{2}N. \tag{5.95}$$

If all the one-spin excitations like (5.81) were independent this would be tantamount to saying that the antiferromagnetic ground state contains equal contributions from all wave numbers in the spin-wave spectrum. Indeed, it is well known that the ground-state energy of a two-dimensional or three-dimensional antiferromagnet is approximately equal to the zero-point energy of a half quantum in each of the *antiferromagnon* modes – these being, of course, waves of spin deviation from the ordered antiferromagnetic array. In the present case, however, the many-spin excitations (5.91) are deviations from the ordered *ferromagnetic* state $|0\rangle$, and are strongly coupled by the phase relations (5.93) and (5.94).

To calculate the effects of these relations on the spectral density, let us introduce a variable $x = 2j/N$, which we treat as continuous, and make the transformation

$$\sum_j \to \tfrac{1}{2} N \int_0^1 dx \qquad\qquad (5.96)$$

in (5.92)–(5.95). The relationship $k(x)$ between the actual wave number and this variable derives from (5.94), i.e.

$$k(x) = 2\pi x + \tfrac{1}{2} \int_0^1 \phi\{k(x), k(x')\}\, dx', \qquad\qquad (5.97)$$

where the kernel is defined by (5.93),

$$\cot \tfrac{1}{2}\phi\{k, k'\} = \cot \tfrac{1}{2}k - \cot \tfrac{1}{2}k'. \qquad\qquad (5.98)$$

Differentiating (5.97) with respect to k, and using the fact that ϕ jumps from π to $-\pi$ as k' passes through k, we get the integral equation

$$A(k) = \frac{1}{\pi} - \frac{1}{\pi}\{1 + \cot^2 \tfrac{1}{2}k\} \int_0^{2\pi} \frac{A(k')\, dk'}{4 + \{\cot \tfrac{1}{2}k - \cot \tfrac{1}{2}k'\}^2} \qquad (5.99)$$

to be satisfied by the 'wave number density' $A(k) = dx/dk$. This integral equation looks rather formidable but is reduced to simpler terms by the substitution $\xi = \cot \tfrac{1}{2}k$, when it may be solved as an infinite sum or as an integral.

To determine the energy of the antiferromagnetic ground state we have to evaluate the sum (5.92). Using the spectral transformation (5.96) and the solution of (5.99), we get

$$E_0 = -NJ\{1 - \int_0^{2\pi} (1 - \cos k)A(k)\, dk\}$$
$$= -NJ(1 - 4 \ln 2). \qquad\qquad (5.100)$$

Since J is negative, this lies somewhat above the value NJ which would be the energy of an ordered antiferromagnetic array of *Ising* spins. This demonstrates the disordering effect and extra 'zero-point energy' produced by the off-diagonal components of the Heisenberg spin operators, which exchange spins along the chain. The transition from the Heisenberg to the Ising model by reducing the strength of the interaction between the off-diagonal components in $\mathscr{H}$ has been discussed at length by Bonner & Fisher (1964), who showed that the ground state of an antiferromagnet lacks long-range order only for the fully isotropic Heisenberg model (5.73). Various exact theorems for various models of this kind have been proved by Yang & Yang (1966). It is interesting to note (Des Cloiseaux & Pearson 1962) that the spin-wave spectrum of the antiferromagnetic linear chain with $S = \tfrac{1}{2}$ can be constructed as a simple generalization of the Bethe–Hulthén theory: for an excitation of total wave number q the energy is exactly

$$\mathscr{E}(q) = \tfrac{1}{2}\pi |\sin q|. \qquad\qquad (5.101)$$

But these states, like the ground state, are without long-range order; although there is no exact formula for the partition function of this system we can be quite sure that it exhibits no order–disorder phenomena even in the limit as $T \rightarrow 0$.

5.7 The Onsager solution of the two-dimensional Ising problem

The partition function for a planar lattice of Ising spins with nearest neighbour interaction can be evaluated exactly in the absence of an external 'magnetic field' (Onsager 1944). The mathematical argument is elaborate and subtle (see e.g. Newell & Montroll 1953; Domb 1960; Huang 1963; Temperley 1972) but the essential elements are not beyond ordinary comprehension.

The starting point is the transfer matrix introduced in § 5.5. In (5.56), for example, we represent the partition function for a linear chain of N Ising spins (with $H = 0$) as the trace of the Nth power of a 2×2 matrix, whose elements are

$$V_{\sigma\sigma'} \equiv \exp\{\beta J \sigma \sigma'\}. \tag{5.102}$$

This matrix operates in the space of the values $\sigma = +1, -1$ of a given Ising variable, relating it to the similar variable σ' on a neighbouring site. For abstract algebraic reasons, there is an advantage in representing such a two-dimensional operator in terms of the canonical Pauli matrices $\mathbf{1}, \tau^x, \tau^y, \tau^z$ in the same space, for example

$$V_{\sigma\sigma'} \equiv \begin{pmatrix} e^{\beta J} & e^{-\beta J} \\ e^{-\beta J} & e^{\beta J} \end{pmatrix} \equiv e^{\beta J}\begin{pmatrix} 1 & \cdot \\ \cdot & 1 \end{pmatrix} + e^{-\beta J}\begin{pmatrix} \cdot & 1 \\ 1 & \cdot \end{pmatrix}$$
$$\equiv e^K \mathbf{1} + e^{-K}\tau^x, \tag{5.103}$$

where $K \equiv \beta J$ as in (5.61).

This can be written even more compactly by using the identity $[\tau^x]^2 = \mathbf{1}$. By expanding in power series we verify that

$$V_{\sigma\sigma'} \equiv (2 \sinh 2K)^{\frac{1}{2}} \exp(K^*\tau^x) \tag{5.104}$$

where

$$\tanh K^* = e^{-2K}. \tag{5.105}$$

This very abstract formalism has the great advantage that a *product* of transfer matrices involving various different Ising variables becomes simply an exponential of the *sum* of the corresponding Pauli matrices. In the same notation, moreover, any operator referring to the values of a single Ising spin can be represented (as in 5.64) by the diagonal Pauli matrix

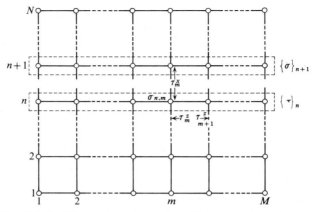

Fig. 5.8. Building up two-dimensional Ising model row by row.

τ^z. In (5.60), for example, the influence of an external field H is described by a matrix

$$\exp(\beta H \sigma)\delta_{\sigma\sigma'} \equiv \begin{pmatrix} e^{\beta H} & \\ & \cdot \\ & & e^{-\beta H} \end{pmatrix} \equiv \exp(K\tau^z). \qquad (5.106)$$

Now let us build up a two-dimensional square lattice row by row (fig. 5.8). Thus, the nth row contains M spins, having some configuration

$$\{\sigma\}_n \equiv \{\sigma_{n,1}, \sigma_{n,2}, \ldots, \sigma_{n,m}, \ldots, \sigma_{n,M}\}. \qquad (5.107)$$

This may be taken as the label for a basis vector in a space of 2^M dimensions, each corresponding to one of the 2^M different configurations of the spins in the row. The interaction of this whole row with the neighbouring row thus has the effect in the partition function of multiplying by a generalized transfer matrix

$$\mathbf{V}_1 \equiv V_{\{\sigma\}_n, \{\sigma\}_{n+1}} \equiv \exp\Big\{\beta J \sum_{m=1\ldots M} \sigma_{n,m}\, \sigma_{n+1,m}\Big\}$$

$$\equiv (2 \sinh 2K)^{\frac{1}{2}M} \exp\Big(K^* \sum_m \tau_m^x\Big). \qquad (5.108)$$

The interactions along the m column of the lattice are thus taken into account by defining a special Pauli operator τ_m for this column, just as in (5.104).

But we must also include the interactions between neighbouring spins in the same row. These introduce another matrix,

$$\mathbf{V}_2 \equiv \exp\left\{\beta J \sum_{m=1\ldots M} \sigma_{n,m}\, \sigma_{n,m+1}\right\}$$

$$\equiv \exp\left\{K \sum_{m=1\ldots M} \tau_m^z\, \tau_{m+1}^z\right\}, \tag{5.109}$$

by a simple extension of (5.106). In both (5.108) and (5.109) we must, of course, make special provision for conditions at the edges and ends of the carpet of spins. The simplest convention is to wrap the whole lattice around the surface of a torus, and to apply cyclic boundary conditions for the labels $m = 1 \ldots M$ and $n = 1 \ldots N$.

The partition function can now be expressed, as in (5.56), as the trace of a product of transfer matrices. The operators $\mathbf{V}_1$ and $\mathbf{V}_2$ are applied alternately to build up the lattice row by row, counting first column-wise and then row-wise interactions:

$$Z = \mathrm{Tr}\{\mathbf{V}_1\ \mathbf{V}_2\ \mathbf{V}_1\ \mathbf{V}_2 \ldots\}$$

$$= \mathrm{Tr}\{\mathbf{V}_1\ \mathbf{V}_2\}^N \tag{5.110}$$

if there are N such rows. By analogy with (5.59), we have only to evaluate λ_M, the maximum eigenvalue of the $2^M \times 2^M$ matrix

$$\mathbf{V} = \mathbf{V}_1\ \mathbf{V}_2 \tag{5.111}$$

to find the free energy and other thermodynamic properties of the system.

But this is obviously not a trivial problem! The fact that it can be solved at all seems a mathematical accident. Onsager studied the commutation relations of the operators

$$\mathbf{A}_0 \equiv \sum_{m=1}^{M} \tau_m^x \quad \text{and} \quad \mathbf{A}_1 = \sum_{m=1}^{M} \tau_m^z\, \tau_{m+1}^z \tag{5.112}$$

whose exponentials occur in $\mathbf{V}_1$ and $\mathbf{V}_2$ respectively. By constructing the commutator $[\mathbf{A}_0, \mathbf{A}_1]$ and then forming the commutators of this with $\mathbf{A}_0$ and $\mathbf{A}_1$, and so on, he systematically generated the *Lie algebra* of the problem. This turns out to have only $3M - 1$ linearly independent elements (instead of a possible 4^M) and as a consequence of the translational symmetry along the rows can be decomposed by Fourier transformation into very simple sub-algebras. But if an external magnetic field, or next-nearest neighbour interactions had been included in the partition function, the Lie algebra generated by the additional operators would have been much too complicated to permit this decomposition. For this fundamental reason, the Onsager method cannot solve the Ising problem in three dimensions.

The details of the Onsager technique can be expressed somewhat more concisely in the language of spinor representations of the rotation group in 2^M dimensions (Onsager & Kaufman 1949), although this also demands some very abstract algebra. The most elementary algebraic derivation of the same results (Schultz, Mattis & Lieb 1964) uses only the standard field theoretical language of boson and fermion creation and annihilation operators. The Pauli matrices τ_m, for example, can be transformed into the fermion operators (cf. the spin deviation operators of § 1.8)

$$\tau_{\bar{m}}^{\pm} = \tfrac{1}{2}(\tau_m^z \pm i\tau_m^y), \tag{5.113}$$

which satisfy the anticommutation relations

$$\{\tau_m^+, \tau_m^-\} = 1; \quad (\tau_m^+)^2 = (\tau_m^-)^2 = 0. \tag{5.114}$$

In terms of these, the operators (5.112) are both bilinear

$$\mathbf{A}_0 = -2\sum_{m=1}^{M}(\tau_m^+\tau_m^- - \tfrac{1}{2}); \quad \mathbf{A}_1 = \sum_{m=1}^{M}(\tau_m^+ + \tau_m^-)(\tau_{m+1}^+ - \tau_{m+1}^-). \tag{5.115}$$

Unfortunately, these are still not perfect fermion operators. The algebraic independence of the transfer matrices for adjacent columns of the lattice implies that they commute, i.e.

$$[\tau_{\bar{m}}^{\pm}, \tau_{\bar{m}'}^{\pm}] = 0 \quad \text{if } m \neq m'. \tag{5.116}$$

This boson-like property is turned into the corresponding anticommutation relation by a well-known non-linear transformation:

$$\mathbf{C}_m = \left[\exp(i\pi\sum_{i=1}^{m-1}\tau_i^+\tau_i^-)\right]\tau_{\bar{m}}^-; \quad \mathbf{C}_m^{\dagger} = \left[\exp(i\pi\sum_{i=1}^{m-1}\tau_i^+\tau_i^-)\right]\tau_{\bar{m}}^+. \tag{5.117}$$

The exponential operators here simply provide alternations of sign, according to the parity of the total occupation number of the eigenstates of $\tau_i^+\tau_i^-$, to generate standard anticommutators for the true fermion annihilation and creation operators $\mathbf{C}_m$, $\mathbf{C}_m^{\dagger}$.

Inverting (5.117) and substituting in (5.118) we get

$$\mathbf{A}_0 = -2\sum_m(\mathbf{C}_m^{\dagger}\mathbf{C}_m - \tfrac{1}{2}); \quad \mathbf{A}_1 = \sum_m(\mathbf{C}_m^{\dagger}\mathbf{C}_m)(\mathbf{C}_{m+1} + \mathbf{C}_{m+1}^{\dagger}). \tag{5.118}$$

But if we insist on cyclic boundary conditions and include in $\mathbf{A}_1$ the term for $m = M$ we make a mistake: careful algebraic manipulations show that this irregularity does not introduce a significant error into the final answer. Notice, however, that if we had made transformation (5.117) on a transfer matrix containing an external magnetic field, or with interactions $\sigma_{n,m}$ $\sigma_{n+1,m+1}$, etc. across the diagonal of a lattice cell, then we should arrive at much more complicated formulae than the bilinear expressions (5.118) in the operators $\mathbf{C}_m$, $\mathbf{C}_m^\dagger$. This is the step at which the Onsager technique shows its limitations.

Now, of course, we use the translational symmetry of (5.118) by making a Fourier transformation to the Bloch representation:

$$\mathbf{C}_m = M^{-\frac{1}{2}} e^{-i\pi/4} \sum_q e^{iqm} \, \boldsymbol{\eta}_q \qquad (5.119)$$

(where q runs through allowed wave numbers for the chain of M sites in the range $-\pi < q \leqslant \pi$) defines a new set of fermion operators $\boldsymbol{\eta}_q$, $\boldsymbol{\eta}_q^\dagger$. Substituting into (5.118), we get

$$\mathbf{A}_0 = -2 \sum_{q \geqslant 0} \{\boldsymbol{\eta}_q^\dagger \, \boldsymbol{\eta}_q + \boldsymbol{\eta}_{-q}^\dagger \, \boldsymbol{\eta}_{-q} - 1\};$$

$$\mathbf{A}_1 = 2 \sum_{q \geqslant 0} \{\cos q (\boldsymbol{\eta}_q^\dagger \, \boldsymbol{\eta}_q + \boldsymbol{\eta}_{-q}^\dagger \, \boldsymbol{\eta}_{-q}) + \sin q (\boldsymbol{\eta}_q \, \boldsymbol{\eta}_{-q} + \boldsymbol{\eta}_{-q}^\dagger \, \boldsymbol{\eta}_q^\dagger)\}. \qquad (5.120)$$

The terms in each of these sums contain products of even numbers of fermion operators and, therefore, commute with one another for different values of q. Referring back to (5.108)–(5.112), we see that $\mathbf{V}$ has been decomposed into a product, over all positive q, of matrices such as

$$\mathbf{V}_{1q} = \exp\{-2K^*(\boldsymbol{\eta}^\dagger \boldsymbol{\eta}_q + \boldsymbol{\eta}_{-q}^\dagger \, \boldsymbol{\eta}_{-q} - 1)\};$$

$$\mathbf{V}_{2q} = \exp\{2K \cos q \, (\boldsymbol{\eta}_q^\dagger \, \boldsymbol{\eta}_q + \boldsymbol{\eta}_{-q}^\dagger \, \boldsymbol{\eta}_{-q}) + 2K \sin q \, (\boldsymbol{\eta}_q \, \boldsymbol{\eta}_{-q} + \boldsymbol{\eta}_{-q}^\dagger \, \boldsymbol{\eta}_q^\dagger)\}. \qquad (5.121)$$

Each of these operators may be represented as a 4×4 matrix operating on the four fermion states with occupation numbers n_q, $n_{-q} = 0$, 1 and may, therefore, be diagonalized by elementary algebra. The most convenient procedure is to use the identity (5.103)–(5.104) to express the exponential of a sum of fermion operators as a sum of similar operators with exponential coefficients. The non-diagonal terms in $(\boldsymbol{\eta}_q \boldsymbol{\eta}_{-q} + \boldsymbol{\eta}_{-q}^\dagger \boldsymbol{\eta}_q^\dagger)$ can then be removed by the standard Bogoliubov transformation to new fermion operators

$$\boldsymbol{\xi}_q = \cos\phi_q \, \boldsymbol{\eta}_q + \sin\phi_q \, \boldsymbol{\eta}_{-q}^\dagger; \quad \boldsymbol{\xi}_{-q} = \cos\phi_q \, \boldsymbol{\eta}_{-q} - \sin\phi_q \, \boldsymbol{\eta}_q^\dagger, \qquad (5.122)$$

for a suitable choice of ϕ_q for each value of q. The final result of various straightforward algebraic manipulations is that we may write

$$\mathbf{V} = (2 \sinh 2K)^{\frac{1}{2}M} \exp\left\{-\sum_{\text{all } q} \varepsilon_q (\boldsymbol{\xi}_q^\dagger \, \boldsymbol{\xi}_q - \tfrac{1}{2})\right\} \qquad (5.123)$$

where the eigenvalues of each product $\mathbf{V}_{1q} \mathbf{V}_{2q}$ are exponentials of the roots, ε_q, of

$$\cosh \varepsilon_q = \cosh 2K \cos 2K^* - \sinh 2K \sin 2K^* \cos q$$

$$= \cosh 2K \coth 2K - \cos q \qquad (5.124)$$

(using (5.105)).

Now let us suppose that all values of ε_q are positive (which is indeed the case at high temperatures): the maximum eigenvalue of $\mathbf{V}$ is given at once by the vacuum states of the operators $\xi_q^\dagger \xi_q$ for all q, i.e.

$$\Lambda_M = (2 \sinh 2K)^{\frac{1}{2}M} \exp\left\{\tfrac{1}{2}\sum_q \varepsilon_q\right\}. \tag{5.125}$$

Thus, from (5.59), the thermodynamic free energy (per spin site) is given by the closed analytical formula

$$F = -kT\left[\tfrac{1}{2}\ln(2 \sinh 2K) + \frac{1}{4\pi} \int_{-\pi}^{\pi} \cosh^{-1}\{\cosh 2K \coth 2K - \cos q\}\, dq \right]$$

$$\tag{5.126}$$

where $K \equiv J/kT$. This is the exact Onsager solution.

To demonstrate the existence of a phase transition let us look again at (5.123). If any value of ε_q were to be zero or negative, the maximum eigenvalue (5.125) would no longer be unique; according to (5.66), the system would then show long-range order. Inspection of (5.124) and reference to the rules for choosing the proper root of this equation shows that this can occur, for $q = 0$, at the temperature where $K = K^*$, i.e. where

$$kT_c = 2J/\sinh^{-1}1 = 0.566\, zJ. \tag{5.127}$$

The same result is obtained by differentiating (5.126) twice with respect to temperature: the point where ε_q vanishes produces a logarithmic singularity in the specific heat. Notice that this exact value for the critical temperature is a long way below the mean field approximation (5.6), and the QCA result (5.17), but is not so far from the Kikuchi cluster approximation (5.52).

The degree of spontaneous magnetization below T_c can be calculated by 'perturbing' the transfer matrix by a weak magnetic field H, and then evaluating the change in free energy up to terms in H (Yang 1952). Exactly the same result can be obtained by finding the limit of the correlation function for spins at a great distance apart (Montroll, Potts & Ward 1963). Following the analogy of (5.63), we can write down the correlation between spins in the same row of the lattice:

$$\begin{aligned}
\langle \sigma_{n,m}\, \sigma_{n,m'} \rangle &= \operatorname*{Lt}_{M,N\to\infty}\ \{\operatorname{Tr}(\tau_m^z\, \tau_{m'}^z\, \mathbf{V}^N)\big/ \operatorname{Tr}(\mathbf{V}^N)\}\\
&= \operatorname*{Lt}_{M,N\to\infty}\ \left\{\sum_\alpha \langle \Psi_\alpha | \tau_m^z\, \tau_{m'}^z | \Psi_\alpha \rangle \Lambda_\alpha^N \Big/ \sum_\alpha \Lambda_\alpha^N \right\}\\
&\to \operatorname*{Lt}_{M\to\infty}\ \tfrac{1}{2}\{\langle \Psi_M^+ | \tau_m^z\, \tau_{m'}^z | \Psi_M^+ \rangle + \langle \Psi_M^- | \tau_m^z\, \tau_{m'}^z | \Psi_M^- \rangle\},
\end{aligned} \tag{5.128}$$

where we represent the matrix $\mathbf{V}$ in terms of its eigenvectors $|\Psi_\alpha\rangle$ and keep only the contributions from the eigenvectors $|\Psi_M^\pm\rangle$ of the two degenerate maximal eigenvalues $\Lambda_M^\pm$ given by (5.125). This generalization of (5.66) can be transformed into a determinant of order $(m-m')$ by following through the various steps (5.117), (5.119), (5.122) and using Wick's theorem concerning multiple products of fermion field operators. In the limit of very large values of $m-m'$, this determinant can be evaluated analytically by some specialized techniques of no obvious physical significance. This yields the 'Onsager cryptogram' (see Montroll, Potts & Ward 1963) for the spontaneous long-range order of the two-dimensional Ising lattice:

$$\langle\sigma\rangle \equiv \mathop{\mathrm{Lt}}_{(m-m')\to\infty} \langle\sigma_{n,m}\,\sigma_{n,m'}\rangle = \{1-\mathrm{cosech}^4(2J/kT)\}^{\frac18}. \qquad (5.129)$$

This function evidently goes to zero as T approaches the critical temperature (5.127) but otherwise looks very much like the approximate results obtained by the MFA, QCA, and other methods. Although (5.128) only considers correlations along the rows of the lattice, it can be shown quite generally (Schultz, Mattis & Lieb 1964) that various alternative definitions of the 'intrinsic magnetic moment' of this model are all equivalent.

Considered as a function of the distance $R=m-m'$ along a row, the spin correlation function (5.128) decays exponentially, as suggested in (1.37). The correlation lengths ξ and ξ' are proportional to $|T-T_c|^{-1}$ above and below the critical temperature, showing the deficiencies of the MFA results (5.29) and (5.30) compared with the exact result. For $T>T_c$, the 'pre-factor' R^{-n} of the exponential decay behaves as predicted by the Ornstein–Zernike theory (§§ 4.6, 5.3), where $n=\frac12(d-1)=\frac12$ for a system of dimensionality $d=2$. But *below* the critical temperature the exact result corresponds to an exponent $n=2$, whilst at $T=T_c$, where the correlation length ξ is infinite, a factor $R^{-\frac14}$ describes the fall off of the correlations with distance. These results are important in the general theory of critical phenomena (§ 5.12).

The Onsager method is not entirely restricted to the case of the planar square lattice with isotropic nearest neighbour interactions. The generalization to the *anisotropic exchange* model, where the parameters J_1 and J_2, for interactions along columns and rows respectively, are not equal, is discussed in all the review articles. The properties of this model are, indeed, almost identical with those of an isotropic model with exchange parameter J given by

$$\sinh(2J/kT) = \{\sinh(2J_1/kT)\sinh(2J_2/kT)\}^{\frac12}. \qquad (5.130)$$

Various topological transformations of the square lattice have also provided the thermodynamic properties (in zero field) of a number of exotic planar lattices – 'honeycomb', 'kagomé', 'diced', etc. (see e.g. Syozi 1972). But the theory of these transformations is of more significance for the mathematical analysis of *order* than for the study of *disorder*. For the moment we also defer discussion of graphical treatment of the Ising model in two dimensions, such as the derivation of the Onsager formulae by Kac & Ward (1952), since this technique is also the basis of a variety of series expansions in the more general Ising problem in three dimensions (see § 5.10).

5.8 Ferroelectric models in two dimensions

Exact formulae for the thermodynamic properties can be derived for many members of another class of two-dimensional systems – the square lattice 'ferroelectric' models of the tetrahedral hydrogen-bonded materials discussed in § 1.4. Many of these models are mathematically complicated and of little physical relevance (see e.g. Lieb 1971; Lieb & Wu 1972): all the exact solutions are, in fact, elaborations of the technique used originally by Lieb (1967a) to calculate the residual entropy of 'square ice'.

We start, once more, with the transfer matrix of §§ 5.5, 5.7. As with the two-dimensional Ising model (cf. fig. 5.8) we build up an $M \times N$ (toroidal) lattice, row by row (fig. 5.9). Consider the nth row of 'vertical' bonds. In a given configuration there may be 'down' arrows on just r sites, labelled m_1, $m_2, \ldots, m_r$ along the row. In other words, this configuration would be represented by a 'state vector' $|m_1, m_2, \ldots, m_r\rangle$, whose arguments are always in ascending order:

$$1 \leqslant m_1 < m_2 < \ldots < m_r \leqslant M. \tag{5.131}$$

The transfer matrix $\mathbf{V}$ links this with a configuration $|m'_1, m'_2, \ldots, m'_{r'}\rangle$ of the arrows on the vertical bonds of the next row – and so on.

Since we are considering the ice model, all configurations that are allowed by the ice vertex conditions (fig. 1.7) are of equal thermodynamic weight: the matrix elements of $\mathbf{V}$ are either 1 or 0, depending on whether or not the configurations $|m_1, m_2, \ldots, m_r\rangle$ and $|m'_1, m'_2, \ldots, m'_{r'}\rangle$ satisfy these constraints. One might imagine that a further combinatorial factor would arise from different arrangements of the arrows in the 'horizontal' bonds: it is easy to verify, however, that these are now determined uniquely (with trivial exceptions) by the up and down arrows on the vertical bonds and by the ice condition that there should be exactly two arrows entering and two

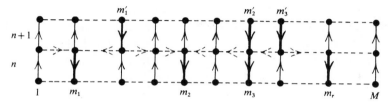

Fig. 5.9. Labelling arrows in two-dimensional ice model.

arrows leaving each vertex. The thermodynamic properties of the model can be deduced from the largest eigenvalue of $\mathbf{V}$, just as in (5.56)–(5.59).

As we try to insert arrows on the horizontal bonds we quickly learn that the proportion of 'down' arrows on the vertical bonds must be the same from row to row. The condition $r' = r$ implies that $\mathbf{V}$ is a diagonal block matrix, operating independently in each of the sub-spaces corresponding to different values of r. Since the maximum amount of disorder occurs when there are equal numbers of up and down arrows in each row, we may assume that the largest eigenvalue Λ_M has its eigenvector in the sub-space $r = \frac{1}{2}M$.

We further discover that the set of 'down' arrows in each new row – i.e., at $m'_1, m'_2, \ldots, m'_r$ – must either coincide with or 'interleave' the ones in the previous row: e.g.

$$1 \leqslant m'_1 \leqslant m_1 \leqslant m'_2 \leqslant m_2 \leqslant \ldots \leqslant m'_r \leqslant m_r \leqslant M$$

or $\qquad 1 \leqslant m_1 \leqslant m'_1 \leqslant m_2 \leqslant \ldots \leqslant m_r \leqslant m'_r \leqslant M. \qquad (5.132)$

These conditions, together with (5.131), define the non-vanishing matrix elements of $\mathbf{V}$ in each sub-space.

Now suppose that an eigenfunction Ψ is constructed as a linear combination of the configuration functions in the sub-space $r = \frac{1}{2}M$

$$\Psi = \sum_{1 \leqslant m_1 < m_2 < \ldots < M} a(m_1, m_2, \ldots, m_r)|m_1, m_2, \ldots, m_r\rangle. \qquad (5.133)$$

The eigenvalue equations simply read

$$\Lambda a(m_1, \ldots, m_r) = \sum_{m'_1 \geqslant 1}^{m_1} \sum_{m'_2 \geqslant m_1}^{m_2} \ldots \sum_{m'_r \geqslant m_r - 1}^{m_r} a(m'_1, \ldots, m'_r) D_1$$

$$+ \sum_{m'_1 \geqslant m_1}^{m_2} \sum_{m'_2 \geqslant m_2}^{m_3} \ldots \sum_{m'_r \geqslant m_r}^{M} a(m'_1, \ldots, m'_r) D_2, \qquad (5.134)$$

where the regions of summation are, of course, restricted by the condition that no two successive labels, m'_j, m'_{j+1} should be equal.

In the ice problem, the factors D_1 and D_2 are both equal to unity. But different energies assigned to various configurations of arrows at each vertex in the KDP and F-models (§ 1.4) merely introduce exponentials of linear functions of $(m_i - m'_j)$, etc., in these factors. The algebra is thereby greatly complicated but because of the cyclic invariance of such terms for translations along the row of bonds the Lieb technique is not hindered.

The cyclic symmetry suggests 'Bloch-like' solutions in which each label m_j would carry a wave-like factor $\exp(ik_j m_j)$. We are at once reminded of the antiferromagnetic ground state of the one-dimensional Heisenberg model (§ 5.6). It can be shown, in fact (see e.g. Lieb & Wu 1972), that the Hamiltonian of this model is closely related to our present transfer matrix, suggesting that *the eigenfunctions of (5.80) generated by the Bethe Ansatz (5.91) are also eigenfunctions of (5.134)*. The algebraic verification of this conjecture is extremely laborious (Lieb 1967), but it turns out that the restrictions on the labels m'_j, etc. lead to precisely the same consistency conditions (5.93) on the relative phase angles ϕ_{jm} as was required in (5.88), for example, to deal with the special case of two reversed spins on adjacent sites in the Heisenberg model.

On the other hand, the *eigenvalues* of (5.80) are quite different from those of (5.134), where we have successive summations over $r = \frac{1}{2}M$ different labels m_j. Each such summation is itself a geometric series in $\exp(ik_j)$ which introduces a factor of the form $\{1 - \exp(ik_j)\}^{-1}$ in the eigenvalue: in place of (5.92), we have

$$\lambda(k_1 \ldots k_r) = \prod_j \{1 - \exp(ik_j)\}^{-1}. \tag{5.135}$$

The 'allowed' values of k_j must, as before, be chosen from the set (5.94); the condition (5.95) for the ground state of the Heisenberg Hamiltonian evidently yields the maximum eigenvalue (5.135) of the transfer matrix. Following the line of argument from (5.96) to (5.100), we evaluate (5.59) for the entropy per vertex:

$$S = -F/MNT = (\pounds/M)\ln \lambda_M$$

$$= -\pounds \operatorname*{Lt}_{M \to \infty} \frac{1}{M} \sum_j \ln\{1 - \exp(ik_j)\}$$

$$= -\pounds \tfrac{1}{2} \int_{-\pi}^{\pi} A(k)\ln(2 - 2\cos k)\, dk, \tag{5.136}$$

where $A(k)$ satisfies the integral equation (5.99). Elaborate functional analysis yields the exact result for the entropy of 'square ice',

$$S/\pounds = \ln(4/3)^{\frac{3}{2}} = \ln(1.5396), \tag{5.137}$$

which is not much larger than the Pauling approximation (1.8). For a real three-dimensional ice model, the best series approximation (Meijering 1957; Nagle 1966) is

$$S/\textit{k} \approx \ln(1.5068) \tag{5.138}$$

which is even closer to $\ln(3/2)$: was our long journey from § 1.4 really necessary?

The eigenvalues for the KDP and F-models are somewhat more complicated than (5.135) so that the free energy integral analogous to (5.136) cannot be evaluated analytically. It turns out, however, as might be expected, that these models exhibit phase transitions to ferroelectric or antiferroelectric order at a temperature comparable with the energy separation of the various vertex configurations, i.e.

$$\textit{k}T_{\text{c}} = \varepsilon/\ln 2. \tag{5.139}$$

The dependence of this transition temperature on an external electric field can also be calculated.

The most remarkable characteristic of these models is that in zero field the polarization (i.e. spontaneous order) below T_{c} is complete: there is no analogue of the temperature-dependent spontaneous magnetization (5.129) of the Ising ferromagnet on the same lattice. But these models are all closely connected, since they are all special cases of the *eight-vertex model* (§ 1.4) with particular values of the interaction parameters J_{ij} and J_7 of the general Ising Hamiltonian (1.26a). The transfer matrix for this model can be expressed in Pauli operator language (cf. (5.109)) and general conditions found for the existence of a matrix with which it commutes – i.e. which has the same eigenvectors. Just as the Bethe Ansatz (5.91) provides eigenvectors of both the Heisenberg chain and the planar ferroelectric models, although with very different eigenvalues, so it proves possible (Baxter 1972) to find a general algebraic scheme where the maximum eigenvalue of the transfer matrix is expressible as a function of the energy parameters assigned to the eight-vertex configurations in fig. 1.10. We then find, for example (Barber & Baxter 1973), that the behaviour of the spontaneous long-range order as a function of the temperature depends on the relative ratios of these parameters, with the Ising and KDP models as special cases. This powerful mathematical technique is obviously of great interest in the general theory of phase transitions and critical exponents, but the various models to which it is applicable, being square planar lattices instead of tetrahedrally coordinated structures in three dimensions, are somewhat remote from physical reality.

5.9 The spherical model of ferromagnetism

For fundamental algebraic reasons the methods of §§ 5.7–8 cannot be generalized to three-dimensional lattices. Even for the Ising model on a simple cubic lattice, itself a greatly simplified representation of any real physical system, there is no exact theory of the phase transition. In the search for a mathematically soluble model we modify still further the assumed properties of the interacting 'spins'.

In § 1.8 we saw the advantages of representing a random scalar variable u_l on a lattice by its Fourier amplitudes $U(\mathbf{q})$, defined as in (1.42). In this representation the near-neighbour interaction Hamiltonian is easily diagonalized:

$$\mathscr{H} = -\tfrac{1}{2} \sum_{l,l'} J_{ll'} u_l^* u_{l'} - H \sum_l u_l$$
$$= -\tfrac{1}{2} \sum_{\mathbf{q}} J(\mathbf{q}) U^*(\mathbf{q}) U(\mathbf{q}) - HN^{\frac{1}{2}} U(0) \qquad (5.140)$$

where the interaction has Fourier transform

$$J(\mathbf{q}) = \sum_{l'} J_{ll'} \exp\{i\mathbf{q} \cdot (l - l')\}. \qquad (5.141)$$

If we were to identify u_l with a spin variable, this would be a way of treating the typical ferromagnetic Hamiltonian (5.1) in a magnetic field.

But this procedure is fatally hindered by the diabolical condition that each spin is quantized or, as in the classical model, must be a vector of constant length, i.e.

$$|\mathbf{S}_l|^2 = S(S+1) \quad \text{for each site } l. \qquad (5.142)$$

As we saw in § 1.4, such conditions impose a Byzantine set of phase relations on the Fourier amplitudes $S(\mathbf{q})$, rendering the transformation (5.140) quite useless.

The central idea of the *spherical model* (Berlin & Kac 1952) is to replace these conditions by the very much looser restraint that the *sum* of the squares of all the spin vectors should be a constant. For our model scalar variable u_l, this overall normalization condition would be written

$$\sum_l |u_l|^2 = N. \qquad (5.143)$$

In other words, the point $(u_1, u_2, \ldots, u_N)$ in an N-dimensional space is constrained to lie on the hypersphere passing through the points $|u_1|^2 = |u_2|^2 = \ldots = |u_N|^2 = 1$.

This condition permits the N Fourier components $U(\mathbf{q})$ to behave as if they were completely random, subject only to the fixed sum

$$\sum_{\mathbf{q}} |U(\mathbf{q})|^2 = N. \qquad (5.144)$$

For simplicity, the magnetic field term is dropped from the Hamiltonian (5.140), which is put into the partition function (5.2). For this 'classical' system the 'Trace' is interpreted as a sum over all values of all the variables $U(\mathbf{q})$ consistent with (5.144) – i.e. as the N-fold integral

$$Z = A_N^{-1} \int_{-\infty}^{\infty} \dots \int_{-\infty}^{\infty} \delta\left\{ N - \sum_{\mathbf{q}} |U(\mathbf{q})|^2 \right\}$$

$$\exp\left\{ \tfrac{1}{2}\beta \sum_{\mathbf{q}} J(\mathbf{q}) |U(\mathbf{q})|^2 \right\} dU(1) \dots dU(N), \qquad (5.145)$$

where the normalization factor A_N is the area of the hypersphere (5.144) of N dimensions.

To evaluate this expression we represent the delta function as an integral over a complex variable s, using the identity

$$\delta\left\{ N - \sum_{\mathbf{q}} |U(\mathbf{q})|^2 \right\} \equiv \frac{1}{2\pi i} \int_{-i\infty}^{i\infty} \exp\left[s\left\{ N - \sum_{\mathbf{q}} |U(\mathbf{q})|^2 \right\} \right] ds. \qquad (5.146)$$

Putting this into (5.145), but postponing to a later stage the contour integration over s, we find that the N-fold integration over the variables $U(\mathbf{q})$ separates into the product of N independent integrals, each of the simple form

$$\int_{-\infty}^{\infty} \exp[-\{s - \tfrac{1}{2}\beta J(\mathbf{q})\} |U(\mathbf{q})|^2] \, dU(\mathbf{q}) = \pi^{\frac{1}{2}} \{s - \tfrac{1}{2}\beta J(\mathbf{q})\}^{-\frac{1}{2}}$$

$$= \pi^{\frac{1}{2}} \exp[-\tfrac{1}{2} \ln\{s - \tfrac{1}{2}\beta J(\mathbf{q})\}]. \qquad (5.147)$$

The product of all these factors becomes a sum of the exponents: the partition function can now be written in the form

$$Z = A_N^{-1} \pi^{\frac{1}{2}N} \frac{1}{2\pi i} \int_{-i\infty}^{i\infty} \exp N\left[s - \frac{1}{2N} \sum_{\mathbf{q}} \ln\{s - \tfrac{1}{2}\beta J(\mathbf{q})\} \right] ds. \qquad (5.148)$$

The integrand of (5.148) contains the exponential of a very large number, N, multiplied by an expression that tends, as $N \to \infty$, to a definite limit

$$\gamma(s) \equiv \operatorname*{Lt}_{N \to \infty} \left[s - \frac{1}{2N} \sum_{\mathbf{q}} \{s - \tfrac{1}{2}\beta J(\mathbf{q})\} \right]$$

$$= s - \frac{1}{(2\pi)^d} \int \ln\{s - \tfrac{1}{2}\beta J(\mathbf{q})\} \, d\mathbf{q}. \qquad (5.149)$$

The integration is over the Brillouin zone of a d-dimensional lattice with unit volume per unit cell. A contour integral of the type

$$I_N = \frac{1}{2\pi i} \int_{-i\infty}^{i\infty} e^{N\gamma(s)} \, ds \qquad (5.150)$$

can be evaluated to order $1/N$ by the *method of steepest descents*. We look for a saddle point s_s of the real part of $\gamma(s)$, and lead the contour through this point in a direction along which the imaginary part of $\gamma(s)$ stays constant. The main contribution to the integral then comes from a small region about s_s, where the exponential factor is large and there are no oscillations due to rapid changes of phase. In other words, provided that $\gamma(s)$ is analytic at s_s, the partition function (5.148), in the limit $N \to \infty$, can be evaluated exactly as an analytic function of the temperature $T (\equiv 1/\mathit{k}\beta)$.

The position of the saddle point is determined by solving $\gamma'(s_s) = 0$. But we see from (5.141) that the largest value of $J(\mathbf{q})$ comes at $\mathbf{q} = 0$; from (5.149) it is evident that $\gamma(s)$ is analytic along the real axis only for $s > \frac{1}{2}\beta J(0)$. It is convenient to transform to a standard complex variable $\xi = 2s/\beta J(0)$; the saddle point occurs at ξ_s, satisfying

$$R(\xi_s) \equiv \frac{1}{(2\pi)^d} \int \frac{d\mathbf{q}}{\xi_s - \lambda(\mathbf{q})} = \beta J(0) \tag{5.151}$$

where

$$\lambda(\mathbf{q}) \equiv J(\mathbf{q})/J(0) \tag{5.152}$$

is a normalized Fourier component of the interaction function $J_{ll'}$. The steepest descents formula is valid only for $\xi_s > 1$ – the branch point of $R(\xi)$ corresponding to the largest value of $\lambda(\mathbf{q})$.

Since $\beta \to \infty$ as $T \to 0$, we can satisfy (5.151) at all temperatures provided that $R(\xi)$ tends to infinity as $\xi \to 1$; whether or not this occurs depends on the dimensionality of the lattice. For $d = 1$ and $d = 2$ it is easy to show that (5.151) can always be satisfied, whatever the value of β; the free energy per spin is thus an analytic function of the temperature down to $T = 0$ and there is no phase transition.

In three dimensions, however, the spherical model always has a phase transition. Taking a simple cubic lattice with nearest neighbour interactions, we get from (5.141) and (5.151) the expression

$$R_3(\xi) = \frac{1}{(2\pi)^3} \int\limits_{-\pi}^{\pi} \int \int \frac{dq_1\, dq_2\, dq_3}{\xi - \frac{1}{3}(\cos q_1 + \cos q_2 + \cos q_3)}, \tag{5.153}$$

which happens to tend to the finite limiting value $R_3(1) = 1.516$ as $\xi \to 1$. The evaluation of Z by the method of steepest descents, therefore, breaks down when $\beta J(0) \geqslant R_3(1)$, i.e. at a critical temperature

$$\frac{\mathit{k} T_c}{zJ} = 0.660. \tag{5.154}$$

It is extraordinary that this is very close to the best estimate (e.g. (5.52)) of the critical temperature of an Ising model on the same lattice.

What happens to the partition function (5.148) below T_c? The contour must be made to pass round the branch point at $\xi = 1$, although this is no longer a saddle point of $\gamma(s)$. But now, instead of smoothing all the terms in the sum in (5.149) into a continuous integral over $\mathbf{q}$, we keep separate the contribution from $\mathbf{q} = 0$. This appears as a special factor $[s - \frac{1}{2}\beta J(0)]^{-1}$ in (5.148) and must, therefore, be treated as a pole contributing its residue to the contour integral. This residue then appears as a large contribution to the logarithm of the partition function and eventually as a finite part of the free energy per spin. What this means, physically, is that a finite fraction of the total sum $\sum_{\mathbf{q}} |U(\mathbf{q})|^2$ is coming from the single mode $\mathbf{q} = 0$. This corresponds to a finite value for the average 'spin' $\langle u_l \rangle$ and, therefore, describes a state with long-range order.

Another way of thinking about the spherical model is to regard $J(\mathbf{q})$ as the energy of a particle of momentum $\mathbf{q}$. If this particle is a boson, then (5.144) merely adds up the occupation numbers of the various levels in a gas of N independent particles. The phase transition is then precisely equivalent (Gunton & Buckingham 1968) to the *Bose–Einstein condensation* of a finite proportion of the particles into the ground state. Below T_c the relative occupation number

$$N^{-1}|U(0)|^2 = \langle u_l \rangle^2 \qquad (5.155)$$

plays the part of an order parameter, which rises linearly to unity as $T \to 0$: the spontaneous magnetization of the spherical model is thus given by

$$M(T) = \langle u_l \rangle = \left(1 - \frac{T}{T_c}\right)^{\frac{1}{2}} \qquad (5.156)$$

which is not nearly as abrupt as, for example, the two-dimensional Ising model (5.129).

On the face of it, the theory of the spherical model looks quite different from the algebraic techniques, based on the concept of a transfer matrix, outlined in the last few sections. It can be shown, however (Thompson 1968), that the analogue of the transfer matrix is a continuous operator whose eigenfunctions satisfy an integral equation and that the phase transition occurs when the largest eigenvalues become degenerate, just as in (5.66), (5.125), etc. This connection confirms that the spherical model is by no means completely unrealistic and artificial as a representation of the key mechanisms of order–disorder transitions on a lattice.

The thermodynamic properties of the spherical model for various types of lattice and various types of interaction $J_{ll'}$ can be worked out in detail (Joyce 1972). The spin–spin correlation function has a particularly simple form, which can be deduced from the partition function (5.145) by introducing the factor (cf. (5.63))

$$N^{-1} \sum_{l'} u_{l'} u_{l'+l} = \sum_{\mathbf{q}} |U(\mathbf{q})|^2 e^{i\mathbf{q} \cdot l}. \tag{5.157}$$

Carrying this through to (5.148), we get

$$\Gamma(l) \equiv \langle u_0 u_l \rangle - \langle u_0^2 \rangle$$
$$= \{1/\beta J(0)\} R(\xi_s, l), \tag{5.158}$$

where

$$R(\xi, l) \equiv \frac{1}{(2\pi)^d} \int \frac{e^{-i\mathbf{q} \cdot l}}{\xi - \lambda(\mathbf{q})} d\mathbf{q} \tag{5.159}$$

is a generalization of $R(\xi)$.

In the limit of large distances, where only small values of q contribute to the integral, we can write

$$\lambda(q) \sim 1 - \tfrac{1}{6} q^2 + \dots \tag{5.160}$$

By comparison with (4.27) we see that $\Gamma(l)$ must have the Ornstein–Zernike form (4.28) with correlation length (in units of the lattice constant)

$$\xi_1 = \{6(\xi_s - 1)\}^{-\frac{1}{2}}. \tag{5.161}$$

Above the critical temperature this can be calculated from (5.151): the correlation length evidently goes to infinity as $T \to T_c$ and $\xi_s \to 1$. Below T_c, where $\xi_s = 1$, the exponential factor does not occur and correlations additional to the spontaneous magnetization fall off with distance as $1/r$. It is interesting to note (Stell 1969) that if we put the Fourier transform of $\Gamma(l)$ into (4.14) in place of $S(\mathbf{q})$, we get the Fourier transform of the direct correlation function in the form

$$C(\mathbf{q}) = 1 - \beta J(0)\xi_s + \beta J(\mathbf{q}). \tag{5.162}$$

In other words, the spherical model represents a system in which the direct correlation between sites at l and l' is precisely $J_{ll'}$, the strength of the actual interaction between these sites, in units of kT.

Although the spherical model seems very artificial, it is not entirely unrealistic. Consider a system with the Hamiltonian (1.16) in which each of the 'spin vectors' $\mathbf{S}_l$ is a classical vector with D components. It is not very difficult to show that the properties of the spherical model are precisely those of such a system in the limit where the *spin dimensionality D* is allowed to tend to infinity (Stanley 1968, 1971). In this sense we may say that the behaviour of the classical Heisenberg model in d lattice dimensions (for which, of course, $D = d$) must be intermediate between the behaviour of the corresponding Ising model $(D=1)$ and the behaviour predicted by the simple exact formulae of the spherical model $(D=\infty)$. Thus, the fact that the spherical model has no phase transition for $D=2$ is consistent with the argument (§ 2.5) of Mermin & Wagner (1966) against long-range order in two-dimensional magnetic systems of higher spin dimensionality than the Ising model (§ 5.7).

5.10 Graphical expansions

In default of an exact formula for the partition function of any realistic three-dimensional model we have a choice between an approximate *compact representation* (e.g. § 5.4) or an expansion in a power series. The sum of a *finite* number of terms of such a series is always well behaved but it is not difficult to extrapolate to the sum of an infinite series, whose divergence betokens a possible phase transition. This brute force approach has developed into a gentle art (see e.g. Domb 1960, 1970a, b; Domb & Green 1974), which has yielded a great deal of important information about such unsolved problems as the order–disorder transition of the three-dimensional Ising model. The representation of the coefficients for such a series as combinatorial factors associated with the enumeration of lattice graphs is also an important unifying principle for the whole of the theory of disordered systems.

The obvious starting point is the general expression (5.2) for the partition function. At high temperatures, where the parameter is small, we write down the *moment expansion*

$$Z = \langle \exp(-\beta \mathcal{H}) \rangle$$
$$= 1 - \beta \langle \mathcal{H} \rangle + \tfrac{1}{2}\beta^2 \langle \mathcal{H}^2 \rangle - \ldots + \frac{1}{n!}(-\beta)^n \langle \mathcal{H}^n \rangle + \ldots \quad (5.163)$$

where $\langle \ \rangle$ denotes a quantum mechanical trace or a classical average over all configurations of the system, normalized to unity. In the Ising model, for example, where (cf. (5.61))

$$\beta \mathcal{H} \equiv -\tfrac{1}{2}K \sum_{l.l'}^{\text{neighbours}} \sigma_l \sigma_{l'} \quad (5.164)$$

this symbol means 'sum over all values ± 1 of each variable σ_l'.

Nothing but the immense labour involved prevents us from evaluating

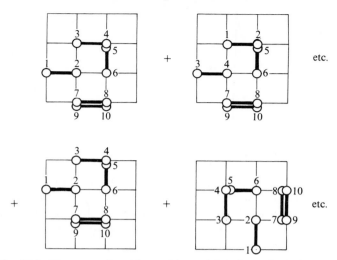

Fig. 5.10. Dimer graph, with combinatorial and topological equivalents.

any term in the series: in practice, elementary algebra becomes inadequate as soon as n exceeds 4. To try to keep account of the various terms we automatically introduce a graphical representation along the following lines: Consider the nth term in (5.163) for the Ising Hamiltonian (5.164). The nth moment $\langle \mathcal{H}^n \rangle$ will evidently contain a factor K^n together with an average $\langle \sigma_1 \sigma_2 \ldots \sigma_{2n} \rangle$ arising from the sum of all possible products of n pairs of factors $\sigma_1 \sigma_2$, $\sigma_3 \sigma_4 \ldots$, etc. The site labels $1, 2, \ldots, 2n$ in any such product can be represented (fig. 5.10) as points on a lattice obeying the condition that they can be paired as neighbours. Any such *dimer graph*, therefore, corresponds to a possible contribution to a term of the moment series and may be evaluated algebraically. But there must be a very large number of equivalent graphs, each making exactly the same contribution. For example, the labels in fig. 5.10 may be permuted in pairs, yielding a factor $n!$, whilst further factors arise from various layouts of the dimers that are topologically equivalent to the chosen graph. The calculation thus reduces to evaluating the average $\langle \sigma_1 \ldots \sigma_{2n} \rangle$ for each distinct type of graph and multiplying by the appropriate combinatorial factor.

Even for small values of n, this remains a formidable task. It is obvious, moreover, that the terms in (5.163) contain successive powers of N, the number of sites in the actual lattice, so that the convergence of the series would appear to depend on this physically irrelevant number. What we are

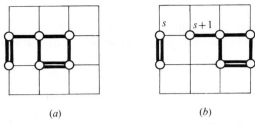

(a)   (b)

Fig. 5.11. (a) Connected, (b) separated graph.

really interested in is the free energy *per atom* which should be a well defined quantity. We thus construct a new series of the form

$$\underset{N\to\infty}{\text{Lt}}\left(\frac{1}{N}\ln Z_N\right)=\sum_{n=1}^{\infty}\frac{1}{n!}\,c_n(-\beta)^n \qquad (5.165)$$

whose coefficients can be derived from those of (5.163) by expanding in a logarithmic series and collecting like powers of β.

But the relationship between (5.163) and (5.165) is a standard transformation in probability theory, from the moment expansion to the *cumulant expansion*. Each coefficient c_n is *defined* as a *cumulant average* of the Hamiltonian, i.e.

$$c_n\equiv\langle\mathscr{H}^n\rangle_{\text{cum}}. \qquad (5.166)$$

The actual formulae for successive cumulants (or, as they are sometimes termed, the *semi-invariants of Thiele*) in terms of the corresponding moments become more and more complicated as n increases, but the first few expressions are easily written down, e.g.

$$\langle X\rangle_{\text{cum}}=\langle X\rangle$$
$$\langle X^2\rangle_{\text{cum}}=\langle X^2\rangle-\langle X\rangle^2 \quad \text{(cf. (1.14))}$$
$$\langle X^3\rangle_{\text{cum}}=\langle X^3\rangle-3\langle X\rangle\langle X^2\rangle+2\langle X\rangle^3$$

etc.   (5.167)

The real power of this transformation of the moment series resides in a simple but very general property of cumulants: *the cumulant of a product* $\langle XY\rangle_{\text{cum}}$ *is zero if the factors X and Y are statistically independent.* This theorem, which can be proved by quite elementary methods (e.g. Kubo 1962) has profound consequences.

Consider, for example, a contribution to $\langle\mathscr{H}^n\rangle_{\text{cum}}$ coming from a term such as $\langle\sigma_1\ldots\sigma_s\,\sigma_{s+1}\ldots\sigma_{2n}\rangle$ corresponding to a *separated graph* (fig. 5.11). The jump from one connected sub-graph to another in going from site s to site $s+1$ divides the product into two independent factors; the cumulant average must therefore vanish. In other words, in evaluating c_n we need include only *connected graphs*, thus enormously reducing the labour.

This sort of simplification arises in many techniques of theoretical physics. The present analysis may be seen, for example, as a special case of the *Mayer cluster expansion* (§ 6.5) for the virial coefficients of a dense gas (see e.g. Hill 1956). Or we recall a similar theorem for the perturbation expansion of the S-matrix in quantum field theory which can be applied to statistical mechanics by the device of *temperature Green functions* (Abrikosov, Gorkov & Dzyaloshinski 1963).

These are all, in fact, examples of a fundamental theorem of *graph theory* which can be expressed as follows (Uhlenbeck 1960; Uhlenbeck & Ford 1962):

Let

$$F(x) = \sum_{N=1}^{\infty} F_N \, x^N / N! \qquad (5.168)$$

where the sum

$$F_N = \sum_{G_N} W(G_N) \qquad (5.169)$$

runs over all N-pointed labelled graphs. Assume that the weight $W(G_N)$ assigned to the graph G_N is independent of the labelling of the points and can be written as a product

$$W(G_N) = \prod_{\text{all } C_l} W(C_l), \qquad (5.170)$$

where C_l is one of the disjoint connected parts of G_N. Then

$$\ln\{1 + F(x)\} = \sum_{l=1}^{\infty} f_l \, x^l / l!, \qquad (5.171)$$

where

$$f_l = \sum_{C_l} W(C_l). \qquad (5.172)$$

Putting it crudely, we may say that the various terms that can be generated from (5.171) by replicating, translating bodily and permuting all the connected graphs in (5.172) merely have the effect of transforming this series into its own exponential (5.168).

In evaluating c_n, the number of graphs can be further reduced by the obvious fact that a 'loose end' or 'odd' vertex j contributes $\sigma_j = \pm 1$ to an odd power and thus averages to zero. The summation therefore includes only *closed* graphs with *even* vertices. But the assignments of weight and the enumeration of these is still very laborious (Domb 1970b) because they may include contributions from any number of internal cycles of a sub-graph in which a given line is traversed many times. This is the type of series that must be used for, say, the calculation of the free energy of a three-dimensional Heisenberg model at high temperatures.

For the Ising model, however, an alternative graphical reduction is often more convenient (see e.g. Newell & Montroll 1958). For $\sigma_i \sigma_j = +1$ or -1, the interaction matrix (5.60) or (5.103) satisfies the identity

$$\exp\{K\sigma_i\sigma_j\} \equiv \cosh K + \sigma_i \sigma_j \sinh K$$
$$= \cosh K \{1 + \sigma_i\sigma_j w\} \qquad (5.173)$$

where

$$w \equiv \tanh K \equiv \tanh(J/kT). \qquad (5.174)$$

The partition function (5.163)–(5.164) now takes the form

$$Z = (\cosh K)^{\mathscr{P}} \sum_{\sigma_i = \pm 1} \prod_{ij}' (1 + \sigma_i\sigma_j w), \qquad (5.175)$$

where

$$\mathscr{P} = \tfrac{1}{2}zN \qquad (5.176)$$

is the number of nearest neighbour pairs on the lattice. The product is taken over all such pairs, and the summation contains 2^N terms. Expanding in powers of w we get

$$Z = (\cosh K)^{\frac{1}{2}zN} 2^N (1 + \langle \sigma_i \sigma_j \rangle w + \langle \sigma_i \sigma_j \sigma_k \sigma_l \rangle w^2 + \ldots). \qquad (5.177)$$

The successive terms in this series evidently correspond to closed graphs with even vertices, but each bond appears only once in the continued product (5.175), unlike the graphs counted in (5.165) where the same adjacent pair of sites may occur many times in $\langle \mathscr{H}^n \rangle_{\text{cum}}$. In other words, the partition function can be expressed in the form

$$Z = 2^N (\cosh K)^{\frac{1}{2}zN} \left[1 + \sum_{n=1}^{\frac{1}{2}zN} p(n)w^n \right] \qquad (5.178)$$

where $p(n)$ is the number of independent *closed, even-vertex* graphs that can be drawn with n *distinct* links on a lattice (fig. 5.12(a)). Unfortunately this series is not quite of the form (5.168), so that it cannot be reduced further by

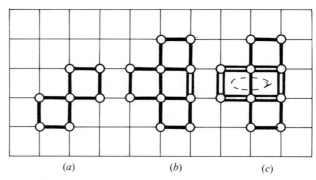

(a)   (b)   (c)

Fig. 5.12. Closed graphs with even vertices; (a) with distinct links; (b) including multiple links; and (c) further cycles on sub-graphs.

considering only connected graphs without reintroducing multiple bonds (Domb 1970b).

This formula holds for an Ising model on a lattice of any number of dimensions. Since there is only one graph of the required type (i.e. with $n = N$) on a cyclic chain, the one-dimensional solution (5.59)–(5.62) is a trivial case. The derivation of the Onsager solution (5.126) for the two-dimensional lattice by combinatorial arguments (Kac & Ward 1952) is also essentially elementary although lengthy and elaborate in detail. The connection between this type of formula and the solution of the *dimer problem* – the computation of the number of distinct ways of placing 'diatomic molecules' on the lattice without overlapping – is discussed in detail by Kasteleyn (1967) and references to the corresponding algebraic theory of *Pfaffians* are given by Temperley (1972): but this takes us a long way from the physics of disordered systems!

The graphical expansions in (5.165) and (5.178) are convergent at high temperatures, where β and w are small. An alternative approach, valid at low temperatures, is to start from the ordered phase, at $T = 0$ and then to consider the effects of reversing more and more spins. The partition function for the Ising model (5.22) can thus be written in the form

$$Z = 2e^{\mathcal{P}J/kT}\left[1 + \sum_{n=1}^{N} v(n)e^{-2nJ/kT}\right] \tag{5.179}$$

where $\mathcal{P}J$ is the energy of the ground state and $v(n)$ is the number of independent spin configurations in which exactly n nearest-neighbour interactions have been reversed in sign.

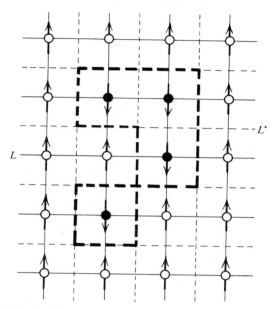

Fig. 5.13. Graph on dual lattice L^* for low-temperature series.

For a two-dimensional system on a lattice L this series can be represented by graphs on the *dual lattice* L^*. For any configuration of reversed spins (fig. 5.13) we connect together neighbouring broken links; we evidently produce a collection of closed polygons of exactly the kind counted in (5.178). But these polygons are drawn on the lattice L^* which is derived from L by applying this geometrical construction t. The combinatorial factor $v(n)$ in (5.179) is thus precisely $p^*(n)$, the number of closed even-vertex graphs with n distinct links on the dual lattice. Introducing a new variable (cf. (5.105))

$$w^* \equiv e^{-2K} \equiv \tanh K^* \qquad (5.180)$$

we can write (5.179) in the form

$$Z = 2e^{\mathscr{P}K}\left[1 + \sum_{n=1}^{\mathscr{P}^*} p^*(n)w^{*n}\right] \qquad (5.181)$$

which is thus a low-temperature expansion (since $w^* \to 0$ as $T \to 0$) in which the coefficients are obtained by counting the same type of graphs as the high-temperature series (5.178).

The advantage of this transformation is that if we already know the exact partition function on L^* by the Onsager method, we can write this in the form (5.178), as if w^* were a high-temperature variable, i.e.

$$Z^*(K^*) = 2^{N^*}(\cosh K^*)^{\mathscr{P}^*}\left[1 + \sum_{n=1}^{\mathscr{P}^*} p^*(n)w^{*n}\right] \qquad (5.182)$$

and by comparing (5.181) with (5.182) obtain a new formula for the partition function Z on the lattice L in terms of K and K^*. This topic is reviewed at length by Syozi (1974).

For a *self-dual* lattice, such as the square planar lattice, where $L^* = L$, the two series (5.178) and (5.181) have identical coefficients. Yet the variables w and w^* are not the same. If the critical point is unique, then it must correspond to the same singularity in both expressions for Z, which means that it can only occur at the temperature where $K = K^*$. This method was originally used by Kramers & Wannier (1941) to locate exactly the temperature of the transition (5.127), before the complete Onsager solution had been discovered.

But for a three-dimensional system all such arguments are useless. The only way to evaluate the partition function and other thermodynamic quantities is to compute as many terms of each series as possible and then to extrapolate to a closed formula for the sum by various algebraic devices (see e.g. Domb 1970a). By assessing the radius of convergence of each of the various series, one can locate the critical temperature for any given model with high accuracy and also determine the behaviour of sensitive properties such as specific heat and susceptibility in this neighbourhood.

The graphical expansions for spin correlation functions can also be written down and evaluated in the same spirit. At high temperatures, for example, we might start from (5.63) or (5.128) and use (5.173) to express the expectation value, in a form analogous to (5.175) and (5.178),

$$\langle \sigma_l \sigma_{l'} \rangle = Z^{-1}(\cosh K)^{\mathscr{P}} \sum_{\sigma_i = \pm 1} {\prod_{ij}}' \sigma_l \sigma_{l'}(1 + \sigma_i \sigma_j w)$$

$$= Z^{-1}(\cosh K)^{\mathscr{P}} 2^N \sum \eta_{ll'}(n)w^n. \qquad (5.183)$$

The factors σ_l and $\sigma_{l'}$ preserve *odd* powers in the averaging procedure so that $\eta_{ll'}(n)$ differs from $p(n)$ by counting graphs that have *odd* vertices at these two sites (fig. 5.14).

When l and l' are actual neighbours, the line joining them (fig. 5.14(b)) gives the *direct correlation* (5.12) between the two spins. More complicated graphs, where, for example, the two sites are joined by a long, rambling path through the lattice (fig. 5.14(c)) give rise to the *indirect correlations*

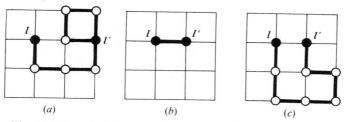

Fig. 5.14. Correlation graphs: (*a*) general; (*b*) direct; (*c*) indirect.

represented in (5.24) by which SRO is propagated through the lattice. This qualitative discussion can be made the basis of much elaborate mathematical theory. We may, for example, expand any of the 'compact' approximate formulae of §§ 5.2–5.4 in a power series, in β, w or w^*, comparing coefficients with the exact results obtained by graphical enumeration. Or we may decide to retain only certain simple types of graph – 'rings', 'bubbles', 'ladders', etc. – in the graphical expansion, which can then be summed approximately by combinatorial topology. The graphical significance of the various compact representations of the order–disorder transition provides valuable information concerning their relative accuracy and interconnection (see e.g. Brout 1965).

Suppose, for example (Domb 1970*b*), that we start from a high-temperature series such as (5.165) and re-expand in powers of the variable w. The resulting series is rather complicated, because multiple-bond graphs are not eliminated and do not all occur with the same positive 'weight'. But many of the most important contributions are counted in the expression that we might expect to get if we could apply the 'cumulant averaging' principle (5.171) to the series (5.178): to this approximation

$$N^{-1} \ln Z \sim \sum_{n=1} p'(n) w^n, \tag{5.184}$$

where the sum is now over *connected*, even-vertex graphs of n distinct links. As a further approximation we might limit ourselves to closed *polygons*, where the path never crosses itself: in other words, we take

$$p'(n) \sim u(n)/2n \tag{5.185}$$

where $u(n)$ is the total number of distinct *self-avoiding random walks* of length n.

On a Bethe lattice this topological parameter is given exactly by (5.35), i.e.

$$u(n) = z(z-1)^{n-1}, \tag{5.186}$$

so that (5.184) would diverge when $w = 1/(z-1)$; this gives precisely the QCA critical temperature (5.17). For a real lattice, the enumeration of self-avoiding random walks can only be done by direct counting; but there is a well-tested conjecture (Domb 1969, see §§ 7.8, 9.10) that for large n this function takes the form

$$u(n) \approx n^{-\alpha} \zeta^n \qquad (5.187)$$

where the exponent α takes the value $3/2$ for all two-dimensional lattices and $7/4$ for all three-dimensional lattices. The *connective constant* ζ depends upon the local coordination of the lattice and has been determined very accurately for a number of standard cases (table 5.1). Since the series

Table 5.1

Lattice	Triangular	Square	Honeycomb	f.c.c.	b.c.c.	s.c.
z	6	4	3	12	8	6
ζ	4.1515	2.6386	1.8484	10.035	6.5288	4.6826

(5.184) would now diverge when $w_i = 1/\zeta$, we obtain yet another estimate of the critical temperature. In the two-dimensional square lattice this gives $\not k T_c/zJ = 0.625$, which is not even as good as the simplest Kikuchi result (5.52). But this estimate of the critical temperature for three-dimensional lattices is very close to the best values obtained by extrapolation from the exact series expansions. The corresponding approximation (Domb 1970b) for the spin correlation function just above the critical point gives a range of order (1.37) behaving like

$$\xi_1 \sim (1 - w/w_c)^{-\frac{1}{2}\theta} \qquad (5.188)$$

where θ is $3/2$ in two dimensions and $6/5$ in three dimensions. In two dimensions the correct exponent deduced from the Onsager solution is 1, but in three dimensions where the best graphical estimate of this exponent is about 0.64, this approximation is obviously much better than the RPA Ornstein–Zernike result (5.29). Indeed, this connection between the self-avoiding walk problem and the critical behaviour of a ferromagnet is exact to leading order in D, where D is the dimensionality of the spin vector (Bowers & McKernell 1973).

The above discussion of graphical methods in the statistical mechanics of substitutional disorder does scant justice to a very large, active and sophisticated sub-field of theoretical physics. On approaching this activity more closely, however (see, especially, the detailed reviews in Domb & Green 1974), one is overwhelmed by a mass of pure mathematical theorems from graph theory, analytical devices for determining coefficients and estimating the radius of convergence of a series, and numerical results for particular physical properties of particular physical models. Much of this material is only of technical interest, being concerned mainly with subsidiary questions that have had to be answered in the attempt to arrive at a theory of the order–disorder transition in these systems, but not contributing directly to our conceptual understanding of the phenomena themselves. This is my excuse for not having plunged further into this formidable subject.

5.11 Order as a thermodynamic variable

For the moment, let us forget about microscopic theories of disorder and introduce the order parameter, $\mathscr{S}$, as a macroscopic variable whose value is determined by the conditions for thermodynamic equilibrium. For example, $\mathscr{S}$ might measure the magnetization of a lattice of spins. Following Landau & Lifshitz (1958), we expand the free energy of the system (per unit volume) as a power series in $\mathscr{S}$, with coefficients depending on the temperature (and, if necessary, other thermodynamic forces such as magnetic field, pressure, etc.). When $\mathscr{S} = 0$, the system is in the disordered phase with free energy $F_0(T)$. In the absence of any external polarizing field the 'state of order' may be assumed to be independent of the sign of $\mathscr{S}$, so that odd powers do not appear in the phenomenological series

$$F(\mathscr{S}, T) = F_0(T) + B(T)\mathscr{S}^2 + D(T)\mathscr{S}^4 + \ldots \qquad (5.189)$$

In thermodynamic equilibrium, the free energy must be a minimum as a function of $\mathscr{S}$. Let us suppose that $D(T)$ is always positive; at high temperatures, where the disordered phase is always observed, the coefficient $B(T)$ must also be positive. But, as the temperature decreases, this coefficient may change sign: we thus discover a critical temperature such that

$$B(T_c) = 0; \quad B(T) < 0 \quad \text{for } T < T_c. \qquad (5.190)$$

In this temperature range the minimum free energy is obtained, as in (5.11), when

$$\frac{\delta F(\mathscr{S}, T)}{\delta \mathscr{S}} = 0, \qquad (5.191)$$

i.e. by elementary algebra, when

$$\mathscr{S}^2 = -\tfrac{1}{2}B(T)/D(T). \qquad (5.192)$$

In other words, at T_c, we get a transition to a phase in which the order increases from zero as the temperature falls.

As far as it goes this argument is unexceptionable. In the original *Landau theory*, however, the function $B(T)$ is assumed to be an analytic function of T in the neighbourhood of T_c, with a Taylor expansion

$$B(T) \approx b_1(T - T_c) + \ldots \tag{5.193}$$

where b_1 must be positive to agree with (5.190). The order parameter ought then to behave like

$$\mathscr{S} \approx \begin{cases} \{b_1/2D(T_c)\}^{\frac{1}{2}}(T_c - T)^{\frac{1}{2}} & \text{for } T \leqslant T_c \\ 0 & \text{for } T > T_c \end{cases}. \tag{5.194}$$

Thermodynamic analysis of (5.189)–(5.194) shows that this describes a typical *second-order phase transition*, in which there is no latent heat but the specific heat is discontinuous at T_c. We may also verify from (5.5) that the mean field approximation yields this type of behaviour of the order parameter in the neighbourhood of T_c.

The Ornstein–Zernike formulae (4.28) and (5.29) can also be derived in the same spirit (Kadanoff *et al.* 1967). The order parameter is now to be treated as a field variable, $\mathscr{S}(\mathbf{r})$, that may fluctuate from point to point through the region. The free energy density (5.189) now takes the form

$$F(\mathbf{r}) = F_0(T) + B(T)\{\mathscr{S}(\mathbf{r})\}^2 + D(T)\{\mathscr{S}(\mathbf{r})\}^4 + \ldots$$
$$- H(\mathbf{r})\mathscr{S}(\mathbf{r}) + \beta(T)\{\nabla \mathscr{S}(\mathbf{r}) \cdot \nabla \mathscr{S}(\mathbf{r})\}. \tag{5.195}$$

In this expression we include a term for any possible polarizing field, $H(\mathbf{r})$, applied externally or arising from fluctuations of $\mathscr{S}(\mathbf{r})$ itself. The term with coefficient $\beta(T)$ is introduced to damp out rapid spatial variations in $\mathscr{S}(\mathbf{r})$; phenomenologically, this is the mechanism by which order is transmitted through the medium.

For the total free energy of a macroscopic specimen, we integrate (5.195) through the whole volume. This integral is a functional of the unknown distribution of order, $\mathscr{S}(\mathbf{r})$, which may be varied for a stationary value of F. The functional derivative (5.191) yields the partial differential equation

$$\{2B(T) + 4D(T)\mathscr{S}^2(\mathbf{r}) - 2\beta(T)\nabla^2\}\mathscr{S}(\mathbf{r}) = H(\mathbf{r}). \tag{5.196}$$

If $H(\mathbf{r})$ is identically zero, this equation has the solution $\mathscr{S}(\mathbf{r}) = \mathscr{S}$, whose value is given by (5.192) or (5.194). But the polarizing field may have small fluctuations $\delta H(\mathbf{r})$ about zero. The local order must then fluctuate in sympathy by an amount $\delta \mathscr{S}(\mathbf{r})$ about the average value $\mathscr{S}$. The fluctuations are linked thermodynamically: to lowest order in $\delta \mathscr{S}(\mathbf{r})$, (5.196) yields

$$\{2B + 12D\mathscr{S}^2 - 2\beta\nabla^2\}\delta\mathscr{S}(\mathbf{r}) = \delta H(\mathbf{r}). \tag{5.197}$$

But the order parameter must be continuous from point to point. As shown in § 3.2, this implies the existence of a correlation function

$$\Gamma(\mathbf{r}-\mathbf{r}') = \langle \delta \mathscr{S}(\mathbf{r}) \delta \mathscr{S}(\mathbf{r}') \rangle, \qquad (5.198)$$

which tends to a finite value as $\mathbf{r}' \to \mathbf{r}$. In classical statistical mechanics this is also a generalized *response function*, linking the fluctuations of local order (e.g. of magnetization) with fluctuations of the polarizing field, i.e.

$$\delta \mathscr{S}(\mathbf{r}) = \frac{1}{\ell T} \int \Gamma(\mathbf{r}-\mathbf{r}') \delta H(\mathbf{r}') \, d\mathbf{r}'. \qquad (5.199)$$

We now substitute for $\delta \mathscr{S}(\mathbf{r})$ from (5.199) into (5.197):

$$\int [\{2B + 12D\mathscr{S}^2 - 2\beta\nabla^2\}\Gamma(\mathbf{r}-\mathbf{r}') - \ell T\delta(\mathbf{r}-\mathbf{r}')]\delta H(\mathbf{r}') \, d\mathbf{r} = 0. \qquad (5.200)$$

Remembering that $\delta H(\mathbf{r})$ is an arbitrary fluctuation field, we arrive at a differential equation for the correlation function:

$$[2B(T) + 12D(T)\{\mathscr{S}(T)\}^2 - 2\beta(T)\nabla^2]\Gamma(\mathbf{R}) = \ell T \, \delta(\mathbf{R}), \qquad (5.201)$$

where the coefficients are simply functions of T defined in (5.189) and (5.195) or given by (5.192). This equation has the Ornstein–Zernike solution (4.28), i.e.

$$\Gamma(\mathbf{R}) = \frac{\ell T}{8\pi\beta(T)} \frac{1}{R} \exp(-R/\xi_1) \qquad (5.202)$$

where

$$\xi_1 = \begin{cases} \{\beta(T)/B(T)\}^{\frac{1}{2}} & \text{for } T > T_c \\ \{-\beta(T)/2B(T)\}^{\frac{1}{2}} & \text{for } T < T_c \end{cases}. \qquad (5.203)$$

If (5.193) is also assumed, then we get

$$\xi_1 \propto |T_c - T|^{-\frac{1}{2}} \qquad (5.204)$$

above and below T_c, just as in (5.29) and (5.30).

If it were not for the exact Onsager solution (5.129) for the spontaneous magnetization of a two-dimensional Ising model and also the very precise estimates of critical exponents such as (5.188) by series expansions, we might easily be persuaded that the behaviour of any system near a second-order phase transition ought to obey the Landau formulae. We know, however, that (5.194) and (5.204) are not correct. The unjustified assumption is (5.193); there is no *a priori* reason why the phenomenological coefficients in (5.189) should be analytic functions of T through the critical point. Detailed study of the Onsager solution shows, for example (Fisher 1967*a*), that the free energy above T_c contains a term in $(T - T_c)^2 \ln(T - T_c)$ which contributes a logarithmic singularity to the specific heat.

It is interesting to note (Kadanoff *et al.* 1967) that this singularity is related to the deviation of the correlation function from the Ornstein–Zernike form at large distances. As was assumed in (5.197), the Landau theory is valid only if the fluctuations of the order parameter over distances of the order of ξ_1 are small compared with the order parameter itself. As we approach the critical point, where the correlation length becomes very large, this assumption becomes more and more implausible and leads to erroneous conclusions. It is not difficult, of course (see e.g. Luban 1976), to generalize the Landau theory phenomenologically by introducing more complicated functionals into the free-energy density (5.195) but the theory provides no guidance as to the precise form of such functions, whose parameters must be adjusted arbitrarily to fit to critical data.

The simplicity and generality of the basic Landau theory, which makes no direct appeal to microscopic models, has much to commend it. In some cases, for example (Landau & Lifshitz 1958), the phase transition is governed by higher-order terms in an expansion of the free energy such as (5.184). Thus, if we introduce pressure P as an independent thermodynamic variable, we can find a critical point (T_c, P_c), where both

$$B(T_c, P_c) = 0 \quad \text{and} \quad D(T_c, P_c) = 0. \tag{5.205}$$

Since the condition for the stability of the transition (5.190) was that $D(T) > 0$, the transition goes over from second order to first order at this point on the critical curve in the (P, T) diagram. This behaviour is, of course, quite familiar in the gas–liquid phase transition (§ 6.2). On the other hand, by introducing more complex order parameters and, considering higher-order terms in the free energy density (de Gennes 1974), one arrives at an excellent formal framework for the representation of all the subtleties and complexities of the spatial order patterns observed in *liquid crystals* (§ 2.14), where the statistical mechanics of the homogeneous order–disorder transition is only of secondary interest.

5.12 Scaling and renormalization of critical phenomena

As we saw in § 5.11, the phase transition near the critical point is characterized by long-range fluctuations of the order parameter. If the correlation length ξ is already quite large, then we can say little more about the individual spins than that they are strongly correlated locally. In other words, inside a block of size L, where $L \ll \xi$, all the spins are so nearly aligned that they tend to behave as a single unit. We could ignore the microscopic internal structure of such a block and think of the phase transition as a collective phenomenon of an assembly of blocks, interacting

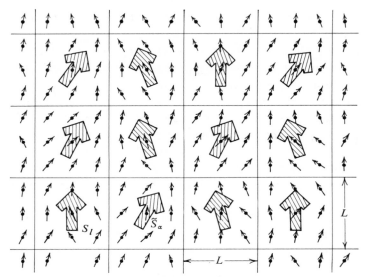

Fig. 5.15. Block spin variables.

via long-range correlations, etc. This general idea can now be put on a firm mathematical basis.

Let us suppose, quite generally, that we have a lattice of dimensionality d, on each of whose sites is a spin variable s_l, which may have Ising, Heisenberg, classical or other components as required. This lattice is mentally sub-divided into a similar lattice of identical blocks each containing L^d spins (fig. 5.15). To the block labelled α we assign a *block spin variable* $\tilde{s}_\alpha$, which describes its average state of polarization. If $\tilde{s}_\alpha$ has the same characteristics as s_l (i.e. the same number of components, commutation relations, etc.) then we may plausibly assert that the correlations between block spins must be of the same mathematical structure as the correlations between the original site spins s_l.

The essential difference must be in the strengths of the interactions between blocks. Suppose we write our original Hamiltonian in 'temperature normalized form'

$$-\beta\mathcal{H}(s) = \tfrac{1}{2}\sum_{ll'}K_{ll'}s_l s_{l'} + h\sum_l s_l, \tag{5.206}$$

where h measures the strength of the external magnetic field. We can write down a corresponding *block Hamiltonian*

$$-\beta\tilde{\mathcal{H}}(\tilde{s}) = \tfrac{1}{2}\sum_{\alpha\beta}\tilde{K}_{\alpha\beta}\tilde{s}_\alpha\tilde{s}_\beta + \tilde{h}\sum_\alpha \tilde{s}_\alpha, \tag{5.207}$$

where the block interaction parameters $\tilde{K}_{\alpha\beta}$ and the effective magnetic field h are defined so as to make the thermodynamic properties deduced from $\tilde{\mathcal{H}}$ identical with those deduced from $\mathcal{H}$. This means, of course, that the interactions in the block Hamiltonian depend in detail on the block size L and must be much more complicated than the original interactions in $\mathcal{H}$.

But suppose that our original Hamiltonian contains only nearest neighbour interactions of strength K; it is tempting to assume (Kadanoff 1966; see e.g. Kadanoff *et al.* 1967; Fisher 1967*a*; Stanley 1971; Kadanoff 1976) that $\tilde{\mathcal{H}}$ retains this simplicity near the critical point – i.e. that block spins interact by a single nearest neighbour parameter $\tilde{\mathcal{H}}$. In this approximation the free energy per *block*, $F(\tilde{K}, \tilde{h})$ would be of the same functional form $F(K, h)$ as the free energy per spin that we might have calculated from the original Hamiltonian (5.206). But there are L^d spins per block; since these are just alternative descriptions of the same system, with the same total free energy, we may write

$$F(K, h) = L^{-d}F(\tilde{K}, \tilde{h}). \qquad (5.208)$$

Similarly, if we had succeeded in calculating the correlation length $\xi(K, h)$ from the original Hamiltonian, at a certain temperature and magnetic field, then our hypothesis would be inconsistent unless we could write

$$\xi(K, h) = L\xi(\tilde{K}, \tilde{h}), \qquad (5.209)$$

in units of the block size L, for the block correlation function $\xi(\tilde{K}, \tilde{h})$.

These approximations are only valid when $L \ll \xi$ – i.e. when h is small, and when T is near the critical temperature T_c. Under these circumstances (5.208) and (5.209) are *homogeneity conditions* on the unknown functions F and ξ, from which the arbitrary *scaling parameter* L can be eliminated. It is convenient to express the interaction parameters in terms of the *reduced temperature*

$$t \equiv \frac{T - T_c}{T_c} \approx \frac{K_c - K}{K_c}. \qquad (5.210)$$

To satisfy (5.208) and (5.209) we may then write

$$\tilde{h} = L^x h; \quad (K_c - \tilde{K})/K_c \equiv \tilde{t} = L^y t, \qquad (5.211)$$

where the exponents x and y are unknown numbers, independent of L. Substituting into (5.208), we find that the free energy per spin near the critical point must have the functional form

$$F(t, h) = t^{-d/y}f(t/h^{y/x}), \qquad (5.212)$$

where f represents a function that could only be determined by a proper statistical calculation. Similarly, the correlation length must behave like

$$\xi(t, h) = t^{-1/y}\phi(t/h^{y/x}) \tag{5.213}$$

where again ϕ would need to be calculated by other methods.

The *Kadanoff construction* thus provides a plausible model from which to deduce the *homogeneity hypothesis* (5.212), first proposed by Widom (1965). These functional relations lead by standard thermodynamic arguments (Kadanoff *et al.* 1967; Fisher 1967; Stanley 1971) to *scaling laws* linking the various *critical exponents* (e.g. (5.29), (5.30)) for the behaviour of the specific heat, susceptibility, correlation functions, etc. in the limits $t \to 0$ and $h \to 0$. All these exponents can thus be expressed in terms of the numbers x and y, introduced in (5.211) and, therefore, ought to satisfy a number of relations amongst themselves which we can attempt to verify experimentally or from calculations on model systems. Notice, in particular, that in the above argument we would not expect to observe any special effects inside a given block as the system passed through the critical point, so that the exponents introduced in (5.211) should be the same for both positive and negative values of t. We thus generate important identities between the critical exponents above and below the critical temperature T_c.

Since the structure of the block spin Hamiltonian (5.207) is not necessarily as simple as we have postulated there is no reason why the scaling conditions should be exactly fulfilled. Nevertheless, the Onsager solution (§ 5.7) for the two-dimensional Ising model scales exactly according to (5.212), with $y = 1$ and $x = 15/8$. The three-dimensional spherical model (§ 5.9) also satisfies (5.212) with $y = 1$. But the various formulae obtained using the mean field approximation (§§ 5.2–5.4, 5.11) are not mutually consistent with the scaling law for $d = 3$. Series expansion of the correlation function for the three-dimensional Ising model, as in (5.188), gives $1/y \approx 0.64$, which is near to, but not quite equal to, the value 0.625 deduced from combinations of other critical exponents. The classical Heisenberg model seems to be fairly consistent with the value $1/y \approx 0.70$ – and so on.

Detailed numerical investigations have also confirmed the *universality hypothesis* (see e.g. Kadanoff 1976): *the critical exponents for a given system depend only on the dimensionality, d, of its lattice, and on the number of components, D, of each spin vector s_l.* That is to say, the critical properties are independent of the lattice type (b.c.c., f.c.c., etc.) and of the ratios of next-nearest neighbour to nearest neighbour interactions, etc.

To see how such a principle might arise, let us look again at the block

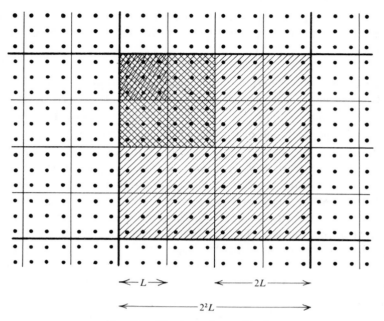

Fig. 5.16. Blocks and superblocks.

Hamiltonian (5.207). Let $\tilde{K}$ stand for the set of interaction parameters $\tilde{K}_{\alpha\beta}$ for a standard sub-division of the system into a standard type of lattice of blocks of size L. Now assemble a fixed number of these blocks (e.g. 2^d in a d-dimensional system) into a 'superblock'; make a standard lattice of superblocks (fig. 5.16). In the superblock Hamiltonian, the interaction parameters $\tilde{K}'$ between 'superblock spins' will depend on the values of $\tilde{K}$ in much the same way as the interactions in (5.207) depend on those in (5.206). But this construction does not involve any reference to the block size L. In other words

$$\tilde{K}' = \tilde{K}'(\tilde{K}) \qquad (5.214)$$

is a standard mapping of $\tilde{K}$ to *renormalized* parameters $\tilde{K}'$ and is independent of L. The *renormalization group* (Wilson 1971a; see e.g. Ma 1973; Wilson & Kogut 1974; Fisher 1974; Wilson 1975) generated by (5.214) must be determined by the dimensionality of the standard block lattice and the number of components of the block spins whose interactions are described by the parameters $\tilde{K}$.

Now suppose that $\tilde{K}$ is close to, but not quite equal to, the critical value $\tilde{K}_c$ for the block lattice. Making superblocks, and super-superblocks, and so on to the pth power, we shall eventually construct a lattice whose spin units are larger than the correlation length ξ. Going from site to site of this lattice we shall find little local order amongst the 'spins', whose mutual interaction parameters $\tilde{K}^{(p)}$ must, therefore, be a long way from $\tilde{K}_c$. In other words, if the whole assembly is to make a phase transition we must go to a temperature where the mapping (5.214) reproduces itself identically; the critical point $\tilde{K}_c$ must be a *fixed point* of the renormalization group, satisfying

$$\tilde{K}_c = \tilde{K}'(\tilde{K}_c). \tag{5.215}$$

If we assume that the whole situation is dominated by a single interaction parameter, then we may readily deduce the Kadanoff relations (5.211), etc. from this powerful theorem. The formal analogy between this approach and the theory of the stability of dynamical systems is lucidly explained by Wilson (1971*a*). In the nomenclature of *global analysis* the critical point corresponds to a *catastrophe* in the evolution of the interaction parameters to infinite scale distance.

Near the critical point, the pattern of fluctuations is thus *scale-independent*; it must look nearly the same, whether observed on the microscopic level or over distances comparable with the correlation length ξ. To appreciate the nature of this type of disorder, consider an Ising model just below T_c. As shown in § 2.4, a region of 'reversed spins' is nearly thermodynamically stable under these conditions so that we naively expect the fluctuations to appear as 'droplets' (fig. 5.17(*a*)) whose size increases as T approaches T_c. But the free energy excess (2.13) does not depend strongly on the size of the region, so that the interior of each large droplet is itself subject to critical fluctuations – and so on, right down to the atomic level. The pattern of reversed regions may thus be of immense topological complexity (Kadanoff 1976) with 'droplets' inside 'droplets' inside 'droplets' (fig. 5.17(*b*)). The renormalization group formalism is the natural mathematical language in which to describe systems whose disorder thus covers a very wide spectrum from microscopic to macroscopic wavelengths (Wilson 1975).

To use (5.215) to determine the critical point and critical exponents we must remember that this is really a matrix relation involving sets of interaction parameters and, therefore, reduces to an eigenvalue problem. This can be tackled directly (see e.g. Wilson 1975; Kadanoff 1976). For example (Niemeijer & van Leeuwen 1973), there is a straightforward mapping relation between the interaction of spins on a triangular lattice (fig. 5.18) and the interactions between similar variables defined for tri-

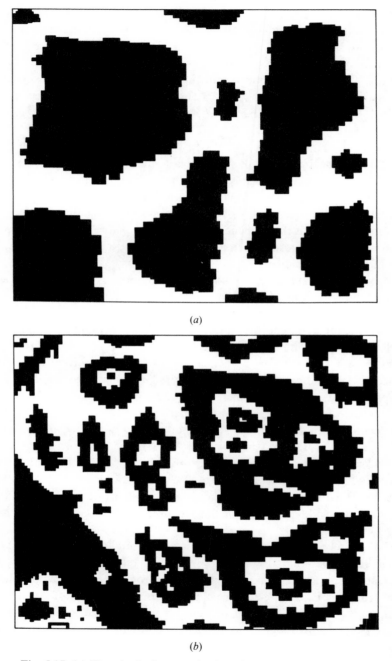

(a)

(b)

Fig. 5.17. (a) 'Droplets' of reversed spins. (b) Critical fluctuations on all scales.

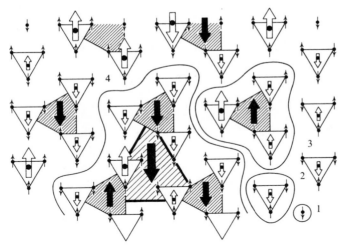

Fig. 5.18. On a triangular lattice, Ising spins (1) can be combined in triangles (2), and similarly into larger blocks (3), (4).

angular blocks of spins on the corresponding block lattice. By counting the contributions of various characteristic types of interaction to the partial partition function for a hexagonal 'cluster' of spins, this mapping matrix can be evaluated. The solution of the eigenvalue equation (5.214) then gives a value of K_c which is almost identical with the exact Onsager solution for this lattice.

The main weakness of the Kadanoff construction is that a single spin variable $\tilde{s}_\alpha$ is supposed to represent the state of a whole block of Ising spins. But, if the block is large enough and we are near enough to the critical point, we are justified in representing the local state of spin polarization by a continuous field variable $s_L(\mathbf{r})$ (cf. § 5.11). The interactions between spins at various points then becomes a Hamiltonian density (5.195), e.g.

$$\mathscr{H}_L = -\tfrac{1}{2}K_L\{\nabla s_L(\mathbf{r})\}^2 - P_L[s_L(\mathbf{r})], \tag{5.216}$$

where the coefficient K_L and the function P_L depend on the scale parameter L. The mapping (5.214) is now to be replaced by a *recursion formula* linking P_{2L} with P_L. The trick is (Wilson 1971b; Wilson & Kogut 1974) to limit the wavelengths in the Fourier representation of $s_L(\mathbf{r})$ to distances greater than L, i.e. to use wave numbers in the range $0 < q < 1/L$. The effect of going from s_L to s_{2L} is thus to integrate the partition function, in its Fourier representation (cf. (15.145)), over the range of wave numbers $1/2L < q < 1/L$. Instead of going to larger and larger 'blocks' we go to smaller and smaller 'boxes' surrounding the origin in reciprocal space, looking for the recursion relation between one shell and the next inside it. These calculations are much too complicated to be summarized. The result is, however, that the critical exponents of the three-dimensional Ising model can be determined numerically to high accuracy. A particularly interesting result obtained by the same method (Wilson & Fisher 1972) is an approximate formula for the critical exponents of the Ising model in '4 − ε dimensions', where ε is small. The point is that the recursion formula is exactly satisfied in four 'dimensions' by the so-called *Gaussian model* in which the magnitude of each 'spin' has a Gaussian distribution. This purely artificial model has the same critical exponents as those obtained by the mean field theory, which is thought to be an exact procedure for $d \geqslant 4$ for all different values of the spin dimensionality D (§ 5.9). But, in the renormalization group formulae, the dimensionality of the system occurs as a factor 2^{-d} measuring the ratio of the volumes of successively smaller 'boxes' in the sub-division of reciprocal space. Approximate solutions of the recursion formula can thus be calculated in terms of the variable $\varepsilon = 4 - d$. The results obtained for $\varepsilon = 1$ are in quite good agreement with the best estimates from series expansions, etc. for the various critical exponents (see e.g. Fisher 1974).

These few pages do scant justice to the power of the renormalization group in the theory of phase transitions and critical phenomena. In just a few years, hundreds of papers have appeared applying this technique to a very wide range of problems concerning a diversity of theoretical models, real or hypothetical. It would be quite impossible to deal adequately with this burgeoning field of research without hopelessly distorting the distribution of subjects in the present work. In any case, many excellent reviews of this subject have already appeared and more are promised. And perhaps it is worth remarking that the study of the statistical mechanics of substitutional disorder cannot be completely reduced to the mathematical theory of scaling transformations and is concerned with more questions than the asymptotic behaviour of the various thermodynamic variables in the immediate neighbourhood of the critical point. For these reasons we must regretfully draw back from this fascinating subject and bring this chapter to a close.

6
Thermodynamics of topological disorder

'So let us melt, and make no noise . . .'
Donne

6.1 The linear gas–liquid–crystal

The theory of topological disorder is very much more difficult than that of substitutional disorder on a regular lattice. As we saw in chapter 2, the most immediate problem of describing a liquid or a glass in analytical terms is still not satisfactorily solved. For this reason, the dream of a rigorous mathematical theory of topological phase transitions, such as melting or evaporation, is very far from realization. Most of the methods to be discussed in this chapter are purely phenomenological or make arbitrary mathematical approximations whose error cannot be checked. These deficiencies are so evident in the theory of the thermodynamics of liquid *mixtures* (see § 2.13) that we shall not consider such systems at all in the present chapter.

As usual, there is an exact solution for the *one-dimensional* case. The classical calculation of the free energy of a linear chain of 'atoms' interacting with their neighbours by a pairwise potential is extremely simple (Takahashi 1942; Gürsey 1950). Following the general remarks of § 2.2, the classical configurations of a closed chain of N atoms are mapped by the values $(\xi_1, \xi_2, \ldots, \xi_N)$ of successive spacings (2.5) between atomic centres (the final gap separating the Nth atom from the first, round the loop). Assigning an energy $\phi(\xi_j)$ to the interaction across the jth gap we can at once write down a partition function which factorizes, as in (2.34), into the perfect gas partition function Z_{gas} arising from the kinetic energy terms and a configurational part of the form

$$Q = \int \ldots \int \exp[-\beta\{\phi(\xi_1) + \phi(\xi_2) + \ldots + \phi(\xi_N)\}] \, d\xi_1 \ldots d\xi_N. \quad (6.1)$$

This expression is very like the partition function (5.71) for the classical Heisenberg chain where the angles between successive spins define the configuration. If the variables ξ_j were truly independent and individually bounded we could express (6.1) as the Nth power of a simple integral. But, to obtain any thermodynamic properties, we must impose a 'volume' constraint; the integral (6.1) must be evaluated for values of $(\xi_1, \xi_2, \ldots, \xi_N)$ adding up to a fixed length L. As in (5.146), we insert into the integral a 'selector function'

$$\delta\left\{L - \sum_j \xi_j\right\} \equiv \frac{1}{2\pi i} \oint \exp[s\{L - (\xi_1 + \xi_2 + \ldots + \xi_N)\}]ds \qquad (6.2)$$

where the contour integral runs to the right of the imaginary axis in the complex s plane. Assuming that the interaction potential $\phi(\xi)$ is of finite range, we may now allow each variable ξ_j to run freely from 0 to ∞. Thus we obtain the desired solution:

$$Q(L, \beta) = \frac{1}{2\pi i} \oint \int_0^\infty \cdots \int_0^\infty \exp[\beta\{\phi(\xi_1) + \phi(\xi_2) + \ldots\} -$$

$$- s\{\xi_1 + \xi_2 + \ldots + \xi_N\}] e^{sL} d\xi_1 \ldots d\xi_N \, ds$$

$$= \frac{1}{2\pi i} \oint e^{sL} [\int_0^\infty \exp\{\beta\phi(\xi) - s\xi\} \, d\xi]^N \, ds, \qquad (6.3)$$

which is easily evaluated analytically in simple cases.

Consider, for example, the most elementary case – an assembly of 'hard' O-spheres of diameter D (§ 2.2). The interatomic force function then has the form

$$\phi(\xi) = \begin{cases} \infty & \xi < D \\ 0 & \xi \geqslant D \end{cases} \qquad (6.4)$$

The integral over ξ becomes

$$\int_0^\infty \exp\{-\beta\phi(\xi) - s\xi\} \, d\xi = \int_D^\infty e^{-s\xi} \, d\xi$$

$$= e^{-sD}/s \qquad (6.5)$$

so that the configurational part of the partition function (6.3) reduces to a standard contour integral

$$Q(L, \beta) = \frac{1}{2\pi i} \oint e^{sL}(s^{-1} e^{-sD})^N \, ds$$

$$= \frac{1}{2\pi i} \oint e^{s(L - ND)} s^{-N} \, ds$$

$$= (L - ND)^{N-1}/(N-1)! \qquad (6.6)$$

provided that $L - ND > 0$. Combining this with Z_{gas} and ignoring the difference between $N - 1$ and N, we arrive at the free energy

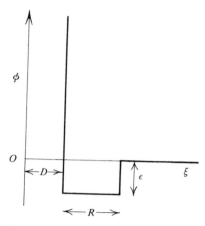

Fig. 6.1. Hard spheres with attractive square-well potential.

$$F = -kT \ln Z = -kT \ln \left[\left(\frac{MkT}{2\pi\hbar^2} \right)^{\frac{1}{2}N} \frac{(L-ND)^N}{N!} \right] \tag{6.7}$$

which is just the standard formula for a one-dimensional perfect gas of N atoms moving on a line of reduced length

$$L' = L - ND. \tag{6.8}$$

In other words, the only effect of giving each atom a finite length D is to exclude this length from the average length $l = L/N$ available for translational motion. The pressure is easily obtained from (6.7) by differentiating with respect to L; the equation of state (Tonks 1936) is simply

$$p(l - D) = kT. \tag{6.9}$$

By arguments analogous to those leading to (5.63) we deduce that the interatomic distances are distributed at random, with precisely the Poisson distribution (2.6).

The hard-sphere model shows no sign of a condensed phase. To encourage condensation, let us introduce a short-range attractive potential of the form shown in fig. 6.1 – i.e. a range $D \leqslant \xi \leqslant D + R$ for which $\phi(\xi) = -\varepsilon$. The analytic form of $Q(L, \beta)$ is now somewhat more complicated, but it is simply a matter of algebra (Gürsey 1950) to derive an equation of state (Herzfeld & Goeppert-Mayer 1934)

$$p(l - D) = kT - pR[e^{pR/kT}\{1 - e^{-\varepsilon/kT}\}^{-1} - 1]^{-1}. \tag{6.10}$$

The isotherms of this equation (in reduced units $T' = kT/\varepsilon$; $p' = pR/\varepsilon$; $l' = (l - D)/R$) are plotted in fig. 6.2. For high temperatures we find nearly

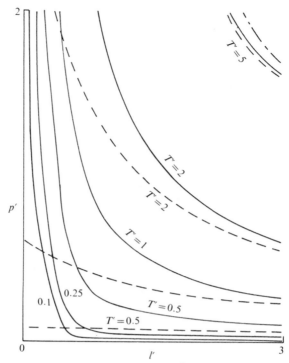

Fig. 6.2. Isotherms of the linear assembly for various values of the reduced temperature T'.

————, Exact solution of Gürsey model.
– – – –, Hyperbolic approximation $l' = T'/p' - \{e^{-1/T'} - 1\}$.
–·–·–·–, Perfect gas law $l' = T'/p'$ for $T' = 5$.

perfect gas behaviour as in (6.9). At low temperatures the isotherms seem to fall into two branches as if describing a condensed phase at high densities which 'vaporizes' below a certain pressure. But the transition between these two regimes is not a mathematical singularity and the analogy with melting, etc. is not valid. As shown more generally by van Hove (1950) any linear gas model with finite range of interaction behaves thermodynamically in just the way we have learnt to expect from §§ 2.4, 5.5; there is no sharp transition to a state of long-range order at any non-zero temperature.

Alternatively (§ 2.2) we may regard the restriction of the atoms to a single dimension, along which they are inevitably ordered in linear succession, as a denial of the possibility of genuine topological disorder. This system is, therefore, always a 'crystal' which can never really melt or evaporate, however hot is may get, however low the pressure.

6.2 The van der Waals approximation

In reality, we observe three distinct 'topological' phases of matter – crystalline, liquid and gaseous. The familiar transitions between these phases cannot be deduced mathematically from a single *a priori* thermodynamic theory. But the various analytic approximations (§ 2.12) for atomic distribution functions and correlation functions used to describe the structure of liquids are all based tacitly upon the assumption that a liquid is a highly condensed gas and, therefore, apply very accurately to low-density vapours. For the moment we consider only the two topologically disordered *fluid* phases, liquid and vapour, and ignore the possibility of a transition to the highly ordered crystalline phase.

From the theory of liquid structure (§§ 2.11–2.12) we proceed to the calculation of the corresponding thermodynamic properties. This looks straightforward: evaluate the partition function (2.33) to obtain the free energy and hence all other thermodynamic variables. But although this general formula was the starting point for the derivation of various relations, such as (2.40), involving the interatomic potential $\phi(1, 2)$ and the successive distribution functions $g(1, 2)$, $g(1, 2, 3)$, etc. we did not actually evaluate Z as such. This calculation (see § 6.4) is much more laborious and uncertain than the use of certain identities that can easily be deduced from (2.34) and (2.35) by differentiation with respect to the macroscopic variables T and V (see e.g. Hill 1950). If, as in (2.32), only two-body interactions are important, then only the radial distribution function $g(R)$ appears explicitly. The *internal energy E*, per atom of an isotropic fluid, is given by

$$E = \frac{kT^2}{N}\left[\frac{\partial \ln Z}{\partial T}\right]_{N,V} = \tfrac{3}{2}kT + \tfrac{1}{2}n \int_0^\infty \phi(R)g(R)4\pi R^2 \, dR, \qquad (6.11)$$

whilst the *virial* defines the pressure in terms of the *force* $\partial\phi(R)/\partial R$ between atoms;

$$P = kT\left[\frac{\partial \ln Z}{\partial V}\right]_{N,T} = nkT - \tfrac{1}{6}n^2 \int_0^\infty R\frac{\partial\phi(R)}{\partial R} g(R)4\pi R^2 \, dR. \quad (6.12)$$

Thus, any formula that might already have been obtained for $g(R)$ as a function of temperature and density $n(=N/V)$, can be used to derive an *equation of state* from (6.12). Another useful relation between macroscopic thermodynamic variables and microscopic statistical functions follows from (4.21) and (4.22): the *isothermal compressibility* is given by

$$\kappa_T \equiv \frac{1}{n}\left(\frac{\partial n}{\partial p}\right)_T = \frac{1}{nkT} + \frac{1}{kT}\int_0^\infty \{g(R) - 1\}4\pi R^2 \, dR. \qquad (6.13)$$

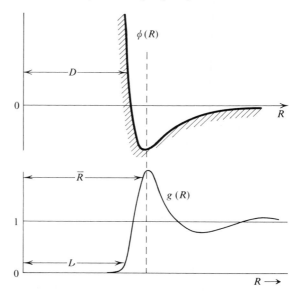

Fig. 6.3. Characteristic behaviour of pair potential $\phi(R)$ and radial distribution function $g(R)$ of a liquid.

But these three relations are not, strictly speaking, independent. For example, if we differentiate the equation of state, we get κ_T: the mutual consistency of the results obtained from (6.12) and (6.13) is a valuable check on the actual accuracy of any approximate formula that may have been assumed for $g(R)$.

A phenomenological equation of state for a fluid can be derived by putting into (6.12) our general knowledge of the behaviour of the radial distribution function $g(R)$. As we saw in §§ 2.6, 2.12, this rises rapidly from zero at some distance of closest approach or 'diameter', D, and attains a substantial peak at the typical nearest neighbour distance $\bar{R}$(fig. 6.3). This peak must come very near the point where the interatomic potential $\phi(R)$ has a minimum. We can thus conveniently divide the range of the integral in (6.12) at $\bar{R}$:

$$p = nkT - \tfrac{1}{6}n^2 \int_D^{\bar{R}} R \frac{\partial \phi(R)}{\partial R} g(R) 4\pi R^2 \, dR - \tfrac{1}{6}n^2 \int_{\bar{R}}^{\infty} R \frac{\partial \phi(R)}{\partial R} g(R) 4\pi R^2 \, dR.$$

(6.14)

Since the interatomic forces in a liquid are usually attractive beyond $\bar{R}$, whilst $g(R)$ oscillates through several shells of neighbours and then tends to unity, the expression

$$a \approx \tfrac{1}{6} \int_R^\infty R \frac{\partial \phi(R)}{\partial R} g(R) 4\pi R^2 \, dR \qquad (6.15)$$

should be positive and practically independent of density and temperature. The other part of the integral is much more difficult to take account of because it describes the effects of the strong repulsive forces that keep the atoms apart. The rise in the radial distribution function from D to $\bar{R}$ is combined with a strong negative function $\partial \phi / \partial R$, so that the whole expression would vary rapidly with density and temperature. What is really important, however, is that the atoms are excluded from one another – that there is a maximum density of packing, or minimum average atomic volume, b, which cannot be passed, whatever the external pressure applied to the fluid. The one-dimensional analogue (6.9) suggests that this *excluded volume* effect can be expressed to the full by writing

$$n\not{k}T - \tfrac{1}{6}n^2 \int_D^R R \frac{\partial \phi}{\partial R} g(R) 4\pi R^2 \, dR \approx \frac{n\not{k}T}{1 - nb}. \qquad (6.16)$$

From (6.14)–(6.16) we immediately derive the famous *van der Waals equation of state* for a fluid:

$$(p + an^2)(1 - nb) = n\not{k}T. \qquad (6.17)$$

Since this formula was first proposed in 1873, it has retained its place as the simplest and most plausible semi-empirical formula for the properties of a fluid in the liquid–vapour regime. The coefficients a and b are easily fitted to experiment over a wide range of densities and temperatures.

It would be wrong, however, to interpret relations such as (6.15) and (6.16) too literally. The same type of equation of state can be derived equally plausibly from (6.11) or (6.13), with somewhat different definitions of the parameters. Various other semi-empirical equations of state have been suggested (see e.g. Fowler & Guggenheim 1949) which fit the observations even better than the van der Waals equation. Discussion of such equations adds little to our understanding of the underlying phenomena.

The van der Waals equation gives a good qualitative description, in elementary algebraic form, of the liquid–vapour phase transition. It is, in fact, closely analogous to the mean field approximation (§ 5.2) for substitutional disorder. As pointed out by Yang & Lee (1952) we can re-interpret an Ising Hamiltonian as the energy of a *lattice gas* (§ 1.5) in which 'up spins' stand for atoms and 'down spins' for vacancies on the sites of a regular lattice. The 'order parameter' $\mathscr{S}$ measures condensation to a phase of atomic density

$$n = \tfrac{1}{2}(1 + \mathscr{S}) \qquad (6.18)$$

under the influence of an energy $-\varepsilon/z$ between pairs of neighbouring atoms. Because the actual density may vary over a wide range, we must keep an 'external magnetic field' H, under the guise of a chemical potential μ. The self-consistency equation (5.5) takes the form

$$(2n-1)=\tanh\{(n\varepsilon+\mu)/2\mathit{k}T\} \qquad (6.19)$$

from which, by standard thermodynamics (e.g. Brout 1965), we derive an equation of state

$$(p+\tfrac{1}{2}\varepsilon n^{2})\left\{\frac{n}{-\ln(1-n)}\right\}=n\mathit{k}T. \qquad (6.20)$$

Interpreting $\tfrac{1}{2}\varepsilon$ as the attraction parameter, a, and noting that the maximum density of packing on the lattice is unity, by definition, we see that the lattice gas equation of state (6.20) is very similar to the van der Waals equation (6.17).

As discussed in detail in standard treatises on statistical physics (e.g. Landau & Lifshitz 1958) the isotherms of (6.17) in the (p, V) plane are apparently continuous for all values of T. But when

$$\mathit{k}T<\mathit{k}T_{\mathrm{c}}=8a/27b, \qquad (6.21)$$

parts of these curves have unphysical thermodynamic properties and must be replaced by straight lines of constant pressure corresponding to gross mixtures of two distinct phases (fig. 6.4). In other words, the system undergoes a transition from a vapour to a liquid phase.

This transition shows up clearly in the lattice gas model. Starting from a low density, and increasing the pressure, we reach a value of the chemical potential μ for which (6.19) has two distinct roots for n, corresponding to two distinct phases in equilibrium. The transition between these phases is mathematically identical with the switch in the sign of the spontaneous magnetization, $\mathscr{S}$, of an Ising ferromagnet as the external field H goes through zero. The *condensation* of a vapour to a liquid can thus be attributed to the attractive forces between the atoms or molecules, without regard to the particular arrangement that these atoms take up in the high-density phase. This point of view is expressed with great clarity by Widom (1967).

Above the critical temperature (6.21) there is no sharply defined point of condensation from vapour to liquid: this continuity of phase for $T>T_{\mathrm{c}}$ is also a well-known physical phenomenon. From this point we ought perhaps to embark upon a detailed description of *liquid–vapour critical phenomena*, which are well represented qualitatively by the van der Waals

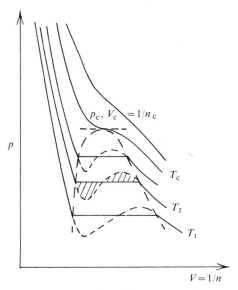

Fig. 6.4. Van der Waals equation of state.

equation. To eliminate arbitrary differences in the properties of particular liquids, we would write down the *critical pressure* and *critical density*

$$p_c = a/27b^2; \quad n_c = 1/3b, \tag{6.22}$$

which provide excellent scale parameters for *reduced* temperature, pressure and density

$$T' = T/T_c; \quad p' = p/p_c; \quad n' = n/n_c. \tag{6.23}$$

In terms of these variables, (6.17) takes the dimensionless form

$$(p' + 3n'^2)(3 - n') = 8n'T', \tag{6.24}$$

which is a rough approximation to the actual behaviour of a wide range of fluids. Quite apart from the precise applicability of the *reduced van der Waals equation of state,* it is observed that most fluids do show very nearly the same thermodynamic behaviour when expressed in terms of the reduced variables (6.23). This *Law of Corresponding States* is a fundamental experimental fact justifying the search for a general theory of fluids that is more or less independent of the details of the shapes and interactions of the constituent molecules.

Since the more striking critical phenomena (density fluctuations, critical opalescence, specific heat anomalies, etc.) are observed only in the neighbourhood of the critical point, it is natural enough to measure the corresponding reduced variables from this point and to determine the appropriate critical exponents. All this is discussed at length in the standard texts (e.g. Fisher 1967*a*; Egelstaff & Ring 1968; Stanky 1971). Indeed, with the lattice gas model as a guide, it is quite easy to write out a list of thermodynamic analogues of magnetization and to discuss critical phenomena in fluids in the same terms as those of Ising ferromagnets and similar substitutionally disordered systems. The critical exponents of fluids are observationally well defined and obey typical scaling laws (§ 5.12) which are very close to those characteristic of magnetic systems (see e.g. Vicentini-Missoni 1972).

It must be emphasized, however, that this parallelism does not imply that we have deduced from first principles a general theory of the condensation of a topologically disordered fluid. The van der Waals equation itself is frankly phenomenological; if it did not, in fact, describe the behaviour of real fluids remarkably well, there would be little reason to accept it. From it we can deduce plausible critical exponents, in the spirit of the classical Landau theory (§ 5.11), but these do not agree with precise experiment. It may turn out, eventually, that the universality hypothesis (§ 5.12) can be trusted and that exact results deduced for, say, the long-range critical behaviour of the lattice gas condensation apply also to proper fluids, but the necessary theorems on this point have not been proved and even the application of the scaling hypothesis is not unequivocal (Nicoll, Chang, Hankey & Stanley 1975).

6.3 The Percus–Yevick approximation

All that we really need to write down the equation of state of a fluid is a structure factor $g(R)$ to be substituted in one of the thermodynamic identities (6.11)–(6.13). In § 2.12 we discussed various analytical approximations for the correlation functions in a liquid and came to the conclusion that the best of these was given by the Percus–Yevick approximation (2.44). In particular, the total correlation function for a *hard-sphere* fluid is given by a simple algebraic formula (2.46), which is related to $g(1, 2)$ by means of (2.41) and (2.42). It is quite easy, therefore (see e.g. Kovalenko & Fisher 1973), to deduce from the virial formula (6.12) a simple equation of state,

$$(p/n \hbar T)_{\text{vir}} = \frac{1 + 2\eta + 3\eta^2}{(1 - \eta)^2}, \tag{6.25}$$

where η is the packing fraction, $\frac{1}{6}\pi d^3 n$, for spheres of diameter d. Alternati-

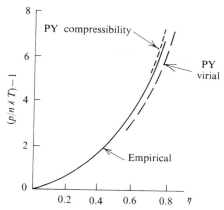

Fig. 6.5. Equations of state for hard spheres.

vely we may substitute in the compressibility identity (6.13), and integrate to get

$$(p/nk\mathcal{T})_{comp} = \frac{1+\eta+\eta^2}{(1-\eta)^3}. \qquad (6.26)$$

Because the PY structure factor is not really an exact solution of the statistical mechanical problem in a liquid, these two expressions are not identical. An intermediate formula (Carnahan & Starling 1969),

$$(p/nkT)_{emp} = \frac{1+\eta+\eta^2-\eta^3}{(1-\eta)^3}, \qquad (6.27)$$

has some merits in terms of the virial series (§ 6.5) and agrees very well with the equation of state deduced by molecular dynamics calculations (fig. 6.5).

But, of course, the thermodynamic properties of the hard sphere gas are not so interesting as those of a real fluid. Comparing with the van der Waals equation (6.17), we note that the parameter $a = 0$, because there is no attractive force in the integral (6.15). This means, simply, that the hard-sphere fluid does not have a gas–liquid phase transition and exhibits none of the critical phenomena discussed in § 6.2. Attempts have sometimes been made to find 'solid-like' (i.e. locally crystalline) solutions of the PY equation or its more complicated generalizations, but these do not seem to be physically meaningful (Katz & Chandler 1972).

The change of free energy along an isotherm can easily be calculated thermodynamically by integrating the pressure with respect to density. In the limit of low density, $\eta \rightarrow 0$, the hard sphere fluid tends to a perfect gas, with the standard entropy

$$S_{gas} = \tfrac{3}{2}k + k \ln\left\{\frac{e}{n}\left(\frac{MkT}{2\pi\hbar^2}\right)^{\frac{3}{2}}\right\}. \qquad (6.28)$$

An analytical expression for the *excess entropy*, S_{exc} can thus be obtained by integrating the right-hand side of (6.25), (6.26) or (6.27) with respect to η. Thus, from the compressibility equation of state (6.26) we get (Lebowitz & Rowlinson 1964; Stroud & Ashcroft 1972)

$$S_{exc} \equiv S_{fluid} - S_{gas} = k\left\{\ln(1-\eta) + \frac{3\eta(\eta-2)}{2(1-\eta)^2}\right\}. \qquad (6.29)$$

Since the hard sphere system has no potential energy this is wholly *configurational entropy*, associated with the number of distinct atomic configurations that may occur in the fluid phase. As defined by (6.29), this is necessarily negative, since a highly condensed fluid is much more 'ordered' by the constraints of random packing than is a gas of point particles.

For a more realistic model of a fluid with a 'soft' interatomic potential $\phi(1, 2)$, the PY equations (2.42), (2.44) must be integrated numerically. The structure factor deduced, for example, from the Lennard-Jones '6:12' potential (2.47) can be put into (6.12) or (6.13) to produce an equation of state (see e.g. Kovalenko & Fisher 1973). The isotherms thus derived look very like the van der Waals isotherms (6.17) and exhibit the typical characteristics of a critical liquid–vapour phase transition. It turns out, for example, that the PY equations have no solution in a region of temperature and density that we should suppose to be the phase separation region, where no uniform fluid state can exist. At low temperatures there is considerable deviation between the 'compressibility' and 'pressure' results, but the mean of these is close to the equation of state for an ideal LJ system computed by molecular dynamics. Even better agreement is obtained (Henderson, Barker & Watts 1970) by using (6.11) to calculate the internal energy, from which the free energy and pressure can be obtained by integration with respect to T and differentiation with respect to n. The significance of this empirical agreement with real liquid properties is obscured by the phenomenological status of the PY theory itself, whose *a priori* justification still leaves much to be desired (§ 2.12).

6.4 Perturbation methods

We already know a great deal about the structure and thermodynamic properties of certain model systems such as the hard-sphere fluid. These provide a starting point for a convergent series expansion of the same properties for a real fluid with more complex interatomic forces. This is the essence of the *perturbation method* introduced by Zwanzig (1954).

The idea is simple. Suppose that the Hamiltonian of an assembly of N atoms may be written as the sum of two parts,

$$\mathcal{H}(1 \ldots N) = \mathcal{H}^{(0)}(1 \ldots N) + \mathcal{H}^{(1)}(1 \ldots N). \qquad (6.30)$$

Suppose, moreover, that $\mathcal{H}^{(0)}$ is the Hamiltonian of a simple 'unperturbed' fluid whose structure and thermodynamic properties are well known. Thus, for example, we know the partition function

$$Z^{(0)} = \frac{1}{V^N} \frac{1}{N!} \int \ldots \int \exp\{-\beta \mathcal{H}^{(0)}(1 \ldots N)\} \, \mathrm{d}1 \ldots \mathrm{d}N, \qquad (6.31)$$

and can thus easily calculate the probability distribution of other configurations of this *reference fluid* over an equilibrium ensemble.

For the 'perturbed' system, therefore, we may write

$$Z = \frac{1}{V^N} \frac{1}{N!} \int_V \ldots \int_V \exp\{-\beta \mathcal{H}^{(0)}(1 \ldots N)\}\exp\{-\beta \mathcal{H}^{(1)}(1 \ldots N)\} \, \mathrm{d}1 \ldots \mathrm{d}N$$

$$= Z^{(0)}\langle \exp\{-\beta \mathcal{H}^{(1)}\}\rangle, \qquad (6.32)$$

where the average is taken over this reference ensemble. But this is merely a slight modification of the formalism for the moment expansion (5.163). Taking the logarithm of (6.32) we obtain the free energy $F^{(0)}$ of the unperturbed system, together with an expression which can be expanded as a cumulant series (5.165):

$$F = F^{(0)} - kT \sum_j \frac{(-\beta)^j}{j!} \alpha_j, \qquad (6.33)$$

where, as in (5.166) and (5.167),

$$\alpha_j \equiv \langle \{\mathcal{H}^{(1)}\}^j \rangle_{\mathrm{cum}} \qquad (6.34)$$

is a cumulant average over the reference ensemble.

For the usual pairwise interatomic forces, the leading term α_1 in (6.33) is simply an average of the perturbing potential $\Phi^{(1)}(R)$ over the unperturbed radial distribution function $g^{(0)}(R)$,

$$F = F^{(0)} + \tfrac{1}{2}n^2 V \int \Phi^{(1)}(R) g^{(0)}(R) 4\pi R^2 \, \mathrm{d}R. \qquad (6.35)$$

We usually take as our reference system a hard-sphere fluid. If $F^{(1)}$ represents simply the attractive long-range forces between the atoms, this integral, like the van der Waals integral (6.15), should be more or less independent of temperature. Indeed, if we approximate crudely to $F^{(0)}$ by an excluded volume term like (6.7), we arrive at the van der Waals equation (6.17).

But this formula can be made quite realistic by using for $g^{(0)}(R)$ any good representation of the RDF of a hard-sphere gas, such as the PY formula (2.46) and a corresponding approximation, such as (6.29), for the 'unperturbed' free energy. The results thus obtained for a 'square well' attractive potential outside hard cores (Barker & Henderson 1967) agree excellently with molecular dynamics computations over a wide range of densities and temperatures.

It is not difficult to write down a formal expression for the next term in the cumulant series (6.33), in terms of the correlation functions $g^{(0)}(1, 2)$, $g^{(0)}(1, 2, 3)$ and $g^{(0)}(1, 2, 3, 4)$ of the reference fluid. Since these are not known accurately, the contribution of this term can only be evaluated approximately using, for example, the superposition approximations (2.27) and (2.28) (Smith, Henderson & Barker 1970). It is perhaps worth noting, however, that higher-order correlation functions probably contribute very little to the higher terms in the perturbation series, by virtue of the fundamental theorem (§ 5.10) on the vanishing of the cumulant average of the product of statistically independent variables. This would be true, for example, of $g^{(0)}(1, 2, 3, 4)$ except when all four atomic centres are clustered closely together.

The problem of applying the perturbation method to 'softer' interatomic potentials, such as LJ '6:12', is more subtle, because of the arbitrariness that now arises in the choice of a hard-sphere diameter for the reference fluid. Barker & Henderson (1972) review the various devices that have been used to separate the actual potential into unperturbed and perturbing parts to ensure the best convergence in the computation. Many of these calculations give results that are numerically in excellent agreement with model computations, but we should learn little more about the theory of disordered systems by going into technical details.

The validity of a perturbation approximation such as (6.35) can be assessed by reference to the *Gibbs–Bogoliubov inequality* (Isihara 1968) which provides general variational principles for estimates of the free energy or entropy of any statistical–mechanical system. This theorem has been applied, for example, by Watabe & Young (1974) to obtain an equation of state for liquid metals that does not depend explicitly on the hard-sphere pressure (6.25–6.27) whilst using the hard-sphere distribution function (2.46) as a parametric representation of $g(R)$. This method also yields an experimentally verifiable relationship between the entropy and the structure factor for many liquid metals (Silbert, Umar, Watabe & Young 1975).

6.5 The virial series

In the theory of substitutional disorder, an important part is played (§ 5.10) by graphical expansions of the free energy at high and low temperatures. The corresponding technique for a topologically disordered system is not so powerful: the successive terms of the high-temperature/low-density series are much more difficult to evaluate and there is no convergent expansion for the properties of the liquid phase. Nevertheless, the attempt to extend the thermodynamic properties of an *imperfect gas* into the regime of a condensed fluid is a standard constituent of the statistical theory of liquids and is often used to provide reference formulae for the assessment of more heuristic theories.

We might, for example, use the series (6.33), treating the whole of the interatomic potential $\sum_{i<j} \phi(i,j)$ as the 'perturbing Hamiltonian', $\mathscr{H}^{(1)}$ acting on the perfect gas itself as reference fluid. This is entirely analogous to the high-temperature cumulant expansion (5.165) for the Ising model, valid at high temperatures. Each coefficient α_j defined by (6.34) would be a sum of integrals of the form

$$\int \ldots \int \phi(1,2)\phi(2,3)\ldots \mathrm{d}1 \ldots \mathrm{d}l \qquad (6.36)$$

where the l points $\mathbf{1} \ldots \mathbf{l}$ form a *connected graph* with j links, possibly repeated.

A more compact series is provided, however, by the *Ursell–Mayer representation* (see e.g. Mayer & Mayer 1940; Hill 1956), where the Boltzmann factor in the partition function is written in a form analogous to (5.175):

$$\exp\left\{-\beta \sum_{i<j} \phi(i,j)\right\} = \prod_{i<j}(1 + f_{ij}), \qquad (6.37)$$

where

$$f_{ij} \equiv \exp\{-\beta\phi(i,j)\} - 1. \qquad (6.38)$$

The advantage of this representation, as in (5.177), is that a given link can occur at most once in any product of the f_{ij} generated by expanding (6.37). The potential multiplicity of graphs is thus greatly reduced: going from $\phi(i,j)$ to f_{ij} we automatically sum all ladder diagrams (Brout 1965).

The conventional derivation of the coefficients in the series expansion of the free energy is quite complicated but can be bypassed, once more, by

quite elementary applications of the theory of cumulants (Kubo 1962). To
go from (5.163) to (5.165) we actually used a defining principle:

$$\left\langle \exp\left(\sum_j X_j \right) \right\rangle = \exp\left\langle \exp\left(\sum_j X_j \right) - 1 \right\rangle_{\text{cum.}}$$

This principle applies to *levelled exponentials* such as (6.37), which we write

$$\prod_{i<j} (1 + f_{ij}) \equiv \exp_{\text{L}}\left(\sum_{i<j} f_{ij} \right), \tag{6.39}$$

on the understanding that the 'levelling' operation erases terms in which
any f_{ij} occurs to a higher power than unity. The configurational free energy
can thus be expressed in the form

$$\begin{aligned}
F_1 &\equiv -\ell T \ln \left\langle \exp\left\{ -\beta \sum_{i<j} \phi(i,j) \right\} \right\rangle \\
&= -\ell T \ln \left\langle \exp_{\text{L}}\left(\sum_{i<j} f_{ij} \right) \right\rangle \\
&= -\ell T \ln \exp\left\langle \exp_{\text{L}}\left(\sum_{i<j} f_{ij} \right) - 1 \right\rangle_{\text{cum}} \\
&= -\ell T \left\langle \prod_{i<j} (1 + f1_{ij}) - 1 \right\rangle_{\text{cum}}. \tag{6.40}
\end{aligned}$$

The general term in the expansion of this product is a cumulant average
of a product of the f_{ij} over a set of pair bonds. By the fundamental cumulant
theorem on statistical independence, this average vanishes unless these
bonds form a *connected* graph. Consideration of the fact that each f_{ij}
depends only on the relative positions of the two atoms shows that we need
only count graphs that are *irreducible* – i.e. that cannot be separated by
cutting a single bond. Counting the different ways of choosing $(k+1)$
atoms to put on the $(k+1)$ nodes of such a graph, we may expand (6.40) in
the form

$$F_1 = -\ell T \sum_{k-1}^{N-1} \binom{N}{k+1} \sum_{\text{irreducible}} \left\langle \prod_{(k+1)} f_{ij} \right\rangle_{\text{cum}}. \tag{6.41}$$

But the ordinary ensemble average associated with such a graph is a
multiple integral of the form

$$\left\langle \prod_{(k+1)} f_{ij} \right\rangle = V^{-k} \int \ldots \int \prod_{(k+1)} f_{ij} \, \mathrm{d}\mathbf{1} \ldots \mathrm{d}\mathbf{n} \tag{6.42}$$

since there are only k relative coordinates for a cluster of $(k+1)$ atoms.
From the form of (6.38) we see that the integrand vanishes unless the atoms
are all quite close together, so that this power of $1/V$ is dominant. The
contributions to the corresponding cumulant average from lower moments
(cf (5.167)) are also of higher order in $1/V$ and may, therefore, be dropped

in the limit as $V \to \infty$. In the thermodynamic limit, where we also let $N \to \infty$, keeping $n = N/V$ constant, we get a simple expression for (6.41):

$$F_1 = -N \not k T \sum_{k=1}^{\infty} \frac{\beta_k}{k+1} n^k, \tag{6.43}$$

where the *irreducible cluster integral* is defined by

$$\beta_k = \frac{V^k}{k!} \sum_{\text{irreducible}} \left\langle \prod_{(k+1)} f_{ij} \right\rangle. \tag{6.44}$$

This is the well-known *Ursell–Mayer expansion*. The same result may be obtained by graph-theoretical methods from the Uhlenbeck theorem (5.168)–(5.172), which is deliberately tailored to produce the Mayer theory as a special application (Uhlenbeck 1960; Uhlenbeck & Ford 1962).

For practical purposes it is convenient to differentiate the total free energy with respect to the density, generating the *virial series*

$$\frac{p}{n \not k T} = 1 - \sum_{k=1}^{\infty} \frac{k}{k+1} \beta_k n^k. \tag{6.45}$$

This expansion in powers of the density measures the deviation from perfect gas behaviour as we go away from the limit $n \to 0$. The cluster integrals (6.44) are thus explicit formulae for the *virial coefficients*

$$B_k = -\frac{k}{k+1} \beta_k. \tag{6.46}$$

Although very clearly defined by (6.44), the virial coefficients are not easy to calculate (see e.g. Alder & Hoover 1968). The number of distinct irreducible graphs goes up very rapidly as k exceeds 4 or 5 and each graph represents a complicated multi-dimensional integral (6.42) whose integrand depends on the interatomic potential and on the temperature. The only system that has been studied in detail is our old friend, the hard-sphere gas, where each f_{ij} equals -1 or 0 according as R_{ij} is less than or greater than the hard-sphere diameter. The virial coefficients are thus independent of the temperature and have been computed exactly up to B_7 for both the three-dimensional hard-sphere gas and its two-dimensional analogue, an assembly of *hard discs* (Ree & Hoover 1964). These coefficients can then be compared numerically with those obtained by expanding one or other of the compact analytical representations of § 6.3 in powers of n, although there is no definite principle by which the result of such a comparison can be regarded as a mathematical test of the accuracy of the approximation in question. Nor is it clear what precisely is proved by continuing the series arbitrarily as an expansion of a Padé approximant.

For a genuine interatomic potential with attractive forces outside a repulsive core, the virial coefficients depend on the temperature. The condition $(\partial p/\partial n)_T > 0$ for thermodynamic stability becomes

$$1 - \sum_k k\beta_k(T)n^k > 0. \tag{6.47}$$

At low temperatures this condition can only be satisfied up to some density $n_c(T)$ for which the left-hand side of (6.47) vanishes; this we interpret as the density at which condensation occurs at the temperature T. As shown by Mayer & Mayer (1940), this is also the density at which a cluster expansion of p/nkT in powers of the fugacity diverges – i.e., where most of the fluid is to be found in very large, condensed, clusters. At high temperatures, however, the higher virial coefficients $\beta_k(T)$ can become negative, so that the condition (6.47) can be satisfied for all values of n and there is no condensation. The temperature T_c separating these two regimes is then identified as the critical temperature of the fluid. But the question of the convergence of the virial expansion seems obscure. Estimates of the radius of convergence of the fugacity series for various types of potential have been derived by Groenveld (1962), Lieb (1963) and Penrose (1963) without establishing a connection with any thermodynamic singularity such as condensation to a liquid or solid.

In the original exposition of the Mayer theory it was argued that the cluster expansion ought also to be valid in the high-density liquid phase. The general theory of Yang & Lee (1952) showed that this was not compatible with the analytical properties of the grand partition function considered as a function of a complex fugacity variable. The virial series is meaningless beyond the vapour/supercritical region of the phase diagram and provides no information about the liquid state.

6.6 Computer simulation methods

As emphasized in § 2.11, the study of the properties of computer models plays a very important part in the theory of the liquid state. Such models bridge the gap between the phenomenological analytical theories discussed in §§ 6.3, 6.4 and observed physical reality. Thus, simple approximations such as the PY formula for a hard-sphere fluid can be tested against the 'experimental' properties of the corresponding computer model, whilst the effects of more complicated interatomic forces can be simulated to a high degree of accuracy on the computer and compared directly with, say, the actual behaviour of liquid argon. The very considerable effort involved in constructing and running such computer programs is more than justified by their results, checking the validity of microscopic models, theoretical assumptions and empirical approximations in this very uncertain field.

Computer simulation is particularly valuable in the theory of *melting* (§ 6.7). As we saw in § 2.7, the distinction between a topologically disordered system such as a liquid or a glass and an assembly of ordered microcrystals does not show up at all clearly in the two- or three-body atomic distribution functions, so that the attempt to describe this order–disorder transition in the standard statistical language faces formidable obstacles. The subtle geometrical differences between random close packing and local crystalline order are properly weighted in the real-space computer model, so that the phenomenon takes place naturally in this representation and can be studied in 'microscopic' detail.

The most direct approach is the method of *molecular dynamics,* where the individual trajectories of a few hundred molecules are followed over a period of time, making full allowance for their mutual interactions and collisions. In the pioneering work of Alder & Wainwright (1957, 1959, 1960) the main emphasis was on assemblies of hard spheres and hard discs, with some applications to the case of a square-well potential outside a hard core. To start with, the molecules are supposed to be arranged on a regular lattice, but moving in random directions, each with the same kinetic energy. The procedure is to search through all possible pairs of molecules looking for the most imminent collision. All the molecules are now allowed to move with constant velocities until this collision takes place. After an appropriate change in the velocities of the colliding pair the program looks for the next collision – and so on. Within the limits of numerical accuracy, therefore, the computation is exact and may be allowed to proceed for many thousands of collisions by each molecule, corresponding perhaps to a fraction of a nanosecond in the history of a real fluid.

For softer interatomic potentials, such as $\phi_{LJ}(R)$, (2.47), the computation cannot be so exact. A short step time, h, is chosen, during which each particle moves in accordance with its initial momentum in response to the accelerating forces from other particles. Thus, the position of the ith molecule at time $(t+h)$ is given (approximately) by

$$\mathbf{R}_i(t+h) = \mathbf{R}_i(t) + \{\mathbf{R}_i(t) - \mathbf{R}_i(t-h)\}h - \tfrac{1}{2}h^2 \sum_{j \neq i} \nabla \phi_{ij}\{\mathbf{R}_{ij}(t)\}. \qquad (6.48)$$

This program (Rahman 1964, 1966) can be speeded up considerably (Verlet 1967, 1968) by ignoring the forces from molecules beyond a short distance. Any such computation is limited in accuracy by the size of the sample that can be studied. To avoid boundary effects, the assembly of N molecules is treated as if it were in one cubic cell of a periodic array of identical cells, in each of which the molecules go through the same motions. That is, the program ensures that whenever the force term in (6.48) includes contributions from the 'ghost' molecules in neighbouring cubes and a molecule is ejected from the basic cube, a new molecule automatically enters through the opposite face. By this device, even so small a sample as $N = 32$ hard spheres can reproduce many of the properties of an infinite system (fig. 6.6). The models of Rahman and Verlet, with $N \sim 1000$, must be very close indeed to the thermodynamic limit $N \rightarrow \infty$ (see e.g. Hubbard 1971).

The most elaborate application of molecular dynamics (Rahman & Stillinger 1971; Stillinger & Rahman 1972, 1974) is to the structural and thermodynamic properties of water (§ 2.8). The program proceeds essentially as in (6.48) with the addition of rotational motion of each charge tetrahedron under the influence of its neighbours.

The great advantage of the molecular dynamics method is that the motion of individual molecules is simulated, thus yielding important information about self-diffusion, auto-correlation of molecular velocities, correlation of motion of neighbouring molecules, etc. But for the calculation of thermodynamic properties one simply takes time averages along the dynamical path, which are equated, by the ergodic hypothesis, to the corresponding equilibrium ensemble averages. The validity of this procedure is not in serious doubt, though its accuracy for small values of N has to be considered carefully.

From this point of view, molecular dynamics is simply a means of constructing an ergodic path through a microcanonical ensemble in the phase space of the model assembly. Any equilibrium property could equally well be computed as an average over the canonical *configurational ensemble* (2.36), with probability distribution

$$g(1 \ldots N) \propto \exp\{-\beta U(1 \ldots N)\}, \tag{6.49}$$

without necessarily obeying the equations of motion along the path. For a small specimen, such an average can be calculated very efficiently by the *Monte Carlo method* (Metropolis *et al.* 1953; see Wood 1968 for an extensive review).

One might at first think that one should generate a random sample of

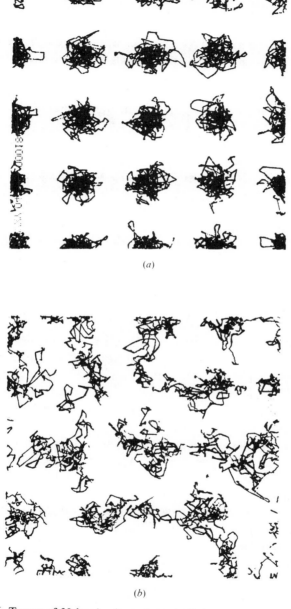

(a)

(b)

Fig. 6.6. Traces of 32 hard spheres in periodic boundary conditions:
(a) solid phase; (b) fluid phase (Alder & Wainwright 1959).

points in the N-dimensional space of the variables $(\mathbf{R}_1 \ldots \mathbf{R}_N)$, and compute the weight factor (6.49) for each one. But, as can easily be seen in the extreme case of a hard-sphere fluid, this is very inefficient. Almost every configuration that might thus be generated would have at least two spheres overlapping, making $u(1 \ldots N)$ infinite and thus contributing nothing to the ensemble. It is much more efficient to start from some 'allowed' configuration and to make a succession of random moves, one molecule at a time, rejecting configurations that violate the constraints as soon as they occur.

Suppose, then, that we have reached the rth configuration, $|r\rangle$, where the molecules are at $\mathbf{R}_1, \mathbf{R}_2, \ldots, \mathbf{R}_N$. The heart of the program is a *random step* generator that moves the ith molecule by the vector distance

$$\mathbf{R}'_i - \mathbf{R}_i = \delta(\xi_1, \xi_2, \xi_3) \qquad (6.50)$$

where the components are random numbers in the range $-1 \leqslant \xi_1, \xi_2, \xi_3 \leqslant 1$. The step length δ is chosen arbitrarily to optimize the whole procedure: it must be large enough to traverse the whole ensemble in a reasonable number of steps, but must not carry the molecule into a region where it is very unlikely to find room. The molecule to be moved at each step can either be decided by regular cycling through 1 to N or may also be chosen at random.

In the hard-sphere case, if the configuration $|r\rangle$ produced by the move (6.50) does not infringe the geometrical constraints, it becomes the next number $|r+1\rangle$ in the ensemble. But, if the moved molecule overlaps other molecules, the step is rejected and we take $|r+1\rangle$ to be identical with the unmoved configuration $|r\rangle$. In this manner, over very large numbers of trials, the sample assembly makes a *random walk* through the space of allowed configurations, all with equal weight, and thus generates a statistical distribution over which structural and thermodynamic properties can be deduced by simple averaging.

In the more realistic case where U is a continuous function of the molecular coordinates, a more complex decision procedure is called for. Having made the tentative step (6.50), we compute the change of energy, ΔU, in going from $|r\rangle$ to $|r'\rangle$. We are then faced with the following alternatives:

(i) If $\Delta U \leqslant 0$ we take $|r+1\rangle = |r'\rangle$: the new configuration, being of lower energy, is suitable as the next point along our walk.

(ii) If $\Delta U > 0$, we move to $|r'\rangle$ with probability $\exp\{-\beta \Delta U\}$. This is done by drawing a random number in the range $0 \leqslant \xi \leqslant 1$ and setting

$$|r+1\rangle = \begin{cases} |r'\rangle & \text{if } \xi \leqslant \exp\{-\beta\Delta U\} \quad (6.51a) \\ |r\rangle & \text{if } \xi > \exp\{-\beta\Delta U\}. \quad (6.51b) \end{cases}$$

Thus, the probability of moving out of the low-energy valleys in configuration space is automatically kept down at each successive step. Notice the accountancy rule that the unmoved state $|r\rangle$ is to be counted as 'new' configuration $|r+1\rangle$ even if the move is rejected as in (6.51b).

It is easy to prove that the random walk generated by this procedure samples a canonical ensemble, each particular configuration having probability (6.49) of being counted. Suppose, for example, that $|l\rangle$ and $|m\rangle$ are two 'adjacent' configurations that can be reached from one another by the step generator (6.50) with equal *a priori* transition probabilities $P_{lm} = P_{ml}$. Suppose, however, that the energy difference $\Delta U_{lm} = U_m - U_l$ is positive. If these configurations occur with relative frequencies v_l and v_m in the ensemble then the rules (6.51) will give rise to a net number of transitions between them

$$T_{lm} = v_l P_{lm} \exp(-\beta\Delta U_{lm}) - v_m P_{ml}$$
$$= P_{lm} \exp(-\beta U_m)\{v_l \exp(\beta U_l) - v_m \exp(\beta U_m)\}. \quad (6.52)$$

These transitions will tend to restore the distribution to an equilibrium canonical ensemble where

$$v_l/v_m = \exp(-\beta U_l)/\exp(-\beta U_m). \quad (6.53)$$

The procedure is thus quite straightforward in principle although, of course, demanding considerable expertise in practice. Naturally, as in the molecular dynamics method, the starting point is a volume containing a regular lattice of N molecules with periodic boundary conditions. If $U(1 \ldots N)$ is the usual pairwise potential energy, the calculation at each step is helped by curtailing the range of the interatomic potentials; but there is no obstacle in principle to the inclusion of triplet potentials in the total energy. The accuracy of the results depends on N, but is quite satisfactory in typical programs with $N \sim 1000$. The Monte Carlo method has now been used to determine the thermodynamic properties of such a wide variety of fluid and solid model systems (see e.g. McDonald & Singer 1970; Ree 1971) that a list of the applications would not be very instructive.

According to the expert practitioners (e.g. Alder & Hoover 1968; Wood 1968) there is little to choose between the Monte Carlo and molecular dynamics methods in terms of computational efficiency. But with both methods a serious issue of principle cannot be avoided at the high-density limit. Consider an FCC lattice of hard spheres, nearly in contact. There is no escape, by a dynamical path or by a random walk, to any other configuration where, say, some spheres have changed sites or where the system has taken up any other equally close-packed arrangement such as HCP. In these circumstances the full equilibrium ensemble is not being sampled; the path through phase space is only *quasi-ergodic*. This difficulty is intimately connected with the melting phenomenon and with the physical existence of metastable states in quasi-equilibrium and has not been resolved within the sphere of model simulations of order–disorder phenomena.

6.7 Melting

The most notable achievement of computer simulation techniques has been to clarify the circumstances of *melting* or *fusion*. As pointed out in § 2.11, an assembly of mutually impenetrable objects may be arranged as a regularly ordered 'solid' or it may form a topologically disordered, random, close-packed 'liquid' at rather lower density. Computer 'experiments' have shown unequivocally that these two states do not merge continuously into one another, but are nearly stable alternatives over finite range of densities and temperatures (fig. 6.7). This was shown originally for quite small assemblies of hard spheres, by both molecular dynamics (Alder & Wainwright 1957) and by Monte Carlo computations (Wood & Jacobson 1957).

For direct comparison with experiment a more realistic model is required. Computations with the LJ 6:12 interaction (2.47) (Ross & Alder 1966; Hansen & Verlet 1969; Hansen 1970) are in satisfactory agreement with the melting and critical phenomena of argon, thus confirming the physical validity of the basic model. These computations also demonstrate that melting is a topological order–disorder transition associated with the mutual impenetrability of the atomic cores. The attractive part of the interaction at greater distances merely produces condensation of the fluid to a dense liquid (§ 6.2), thus bringing the minimum pressure required to observe the melting–freezing transition down to a low value (i.e. to the vapour pressure at the triple point). From a theoretical viewpoint the melting process is best observed, in all its complexity and all its purity, in the hard-sphere model.

Careful studies (see e.g. Wood 1968) have shown, however, that a finite computer assembly of hard spheres ($N < 1000$) does not simulate all the well-known features of melting (fig. 6.8). As in § 6.3, for example, the equation of state in each phase can be represented as a single curve relating the *reduced pressure* $\phi = p/nkT$ to the packing fraction η. Suppose we move

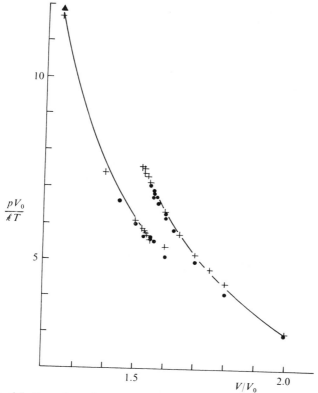

Fig. 6.7. Equation of state for hard spheres from computations of Alder & Wainwright (1957) (− and +) and Wood & Jacobson (1957) (•). V_0 is close-packed (crystalline) volume.

slowly from the close-packed solid phase to lower densities; the system does not 'jump' to the liquid line at a precisely defined density, but may make this transition, apparently at random, at any point in a wide range. As remarked by Alder & Wainwright (1960), 'the accessible region of phase space might be considered as consisting of two pockets, one solid, the other liquid, connected by a narrow passage. The narrow passage represents the relatively improbable cooperative motion among particles necessary to let a particle escape out of a cell and turn a solid into a liquid.' The computer model thus lacks the sharp *melting point* which is so characteristic of almost all crystalline materials.

The converse process of *freezing* to a regular solid is even more haphazard: the liquid phase of the model can be slowly 'compressed' or 'cooled' to

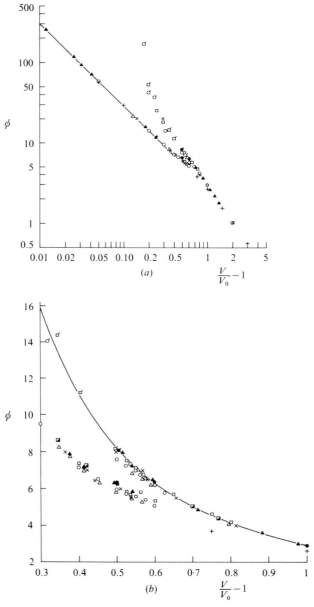

Fig. 6.8. Equation of state for hard spheres: (*a*) over a wide range of densities; (*b*) over the 'melting' region. Data from various computations (Wood 1968).

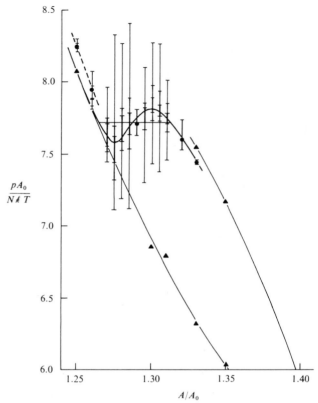

Fig. 6.9. Equation of state for hard discs (Alder & Wainwright 1962).

a glassy amorphous material which seldom 'crystallizes' spontaneously even after a large number of computational steps (Hoover & Ree 1967). This difficulty of nucleating a crystal out of a metastable liquid phase is well known physically and explains, for example, the apparent stability of glassy alloys (§ 2.13).

It is interesting to note, however, that the liquid–solid transition for a two-dimensional model assembly of hard discs (Alder & Wainwright 1962; Hoover & Alder 1967; Wood 1968) is fairly well defined and approximately reversible. For $N \sim 1000$, the liquid and solid portions of the equation of state appear to be continuously linked by a line that lies near to a characteristic 'van der Waals loop' (fig. 6.9). Although this loop is probably an artefact of the finite size of the model, it seems that the topological and geometrical dividing line between a 'liquid' and a 'solid' is much less sharp

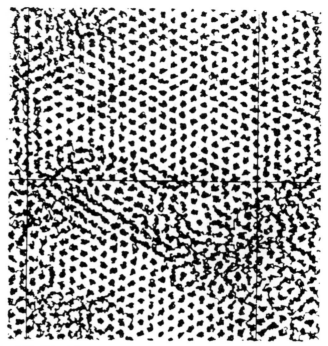

Fig. 6.10. 'Liquid' and 'solid' regions coexisting in a hard-disc assembly (Alder & Wainwright 1962).

in two dimensions than in three. The liquid already contains a high density of nucleation centres for crystallization into a hexagonal network (§ 2.11), whilst the solid can contain much more microcrystalline or dislocation disorder (§§ 2.5, 2.6) than is permissible in a three-dimensional crystal. It is possible, indeed (§ 2.4), that this 'solid' phase does not have long-range order. Along the *tie-line* between liquid and solid, therefore, we may find a mixed phase (fig. 6.10) where liquid and solid regions coexist in equilibrium within the same sample. These features are governed, of course, entirely by the dimensionality of the assembly and would not be expected in a real three-dimensional system.

Returning again to the hard-sphere model, we observe (fig. 6.8) that transitions from the solid to liquid phase occur spontaneously over an 'interval of confusion' in the packing density ranging from about $\eta \approx 0.50$ to $\eta \approx 0.45$. At constant pressure this melting transition is accompanied by an expansion of about 10 per cent in the volume. Notice at once that the density in this range is only about $\frac{2}{3}$ the regular close-packed density ($\eta_{CP} = 0.74$) and is well below the maximum random close packed density, $\eta_{RCP} = 0.63$–0.64, found for the Bernal model (§ 2.11). The pressure certainly rises steeply as we approach η_{RCP} along the liquid line but the Bernal limit does not, apparently, play an important part in determining the density of melting: the regular solid phase is favoured thermodynamically at a much lower density than that at which it becomes necessary for geometrical reasons.

The size of sample that can be treated in a molecular dynamics or Monte Carlo computation in three dimensions (i.e. $N \sim 1000$) is not sufficient to locate the melting point precisely. To draw the tie-line between the two phases we must fall back upon thermodynamic arguments. The elementary condition for a *first-order* phase transition is that the two phases must have the same free energy – i.e. in the hard-sphere case, the same entropy. For the fluid phase we have already seen that S_{fluid} can be calculated from the PY formula as in (6.29) or from any similar semi-empirical formula, such as (6.27), that gives an equation of state in good agreement with the computer simulations over the range of densities in which we are interested.

But the calculation of the absolute entropy of a *crystalline* assembly of hard spheres is made difficult by the necessity of tying this to the entropy of some standard reference state, such as the perfect gas entropy (6.28). This is achieved very directly in the *single-occupancy model* of Hoover & Ree (1967, 1968). In essence this is a straightforward hard-sphere assembly with the additional constaint that each atomic centre may not leave its assigned cell in an underlying close-packed lattice (fig. 6.11). Note that this does not prevent atoms in neighbouring cells from bumping into one another through the cell walls – indeed, at high packing density, the atoms are constrained mainly by their neighbours rather than by the background lattice. Thus, by expanding the lattice constant relative to the atomic diameter, we carry the system continuously from a close-packed crystal to something very like a gas whose absolute entropy S_{SO} at low densities can be computed by cluster expansion techniques (§ 6.5). The equation of state for this model can be computed quite accurately over the whole range of densities by the Monte Carlo method (fig. 6.12). We can thus calculate the absolute entropy of the close-packed, hard-sphere solid by integrating the pressure along this curve (cf. § 6.3).

From this calculation, the melting point of the hard-sphere assembly can be located with great accuracy. At a packing density

$$\eta_L = 0.494 \pm 0.002, \tag{6.54}$$

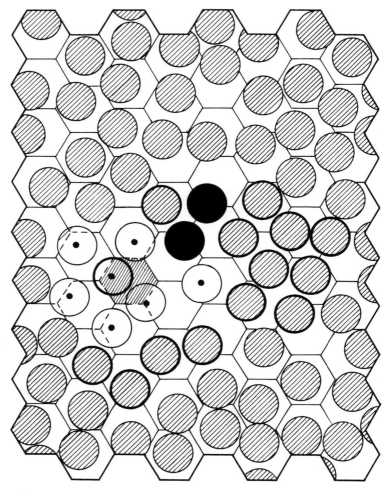

Fig. 6.11. Single-occupancy model. Although the centre of each atom is confined to its own cell, atoms in neighbouring cells can interact.

the liquid is in thermodynamic equilibrium with the close-packed solid at density

$$\eta_S = 0.545 \pm 0.002. \tag{6.55}$$

The melting temperature and pressure are related by

$$p_M = (8.27 \pm 0.13)n_0 \, \text{\textit{k}} T_M \tag{6.56}$$

where n_0 is the particle density per unit volume at regular close packing.

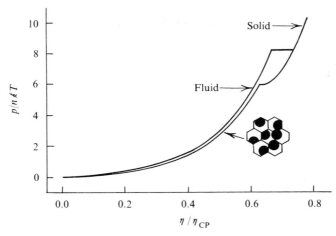

Fig. 6.12. Hard-sphere equations of state. The single-occupancy isotherm extends the solid phase isotherm smoothly to low densities (Hoover & Ree 1968).

These are fundamental numbers in the microscopic theory of melting considered as a topological order–disorder transition.

Notice, for example, that η_L is at the high-density end of the 'interval of confusion' in fig. 6.8. This shows that melting becomes thermodynamically favourable almost as soon as a passage in configuration space opens up between the solid and liquid phases. Perhaps this explains the well-known physical fact that a crystalline solid cannot be superheated above its regular melting point, whereas a liquid can easily be supercooled.

At a packing density around $\eta = 0.45$, the pressure of the single-occupancy model (fig. 6.12) shows something like a cusp. This is interpreted by Hoover & Ree (1968) as the point at which the solid would become *mechanically* unstable; if it were not for the artificial constraints of the cell walls, the crystal structure would disintegrate. In a theory of melting based upon the properties of the solid phase alone this would probably be regarded as the onset of melting. As we see from (6.54)–(6.56), however, this instability has no physical relevance since it would occur at a lower density and pressure, or higher temperature, than the actual thermodynamic transition. This is a very powerful argument against theories of melting based upon the catastrophic multiplication of defects such as vacancies or dislocations (§ 2.5) and also suggests that the search for *premelting phenomena* in the solid phase just below the melting point is likely to prove abortive. Notice, incidentally, that this separation of mechanical from thermodynamic instability does *not* seem to be found in the two-dimensional hard-disc model. This is consistent with the point already noted that this system does not seem to freeze properly into a phase with long-range order and with the observation (§ 2.5) of high densities of dislocation disorder at the transition point.

The method of Hoover & Ree has been applied to the LJ 6 : 12 model with satisfactory results (Hansen & Verlet 1969; Hansen 1970). In their calculation of the melting curve for sodium, Stroud & Ashcroft (1972) used the PY hard sphere entropy (6.29) for the liquid phase, but calculated the entropy of the solid by a self-consistent phonon representation (§ 1.8) of the deviation from a perfect lattice. In the case of a metal there are so many different contributions to the free energy from the self-energy of the electron gas, electronic band structure terms, ion–ion interactions, etc. that the fundamental statistical characteristics of the melting transition are completely obscured.

These computer simulations provide valuable test beds for more phenomenological theories of melting. It turns out, for example, that the well-known *Lindemann criterion* for melting – that the root mean square deviation of an atom from its lattice site should attain a definite fraction of the lattice constant – is very well satisfied for the LJ 6 : 12 model for argon over a wide range of densities (Ross & Alder 1966). A conjecture that solidification should occur when the peak of the liquid structure factor (4.8) reaches the value $S(q_m) = 2.85$ has also been confirmed for this model (Hansen & Verlet 1969). Since this is the value attained in the ideal hard-sphere liquid at $\eta_L = 0.49$, it is tempting to suggest that a liquid ought to freeze when its underlying hard-sphere assembly reaches the point of thermodynamic instability (6.54). It must be admitted, however, that none of the many suggested numerical criteria for the melting of real liquids have been shown to follow *rigorously* from a fundamental statistical theory of the topological order–disorder transition.

6.8 Entropy and free volume

The spatial order of a condensed system is dramatically reduced when it melts; the change of entropy in this process is, therefore, worth special study. For theoretical guidance on this topic we turn again to the hard-sphere model where, by definition, all the *excess entropy* (6.29) beyond the entropy of a perfect gas is wholly configurational. Since the thermodynamic properties of this model are now known with great accuracy (§§ 6.3, 6.7) certain controversial issues can be firmly settled.

It is well known, for example (Faber 1972), that the latent heat of melting for a wide range of materials corresponds to an increase of entropy ΔS_M lying between 0.8 and 1.7 k per atom. It is tempting to regard this quantity as a measure of the *configurational entropy* associated with the increase of disorder as we go from a crystal to a fluid.

The melting point (6.56) for the hard-sphere model is determined for conditions of constant pressure. By equating the Gibbs free energy of the liquid and solid phases we easily deduce that the change of density from η_S to η_L ((6.54) and (6.55)) can only be balanced by an *entropy of melting*

$$\Delta S_M = T_M^{-1} P_M \Delta V_M$$
$$\approx 1.0 \ Nk. \tag{6.57}$$

Allowing for all the consequences of long-range attractive forces, the 'softness' of the atomic repulsions, contributions from any free electrons (as in liquid metals), etc. this is quite a reasonable *a priori* estimate for the observed quantity. But, because of the relatively large (10 per cent) volume change on melting with which this entropy is associated, it is doubtful whether it can be ascribed simply to the loss of 'order' in the process.

Another quantity that we might study is the *communal entropy* (e.g. Kirkwood 1950) implicit in delocalizing the atoms as we go from a crystal to a disordered fluid. In a crystal each lattice cell contains no more than one atom; in the limit of an ideal gas the number of atoms in any small test volume fluctuates over a Poisson distribution (2.6). In the low-density limit the entropy S_{SO} of the single occupancy model (§ 6.7) should differ from that of the corresponding fluid by

$$\delta S_{com} \equiv S_{fluid} - S_{SO} \to Nk. \tag{6.58}$$

The variation of ΔS_{com} as a function of density, computed by Hoover & Ree (1968) for the hard-sphere model, is shown in fig. 6.13. The communal entropy does not jump to its maximum value at the melting point but increases more or less linearly as the density decreases. This contradicts the intuitive notion of a liquid as being so disordered that the main part of the entropy of disorder should appear as soon as a crystal melts.

The quantity plotted in fig. 6.13 does seem to increase by about 0.2 Nk as we go from the solid to the liquid phase. It must be remembered, however, that this calculation was carried out at constant pressure and does not, therefore, make a direct comparison of the two phases at the same density. The actual behaviour of the excess entropy of the fluid and of the single-

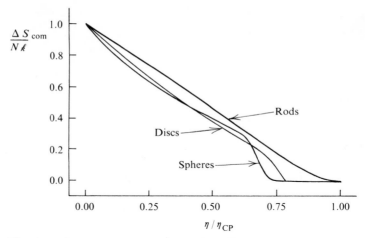

Fig. 6.13. Communal entropy for 'hard' molecules in one, two and three dimensions as a function of density (Hoover & Ree 1968).

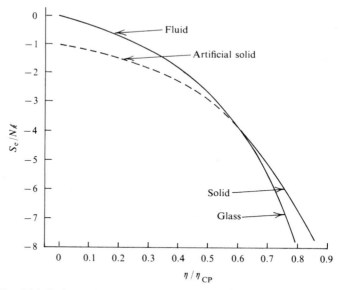

Fig. 6.14. Estimated excess entropy for hard-sphere system in ordered 'solid' or single-occupancy 'artificial solid' and in disordered 'fluid' or 'glass' phases (Hoover & Ree 1967).

occupancy model as a function of density is shown in fig. 6.14. These curves cross at $\eta_M = 0.515$. Since the hard-sphere assembly has no internal potential energy, this is the condition for phase equilibrium under constant volume conditions. In this approximation, by definition, $\delta S_{com} = 0$ at the melting point. In a more complete calculation of the theoretical properties of the hard-sphere assembly we should need to correct the entropy of the solid phase by allowing for some multiple occupation of lattice cells. The true excess entropy for the solid would then be slightly above the single-occupancy entropy so that the value of δS_{com}, as defined by (6.58), would not necessarily vanish at the constant volume melting point. But this has nothing to do with the change of ΔS_{com} in the melting region observed in fig. 6.13, and can scarcely be of large magnitude.

The loss of order when a crystal melts can be considered as loss of *information* about the positions of the atoms. Quite generally (Brillouin 1962), this uncertainty can be assigned a measure S_{inf} that is equivalent to the thermodynamic entropy change in the process. For any variable q that can range over a variety of discrete values with probability p_q, the *informational entropy* per trial is just

$$S_{inf}/N = -k \sum_q p_q \ln p_q. \tag{6.59}$$

For an atomic assembly the constant k is identified with Boltzmann's constant.

Suppose, for example (see Collins 1972), that we regarded the value of the coordination number q of each atomic cell of the structure as significant information about the arrangement of atoms. As we saw in § 2.11, this quantity goes from a single definite value in the perfect crystal to a statistical variable with a range of values in the Bernal liquid or perfect gas. From the data of fig. 2.4, it is easy to deduce values of (6.59): for the 'hot solid', $S_{inf} \approx 1.4\ Nk$; for the liquid model $S_{inf} \approx 1.75\ Nk$; for a perfect gas $S_{inf} \approx 2.5\ Nk$. This term alone, if it were a genuine thermodynamic quantity, would supply a considerable part of the 'configurational entropy of disorder' that we are seeking. Notice, in particular, that the change from the 'hot solid', with considerable fluctuations in local cellular shapes and sizes, to the liquid arrangement does not bring a large change in this measure of disorder – a result consistent with the properties of the communal entropy (6.58).

But the coordination number is not really a well-posed geometrical variable carrying precise information about the arrangement of atoms. The only proper variables for the description of a statistical assembly are the canonical atomic distribution functions defined in § 2.7. These provide (Fisher 1964) successive terms in a cluster expansion of the excess entropy, in the form

$$S_E/N k = -\tfrac{1}{2}n \int g_2(1, 2) \ln g_2(1, 2) \, d2$$

$$-\tfrac{1}{6}n^2 \iint g_3(1, 2, 3) \ln \left\{ \frac{g_3(1, 2, 3)}{g_2(1, 2)g_2(1, 3)g_2(2, 3)} \right\} d2 d3 - \dots \quad (6.60)$$

The first term is obviously the analogue of (6.59), with the variable q standing for the interatomic distance $|\mathbf{R}_1 - \mathbf{R}_2|$. This integral oscillates and converges rather slowly as we go out to large distances but the numbers calculated for typical liquid metals (Faber 1972) account for a considerable proportion of the observed excess entropy near the melting point.

In each of the successive terms of (6.60) the logarithm is taken of the ratio of an exact distribution function to its best 'superposition approximation'. Thus, the second term would vanish in the BBGKY approximation (2.27). The term in g_4 is the higher-order superposition approximation (2.28) which we know to be a good approximation for both liquid and solid phases (§ 2.7): it is possible, therefore, that only the first two terms of (6.60) need be considered. But we know that the BBGKY approximation (2.27) is quite incorrect for a locally crystalline assembly whereas the observed deviations (2.52) from this approximation are not enormous in the liquid state. This term may thus contain, in coded form, the entropy associated with the breakdown of 'long-range' crystalline order at the melting point.

Direct computations of the entropy of the solid phase usually rely on the phonon representations (§ 1.8) of the disorder and depend, therefore, on such parameters as the elastic force constants, Debye temperatures, etc. A much simpler approach (Lennard-Jones & Devonshire 1937) is to treat each atom as if it were moving in the field of its neighbours, which are assumed to be at rest on their equilibrium sites. For an assembly of hard spheres this yields a particularly simple formula, valid when the density, n, approaches the close-packed density n_0. Suppose that the atoms lie on a regular lattice (in v dimensions, for generality), where the distance R between adjacent sites is a little larger than the atomic diameter D. The centre of the atom can trace out a *free volume*

$$v_f = \alpha(R - D)^v$$
$$= \alpha R^v (1 - D/R)^v$$
$$= \alpha'(1/n)\{1 - (n/n_0)^{1/v}\}^v, \quad (6.61)$$

where α and α' are geometrical constants. Going back to the partition function (2.34) we see that the configuration integral (2.35) behaves like $(n v_f)^N$, representing an excess entropy

$$S_f/N\mathscr{k} \approx v \ln \{1 - (n/n_0)^{1/v}\} - 1 \qquad (6.62)$$

(including a communal entropy correction (6.58) for multiple occupancy of cells in the ideal gas limit).

This elementary formula represented the excess entropy of a nearly close-packed assembly due to the localization of each atom near the centre of its atomic cell. For $v = 1$, it obviously leads to the exact formula (6.7) for the free energy of a linear gas. For $v = 2$ and $v = 3$, the equation of state deduced from (6.62) lies quite close to the empirical curves obtained by computer simulation (Wood 1968) for hard discs and hard spheres, right down to the density of melting (fig. 6.8). At first sight this is surprising, since the Lennard-Jones–Devonshire model ignores correlations in the movements of neighbouring atoms; this effect is apparently compensated by other terms that are ignored in this phenomenological approach (Alder, Hoover & Young 1968). Cluster methods (Rudd, Salsburg, Yu & Stillinger 1968), yield further terms in an expansion of the entropy in powers of $\{(n_0/n) - 1\}$ but this does not seem to be a canonical high-density series comparable to the virial expansion (6.45).

It is interesting to observe (fig. 6.8) that the pressure in the liquid phase of the hard-sphere assemblies rises rapidly as we approach the packing density $\eta_{RCP} = 0.63$ of the Bernal RCP structure (§ 2.11), rather than the singularity at the unphysical density $\eta = 1$ of the PY formula (6.29). This suggests the possibility of constructing a similar free volume approximation for the entropy in the fluid regime, behaving like $\ln (1 - n/n_{RCP})$. Putting such an expression into a thermodynamic equilibrium relation such as fig. 6.14 we would have a simple recipe for the melting of a hard-sphere assembly based solely on the geometrical and combinatorial consequences of interatomic repulsion.

7

Macromolecular disorder

—

'O what a tangled web we weave . . .'
Scott

7.1 Regular solutions

An enormous amount of experimental information has been accumulated concerning the thermodynamic properties of liquid mixtures. The interpretation of this mass of data in terms of microscopic molecular parameters is gravely hindered by the fundamental problem (§ 2.13) of calculating partial structure factors for all pairs of constituents. Except for the very simplest case of mixtures of hard spheres (e.g. (2.56)–(2.58)) this can only be done by computer simulation, starting from a suitably simplified prescription for the molecules and their interactions. This technique is so similar to the corresponding computation for a pure fluid (§ 6.6) that we need not discuss the matter further.

In practice, however, the microscopic parameters of a liquid mixture are seldom well enough known to justify a calculation of the topologically disordered structure from first principles. The observed properties of mixed fluids are usually interpreted phenomenologically (see e.g. Rowlinson 1969) by averaging over the molecular parameters of the various components. From the point of view of the general theory of disordered systems, these are all variants of the mean field approximation (§ 5.2) where we ignore local correlations in the distribution of the constituent molecules.

For a simple qualitative interpretation of the behaviour of liquid mixtures we may turn, once more, to the Ising model for substitutional disorder (§1.5). The theory of *regular solutions* is based upon the assumption that the constituent molecules of species A and B are geometrically interchangeable on the sites of a crystalline or liquid assembly, subject only to different intermolecular energies, ϕ_{AA}, ϕ_{AB}, etc., as in (1.19)–(1.25). For simplicity we assume that these potentials act only between nearest neighbours and that the coordination number, z, is the same for every site. In a genuine liquid (cf. § 2.11) these assumptions are obviously very crude but not entirely unrealistic.

This model does not deserve more sophisticated mathematical analysis than is provided by the mean field approximation (§ 5.2). Distributing the atoms at random over the assembly of sites, we derive from (5.8) the *entropy of mixing*

$$\Delta S_{\text{mix.}} = \mathit{k} \ln (N!/N_A!N_B!)$$
$$= -N\mathit{k}(c_A \ln c_A + c_B \ln c_B). \tag{7.1}$$

The internal energy of the mixture is greater than that of the pure condensed phases of the separate components by

$$\Delta \mathscr{E} = NzJc_Ac_B, \tag{7.2}$$

where the 'exchange parameter' is defined as in (1.24):

$$J = \tfrac{1}{2}(\phi_{AB} - \tfrac{1}{2}\phi_{AA} - \tfrac{1}{2}\phi_{BB}). \tag{7.3}$$

Other thermodynamic variables, such as the *free energy of mixing,* can be derived from (7.1) and (7.2).

In the discussion of the Ising model (chapter 5) we focussed mainly on the critical properties of the system. Thus, if J is positive, favouring clusters of like atoms, we introduce an 'order parameter' describing phase separation and show that this would be thermodynamically favoured below a critical temperature T_c. But this temperature may not lie within the limited range, between freezing and vaporizing, over which the mixture can exist as a liquid, so that dramatic critical phenomena of mixing may not be directly observable. At temperatures below T_c, for example, the model simply tells us that the two species are only *partially miscible* – a well known physical phenomenon. To test the model at temperatures above T_c, where A should mix with B in all proportions, the thermodynamic properties of the mixture must be measured. For example, (7.1) and (7.2) lead immediately to a formula for the *partial vapour pressure* of each constituent, relative to the corresponding pure condensed phase:

$$P_A/P_A^0 = C_A \exp\left(\frac{zJ}{\mathit{k}T}c_B^2\right). \tag{7.4}$$

In an *ideal solution,* where the molecular constituents are geometrically and energetically interchangeable (i.e. $J=0$), the partial vapour pressure of each constituent should be directly proportional to its molar concentration. A positive deviation from *Raoult's law* (fig. 7.1) thus indicates that the intermolecular forces favour separation of the components, whereas a

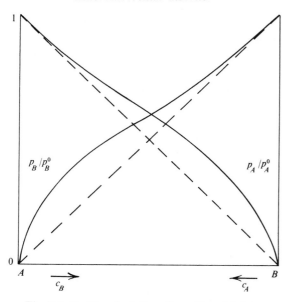

Fig. 7.1. Positive deviations from Raoult's law.

negative deviation might be associated with a tendency towards the formation of AB-type compounds. Similar information may be derived from, say,
the heat of mixing as a function of relative concentration.

In practice, of course, observations on such diverse systems as mixtures
of rare gases, organic and aqueous solutions, or metallic alloys show a
much greater variety of behaviour than can be described by so simple a
formula as (7.4). This is not so much because the MFA solution for the
Ising model needs to be corrected for the effects of short-range order (§ 5.3)
as because the basic assumptions of the model itself are hopelessly over-
simplified. Molecular size and shape, chemical bonding, electronic band
structure, etc. may all contribute to the energetics of mixing, thus influencing
the local structure, the entropy, and other thermodynamic properties. The
regular solution model describes only the most general qualitative features
of real mixtures, and is not to be taken too seriously. But it does provide a
starting point for the study of the statistical thermodynamics of *macromolecular solutions* to which we now turn.

7.2 Entropy of macromolecular solutions

Polymeric materials contain molecules consisting of very large numbers of

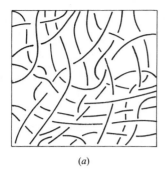
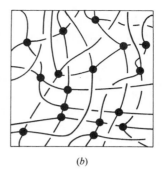

(a)   (b)

Fig. 7.2. (a) Unlinked, (b) cross-linked polymer.

identical *segments* chemically bonded end to end. Each *macromolecule* may constitute a single *unbranched chain,* or it may branch out into shorter or longer *side chains.* The separate chain molecules are usually very flexible (cf. liquid crystals, § 2.14), but a condensed polymer phase may be rendered more or less rigid by the *cross-linking* of one chain with another into a single, multiply-connected network (fig. 7.2) analogous to a glass (§ 2.8). Generally speaking, macromolecular systems do not readily form large regular crystals, but exhibit various complex types of topological disorder which are almost impossible to characterize analytically. The physics of these important materials, therefore, depends mainly on phenomenological theories based on the mathematical properties of a few highly idealized models. For the realities of polymer physics and chemistry, the reader must turn to such treatises as those of Volkenstein (1963) and Flory (1969), or to the voluminous secondary and primary literature on the subject.

The simplest case in principle is that of a solution of perfectly flexible chain molecules dissolved in a liquid. In the spirit of the theory of regular solutions (§ 7.1) we assume that the molecules of the solvent form a 'lattice' (whose geometry need not be regular) whose sites can be occupied by successive segments of the chains, linked neighbour to neighbour. Perfect flexibility implies that there are no constraints on the relative orientations of successive segmental links, except that no site of the lattice may be occupied by more than one segment of the same or different chains. It is convenient to assume that the solution is 'ideal', in the sense that a solvent molecule may be displaced by a chain segment without change of energy. These simplifying assumptions scarcely affect the basic theoretical properties of the model.

A solution of N_p polymer molecules, each of n segments, dissolved in N_s solvent molecules occupies

$$N = N_s + nN_p \tag{7.5}$$

sites of the 'lattice'. To estimate the total number W of configurations of this system (Flory 1941; Huggins 1941; see e.g. Volkenstein 1963; Flory 1970) we need concentrate our attention only on the polymer molecules since we can fill up the unoccupied sites with solvent molecules in only one way for each configuration of the polymer chains.

The argument is purely combinatorial. Suppose that we are about to add the qth polymer molecule to the assembly. Previously added chains occupy $(q-1)n$ sites so that the starting segment may be placed, at random, on any of $N - (q-1)n$ sites. The chain is now built up, segment by segment, in a succession of random steps from site to site. At each step we have $z - 1$ neighbouring sites (that is, excluding the site occupied by the immediately preceding segment of the chain) to which we might go. But some of these will already be occupied by segments of the $q - 1$ polymer molecules that have previously been placed on the lattice. In the spirit of the mean field approximation we assume that *these excluded sites are randomly distributed through the whole volume*. On the average, therefore, only the fraction $\{N - (q-1)n\}/N$ of neighbouring sites will be open for the next step. The number of ways that we can add the qth chain is thus given approximately by

$$v_q \approx \{N - (q-1)n\}[(z-1)\{N - (q-1)n\}/N]^{n-1}. \tag{7.6}$$

For the total number of distinct configurations we take the product of these factors from $q = 1$ to $q = N_p$ and then divide by the $N_p!$ ways of ordering the choice of successive chains:

$$W \approx \{(z-1)^{N_p(n-1)}/N_p!\} \prod_{q=1}^{N_p} \{N - (q-1)n\}^n/N^{n-1}$$

$$= (z-1)^{N_p(n-1)} n^{N_p n} \left[\left(\frac{N}{n}\right)! \Big/ \left(\frac{N}{n} - N_p\right)! \right]^n \Big/ N_p! N^{N_p(n-1)}. \tag{7.7}$$

The logarithm of this expression can be evaluated by Stirling's formula, giving the *Flory–Huggins formula* for the configurational entropy of a macromolecular solution:

$$\Delta S_{\text{config.}} = \Delta S_{\text{disorient.}} + \Delta S_{\text{mix.}}. \tag{7.8}$$

In this theory we treat separately the *entropy of disorientation*,

$$\Delta S_{\text{disorient.}} \approx k N_p \left\{ (n-1) \ln \left(\frac{z-1}{e}\right) + \ln n \right\}, \tag{7.9}$$

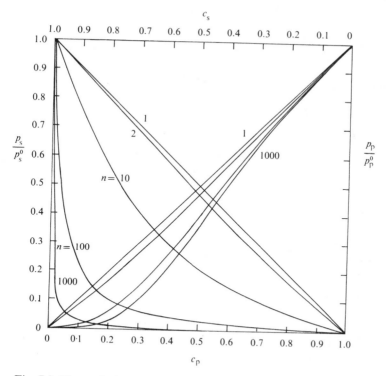

Fig. 7.3. Theoretical deviations from Raoult's law for polymer solutions, for various values of the chain length (Huggins 1958).

associated with the randomness of the arrangement of the segments in the individual polymer molecules, and derive the thermodynamic properties of the solution from

$$\Delta S_{\text{mix.}} = - k\{N_s \ln (N_s/N) + N_p \ln (nN_p/N)\}. \qquad (7.10)$$

Notice that (7.10) contains no reference to the 'lattice' on which solvent and polymer molecules are supposed to be disposed. All we know is that each polymer molecule behaves as a single entity whose segments (however arranged) occupy the same total number of 'cells' of the lattice as n molecules of the solvent. Both (7.1) and (7.10) are special cases of a general expression

$$\Delta S_{\text{mix.}} = - k \sum_j N_j \ln v_j \qquad (7.11)$$

for the entropy of an *athermal* mixture in which N_j molecules of type j

occupy the fraction v_j of the total volume, without other intermolecular forces favouring like or unlike pairs.

From (7.10) or (7.11) we can easily deduce such thermodynamic properties as the partial vapour pressure of the solvent: at low concentrations of polymer

$$\ln (p_s/p_s^0) = \ln v_s + (1 - 1/n)v_p$$
$$\approx -\frac{1}{n}v_p = -N_p/N. \tag{7.12}$$

The very large deviations from Raoult's law observed in macromolecular solutions (fig. 7.3) is a genuine statistical phenomenon. In effect, the addition of a few large polymer molecules provides a great deal of extra volume into which the solvent molecules may move, with considerable increase of entropy and reduction of vapour pressure. Comparison of (7.12) with (7.4) suggests that further improvements to the model by including intermolecular pair interactions (see e.g. Flory 1970) will not give very significant effects and can be dealt with quite adequately by the usual mean field approximation.

7.3 Model chains

The randomness of orientation of successive segments of a chain molecule gives rise to a thermodynamic entropy whose magnitude was calculated approximately in (7.9). For a very long chain, each link contributes

$$\Delta S_{\text{disorient.}}/(n-1)N_p \approx \ln \left(\frac{z-1}{e}\right). \tag{7.13}$$

If each segment were permitted to orientate itself independently of the others in any one of $z - 1$ directions this quantity would be exactly $\ln (z - 1)$; the reduction in these possibilities by the factor $1/e$ allows crudely for physically forbidden configurations where the chain intersects itself. The mathematical problem of calculating the *excluded volume* associated with a *self-avoiding* random chain will be discussed further in § 7.7.

From a practical point of view, the introduction of a liquid lattice with a definite coordination number z is unsatisfactory. It is more realistic to regard each macromolecule as a sequence of segments, with characteristic degrees of freedom and geometrical constraints at each link, freely suspended in a structureless continuous medium. Many properties of real polymer solutions can be compared directly with statistical averages of

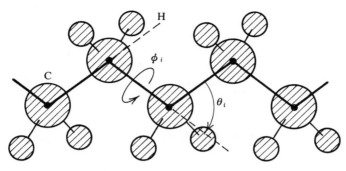

Fig. 7.4. Skeletal carbon chain.

configurational quantities over an ensemble of independent model molecules of the appropriate type.

To set up such a model we must refer to actual chemical structures. The simplest model concentrates on the regularity of the length of monomer segments: in the *random flight* or *freely jointed* chain each segment is of fixed length *l*, but the directions of successive segments are entirely uncorrelated. But most organic macromolecules have a *chain skeleton* (fig. 7.4) whose 'segments' are the covalent bonds between carbon atoms. As we saw in § 2.8, the tetrahedral configuration of bonds about such an atom is strongly constrained so that each *bond angle* θ_i may be assumed constant for all configurations of the chain. In the *free rotation* model we assume that the *bond rotation angle* θ_i, measuring the angle between the plane of segments $(i-1, i)$ and the plane of segments $(i, i+1)$, is an independent random variable.

In practice, however, bond rotation is hindered by a potential energy $U_{\text{rot.}}(\phi)$ which varies considerably with ϕ. The configurational partition function of the chain can be deduced from the total energy

$$\mathscr{H}_{\text{config.}} = \sum_{i=1}^{n} U_{\text{rot.}}(\phi_i). \qquad (7.14)$$

By direct analogy with the case of the one-dimensional classical Heisenberg model (5.72) we immediately verify that this factorizes into the *n*th power of a segmental partition function

$$Z_{\text{segment}} = \frac{1}{2\pi} \int_{-\pi}^{\pi} \exp[-\beta U_{\text{rot.}}(\phi)] d\phi \qquad (7.15)$$

from which the disorientational entropy (7.13) can easily be calculated.

This can be further simplified by the justifiable assumption (§ 2.10) that

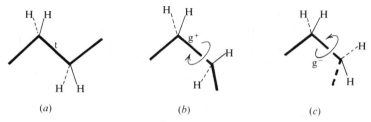

Fig. 7.5. (*a*) Transverse ($\phi=0$), (*b*) gauche ($\phi=2\pi/3$), (*c*) gauche
($\phi=-2\pi/3$) conformations of bonds.

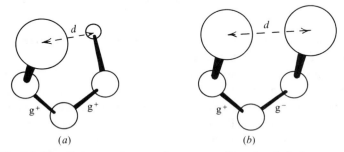

Fig. 7.6. The distance to later carbon atoms along the chain is greater
in the (g^+g^+) conformation (*a*), than in the (g^+g^-) sequence (*b*).

$U_{\text{rot.}}(\phi)$ has deep minima at the 'staggered' configurations about the common bond axis (fig. 2.33). Each ϕ_i is thus treated as a discrete variable,
taking only the three values 0, $2\pi/3$, $-2\pi/3$, corresponding respectively to
the *trans* (t) and two *gauche* (g^+, g^-) configurations of three successive
bonds along the skeleton chain (fig. 7.5). If the energy of a gauche configuration exceeds that of a trans configuration by U_g, the segmental partition
function (7.15) can be reduced to

$$Z_{\text{segment}}=1+2\exp(-\beta U_g) \qquad (7.16)$$

In many actual polymers, however (Flory 1969), the configurational
energy of the chain is not *separable* as in (7.14). It turns out, for example,
that the bond succession g^+g^+ is of lower energy than g^+g^-, which tends to
bring close together carbon atoms separated by three segments of the chain
(fig. 7.6). We must, therefore, write the total energy in the form

$$\mathscr{H}_{\text{config.}}=\sum_i U_{\text{seq.}}(\phi_i,\phi_{i+1}) \qquad (7.17)$$

where $U_{\text{seq.}}$ depends on the sequence of configurational pairs along the molecular chain. This complication can, however, be dealt with by the *Markov method* (see e.g. Lowry 1970) by the introduction of a *transfer matrix*

$$V_{\alpha\alpha'} \equiv \exp[-\beta U_{seq.}(\alpha, \alpha')] \qquad (7.18)$$

where α, α' each run through the three bond rotation angles corresponding to the configurations t, g^+, g^-. Exactly as in §§5.5 and 5.7 the partition function corresponding to the Hamiltonian (7.17) can be represented as the trace of the nth power of the transfer matrix. For a long chain this tends rapidly to the corresponding power of the largest eigenvalue of $V_{\alpha\alpha'}$, which thus provides us with an excellent approximation for the segmental partition function. Since this is a simple 3×3 matrix the thermodynamic properties of the *rotational isomeric chain* model can be evaluated by elementary algebra.

The actual chemistry of organic macromolecules is very much more complicated than these models would suggest (Volkenstein 1963; Flory 1969). A 'segment' may be quite a long sequence of carbon, nitrogen or oxygen atoms, bonded rigidly or flexibly. An interesting phenomenon, of great biological importance, is the *helix–coil transition*. Under appropriate chemical and thermal conditions, the chain arranges itself into a helical structure (e.g. the *alpha-helix* of polypeptide chains) corresponding to a regular succession of bond configurations, such as $tg^+tg^+tg^+ \ldots$ An ordered arrangement of this kind can only be stabilized by interactions between more distant links of the chain – e.g. by hydrogen bonds between skeletal atoms brought into proximity by successive terms of the helix. But, unless these interactions are of 'infinite' range the molecule must be considered one-dimensional, so that (§§ 2.4, 5.5) the observed conformational change cannot be a discontinuous cooperative transition to a phase with long-range order. This does not, of course, preclude the appearance of very long stretches of perfect helix when the chemical conditions are favourable.

7.4 Random coils

Considered as a whole, a three-dimensional random chain of finite length is a globular object floating in the supporting medium. Such an object has overall geometrical properties, such as a characteristic radius, which can be directly associated with measurable physical properties of macromolecules in dilute solution (see e.g. Flory 1953; Huggins 1958; Berry & Casassa 1970). For example, the small-angle scattering of light or X-rays (§ 4.4) provides direct information about the *radius of gyration, D*, of the scattering objects. The dramatic effect of small concentrations of polymeric molecules on the *viscosity* of the solvent can be explained upon the simple assumption that each macromolecular globule behaves as a 'non-draining' sphere, within which the solvent molecules are entrained. Experiment has shown that the apparent radius of this sphere, for a given solute and solvent, is a standard multiple of the radius of gyration found by light scattering on the same solution. The theory of the geometrical properties of a *random coil* is thus of immediate physical interest.

The simplest length parameter is the mean-square distance between the ends of the chain. This can be written in the form

$$\langle R^2 \rangle \equiv \left\langle \left(\sum_i \boldsymbol{l}_i \right)^2 \right\rangle$$

$$= \left\langle \sum_i \boldsymbol{l}_i^2 \right\rangle + \left\langle \sum_{i \neq j} \boldsymbol{l}_i \cdot \boldsymbol{l}_j \right\rangle, \tag{7.19}$$

where $\boldsymbol{l}_i$ is the vector representing the ith segment of the chain and the average $\langle \ \rangle$ is taken over the ensemble of allowed chains. For a *freely jointed chain*, where there is no correlation between the directions of successive segments, this gives

$$\langle R^2 \rangle = nl^2. \tag{7.20}$$

It is easy to prove (see e.g. Flory 1969) a theorem due originally to Lagrange: for large values of n the radius of gyration of the chain (i.e. assigning equal mass to each segment) is proportional to this *end-to-end distance*, i.e.

$$D^2 \to \frac{1}{6} \langle R^2 \rangle. \tag{7.21}$$

Optical and viscosity experiments confirm very satisfactorily the general principle that the apparent radius of a macromolecular globule increases as the square root of the number of segments in the chain.

To evaluate $\langle R^2 \rangle$ for the more complicated molecular chain models of §3.4, it is convenient to introduce a *transformation tensor* $\mathbf{T}_j$ that rotates the jth segmental vector $\boldsymbol{l}_j$ into the direction of its successor, i.e.

$$\boldsymbol{l}_{j+1} = \mathbf{T}_j \cdot \boldsymbol{l}_j. \tag{7.22}$$

Thus, for a typical term in (7.19), we write (for $j > i$)

$$\boldsymbol{l}_i \cdot \boldsymbol{l}_j = \boldsymbol{l}_i \cdot \mathbf{T}_{j-1} \boldsymbol{l}_{j-1}$$

$$= \boldsymbol{l}_i \cdot \mathbf{T}_{j-1} \cdot \mathbf{T}_{j-2} \dots \mathbf{T}_i \cdot \boldsymbol{l}_i. \tag{7.23}$$

In general, each $\mathbf{T}_j$ is a stochastic matrix, about which we only have statistical information. But, if the successive transformations are uncorrelated, we can factorize the ensemble average of the product of tensors in (7.23) into a power of the average transformation tensor at each link:

$$\langle \boldsymbol{l}_i \cdot \boldsymbol{l}_j \rangle = \boldsymbol{l}_i \cdot \{\langle \mathbf{T} \rangle\}^{|j-i|} \cdot \boldsymbol{l}_i. \tag{7.24}$$

Putting (7.24) into (7.19), and assuming that the chain is long enough for end effects to be ignored, we get

$$\langle R^2 \rangle = \sum_i [l_i^2 + 2\boldsymbol{l}_i \cdot \overset{(\infty)}{\underset{k=1}{\sum}} \{\langle \mathbf{T} \rangle\}^k \cdot \boldsymbol{l}_i]$$

$$= \sum_i [\boldsymbol{l}_i \cdot \mathbf{1} \cdot \boldsymbol{l}_i + 2\boldsymbol{l}_i \cdot \langle \mathbf{T} \rangle \{\mathbf{1} - \langle \mathbf{T} \rangle\}^{-1} \cdot \boldsymbol{l}_i]$$

$$= n\boldsymbol{l} \cdot \left[\frac{1 + \langle \mathbf{T} \rangle}{1 - \langle \mathbf{T} \rangle} \right] \cdot \boldsymbol{l}, \tag{7.25}$$

where *l* is a standard segment vector. For a freely jointed chain, where $\langle \mathbf{T} \rangle$ is zero by definition, this gives (7.20). Indeed, the whole effect of the geometrical constraints at each joint is to multiply the random flight formula by a constant factor: in general

$$\langle R^2 \rangle = C_\infty nl^2, \tag{7.26}$$

where C_∞ can be determined, for any particular model, by reference to (7.25). In the case, for example, where the bond angle θ is fixed but the bond energy is separable as in (7.14), we may write

$$\langle \mathbf{T} \rangle = \int_{-\pi}^{\pi} \mathbf{T}(\phi) \exp[-\beta U_{\mathrm{rot.}}(\phi)] \mathrm{d}\phi \Big/ \int_{-\pi}^{\pi} \exp[-\beta U_{\mathrm{rot.}}(\phi)] \mathrm{d}\phi, \tag{7.27}$$

From which, by elementary geometry, we get

$$C_\infty = \left(\frac{1 + \langle \cos \theta \rangle}{1 - \langle \cos \theta \rangle} \right) \left(\frac{1 + \langle \cos \phi \rangle}{1 - \langle \cos \phi \rangle} \right). \tag{7.28}$$

For the free rotation model where, of course $\langle \cos \phi \rangle = 0$, the second factor reduces to unity.

To apply the same method to the rotational isomeric chain, whose conformational energy (7.17) depends on successive pairs of bond rotational configurations, we may feel obliged to introduce a somewhat more complex notation (Flory 1969) to describe ensemble averages of the correlated pairs of neighbouring transformation tensors into which (7.23) would be factorized. But the one-dimensional topology of the chain ensures that no essential difficulties arise in evaluating C_∞ for any specific model. It is obvious also, from the working leading to (7.25), that formulae correct to higher orders in n can be written down by truncating the geometrical series in $\langle \mathbf{T} \rangle$.

The mean square end-to-end distance in a random coil is only an average over a statistical ensemble. But since the chain consists of many segments, whose directions are uncorrelated beyond a small number of links, the conditions are satisfied for the Central Limit theorem of mathematical probability (e.g. Feller 1967). From this theorem, or by more specialized methods, it is easy to deduce that the components of the end-to-end vector **R** must be *normally distributed*. In other words, the probability of finding this vector in an element of volume d**R** must be a three-dimensional *Gaussian* function (cf. (3.12))

$$P(\mathbf{R})d\mathbf{R} = A e^{-b^2 R^2} d\mathbf{R}, \tag{7.29}$$

where A is a suitable normalizing factor. From our knowledge of the *second moment* $\langle R^2 \rangle$ of this distribution, we can deduce the value of b, i.e.

$$b^2 = \frac{3}{2\langle R^2 \rangle}. \tag{7.30}$$

The probability distribution for the *length* of this vector must be of the form

$$P(R) = 4\pi A R^2 e^{-b^2 R^2} \tag{7.31}$$

which has a well-defined maximum at a radius equal to $2D$; the most likely value of the end-to-end distance of a random coil is just twice its radius of gyration.

This derivation of (7.29) makes no reference to the detailed geometrical conditions at each joint. For a long enough chain, the statistical properties are independent of the local chemical structure. For this reason the simplest freely jointed chain model (Kuhn 1934) contains much of the physics of a real polymeric material. All we need to do is define an *equivalent chain,* of n' freely jointed segments, each of length l', having the same stretched length

$$n'l' = nl, \tag{7.32}$$

and the same radius of gyration or end-to-end distance as the real macro-molecule. From (7.20) and (7.26) we readily verify that each *equivalent random link* must have length

$$l' = C_\infty l. \tag{7.33}$$

The number C_∞ defined from (7.25) or (7.28) simply represents the minimum number of 'chemical' segments needed to produce a perfectly flexible stretch of chain (fig. 7.7).

Strictly speaking (7.29) and (7.31) represent the distribution of end-to-end distances in an ensemble of similar chains, which is not quite the same thing as the density distribution of the segments about the centre of mass of a typical random coil; the difference between these two quantities can be

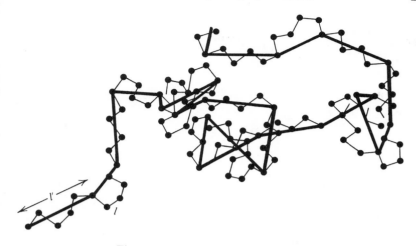

Fig. 7.7. Equivalent freely jointed chain.

calculated and is not large (Isihara 1950; Debye & Bueche 1952). The optical density profile of each macromolecule is thus effectively Gaussian.

7.5 Branching and gel formation

In a concentrated solution or condensed phase, the individual molecules of a polymeric substance come into intimate contact and may interact in a variety of ways. If the material, whether liquid or solid, is genuinely amorphous, and if each macromolecule is a finite unbranched chain, then its configurations may reasonably be supposed (e.g. Flory 1970) to approximate to those of a random coil. In these circumstances, the effects of entanglements must be reckoned with (§ 7.10) and the multitudinous possibilities of local crystalline order.

In many polymeric materials, however, the macromolecules cannot be described as independent linear chains. It is possible, for example (see e.g. Flory 1953; Volkenstein 1963), that they have been produced by chemical condensation from *monomer* molecules with more than two active sites (fig. 7.8(*a*)). Or a small quantity of such a monomer may have been *copolymerized* with a compound that normally produces unbranched chains (fig. 7.8(*b*)). Or, as in the *vulcanization* of a rubber (§ 7.6), chemical bonds are produced at random between pre-existing macromolecular chains (fig. 7.8(*c*)). In each case, *branching chain* molecules are produced, with characteristic physical properties.

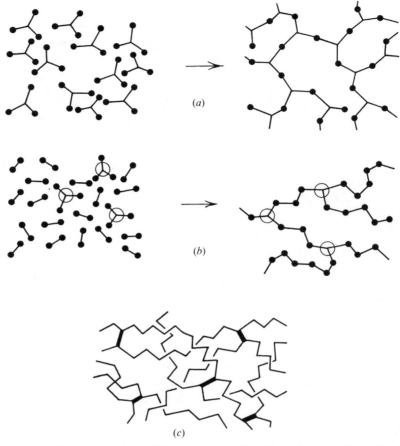

Fig. 7.8. Branching chains produced by (a) polymerization of monomer with functionality $z > 2$, (b) copolymerization with branching species, (c) vulcanization.

The most striking effect of branching is to change a liquid polymer solution into a *gel*. This transformation is due to the appearance of macromolecules of 'infinite' molecular weight – i.e. that extend throughout the whole volume of the specimen. The theory of branching processes is of interest, not only as an example of a special type of homogeneous disorder, but also for its mathematical simularity to other, apparently unrelated, phenomena (Harris 1963).

For simplicity, let us assume that the polymer is chemically condensed

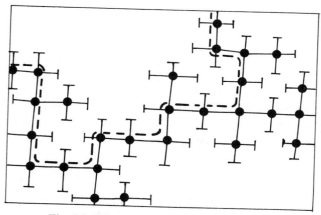

Fig. 7.9. Branching process with $z = 4$, $p > p_c$.

from a solution of identical monomeric units of *functionality* z – i.e. each molecule has z equivalent bonding sites. In the chemical reaction these bonds form at random, until a fraction p of all the sites have actually been linked; under what circumstances will an *infinite* molecule be produced?

It is easy to see that this must occur when p exceeds a critical value

$$p_c = 1/(z-1). \qquad (7.34)$$

The number of possible new branches arising from an old branch in each generation is $(z-1)$ (fig. 7.9); unless p exceeds p_c, the average number of new descendants will decrease with each generation, so that the molecule must eventually stop growing.

But, even when $p > p_c$, the whole system does not necessarily constitute a single, heavily branched macromolecule. Statistical fluctuations leave many of the monomer units in finite molecular clusters; for example, the proportion of entirely isolated single monomers is $(1-p)^z$. These finite *n-mers* – linked clusters of molecular weight n relative to a single monomer – constitute the *sol fraction*, which can be physically separated from the gel network.

The mathematical theory of branching or *cascade* processes provides exact analytical formulae for the sol fraction and similar statistical properties of the system (Gordon 1962; Good 1963). The method depends on the properties of the *weight fraction generating function*

$$W(p, \theta) = \sum_{n=1}^{\infty} w_n(p)\theta^n, \qquad (7.35)$$

where the coefficient of θ^n is defined as the fraction of all monomer units that are condensed into n-mers. By definition, therefore, we must have

$$\sum_{n=1}^{\infty} w_n(p) = 1. \tag{7.36}$$

But when $p > p_c$, this sum includes a highly singular term – a contribution from the gel network itself, for which $n = \infty$. If this term can be excluded we can find the sum of the weights of all *finite* n-mers, converging on

$$S(p) = W(p, 1), \tag{7.37}$$

which is not identically unity for all values of p. By a mathematical device that depends, in effect, on taking the limit $n \to \infty$ before the limit $\theta \to 1$, we thus determine the sol fraction $S(p)$.

But the generating function (7.35) can be constructed directly by successive iterations of the branching process. Taking p to be fixed, we write

$$W(\theta) = \theta F_0(\theta F_1(\theta F_2(\theta F_3(\ldots \theta F_s(\ldots) \ldots)))). \tag{7.38}$$

For the zeroth generation (where we have z functionalities to choose from) the generating function is simply

$$F_0(\zeta) \equiv (1 - p + p\zeta)^z. \tag{7.39}$$

For all later generations, with $(z - 1)$ free branches, the generating functions are identical:

$$F_1(\zeta) \equiv F_2(\zeta) \equiv \ldots \equiv F_s(\zeta) \equiv \ldots \equiv (1 - p + p\zeta)^{z-1}. \tag{7.40}$$

This equivalence (7.40) permits us to write (7.38) in the form

$$W(\theta) = \theta F_0(u(\theta)). \tag{7.41}$$

The auxiliary function $u(\theta)$ is completely self-iterating: it must satisfy

$$u(\theta) \equiv \theta F_1(\theta F_1(\theta F_1(\ldots)))$$
$$= \theta F_1(u(\theta)). \tag{7.42}$$

This fundamental self-consistency condition is the analogue of Dyson's theorem in the graphical representation of perturbation theory. It can now be treated as a functional equation from which we can deduce $u(\theta)$ and hence, by (7.41) and (7.35), the statistical distribution of all finite n-mers. This provides, therefore, the exact solution of the branching problem.

As an example of the power of the method let us evaluate $S(p)$, the sol fraction. Using (7.39) and (7.40), we can eliminate $u(\theta)$ between (7.41) and (7.42) and then take the limit $\theta \to 1$ in (7.37). This yields the following simple algebraic relation:

$$p = (1 - S^{1/z})/(1 - S^{(z-1)/z}), \tag{7.43}$$

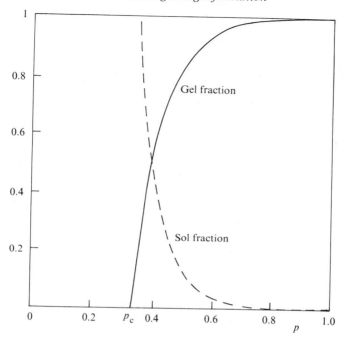

Fig. 7.10. Gel formation for a tree model with $z = 4$.

which is satisfied for all p by

$$S = 1, \tag{7.44}$$

or by the more restrictive condition

$$p = 1 \Big/ \sum_{j=0}^{z-2} S^{j/z} \tag{7.45}$$

Since S, by definition, cannot exceed unity, (7.45) cannot be satisfied for $p < (z-1)^{-1}$: in this case (7.44) merely tells us that all the monomers are in the finite n-mers and there are no infinite clusters. But for $(z-1)^{-1} \leqslant p \leqslant 1$, the root (7.45) provides a unique function $S(p)$ which falls monotonically from 1 to 0 (fig. 7.10). This is the sol fraction. The *gel fraction*

$$G(p) = 1 - S(p) \tag{7.46}$$

is of measure zero so long as the degree of polymerization, p, is less than the critical value (7.34) and then rises abruptly, with finite slope, for $p \geqslant p_c$. *Gelation* is thus a well-defined physical transformation with a sharp critical point as a function of p.

Much more complicated branching processes, such as the copolymerization of several constituents, can be described exactly by a systematically elaborated version of the same mathematical device (Dobson & Gordon 1964; Gordon & Malcolm 1966). This method generates various formulae that had previously been deduced for special cases (e.g. Flory 1953) and also provides useful expressions for a variety of other statistical parameters, such as the average degree of polymerization, the mean square radius of the *n*-mers, etc.

The key to its success is that the branching process generates a *tree* or *Bethe lattice* (§ 5.4). On such a lattice, the Bethe self-consistency condition (5.34) for the average spin – a mathematical analogue to the condition (7.42) for the branch generating function $u(\theta)$ – is also exactly true. The present theory also provides an exact solution (Fisher & Essam 1961) for *percolation* (§ 9.10) on a Bethe lattice; the gel fraction (7.44) is obviously related to the probability that an arbitrarily chosen branch lies on an infinite path through a lattice of this type with a proportion p of 'open' branches.

But this topological property of the model is fundamental; once we allow possibilities of multiple connectivity into the network, the whole problem becomes far more complex and difficult. In the case of gel formation by polymerization this may not be a significant physical possibility, but for the problem of vulcanization (cf. Dobson & Gordon 1965) this is of crucial importance. We may start with the idea that each long chain, with a large number, n, of potential sites for cross-linking to another chain, is a 'monomer' of functionality n. Gelation should then occur for very small values of p, exceeding $1/(n-1)$. Under these conditions, we may expand the right-hand side of (7.43) in powers of $1/n$, i.e.

$$(1 - S^{1/n})/(1 - S^{(n-1)/n}) \rightarrow \frac{-1}{n(1-S)} \ln S. \tag{7.47}$$

The root (7.45) tends rapidly to the form

$$S \rightarrow \exp\{-np(1-S)\} \rightarrow \exp(-np). \tag{7.48}$$

The sol fraction thus falls rapidly to zero as vulcanization proceeds so that the whole effect of additional cross-links is to add further connections between chains that are already attached to the gel assembly. Structurally speaking, a *rubber* must be considered a *random network* (fig. 7.11) whose vertex points (*cross-links*) are joined by polymer chains of variable length. Except for a small proportion of 'loose ends' we lose evidence of the individuality of the original, very long macromolecules from which the

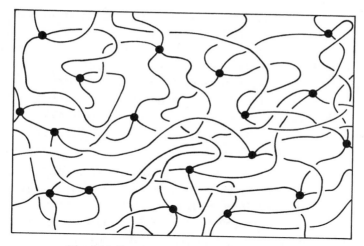

Fig. 7.11. Random network model of rubber.

network was produced. In such a material, each length of linear polymer between cross-links is counted as a distinct chain.

7.6 Rubber elasticity

The extraordinary elastic properties of rubbers can be explained (see e.g. Treloar 1958) by the idea that each chain is free to take up any random configuration between its cross-linked end points. Apart from these constraints, the molecular segments are as mobile as they would be in a liquid. The specimen resists change of shape, not because this does work against intermolecular forces, but because it decreases the disorientational entropy of the chains.

Instead of calculating this entropy directly, as in (7.13) or (7.15), by counting configurations of the segments, we refer to the probability distribution (7.29) for the vector distance between the ends of a long random chain. By definition, $P(\mathbf{R})$ measures the proportion of chain configurations out of a whole ensemble of similar chains in which this quantity has the value $\mathbf{R}$. If we change this distance to $\mathbf{R}'$ we change the relative number of available configurations to $P(\mathbf{R}')$. By the Boltzmann principle, this is equivalent to a change of disorientational entropy (Kuhn 1936)

$$\Delta S_{\text{disorient.}} = k \ln P(\mathbf{R}') - k \ln P(\mathbf{R})$$
$$= k \ln (Ae^{-b^2 R'^2}) - k \ln (Ae^{-b^2 R^2})$$
$$= -kb^2(R'^2 - R^2). \tag{7.49}$$

Thus, stretching the chain decreases its entropy. This, in turn, increases the free energy, producing a restoring force.

When a multiply-connected network of such chains is subjected to a shear stress it deforms homogeneously. That is to say, the cross-link points move as if embedded in a medium which undergoes a homogeneous strain. If this *affine deformation* is represented by a diagonal strain tensor, the vector $\mathbf{R}$, with Cartesian components X, Y, Z is transformed into $R' = (\lambda_1 X, \lambda_2 Y, \lambda_3 Z)$. Putting this into (7.49), we get

$$\Delta S_{\text{disorient.}} = -\hbar b^2 \{(\lambda_1{}^2 - 1)X^2 + (\lambda_2{}^2 - 1)Y^2 + (\lambda_3{}^2 - 1)Z^2\} \quad (7.50)$$

for the contribution to the change of entropy from a chain with these end points. For the N_c different chains in the network we must take an average of (7.50) over the orientations of $\mathbf{R}$ and then use the connection (7.30) between the parameter b^2 for a given type of chain and its equilibrium mean square end-to-end distance $\langle R^2 \rangle$. For the total change of entropy we get

$$
\begin{aligned}
\Delta S_{\text{def.}} &= -N_c \hbar \langle b^2 \{(\lambda_1{}^2 - 1)X^2 + (\lambda_2{}^2 - 1)Y^2 + (\lambda_3{}^2 - 1)Z^2\} \rangle \\
&= -N_c \hbar \langle b^2 \tfrac{1}{3} R^2 \rangle \{\lambda_1{}^2 + \lambda_2{}^2 + \lambda_3{}^2 - 3\} \\
&= -\tfrac{1}{2} N_c \hbar \{\lambda_1{}^2 + \lambda_2{}^2 + \lambda_3{}^2 - 3\}
\end{aligned}
\quad (7.51)
$$

which is equivalent to changing the free energy by

$$\Delta F_{\text{def.}} = -\tfrac{1}{2} N_c \hbar T \{\lambda_1{}^2 + \lambda_2{}^2 + \lambda_3{}^2 - 3\}. \quad (7.52)$$

This formula (Wall 1942; Treloar 1943) is sufficient to explain most of the elastic and thermodynamic properties of a typical rubber. If we take care not to make volume changes, which would be resisted by the cohesive and repulsive forces between the molecular segments (as in an ordinary liquid), we can ignore internal energy terms. For example, if we stretch the specimen along the X axis by an amount $\lambda_1 = \lambda$, and satisfy the volume condition

$$\lambda_1 \lambda_2 \lambda_3 = 1 \quad (7.53)$$

by putting $\lambda_2 = \lambda_3 = \lambda^{-\frac{1}{2}}$, we deduce from (7.52) a restoring force

$$f = N_c \hbar T \left(\lambda - \frac{1}{\lambda^2} \right). \quad (7.54)$$

The beauty of this formula is that the number N_c is simply the total number of chains – that is, twice the number of cross-links – per unit volume of the material and does not depend on the distribution of chain lengths nor on the chemical structure and geometry of the molecular segments that make up these chains. The elastic tension of the stretched rubber is derived from the classical kinetic energy of transverse Brownian motion of the molecular chains; the similarity of (7.54) to the familiar equation of state of a perfect gas, as derived from kinetic theory, is no accident.

In deriving (7.52), of course, a number of dubious assumptions were made. For example, the actual vulcanization process may not produce a network for which each chain is in statistical equilibrium; the elastic coefficients may then be multiplied by a numerical factor such as $\frac{2}{3}$ (James & Guth 1947). Some chains may be too short to justify the Gaussian formula (7.29), although it is well known, from the mathematical theory of probability, that this is a very good approximation, except in the tails of the curve, down to a very small number (e.g. 5) of independent links. The assumption of uniform strain of the cross-link points may also be invalid when these are connected by short chains; such chains may tend to rotate, rather than stretch, when the material is stressed (Dobson & Gordon 1965; Gordon, Love & Pugh 1969). The calculation of such effects is clearly much more complicated, depending on the details of the distribution of chain lengths and involving very uncertain assumptions concerning the local deformational characteristics of the material.

The most serious deficiency of the simple formula (7.52) is at large extensions, where many chains have to be stretched from their equilibrium end-to-end distance $n^{\frac{1}{2}}l$ to something approaching their total length nl. This can easily be allowed for semiquantitatively (James & Guth 1943) by a direct thermodynamic calculation of the tension of a freely jointed chain. Since the segments are independently oriented, each segment vector l_i behaves like the spin vector S_i in a classical paramagnet (cf. § 5.6). The stretching force $\mathbf{f}$ along the direction $\mathbf{R}$ is then analogous to a 'magnetic field' partially aligning the spins. Thus, the length R, being the sum of the projections of the segment vectors l_i along $\mathbf{f}$, is mathematically equivalent to the magnitude of a 'magnetic moment'. The most elementary classical calculation then yields

$$R/n = \int_0^\pi l \cos\theta \, e^{\beta fl\cos\theta} \sin\theta d\theta \Big/ \int_0^\pi e^{\beta fl\cos\theta} \sin\theta d\theta$$

$$= l\mathscr{L}(fl/kT) \tag{7.55}$$

where the *Langevin function*

$$\mathscr{L}(v) \equiv \coth\gamma - 1/\gamma \tag{7.56}$$

saturates to unify as $v \to \infty$. In other words, the tensile force is related to the length by the *inverse Langevin formula*

$$f = (kT/l)\mathscr{L}^{-1}(R/nl) \tag{7.57}$$

268 *Macromolecular disorder*

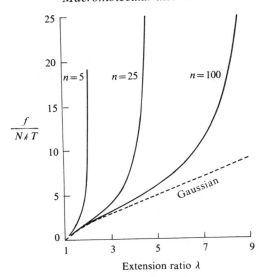

Fig. 7.12. Non-Gaussian force-extension curves for three-chain model (Treloar 1958).

which is linear in R for small values of R/nl, but tends to infinity as R is stretched to nl, the full length of the chain.

A formula equivalent to (7.49) can be written down for the disorientational entropy of the stretched chain, but unfortunately the averaging process (7.51) does not give such a nice result. We have to use a cruder model in which, for example, three independent sets of chains of equal length run parallel to each of the three coordinate axes. We assume also, that the end-to-end distance for each chain in the equilibrium unstrained material is $n^{\frac{1}{2}}l$. For an extension, keeping the volume constant, as in (7.51) we then require a force

$$f = \tfrac{1}{3} N_c k T n^{\frac{1}{2}} \left\{ \mathscr{L}^{-1}\left(\frac{\lambda}{n^{\frac{1}{2}}}\right) - \lambda^{-\frac{3}{2}} \mathscr{L}^{-1}\left(\frac{1}{\lambda^{\frac{1}{2}} n^{\frac{1}{2}}}\right) \right\} \tag{7.58}$$

which reduces correctly to (7.54) for values of λ not much larger than unity. But as λ^2 approaches n, the number of equivalent random segments in the molecular chains, the elastic force increases rapidly (fig. 7.12). This expression, however, is only a semiquantitative approximation which can be refined phenomenologically (Treloar 1958) but not made more precise without further information about the distribution of chain lengths, etc. The effect of large extensions on the deviations of the cross-link positions from affine deformation must also be considered (Edwards 1969).

7.7 Excluded volume

Each segment of a macromolecule occupies space from which other molecules or molecular segments are rigorously excluded. The mutual impenetrability of the segments of two *different* macromolecules was allowed for, approximately, in deriving the thermodynamic properties of polymer solutions (§ 7.2), by reducing the number of segment sites available to each newly added polymer molecule in proportion to the number of previously added molecules of the same type. This approximation does not, however, take account of the fact that no two segments of the *same* chain may occupy the same site; a single, isolated chain is rigorously restricted to *self-avoiding* configurations.

In view of the experimental difficulties of producing precisely characterized macromolecular assemblies, the physical effects of this *excluded volume* restriction are not of great practical significance. But because of its mathematical connections with other statistical problems, such as the properties of the Ising model (see Domb 1969), the problem of calculating these effects for simple theoretical models has been pursued to some depth (see e.g. Yamakawa 1971). In fact, despite its apparent simplicity, no exact analytical solution for this problem has yet been obtained.

The discussion centres on the simplest type of model – a random chain on a regular lattice of coordination number z. At first sight, one might suppose that most of the effect could be allowed for by explicitly prohibiting the superposition of segments lying close to one another along the chain. Small cycles, in which the chain returns to the same site after 3, 4, 5, . . ., s steps are forbidden by assigning a very large positive energy to such superpositions of individual segments up to s units apart. By analogy with (7.18), we may introduce a transfer matrix for the statistical (Markov) process of stepping along the chain from one 'block' of s segments to its neighbour. As in §§ 5.5, 5.7 and 7.3, the thermodynamic properties of the quasi-infinite molecule can be deduced from the maximum eigenvalue of this matrix.

It is immediately obvious that this method yields no new 'qualitative' results. Since blocks of chain separated by more than p segments are assumed to be beyond the range of the interaction, and thus statistically independent, we merely reproduce the theory of a random coil (§ 7.4) with longer equivalent random links (7.33) than the actual segmental length l. For example, the radius of gyration of the molecule would be increased, but would still be proportional to $n^{\frac{1}{2}}$ as in (7.26).

But the plausible procedure of calculating these corrections for successively larger values of s does not converge on the correct answer. The
excluded volume interaction is not confined to nearly adjacent segments; it
has 'long-range' effects, due to interference between segments that happen
to come close to one another even though they are far apart along the chain.
In other words, the analysis must include the small but finite probability
that the trajectory of the chain may return to the origin after a very large
excursion. In the language of probability theory, the model molecule is no
longer a *finite Markov chain,* and the standard procedure breaks down.
Thus, for example (Mazur 1970), the dimensionality of the transfer matrix
goes up as $(z-1)^{s-2}$ or some similar exponential of s. As we try to make s
tend to n, we find that the number of eigenvalues of the transfer matrix
greatly exceeds the total number of segments in the chain, so that the trace
(5.58) can no longer be adequately approximated by the contribution of the
largest eigenvalue.

No exact solution, comparable to the Onsager algebraic reduction of the
two-dimensional Ising problem (§ 5.7), has been discovered for this system.
But the effect of excluded volume on the dimensions of a random coil can
be estimated by an elementary thermodynamic argument (Flory 1949). The
basic idea is that the thermodynamics of solutions (§ 7.2) applies locally
within the region occupied by a single macromolecular chain. From the
point of view adopted in deriving the Flory–Huggins formula (7.10),
interference between very distant stretches of the same long chain looks no
different from interaction between stretches of different chains. In either
case, according to (7.11), if N_s solvent molecules occupy the volume
fraction v_s within a given region, they give rise to an *entropy of dilution*

$$\Delta S_{\text{dil.}} = -N_s \textit{k} \ln v_s \qquad (7.59)$$

regardless of the configurational entropy of the polymer chain or chains. In
our simple model, each solvent molecule occupies a lattice site. If the region
V contains just one macromolecule of n segments, then

$$v_s = N_s/(N_s+n), \qquad (7.60)$$

and for small values of n/N_s the entropy behaves like

$$\Delta S_{\text{dil}}/\textit{k} \approx n - n^2/N_s. \qquad (7.61)$$

In the absence of excluded volume interactions, the macromolecule would
mostly lie within a sphere whose diameter would be about equal to the
root-mean-square, end-to-end distance (7.20). We assume that the effect of
excluded volume is to expand this sphere homogeneously by a radial factor

α. The number of solvent molecules within this region must be something like

$$N_s = (1/C)\alpha^3 n^{\frac{3}{2}},\tag{7.62}$$

where C is a geometrical factor of the order of unity. Putting this into (7.61), we get

$$\Delta S_{\text{dil.}}/\cancel{k} = n - Cn^{\frac{1}{2}}/\alpha^3,\tag{7.63}$$

showing that the entropy of dilution increases as the macromolecule expands.

This expansion is restrained by the kinetic tension of the random coil. As in the theory of rubber elasticity (§ 7.6), the probability distribution of the end-to-end length of the chain is changed from (7.31) to

$$P'(R)\mathrm{d}R = \frac{4\pi A}{\alpha^3} R^2 \mathrm{d}R\, \mathrm{e}^{-b^2 R^2/\alpha^2};\tag{7.64}$$

this function is correctly normalized to give a mean square radius of gyration equal to α^2 times the standard value $2/(3b^2)$ that we would obtain for an 'unexpanded' chain. Exactly as in (7.49), we estimate the average change of entropy due to this expansion

$$\begin{aligned}\Delta S_{\text{exp.}}/\cancel{k} &= b^2(1/\alpha^2 - 1)\langle R^2\rangle + 3\ln\alpha \\ &= 3\ln\alpha - \tfrac{3}{2}(\alpha^2 - 1).\end{aligned}\tag{7.65}$$

The equilibrium configuration of minimum free energy is obtained by adding (7.63) to (7.65) and differentiating with respect to α: we thus derive the *Flory formula*

$$Cn^{\frac{1}{2}} = \alpha^5 - \alpha^3\tag{7.66}$$

for the relationship between the length n of a polymer chain and its expansion factor α under the influence of the mutual exclusion of its segments.

This formula asserts, very specifically, that the mean square radius of gyration of a self-excluding chain should not follow the Gaussian law (7.21), (7.26), but should tend, for large n, to the form

$$D^2 \sim n^{\frac{6}{5}}.\tag{7.67}$$

For a two-dimensional system it is easy to show that the corresponding exponent would be $\tfrac{3}{2}$. Apart from the value of the coefficient C, which depends upon the geometry of the underlying solvent 'lattice' and the physics of the exclusion interaction between the macromolecular segments, the derivation of (7.66) is quite general. From the point of view of formal statistical mechanics, the use of thermodynamic arguments shows that this is essentially a mean field approximation (§ 5.2), analogous to the Landau theory of critical phenomena (§ 5.11) and van der Waals equation (§ 6.2).

It is easy to be sceptical concerning the validity of the phenomenological assumptions made in deriving (7.66) and to suggest alternative hypotheses that might be expected to yield somewhat better results (Flory & Fisk 1966). The excluded volume problem has also been tackled by all the methods familiar in the theory of order–disorder phenomena – for example by a 'virial expansion' (§§ 5.10, 6.5) in powers of the 'strength' of the exclusion interaction (Zimm 1946), by graphical summation (§ 5.10) of the generating function for random walks with mutual interactions (Domb & Joyce 1972), by the renormalization group (§ 5.12) for the critical exponents as a function of the dimensionality of the system (de Gennes 1972) and other complex algebraic procedures (e.g. Fixman 1966). What is surprising, however (Domb 1969; Domb, Barrett & Lax 1973), is the precision with which the best analytical approximations and the numerical data obtained by Monte Carlo and other direct computation, agree with the simple Flory formula (7.66). It is also reassuring that experimental observations of viscosity and light scattering (§ 7.4) are consistent with the exponent in (7.67) (Berry & Casassa 1970; Yamakawa 1972).

7.8 Random walks on a lattice

Although not a reliable quantitative representation of a macromolecule, the simple model of a *random walk* on a regular lattice reproduces most of the important physical properties of the real system. This model plays the same role in the present chapter as the Ising model for substitutional disorder in chapter 5. Many physically significant properties, such as the probability distribution of end-to-end distances (7.31), have already been deduced by more general models such as the freely jointed chain and need not be discussed further. The importance of this model in general probability theory (see e.g. Feller 1967) needs no emphasis. But the mathematical theory of random walks is so closely connected with graphical representations and other statistical features of disordered systems that it deserves some discussion in its own right (see e.g. Kasteleyn 1967; Domb 1969).

The theory of *unrestricted* random walks on a lattice of d dimensions can be deduced from the trivial one-dimensional case. Consider, for example, the resultant of n successive random steps, with equal probabilities to right or left, along an infinite linear chain: what is the total number $w_n(l)$ of such walks that start from the origin and arrive at the site l? This is a purely combinatorial problem. If x represents a step to the right, and x^{-1} a step to the left, we know that $w_n(l)$ must be the coefficient of x^l in $(x+x^{-1})^n$.

Now suppose we replace the dummy variable x by e^{iq}. By elementary Fourier analysis, the integral

$$w_n(l) = \frac{1}{2\pi} \int_{-\pi}^{\pi} dq (e^{iq} + e^{-iq})^n e^{-iql} \qquad (7.68)$$

'filters out' the desired coefficient of $x^l = e^{iql}$. All such numbers can be represented very compactly by constructing the *generating function*

$$\Gamma(l,t) \equiv \sum_{n=0}^{\infty} w_n(l) t^n. \qquad (7.69)$$

Substituting from (7.68), and summing boldly over n, we get the fundamental formula

$$\Gamma(l, t) = \frac{1}{2\pi} \int_{-\pi}^{\pi} dq \frac{e^{-iql}}{1 - 2t \cos q}. \qquad (7.70)$$

To generalize to a Cartesian lattice of d dimensions, we introduce a vector $\mathbf{q}$ whose components appear in place of the d dummy variables

$$(x_1, x_2, \ldots, x_d) \equiv (e^{iq_1}, e^{iq_2}, \ldots, e^{iq_d}) \qquad (7.71)$$

which now replace x. The generating function for the number of walks from 0 to l can then be written as a simple generalization of (7.70):

$$\Gamma(l, t) = \left(\frac{1}{2\pi}\right)^d \int_{-\pi}^{\pi} dq_1 \int_{-\pi}^{\pi} dq_2 \ldots \int_{-\pi}^{\pi} dq_d \frac{e^{iq \cdot l}}{1 - 2t \sum_{j=1}^{d} \cos q_j}. \qquad (7.72)$$

In the familiar language of solid state theory, this is an average over the Brillouin zone of the basic lattice:

$$\Gamma(l, t) = \int_{BZ} dq \frac{e^{iq \cdot l}}{1 - t\lambda(\mathbf{q})} \qquad (7.73)$$

where

$$\lambda(\mathbf{q}) \equiv \sum_{h} e^{iq \cdot h} \qquad (7.74)$$

This formula for the *lattice Green function* is valid for any Bravais lattice provided that all steps $\mathbf{h}$ to neighbouring lattice sites are of equal probability. The similarity of (5.152) to (7.74) is not an accident; the properties of a 'spherical ferromagnet' can be deduced by summing ladder graphs of unrestricted random walks (Domb & Joyce 1972). This function indeed plays a fundamental role in all statistical lattice problems (see e.g. Montroll & Weiss 1965).

For many purposes it is more convenient to divide each coefficient $w_n(l)$ by z^n, the total number of walks of n steps on a lattice of coordination number z. Thus the *probability generating function*,

$$P(l, t) \equiv \sum_{n=0}^{\infty} p_n(l)t^n = \sum_{n=0}^{\infty} w_n(l)z^{-n}t^n \equiv \Gamma(l, t/z), \qquad (7.75)$$

can be used to derive quantities such as $p_n(l)$, which is the probability of reaching site l after just n steps. For reasonably large values of n, this must, of course, yield the Gaussian approximation (7.29), i.e., in d dimensions,

$$p_n(l) \sim \frac{1}{(2\pi n)^{d/2}} e^{-l^2/2n}. \qquad (7.76)$$

One of the most interesting properties of an unrestricted random walk is the probability that it will eventually return to its starting point. Unfortunately, the infinite sum

$$P(0, 1) = \sum_{n=0}^{\infty} p_n(0) \qquad (7.77)$$

is not quite what we are seeking, since $p_n(0)$ includes the probability of returning to the origin more than once in the course of a walk on n steps. We introduce a more suitable quantity, f_n, which is defined as the probability of returning just *once* to the origin, at the *end* of the walk. Now we can split $p_n(0)$ into the probabilities of making a *first* return after $1, 2, \ldots$ steps, followed by a closed cycle of $n-1, n-2, \ldots$ steps respectively, to yield the recurrence relation

$$p_n(0) = f_1 p_{n-1}(0) + f_2 p_{n-2}(0) + \ldots + f_1 p_0(0). \qquad (7.78)$$

Thus, the generating function

$$F(t) = \sum_{n=1}^{\infty} f_n t^n \qquad (7.79)$$

satisfies

$$P(0, t) - 1 = F(t)P(0, t), \qquad (7.80)$$

for all values of t. The total probability of eventual return is thus

$$R = \sum_{n=1}^{\infty} f_n = F(1) = 1 - \{P(0, 1)\}^{-1}. \qquad (7.81)$$

But from (7.76) and (7.77) we see that

$$P(0, 1) \sim \sum_n \frac{1}{(2\pi n)^{d/2}}, \qquad (7.82)$$

which diverges for $d \leqslant 2$. In that case, $R = 1$; this is Polyá's theorem: *an unrestricted random walk on a lattice of one or two dimensions always returns, eventually, to its starting point.*

To evaluate the probability of return for a three-dimension lattice, we express $P(0, 1)$ as the integral (7.73) for $\Gamma(0, 1/z)$. As in (5.151) it is easy to verify that this integral diverges near $q = 0$ for $d = 1$, 2, but has a finite value for any three-dimensional case, so that there is a finite probability that the walker will 'escape'. The actual value of this integral depends on the lattice type, but it may be worth remembering that for the SC lattice the exact value of R (Domb 1954) is slightly greater than $\frac{1}{3}$.

What are the mathematical effects of restricting ourselves to *self-avoiding* walks? Rigorous analysis shows (e.g. Hammersley 1957) that the total number of such walks of n steps behaves asymptotically like

$$u_n \sim \zeta^n f_1(n) \tag{7.83}$$

where

$$\lim_{n \to \infty} [f_1(n)]^{1/n} = 1. \tag{7.84}$$

This is consistent with (5.187) where $f_1(n)$ was assumed to behave like $n^{-\alpha}$. But the values $\frac{3}{2}$ and $\frac{7}{4}$, respectively, for the exponent α, in two and three dimensions, and the estimates for the *connective constant* ζ for various lattices (table 5.1) have been obtained only by computational enumeration of walks.

As already pointed out, numerical studies also confirm the exponents $\frac{3}{4}$ and $\frac{3}{5}$ for the asymptotic behaviour (7.67) of the radius of gyration, for $d = 2$ and $d = 3$, as predicted by the Flory approximation (7.66). The most significant effect of self-avoidance seems to be to reduce the probability of return to the origin after n steps; in both two and three dimensions, this comes out proportional to n^{-2}, which is much smaller than the form

$$p_n(0) \sim n^{-d/2} \tag{7.85}$$

derived from (7.76) for the unrestricted case. According to (7.77) and (7.81) there is a finite probability of 'escape' from the origin in a two-dimensional self-avoiding walk, thus showing the delicacy of Polyá's theorem.

The statistical properties of the self-avoiding random walk on a lattice are closely related, by graphical approximations (§ 5.10), to the thermodynamic properties of the corresponding Ising model (see e.g. Domb 1969) and other 'magnetic' models (e.g. Bowers & McKerrell 1973). An interesting mathematical question, reminiscent of the theory of critical phenomena, concerns the nature of the transition from behaviour characteristic of the simple unrestricted walk to that characteristic of the self-avoiding case. The prohibition of returns to the same site after only a limited number of steps can be dealt with by the Markov method (§ 7.6) and does not change the long-range asymptotic behaviour of the system. But we may introduce something like a repulsive potential J between all segments of the polymer chain and thus reduce the *a priori* probability of any 'overlap' by a factor

$$1 + w = \exp(-\beta J) \tag{7.86}$$

as if we are dealing with a thermodynamic problem. Thus, for $T \to \infty$, we have $w = 0$, as in the simple unrestricted walk, whilst $T = 0$, $w = -1$, corresponds to the ideal self-avoiding walk. The mathematical analysis of this model is not complete, but the evidence (Domb & Joyce 1972; Domb, Barrett & Lax 1973) is that the expansion factor α^2 (cf. (7.62)–(7.66)), the generating function (7.75) and other statistical properties depend essentially on the single variable $-wn^{\frac{1}{2}}$ and that there is no singularity in these functions at the point of perfect self-avoidance, $w = -1$. The transition to non-Gaussian behaviour thus seems to take place immediately upon the introduction of a repulsive interaction (i.e. as we leave $w = 0$ in the negative direction), but would be apparent only for very long chains if $J/\hbar T$ were very small. In other words, for a given strength of interaction, the coil expands outwards as the temperature falls, without any cooperative order–disorder transition.

Some of the thermodynamic effects of the excluded volume interaction between different macromolecular chains are adequately described by the Flory–Huggins theory (§ 7.2). But this approximation is obviously misleading when it comes to such a topologically constrained phenomenon as the *melting* of a macromolecular crystal. Some of the qualitative features of such phenomena may, however, be deduced from simple lattice models (Nagle 1974). Consider, for example, the arrow representation (§ 1.4) for ferroelectric disorder. Any distribution of the arrows that is consistent with the ice condition can be reinterpreted as a configuration of a dense assembly of polymer chains. All that is necessary (fig. 7.13) is to divide the lattice into two interpenetrating sub-lattices of A- and B-type vertices and to substitute a polymer segment for each arrow that points from an A site to a B site. The different energies assigned to different types of vertex in the various ferroelectric models now refer to bond configurations similar in some respects to the 'trans' and 'gauche' configurations of the rotational isomeric chain (§ 7.3). An order–disorder transition of the ferroelectric system would thus be interpreted as the 'melting' of a macromolecular

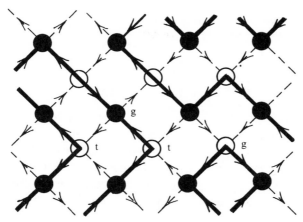

Fig. 7.13. Polymer chain configuration defined uniquely by an arrow diagram (cf. fig. 1.7). Each polymer 'bond' runs *from* a site o of one sub-lattice *to* a site ● of the other sub-lattice. The 'gauche' and 'trans' configurations are assigned according to the 'KDP' energy at each vertex.

'crystal' into a 'liquid' assembly of interpenetrating, conformally disordered chains. Although the defining characteristics of this sort of model are obviously very far from the realities of macromolecular physics, this simple topological transformation from a case of substitutional disorder to a model of long-chain disorder invites closer study.

7.9 Continuum models

On a 'macroscopic' scale (i.e. down to the limits of resolution of a powerful electron microscope!) the chemical segmentation of a macromolecular chain is of little significance. It is physically reasonable, therefore, as well as being mathematically advantageous, to consider the limiting case of a *flexible chain*, or *continuous walk*, whose total length, $L = nl$, is made up of a quasi-infinite number of steps, n, each of 'infinitesimal' length l. From the theory of Brownian motion, it is well known that a random walk tends, in this limit, to a process of *diffusion*.

This can be verified from the Gaussian function (7.29) for the probability density of arriving at **R** on a path of length L;

$$G(\mathbf{R};L) \equiv P(\mathbf{R}) = \left(\frac{3}{2\pi l L}\right)^{\frac{3}{2}} \exp\left(-\frac{3R^2}{2lL}\right). \tag{7.87}$$

This function is a solution of the elementary *diffusion equation*

$$\left\{\frac{\partial}{\partial L}-\frac{1}{6}l\,\nabla_{\mathbf{R}}^2\right\}G(\mathbf{R};\,L)=0 \qquad (7.88)$$

with the boundary condition

$$\lim_{L\to 0}G(\mathbf{R};\,L)=\delta(\mathbf{R}). \qquad (7.89)$$

Thus, if we think of L as a quantity that is increasing steadily with time, the coefficient $\frac{1}{6}l$ is the 'diffusion constant' for the spreading out of the probability density for the position of a particle that started from the point $\mathbf{R}=0$ at the initial time $L=0$. Of course, for a genuine random walk of a finite number of steps (7.87) is not quite correct, but the Gaussian approximation is known to be very good indeed for more than, say, $n=6$.

The great mathematical advantage of this *continuum representation* is that we have all the apparatus of classical and quantum field theory at our command. But since the original general derivation of the Gaussian form (7.87), using the Central Limit theorem, is valid only for Markovian walks, it is necessary to show how the partial differential equation (7.88) can be deduced directly from the definition of such processes without solving explicitly for $G(\mathbf{R};\,L)$.

Consider a finite chain of n steps, l_k. In the absence of correlations between successive steps we can write the probability of reaching the point $\mathbf{R}$ quite formally as a multiple integral over all the n variables l_k:

$$P(\mathbf{R};\,n)=\int \mathrm{d}\{l_k\}\Big[\prod_{j=1}^{n}\tau(l_j)\Big]\delta\Big(\sum_{j=1}^{n}l_j-\mathbf{R}\Big). \qquad (7.90)$$

For the standard *random flight* model (§ 7.3) the *a priori* probability that the jth step is the vector l_j would be

$$\tau(l_j)=\frac{1}{4\pi l^2}\delta(|l_j|-l), \qquad (7.91)$$

because we insist that it should be of exactly length l. If we put this into (7.90) we get rather complicated results that are easily avoided by making a Gaussian approximation for the probability distribution of each link. Thus,

$$\tau(l_j)=\left(\frac{3}{2\pi l^2}\right)^{\frac{3}{2}}\exp\left\{-\frac{3l_j^2}{2l^2}\right\} \qquad (7.92)$$

gives the same mean square step length as (7.91). In any case, since the parameter l would normally stand for the length l' of an *equivalent random link* (7.33) of a real macromolecular chain, there is no significant loss of realism in this modification. Putting (7.92) into (7.90), we get

$$P(\mathbf{R}; n) = \int d\{l_k\} \exp\left[-\sum_{j=1}^{n} \frac{3l_j^2}{2l^2} \right] \delta\left(\sum_{j=1}^{n} l_j - \mathbf{R} \right). \tag{7.93}$$

Now let us think of the successive joints of the chain as falling on a *path* $\mathbf{R}(s)$, defined parametrically as a function of the distance variable s along its length. The interval $\Delta s = l$ between successive joints is supposed to be very small, so that

$$\frac{l_j}{l} = \frac{\mathbf{r}(s + \Delta s) - \mathbf{r}(s)}{\Delta s} \to \dot{\mathbf{r}}(s) \tag{7.94}$$

tends to the derivative along the tangent to the curve. In this limit, therefore, we write the sum in (7.93) as an integral along the same path:

$$\sum_{j=1}^{n} \frac{3l_j^2}{2l^2} \to \int_0^L ds \frac{3}{2l} \dot{\mathbf{r}}^2(s). \tag{7.95}$$

Notice, however, that one power of the 'infinitesimal' l still remains explicit in this expression, showing that the segment length is always a genuine physical parameter of the model even though the abrupt changes of direction at each joint have been smoothed into a continuous path.

Under these conditions, the probability distribution function (7.92) is transformed into a function of the continuous variable L, represented formally as a *functional integral* (see e.g. Freed 1972*a*)

$$G(\mathbf{R}; L) = \int_{\mathbf{r}(0)=0}^{\mathbf{r}(L)=\mathbf{R}} \mathcal{D}[\mathbf{r}(s)] \exp\left[-\int_0^L ds \frac{3}{2l} \dot{\mathbf{r}}^2(s) \right]. \tag{7.96}$$

The integration sign here stands for a summation over every continuous curve $\mathbf{r}(s)$, of length L, joining the origin to the point $\mathbf{R}$. The exact meaning of the symbol $\mathcal{D}[\mathbf{r}(s)]$, as a measure of the relative weights of various alternative paths and their overall normalization, need not concern us here. For all practical purposes (7.96) is what one gets from (7.93) if all the steps are very short compared with the total length of the chain. Expressions of this kind are, of course, quite familiar in the branch of functional analysis known traditionally as the *calculus of variations*.

To arrive at the diffusion equation (7.88) we now make use of what is called the *Smoluchowsky–Chapman–Kolmogorov equation* in the theory of Markov processes, i.e. the *convolution identity* implicit in the definition of (7.96):

$$G(\mathbf{R}; L) \equiv \int d\mathbf{R}' G(\mathbf{R} - \mathbf{R}'; L - L') G(\mathbf{R}'; L'). \tag{7.97}$$

This arises simply from the fact that every continuous curve of length L, running from the origin to the point $\mathbf{R}$, can be constructed out of a curve of length l' ($< L$) running to $\mathbf{R}'$, together with a curve of length $L - L'$ from $\mathbf{R}'$

to **R**. The integration over all positions of the intermediate point **R′** counts all such possibilities with the correct weight.

But suppose that $(L - L')$ is an infinitesimal quantity, ε, comparable to l. We can go back to (7.93) and write

$$G(\mathbf{R} - \mathbf{R}'; \varepsilon) \approx \left(\frac{3}{2\pi l\varepsilon}\right)^{\frac{3}{2}} \exp\left\{-\frac{3|\mathbf{R} - \mathbf{R}'|^2}{2l\varepsilon}\right\}. \tag{7.98}$$

This is a function that falls off very rapidly with $(\mathbf{R} - \mathbf{R}')$, which we can treat as our new variable of integration, $\boldsymbol{\eta}$. In other words (7.97) takes the approximate form

$$G(\mathbf{R}, L) \approx \int d\boldsymbol{\eta} \left(\frac{3}{2\pi l\varepsilon}\right)^{\frac{3}{2}} \exp\left(\frac{-3\eta^2}{2l\varepsilon}\right) G(\mathbf{R} - \boldsymbol{\eta}; L - \varepsilon), \tag{7.99}$$

showing the connection between an increment ε in the length L of the chain and an increment $\boldsymbol{\eta}$ in the position **R** of its end point. Expanding in a Taylor series,

$$G(\mathbf{R} - \boldsymbol{\eta}; L - \varepsilon) \approx G(\mathbf{R}; L) - \varepsilon \frac{\partial G(\mathbf{R}; L)}{\partial L} - \boldsymbol{\eta} \cdot \nabla_{\mathbf{R}} G(\mathbf{R}; L)$$
$$+ \tfrac{1}{2} \boldsymbol{\eta}\boldsymbol{\eta} : \nabla_{\mathbf{R}} \nabla_{\mathbf{R}} G(\mathbf{R}; L) + o(\varepsilon \eta^2) \ldots \tag{7.100}$$

we can carry out the integration in (7.99), and find that the terms of order ε vanish only if $G(\mathbf{R}; L)$ satisfies the diffusion equation (7.88).

The real value of this approach is seen when we want to include any additional fields or interactions in the model. Suppose, for example, that each link of the chain is subject to a local electromechanical or chemical potential $W(\mathbf{r})$. The probability density $P(\mathbf{R}; n)$ will now be interpreted as a partition function, where the *a priori* probability distribution (7.92) at the jth link will have acquired the Boltzmann factor

$$Z_j = \exp\{-\beta W(\mathbf{r}_j)\}. \tag{7.101}$$

The product of all these factors will now appear in (7.93) and in the continuum limit will contribute to the functional integral (7.96) a factor

$$\exp\left\{-\beta \sum_j W(\mathbf{r}_j)\right\} \rightarrow \exp\left\{-\int_0^L ds \beta W[\mathbf{r}(s)]\right\}. \tag{7.102}$$

This chain is still Markovian; it is easy to verify that (7.98) and (7.99) are modified in such a way as to add the 'potential' $\beta W(\mathbf{R})$ to the diffusion equation. It is interesting to observe that this generalization of (7.88), i.e.

$$\left\{\frac{\partial}{\partial L} - \frac{1}{6} l \nabla_{\mathbf{R}}^2 + \beta W(\mathbf{R})\right\} G(\mathbf{R}; L) = \delta(\mathbf{R})\delta(L), \tag{7.103}$$

is analogous to a Schrödinger equation

$$\left\{\frac{\hbar}{i}\frac{\partial}{\partial t} - \frac{\hbar^2}{2m}\nabla_{\mathbf{R}}^2 + V(\mathbf{R})\right\}K(\mathbf{R};\ t) = \delta(\mathbf{R})\delta(t) \qquad (7.104)$$

for the space–time propagator of a quantum-mechanical particle in the potential $V(\mathbf{R})$. This analogy is not accidental, since (7.104) can be derived from a *Feynman path integral* very similar to (7.96).

The physical situation described by (7.103) is not particularly interesting, but this equation is an important ingredient of an approximate solution (Edwards 1965) of the excluded volume problem (§ 7.7). Let us introduce a short-range repulsive interaction $w(\mathbf{r}_i - \mathbf{r}_j)$ between all pairs of segments. Just as in (7.101) and (7.102), the partition function will be a functional integral like (7.96) containing the additional 'Boltzmann factor',

$$\exp\left\{-\tfrac{1}{2}\beta\sum_{i\neq j}w(\mathbf{r}_i-\mathbf{r}_j)\right\}\rightarrow\exp\left\{-\tfrac{1}{2}\beta\int_0^L \mathrm{d}\,s\int_0^L \mathrm{d}s'w[\mathbf{r}(s)-\mathbf{r}(s')]\right\},\ (7.105)$$

to be evaluated along every path.

Unfortunately, this term hinders the transformation of the path integral into a partial differential equation because the system is no longer Markovian and the convolution (7.97) is no longer valid. To proceed further (see e.g. Freed 1972a) one must introduce a *three-point distribution function* $G_3(\mathbf{R}, \mathbf{R}';\ L, L')$ which counts the number of chains of length L going to $\mathbf{R}$ that also pass through $\mathbf{R}'$ after a path length L'. In place of (7.86) or (7.103), we then obtain an integrodifferential equation

$$\left\{\frac{\partial}{\partial L}-\frac{1}{6}l\,\nabla_{\mathbf{R}}^2\right\}G(\mathbf{R};\ L)+\beta\int \mathrm{d}\mathbf{R}'w(\mathbf{R}-\mathbf{R}')\int_0^L \mathrm{d}sG_3(\mathbf{R}, \mathbf{R}';\ L, s)=\delta(\mathbf{R})\delta(L).$$

$$(7.106)$$

But how are we to calculate G_3? It turns out that (7.106) is only the first of a hierarchy of similar equations, involving G_4, G_5, . . . etc. The excluded volume problem is really a many-body problem; (7.106) is the analogue of the relation (2.40) between the two-body and three-body distribution functions in a liquid and of the identity (5.20) linking two-spin and three-spin correlation functions in the Ising model. The hierarchy can only be terminated by some *ad hoc* assumption, such as the superposition approximation (2.17), which leads to the **BBGKY** theory of liquids (§ 2.12) or the analogous approximation (5.23) which leads to the random phase approximation for the Ising model.

The most natural assumption is that G_3 is approximately Markovian, i.e.

$$G_3(\mathbf{R}, \mathbf{R}';\ L, L')\approx G(\mathbf{R};\ L)G(\mathbf{R}';\ L'). \qquad (7.107)$$

which would lead exactly to the convolution (7.97). It is physically reasonable also to assume that the 'self-exclusion interaction' $w(\mathbf{r}_i-\mathbf{r}_j)$ is just a positive delta function of strength w. Under these circumstances (7.106) is reduced to the form of (7.103) with a *self-consistent potential*

$$\beta W(\mathbf{R}) = \beta \int d\mathbf{R}' \int_0^L ds\ w(\mathbf{R} - \mathbf{R}')G(\mathbf{R}'; s)$$
$$= \beta w \int_0^L G(\mathbf{R}; s)ds. \qquad (7.108)$$

It is reasonable to assume that $G(\mathbf{R}; s)$ is spherically symmetrical, but even then the simultaneous solution of (7.103) and (7.108) is not a trivial mathematical problem (Edwards 1965). Quite simple arguments (de Gennes 1969), however, clarify the main conclusion that the mean square radius of an isolated macromolecular chain of length L should increase as

$$\langle R^2 \rangle \sim L^{6/5}, \qquad (7.109)$$

exactly as in (7.67) and as confirmed for lattice models. This analysis thus provides more formal mathematical justification for the Flory phenomenological formula (7.66). The question of whether better solutions of (7.106), etc. can be found (Reiss 1967; Freed 1971; Kyselka 1974) begins to look rather academic.

7.10 Entanglements

The physical properties of dense macromolecular systems (§§ 7.5, 7.6) are dominated by the interactions between many different chains. Consider the case of an unvulcanized rubber – an assembly of long chains without chemical cross-links. It is intuitively obvious that the relatively low density of chain ends in such a system is of no great physical significance, so that we may assume (Edwards & Freed 1969) that these have been connected up into a single superchain of enormous length, filling the whole volume of the specimen. In other words, the chain is confined to a 'box', on whose boundary there is an infinite potential $W(\mathbf{R})$. The solution of (7.103) in this case is just the solution of the simple diffusion equation (7.88), with the additional condition that $G(\mathbf{R}; L)$ must vanish on the boundary of the box. The plausible result is obtained that the molecule exerts a pressure on this boundary, but the whole model is quite unphysical. It is easy to see, for example, that the actual solution of (7.88) in these conditions does not have the polymer spread uniformly throughout the box but makes the density pile up in the centre. It is essential to include the self-avoidance interaction (7.105) if we are to reproduce the incompressibility of such a material beyond the limit of dense random packing.

But such a material would not behave simply like a dense vapour or liquid. It would be a 'bouncing putty': it would respond elastically to

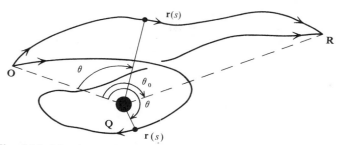

Fig. 7.14. Number of circuits around constraint **Q** is topological invariant of the entanglement.

impulsive strains, just as if it were a rubber (§ 7.6). This resistance to rapid changes of shape is not due to chemical cross-linkings between the chains, but simply to the constraints imposed by their mutual *entanglements*.

A mathematical description of a tangled mass of mutually impenetrable molecular chains is obviously beyond our present reach. But the functional integral formalism provides a possible line of entry into the problem (Edwards 1967, 1968). Suppose, for example, that we consider a chain of length L, going from **O** to **R** in a two-dimensional space, in the neighbourhood of an impenetrable obstacle at **Q** (fig. 7.14). As the representative point $r(s)$ moves along the chain from **O** to **R**, the angle subtended at **Q** changes by an amount

$$\int_0^L \dot\theta \, ds = \theta_q = \theta_0 + 2\pi q \tag{7.110}$$

where θ_0 is the angle **OQR** and q is the number of times the path loops around **Q**. It is evident that the value of q is a *topological invariant* of the model; all configurations available to the chain by continuous deformation must have the same value of this integer. The distribution function $G(\mathbf{R}, q; L)$ would thus be defined subject to this constraint.

A functional integral representation of this distribution would have to contain a selection factor

$$\delta(\theta_q - \int_0^L \dot\theta \, ds) = \delta(\theta_q - \int_0^L \mathbf{A} \cdot \dot{\mathbf{r}} \, ds), \tag{7.111}$$

where the vector field

$$\mathbf{A}(\mathbf{r}) = \mathbf{r} \wedge \boldsymbol{\xi}/r^2 \tag{7.112}$$

is generated from a unit vector $\boldsymbol{\xi}$, normal to the plane at **Q**. We use the identity (5.146), i.e.

$$\delta(x) \equiv \frac{1}{2\pi} \int_{-\infty}^{\infty} \exp(i\lambda x) \, d\lambda, \tag{7.113}$$

to insert this factor into the two-dimensional form of (7.96), i.e.

$$G(\mathbf{R}, q; L) = \frac{1}{2\pi} \int_{-\infty}^{\infty} d\lambda \, \exp(i\lambda\theta_q) \int_0^{\mathbf{R}} \mathscr{D}[\mathbf{r}(s)] \exp\left\{ -\frac{1}{l} \int_0^L ds(\dot{\mathbf{r}}^2 + il\lambda\mathbf{A}\cdot\dot{\mathbf{r}}) \right\}$$

$$= \frac{1}{2\pi} \int_{-\infty}^{\infty} d\lambda \, \exp(i\lambda\theta_q) \, \Gamma(\mathbf{R}, \lambda; L). \tag{7.114}$$

The standard procedure (7.95)–(7.100) allows us to transform the functional integral into a partial differential equation for $\Gamma(\mathbf{R}, \lambda; L)$. But the analogy between the diffusion equation (7.103) and the Schrödinger equation (7.104) gives a clue to the result. The function $(\dot{\mathbf{r}}^2 + il\lambda\mathbf{A}\cdot\dot{\mathbf{r}})$ is analogous to the Lagrangian of a charged particle moving in the field of a vector potential $il\lambda\mathbf{A}$; we deduce that (7.114) contains the path integral representation of the 'propagator' of the corresponding 'Schrödinger equation', i.e.

$$\left[\frac{\partial}{\partial L} - \frac{1}{4} l\{\nabla_{\mathbf{R}} - i\lambda\mathbf{A}(\mathbf{R})\}^2 \right] \Gamma(\mathbf{R}, \lambda; L) = \delta(\mathbf{R})\delta(L). \tag{7.115}$$

From known forms of the solution of this equation in terms of Bessel functions, etc. we can construct analytic expressions for $G(\mathbf{R}, q; L)$ and hence determine the effect of the constraint on the statistical properties of the chain.

What we have in mind, of course, is that our polymer chain is part of a rubber network and that the obstacle at $\mathbf{Q}$ is actually another polymer chain with which it is entangled. The formal analysis for a three-dimensional model is not difficult to set up, and asymptotic solutions can be obtained in simple cases. We might thus hope to estimate the effects of entanglements on the restoring forces in a rubber, as a modification of the general theory of § 7.6.

But a very serious and fundamental difficulty arises: an invariant such as (7.110) does not discriminate adequately between topologically non-equivalent cases. It is easy to verify, for example, that it would not distinguish between the plain crossing of two chains, as in fig. 7.15(a) and the case where one chain is *knotted* about the other by entanglement with itself, as in fig. 7.15(b). Although it is technically possible (Edwards 1968) to define successively more complex analytical representations of the more complex topological invariants that may arise in this way, the procedure obviously does not converge to any more simple general result. The mathematical theory of disordered systems seems completely open in this direction.

Nor can it be said that the effects of cross-linkages in a rubber network can yet be deduced analytically from a continuum theory. The difficulty is

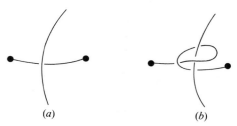

Fig. 7.15. (*a*) Plain and (*b*) 'knotted' crossing of polymer chains with the same 'entanglement' invariant.

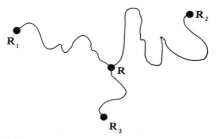

Fig. 7.16. Linkage requiring four-point correlation function.

very similar to the fundamental problem of the theory of liquids and glasses (§ 2.7): many-point correlation functions cannot be neglected. Consider, for example, the very simplest case (fig. 7.16) of three chains, lengths L_1, L_2, L_3, starting from fixed points R_1, R_2, R_3 and tied together at R. The statistical distribution of R for all configurations of the chains is a four-point function $G_4(R, R_1, R_2, R_3; L_1, L_2, L_3)$ which cannot necessarily be represented as the superposition of two-point functions as in (7.106). The field-theoretical formalisms of Edwards & Freed (1970) do not seem to provide for this possibility.

8

Excitations on a disordered linear chain

'Wit will shine
Through the harsh cadence of a rugged line'
Dryden

8.1 Dynamic, magnetic and electronic excitations

In previous chapters we have studied the spatial arrangement of the atomic centres in various types of disordered system; we now consider the characteristic dynamic, magnetic and electronic *excitations* of such structures. These have many mathematical features in common.

Suppose, for example, that the system has a configuration of minimum potential energy in which the *l*th atom would be at R_l. As in the conventional theory of *lattice dynamics* we assume that a small displacement u_l of this atom is subject to restoring forces that are linear in the relative displacements of nearly atoms. In general, therefore, the classical equations of motion for these displacements take the form

$$M_l \ddot{u}_l = -\sum_{l'} \Phi_{ll'} \cdot u_{l'}, \tag{8.1}$$

where M_l is the mass of this atom and the tensor $\Phi_{ll'}$ measures the forces produced at the site R_l by displacements of the atom at $R_{l'}$.

For a regular lattice of identical atoms, the force constants have the translational symmetry

$$\Phi_{ll'} \equiv \Phi(R_l - R_{l'}), \tag{8.2}$$

and the *N* equations (8.1) become equivalent to one another under a lattice translation. This symmetry permits the exact reduction of the whole set by a Fourier transformation (1.42), from which we generate the *phonon* modes of the crystal. For a disordered system, however, where the various coefficients may become random variables, this general symmetry is absent, and the equations (8.1) must be solved by other means.

The simplest case is perfect *isotopic disorder*, where the masses M_l are not the same on every equivalent lattice site. This is obviously a very special case of *substitutional disorder* (§ 1.2), but is often used to describe the effects of mass differences on the lattice dynamics of a substitutional alloy (§ 1.5). Although one ought, in principle, to include short- and long-range order correlations (§ 1.6) in the spatial distribution of the different types of atom, much attention is given to the *binary alloy* model where A- and B-type atoms, of masses M_A and M_B, are distributed at random on a regular lattice.

Since we are mainly interested in the spectrum of the normal modes we assume that each $\mathbf{u}_l$ varies with the same frequency ω, so that (8.1) can be written

$$(\mathbf{\Phi}_{ll} - \omega^2 M_l \mathbf{1}) \cdot \mathbf{u}_l + \sum_{l' \neq l} \mathbf{\Phi}_{ll'} \cdot \mathbf{u}_{l'} = 0. \tag{8.3}$$

It is clear that isotopic disorder affects only the *diagonal* elements of the matrix of these equations. In a genuine alloy, however, 'chemical' differences between the constituents give rise to variations in the force constants, which would then fail to satisfy the symmetry condition (8.2). This is a case where *off-diagonal disorder* cannot be ignored.

More generally, in a liquid or glass, the equilibrium positions $\mathbf{R}_l$ are not topologically isomorphous with a regular lattice. In such a system, not only are all the components of the force tensors $\mathbf{\Phi}_{ll'}$ random variables, but also the very notion of a site l' being 'near' to a given site l can only be defined by statistical terms or by reference to a computer model (§§ 2.10, 2.11). For this reason, the theory of the excitations of topologically disordered systems (chapter 11) is particularly intractable.

The familiar analogy between phonons and *magnons* (§ 1.8) immediately suggests that equations similar to (8.3) could be written down for *spin deviations*. We might start, for example, from the spin Hamiltonian (1.17) under ferromagnetic conditions, where

$$\langle S_l^{(z)} \rangle \lesssim S, \tag{8.4}$$

and write down the equation of motion (in the Heisenberg representation) for each of the spin deviation operators

$$S_l^\pm \equiv S_l^{(z)} \pm i S_l^{(y)}. \tag{8.5}$$

This takes the form

$$i\hbar \partial S_l^{(-)}/\partial t = -2 \sum_{l' \neq l} \{J_\parallel(ll') S_l^{(-)} S_{l'}^{(z)} - J_\perp(ll') S_{l'}^{(-)} S_l^{(z)}\}. \tag{8.6}$$

If we assume that every $S_l^{(-)}$ varies with the same frequency ω, and that the system is very nearly ordered, as in (8.4), we get a set of equations,

$$\left[\left\{ 2S \sum_{l' \neq l} J_\parallel(ll') \right\} - \hbar\omega \right] S_l^{(-)} - 2S \sum_{l' \neq l} J_\perp(ll') S_{l'}^{(-)} = 0, \tag{8.7}$$

Which are essentially of the form (8.3). For a pure perfect crystal these can be transformed into the quasi-classical equations for ferromagnetic spin waves, as in (1.47). But the random mixture of magnetically different atoms in an alloy would introduce diagonal and off-diagonal disorder into the exchange parameters $J_{\parallel}$ and $J_{\perp}$ (§ 1.5). The case of *antiferromagnetic* states in a disordered system is obviously analogous, although very much more complicated because of the difficulty of constructing an ordered ground state (§ 5.6).

The same equations arise quite naturally in the *tight-binding model* for the electron states of a condensed system. We assume, for example, that the *l*th atom, if it were isolated, would have an *atomic potential* $v_l(\mathbf{r})$ in which there are a number of different *atomic orbitals* $\psi_l^{(\alpha)}(\mathbf{r})$ corresponding to the atomic energy levels $\mathscr{E}_l^{(\alpha)}$. If we assume that the potential of the whole system is not too far from a *superposition of atomic potentials,* i.e.

$$\mathscr{V}(\mathbf{r}) \sim \sum_l v_l(\mathbf{r} - \mathbf{R}_l), \tag{8.8}$$

then it is reasonable to suppose that the solution of the Schrödinger equation

$$\left\{ -\frac{\hbar^2}{2m}\nabla^2 + \mathscr{V}(\mathbf{r}) \right\}\Psi(r) = \mathscr{E}\Psi(\mathbf{r}) \tag{8.9}$$

can be written as a *linear combination of atomic orbitals* (LCAO) of the form

$$\Phi(\mathbf{r}) = \sum_{l,\alpha} u_l^{(\alpha)}\psi_l^{(\alpha)}(\mathbf{r} - \mathbf{R}_l). \tag{8.10}$$

Putting this into (8.9), and using the natural properties of the atomic orbitals, we get a set of linear equations for the coefficients $u_l^{(\alpha)}$.

In practice, the various *overlap integrals* that appear in these equations are too complicated and too uncertain to be evaluated from first principles. In the case of a perfect lattice the best we can do is to write down a simplified version, in the form

$$(\mathscr{E}_l^{(\alpha)} - \mathscr{E})u_l^{(\alpha)} + \sum_{l' \neq l}\sum_{\beta} V_{ll'}^{(\alpha\beta)}u_{l'}^{(\beta)} = 0, \tag{8.11}$$

where the matrix elements $V_{ll'}^{(\alpha\beta)}$ of the *model Hamiltonian* are chosen empirically so as to reproduce the electronic band structure of the crystal. These equations, again, are closely analogous to (8.3).

In the theory of disordered alloys it is often assumed that we have only a single atomic orbital $\psi_l(\mathbf{r})$ on each atom, but that this has a different *bound state energy* $\mathscr{E}_l$ for each chemical constituent of the alloy. It is obvious, however, that any realistic *tight-binding alloy* (TBA) model should also have considerable off-diagonal disorder in the *interaction energy* $V_{ll'}$ corresponding to different amounts of overlap of the orbitals on different combinations of constituents on neighbouring sites. This effect is particularly important where the interatomic distances are not all the same, as in the extreme topological disorder of a 'gas' of impurity atoms in a semiconductor (§ 2.15).

Since the theory of *Frenkel excitons* is usually set up in the LCAO representation, the same type of equations will arise when we study these in disordered systems. But for more extended electron–hole pairs, such as the *Wannier exciton*, this type of *local representation* is not satisfactory. This applies particularly to the ordinary *conduction electron* states in metals, which are not correctly described in terms of a finite set of atomic orbitals. It is true that the Bloch states of a regular crystal can always be transformed into a linear combination of localized *Wannier functions*, analogous to the atomic orbitals $\psi_l^{(\alpha)}$ in (8.10), and that the coefficients then satisfy equations like (8.11). Since each band of Bloch states contributes a single Wannier function to each lattice site, it would almost seem as if one could construct the conduction band of a metallic alloy by a simple modification of the TBA model. But the Wannier representation is only valid for a perfect crystal with lattice translational symmetry: there is no *a priori* prescription for the two types of localized functions that would need to be attached to the two different constituents of a binary alloy in such a way as to make (8.11) a reasonable approximation to the Schrödinger equation (8.9). In all such systems, the effects of disorder on the electron states demand a somewhat different scheme of approximation, approached through scattering theory (chapter 10).

With these important exceptions, the theory of the excitations of disordered systems is centred about a standard mathematical model, defined in general by the equations

$$(\mathscr{E}_l - \lambda)u_l + \sum_{l' \neq l} V_{ll'}u_{l'} = 0. \tag{8.12}$$

In discussing the many interesting properties of these equations we may often leave open the question of whether the various symbols are to be interpreted as Cartesian vectors and tensors, as column vectors and square matrices, as quantum-mechanical operators, or as ordinary scalar quantities. All that we need remember is that the variable u_l represents the *amplitude* of the excitation on the *l*th atomic site, whilst the *spectral* *variable*, λ, stands for the square of the dynamical frequency ω^2, or for the quantized energy $\hbar\omega$ of a magnon or exciton, or for the energy $\mathscr{E}$ of an electron eigenstate of the Hamiltonian of the whole system, as the case may be. The meaning and statistical properties of the *diagonal elements* $\mathscr{E}_l$, and of the *off-diagonal elements* $V_{ll'}$, can then be ascertained by comparing (8.12) with (8.3), (8.7) or (8.11). In the lattice dynamical case, the interpretation

$$\mathscr{E}_l \to M_l^{-1}\Phi_{ll}; \quad V_{ll'} \to M_l^{-1}\Phi_{ll'} \tag{8.13}$$

has the unfortunate effect of apparently generating 'off-diagonal disorder' from isotopic randomness of the masses M_l, but no important qualitative features of the system are thereby lost. More complex systems where, for example, there are a number of atoms per unit cell can be handled symbolically by appropriate conventions of matrix algebra.

Where the underlying arrangement of sites forms a regular lattice, we naturally replace the label l by the lattice $l \equiv \mathbf{R}_l$. If the system were, in fact, ordered, with $\mathscr{E}_l$ independent of l, and $V_{ll'}$ satisfying the condition of translational symmetry (8.2), we could immediately solve the equations (8.12) by Bloch's theorem (§1.8). The Fourier transformations (1.42), (1.43) yield

$$\left[(\mathscr{E}_0 - \lambda) + \sum_{\mathbf{h} \neq 0} V(\mathbf{h})e^{i\mathbf{q}\cdot\mathbf{h}}\right]U(\mathbf{q}) = 0 \tag{8.14}$$

showing that λ is an eigenvalue of the matrix [ ]. In the lattice-dynamical case, this is, of course, the familiar dynamical matrix for the squares, $[\omega(\mathbf{q})]^2$, of the normal mode frequencies of wave vector $\mathbf{q}$. In the most elementary tight-binding model (TBM) for a metal, with a single atomic level at each site, we get a typical *band* of states, with energies

$$\lambda(\mathbf{q}) = \mathscr{E}_0 + \sum_{\mathbf{h} \neq 0} V(\mathbf{h})e_{i\mathbf{q}\cdot\mathbf{h}}. \tag{8.15}$$

If the 'overlap integral' $V(\mathbf{h})$ has value $-V$, and extends only to the z nearest neighbours of a given site, we see that this band is centred on $\mathscr{E}_0$, and has a total width

$$2B \leqslant 2zV, \tag{8.16}$$

with lower and upper limits at the centres ($\mathbf{q}=0$) and corners of the Brillouin zone, respectively. The *band width* parameter B thus provides a convenient measure of the energy scale of the system, and of the strength of the interaction between neighbouring sites of the lattice. From the standpoint of conventional solid state physics, these remarks are, of course, completely trivial, and the model of an electron or phonon band implicit in

(8.15) is very far from any real system. In the theory of disordered systems, however, this sort of model is often the only case that can be treated with adequate mathematical success.

8.2 One-dimensional models

Although many characteristic features of disordered systems are not to be found in one-dimensional models (§§ 2.2, 2.4), and despite the extreme artificiality of real systems that simulate such models (§ 2.3), we devote this chapter to a study of the excitations of a disordered linear chain. The theory of the spectral properties of such a chain is by no means as trivial as the classical statistical mechanics of one-dimensional systems (§§ 5.5, 6.1) and some unexpected phenomena make their appearance. It is natural, moreover, to use this simple model, whose behaviour can be calculated exactly or computed to any desired degree of accuracy, as a test-bed for any mathematical technique that is eventually to be applied to more realistic problems. If a new method fails to yield satisfactory answers for a one-dimensional system, then it is unlikely to be a sound approximation in general.

The most important theoretical limitation of one-dimensional models is that they are *always* topologically ordered (§ 2.2). This means, for example, that the 'site label' l in the TBM equations (8.12) is always equivalent to the lattice 'vector' (2.3) of a regular lattice, with the same average spacing as the actual system. Diagonal disorder in the level parameter $\mathscr{E}_l$, or off-diagonal disorder in the interaction energies $V_{ll'}$, may be attributed to physical or chemical differences between the constituents of equally spaced cells of a regular chain; it may, on the other hand, be a secondary consequence of fluctuations in the relative spacing ξ_l of atomic sites along the chain, as in (2.5). Mathematically speaking, there is no distinction between the substitutional disorder of a 'one-dimensional alloy' and the effects of random spacing in a 'linear glass' or 'linear liquid' (§ 2.2). The physical assumptions made about the model merely govern the statistical properties postulated for the diagonal and off-diagonal elements of the equations (8.12).

It turns out, in fact, that the equations for electronic excitations in any one-dimensional potential at *any* energy can be cast into essentially the same form. The easiest way of seeing this (e.g. Hori 1968a) is to follow the example of statistical mechanics in one dimension (§§ 5.5, 6.1, 7.3) by introducing a *transfer matrix* to generate the successive difference equations (8.12). There is no great loss of generality in assuming that the interactions, $V_{ll'}$, are effective only between nearest neighbours. The equations (8.12), i.e.

$$(\mathscr{E}_l - \lambda)u_l + V_{l,l+1}u_{l+1} + V_{l,l-1}u_{l-1} = 0, \tag{8.17}$$

can be written in matrix form:

$$\begin{pmatrix} u_l \\ u_{l+1} \end{pmatrix} = \begin{pmatrix} 0 & 1 \\ -[V_{l,l+1}]^{-1}V_{l,l-1} & -[V_{l,l+1}]^{-1}(\mathscr{E}_l - \lambda) \end{pmatrix} \begin{pmatrix} u_{l-1} \\ u_l \end{pmatrix}. \tag{8.18}$$

This we read as

$$\mathbf{U}_{l+1} = \mathbf{T}_l \mathbf{U}_l, \tag{8.19}$$

showing that the *excitation* $\mathbf{U}_l$ across the lth cell is generated from the corresponding excitation across the previous cell by multiplication by the transfer matrix $\mathbf{T}_l$. In the most elementary case, $\mathbf{T}_l$ is a 2×2 matrix whose components are given by (8.18). But if, as in (8.3), the excitation amplitude u_l is itself a vector with many components, the dimensionality of the transfer matrix (8.18) is governed by the matrix or tensor properties of the symbols $V_{ll'}$.

A transfer matrix is thus associated unambiguously with each cell of the lattice. The propagation of an excitation along the chain is represented as the matrix product of a succession of such matrices: from (8.19) it follows that

$$\mathbf{U}_{l+r} = \mathbf{T}_{l+r-1} \cdot \mathbf{T}_{l+r-2} \cdot \ldots \cdot \mathbf{T}_r \mathbf{U}_r. \tag{8.20}$$

For an *ordered* system, all the transfer matrices are identical, so that we could write

$$\mathbf{U}_{l+r} = [\mathbf{T}]^l \mathbf{U}_r. \tag{8.21}$$

The effect of disorder is to change the transfer matrix from cell to cell because of random variations of the matrix elements (8.18). In other words, the transfer matrix $\mathbf{T}_l$ is a stochastic function of the site label l, with a statistical distribution determined by the physical set-up of the model.

Now consider the problem of electron propagation in a one-dimensional disordered potential. Thinking about substitutional disorder we might set up a *Kronig–Penney alloy* model (fig. 8.1(*a*)), whose lattice sites carry delta-function potentials of varying strengths δ_l. Or we might introduce a *Kronig–Penney liquid* model (fig. 8.1(*b*)), where the random variable is the spacing, ξ_l, of successive delta functions. In either case, reference to the standard theory of the Kronig–Penney model for a regular chain reminds us that the solution of the Schrödinger equation at energy $\mathscr{E} = \kappa^2$ is made up of segments of free electron waves of momentum $\pm \kappa$. Assuming that $0 \leqslant x \leqslant \xi_l$ in the lth 'open stretch', we may write such a function in the form

$$\psi_l(x) = u_l \cos \kappa x + u_l' \kappa^{-1} \sin \kappa x, \tag{8.22}$$

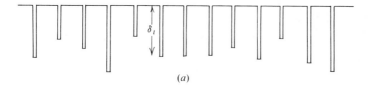

(a)

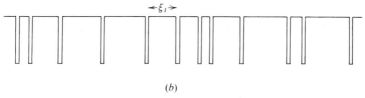

(b)

Fig. 8.1. Kronig–Penney models: (a) 'alloy', (b) 'liquid'.

where the coefficients are arbitrary. But this function is connected to ψ_{l+1} in the next stretch, having passed through the singular potential δ_l. The connectivity conditions at this junction read

$$\psi_{l+1}(0) = \psi_l(\xi_l); \quad \frac{1}{\psi_{l+1}} \frac{\partial \psi_{l+1}}{\partial x}\bigg]_{x=0} - \frac{1}{\psi_l} \frac{\partial \psi_l}{\partial x}\bigg]_{x=\xi_l} = \delta_l. \quad (8.23)$$

Putting (8.22) into (8.23) we get linear equations for the successive excitation amplitudes (u_l, u_l') (u_{l+1}, u_{l+1}') which can be written in matrix form

$$\begin{pmatrix} u_{l+1} \\ u_{l+1}' \end{pmatrix} = \begin{pmatrix} \cos \kappa \xi_l & \kappa^{-1} \sin \kappa \xi_l \\ -\kappa \sin \kappa \xi_l + \delta_l \cos \kappa \xi_l & \cos \kappa \xi_l + \delta_l \kappa^{-1} \sin \kappa \xi_l \end{pmatrix} \begin{pmatrix} u_l \\ u_l' \end{pmatrix}.$$
$$(8.24)$$

Ignoring any question of the physical interpretation of the excitation U_l, this is precisely of the form (8.19). For an ordered system, the transfer matrix T_l would be the same for every cell, as in (8.21). For our disordered models, the elements of T_l would be defined statistically by the distribution of the random variables ξ_l and/or δ_l.

But this again is a special case. Any one-dimensional potential $\mathscr{V}(x)$ may be dissected into a linear succession of 'atomic potentials' $v_l(x)$ separated by (perhaps infinitesimally narrow) regions where the potential is zero (fig. 8.2). Within each of these atomic cells we can construct two independent solutions of the Schrödinger equation, $\phi_l(x)$, $\chi_l(x)$, satisfying the initial conditions

$$\begin{array}{ll} \phi_l(0) = 1; & \phi_l'(0) = 0; \\ \chi_l(0) = 0; & \chi_l'(0) = 1 \end{array} \Bigg\} \quad (8.25)$$

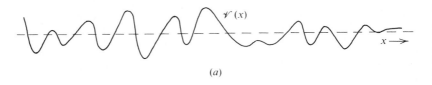

(a)

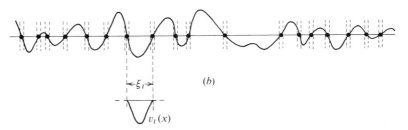

(b)

Fig. 8.2. (a) Continuous random potential $\mathscr{V}(x)$ subdivided at zeros (b) into a linear chain of random cellular potentials v_l with random spacings ξ_l.

at the left-hand end of the cell, at $x = 0$. A linear combination of these functions

$$\psi_l(x) = u_l \phi_l(x) + u_l' \chi_l(x) \qquad (8.26)$$

in the lth cell matches the next such function, in amplitude and slope, at the next cell boundary, where $x = \xi_l$. Just as in (8.24), this gives rise to linear equations coupling successive excitation amplitudes (u_l, u_l') with a transfer matrix

$$\mathbf{T}_l = \begin{pmatrix} \phi_l(\xi_l) & \chi_l(\xi_l) \\ \phi_l'(\xi_l) & \chi_l'(\xi_l) \end{pmatrix}. \qquad (8.27)$$

The mathematical problem is thus reduced, once more, to finding the properties of excitations propagated by the transfer operation (8.19), where the elements of $\mathbf{T}_l$ are stochastic variables.

The statistical distribution of the elements of (8.27) would, however, be somewhat complicated in general, since each of the functions ϕ_l, χ_l is the solution of the Schrödinger equation in a random potential $v_l(x)$, over a length ξ_l that is itself statistically distributed (cf. § 3.4). Each element, moreover, is a function of the energy $\mathscr{E}$ at which the Schrödinger equation has been solved. But the determinant of the matrix (8.27) is the Wronskian of the functions ϕ_l, χ_l and, therefore, remains at unity as x goes from 0 to ξ_l. The three independent elements of (8.27) can be simulated by the corresponding elements of a matrix of the form (8.24) by appropriate choice of κ, ξ_l and δ_l. In other words, the effect of any one-dimensional potential on an electron of given energy $\mathscr{E}$ can be simulated by a disordered Kronig–Penney chain (Borland 1961). For this reason, the Kronig–Penney alloy or liquid is of more theoretical interest than it deserves on 'physical' grounds. Notice, however, that none of these generalizations would be valid in two or three dimensions.

In practice, however, the topological 'order' that can be discerned in a one-dimensional continuous random potential of the type discussed in § 3.2 is arbitrary and artificial and does not really help in the solution of the Schrödinger equation for the electron eigenstates. Although the properties of this model are of some theoretical interest, it seems more convenient to postpone the discussion of this limiting case until § 13.5, where the one-dimensional model is seen as a special case of a general three-dimensional theory, without reference to the linear ordering of hypothetical cells. In the present chapter we confine the discussion to systems where it is natural to use the matrix representation.

8.3 Phase-angle representation

The transfer matrix formalism (8.19) can be used for all theoretical models of excitations on a linear chain. As we saw in § 8.2, the case where the excitation U_l (defined as in (8.18) or (8.24)) has only two components is sufficiently general to include most models of dynamical or electronic excitations on an 'alloy' or 'liquid' chain. The physical problems considered in this chapter are mathematically equivalent to discovering the effects of transforming a two-dimensional vector U by successive multiplications with a set of 2×2 matrices T_l containing stochastic elements.

In studying the *spectral* properties of the system, we look for stationary states of excitation that satisfy prescribed conditions at the ends of the chain. For this problem it turns out *not* to be advantageous to impose cyclic boundary conditions (cf. § 5.5), pretending that the chain has been joined into a closed loop. It is more natural to assume, for example, that the end atoms are fixed (or, in the electronic case, that the wave functions (8.26) vanish at each end) so that the amplitudes u_0 and u_N are zero. This can be considered a special case of a general condition that the *excitation ratio* ζ_l – i.e. the ratio of the two components of U_l – should take prescribed values ζ_0 and ζ_N at $l=0$ and $l=N$ respectively. The general theory of eigenvalues assures us that the spectrum is not significantly affected by the precise values chosen for these boundary conditions, provided that N is large. We may, therefore, assume the simple but slightly unphysical conditions $\zeta_0 = \zeta_N = 0$ without serious error.

Geometrically speaking, this means that the two-dimensional vector $\mathbf{U}_l$ must take prescribed directions at $l=0$ and $l=N$. In the search for a stationary state we can ignore the magnitudes of the successive excitation amplitudes and concentrate our attention on a *phase angle*

$$\theta_l \equiv 2\tan^{-1}\zeta_l = \begin{cases} 2\tan^{-1}(u_l/u_{l-1}) & \text{in (8.18)} \\ 2\tan^{-1}(u_l/u_l') & \text{in (8.24)} \end{cases} \tag{8.28}$$

which varies from cell to cell along the chain, but must satisfy the boundary conditions

$$\tan \tfrac{1}{2}\theta_0 = \tan \tfrac{1}{2}\theta_N = 0. \tag{8.29}$$

Multiplication by the transfer matrix $\mathbf{T}_l$ is equivalent to rotation of the vector $\mathbf{U}_l$; from (8.19) we may calculate the change of phase angle

$$\theta_{l+1} - \theta_l = \eta_l(\theta_{l+1}, \lambda) \tag{8.30}$$

in terms of the elements of $\mathbf{T}_l$. Generally speaking, this *phase shift* depends on the initial phase θ_l (or, as here defined, on the final phase angle θ_{l+1}), so that the product (8.20) of many successive matrix multiplications is not simply equivalent to the sum of a corresponding number of independent phase shifts. Nevertheless, following the algebraic instructions in (8.20) and (8.28), we can start out from the prescribed value $\theta_0 = 0$ at one end of the chain, and calculate the value of θ_N at the other end by successive additions of the form (8.30).

In general, this final phase angle will not be consistent with the final boundary condition (8.29). But each transfer matrix is really a function of the spectral variable λ – either explicitly, as in (8.18), or by definition of λ as the 'free electron' energy κ^2 in (8.24), or as the energy parameter in the Schrödinger equation of which the functions in (8.27) are solutions. In other words, we should write the final phase angle as $\theta_N(\lambda)$, showing that it is a function of λ. The spectrum of stationary states of the system is just the set of values λ_k such that

$$\theta_N(\lambda_k) = 2k\pi \tag{8.31}$$

where k is an integer. Only for these values of λ can we simultaneously satisfy both the boundary conditions (8.29).

The analogy with the theory of excitations on a one-dimensional continuum suggests, and various theorems on difference equations, etc. guarantee, that $\theta_N(\lambda)$ is a monotone function of λ. Starting from a state of

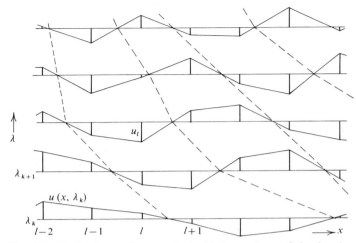

Fig. 8.3. Excitations of successively higher values of λ, showing appearance of new nodes.

very low energy or frequency, we can count λ_k as the kth level in the spectrum. One can also define the phase shifts $\eta_l(\theta, \lambda)$ in such a way that $\theta_l(\lambda)$ increases monotonically with l along the chain. For a given value of λ, the amplitude u_l in each successive cell can thus be regarded as a sample of the amplitude of a continuous wave $u(x, \lambda)$ which goes through zero whenever $\theta_l(\lambda)$ goes through a multiple of 2π. From this picture (fig. 8.3) it is obvious that, for the stationary state of spectral value (energy) λ_k, $u(x, \lambda_k)$ has just $k-1$ nodes. Since $u(x, \lambda)$ changes sign as it passes through a node, the number of *negative factors* in the sequence of excitation ratios $\zeta_l(\lambda)$ also counts the number of spectral levels less than λ.

To show how this works, consider a simple, regular, TBA model where the nearest-neighbour interaction is of strength V. From (8.12) and (8.18) this has the same transfer matrix for every step along the chain:

$$\mathbf{T}_l \equiv \mathbf{T}_0 = \begin{pmatrix} 0 & 1 \\ -1 & -(\mathscr{E}_0 - \lambda)/V \end{pmatrix}. \tag{8.32}$$

In the two-dimensional space of excitation vectors $\mathbf{U}_l$ this operator has two eigenvalues, $\mu^{(+)}$ and $\mu^{(-)}$, with eigenvectors $\mathbf{W}^{(+)}$ and $\mathbf{W}^{(-)}$ respectively. Suppose that we have chosen these so that $|\mu^{(+)}| > |\mu^{(-)}|$. Now any excitation can always be represented as a linear combination of these eigenvectors, i.e.

$$\mathbf{U}_l = u^{(+)}\mathbf{W}^{(+)} + u^{(-)}\mathbf{W}^{(-)}. \tag{8.33}$$

It follows, therefore, from (8.19),

$$\mathbf{U}_{l+1} = \mathbf{T}_0 \mathbf{U}_l = \mu^{(+)} u^{(+)} \mathbf{W}^{(+)} + \mu^{(-)} u^{(-)} \mathbf{W}^{(-)} \tag{8.34}$$

and, *a fortiori*, from (8.21),

$$\mathbf{U}_N = (\mathbf{T}_0)^N \mathbf{U}_0 \rightarrow (\mu^{(+)})^N W^{(+)}, \tag{8.35}$$

whatever the initial choice of $\mathbf{U}_0$ (except the trivial case where $\mathbf{U}_0 = \mathbf{W}^{(+)}$).

This is disappointing. Considered as a function of the spectral variable λ, the 'direction' of $\mathbf{W}^{(+)}$ varies very slowly. There seems no possibility of deducing from (8.35) a whole spectrum of stationary states with rapidly increasing phase angles satisfying (8.31). According to (8.34), the effect of operating on the excitation $\mathbf{U}_l$ by the transfer matrix $\mathbf{T}_0$ is simply to push it 'nearer' to the direction $\mathbf{W}^{(+)}$; there is no mechanism for the introduction of new nodes in the excitation wave $u(x, \lambda)$ by substantial changes in the phase shifts (8.30) with changes in λ. In other words, over any range of λ where these conditions hold, no new spectral levels can arise.

But the whole argument from (8.33) to (8.35) depends upon the eigenvalues $\mu^{(+)}$ and $\mu^{(-)}$ being *real*. From (8.32) we easily verify that

$$\mu^{(\pm)} = \frac{1}{2V} [-(\mathscr{E}_0 - \lambda) \mp \sqrt{\{(\mathscr{E}_0 - \lambda)^2 - 4V^2\}}] \tag{8.36}$$

which has real values only if

$$|\mathscr{E}_0 - \lambda| > 2|V|. \tag{8.37}$$

Comparison with (8.15) shows that this condition is satisfied when λ lies *outside* the band of Bloch-like excitations of the ordered chains. What we have really proved is that a *regular chain has a spectral 'gap' whenever the eigenvalues of the transfer matrix are real.*

This theorem obviously does not depend on the specially simplified form of transfer matrix (8.32). In general, however, every transfer matrix $\mathbf{T}_l$ in a genuine physical model has real trace $\mathrm{Tr}\{\mathbf{T}_l\}$ and is *unimodular*, i.e.

$$\det\{\mathbf{T}_l\} = 1, \tag{8.38}$$

so that the secular equation

$$\mu^2 - \mu \, \mathrm{Tr}\{\mathbf{T}_l\} + \det\{\mathbf{T}_l\} = 0 \tag{8.39}$$

has real roots if

$$\tfrac{1}{2} \mathrm{Tr}\{\mathbf{T}_l(\lambda)\} > 1. \tag{8.40}$$

This is the general condition that λ should lie in a spectral gap.

In the case of the Kronig–Penney model, for example, this condition applied to (8.24) gives

$$\cos \kappa \xi_0 + \tfrac{1}{2} \delta_0 \kappa^{-1} \sin \kappa \xi_0 > 1, \tag{8.41}$$

which is precisely the standard condition that there should *not* be a stationary state of energy κ^2 in a string of delta functions of strength δ_0 spaced a distance ξ_0 apart. Here, of course, the elements of the transfer matrix contain periodic function of the spectral variable κ, so that the regions where (8.41) is satisfied appear only as gaps between the various *allowed bands*. It is obvious, indeed, from the form of (8.41), that these gaps become narrower as κ increases.

To show that spectral levels do indeed appear in regions that are *not* forbidden by the condition (8.40) we observe that (8.39) will then have a pair of complex roots; by (8.38) these must be of the form

$$\mu^{(+)}=e^{2i\beta} \tag{8.42}$$

and

$$\mu^{(-)}=e^{-2i\beta}, \tag{8.43}$$

where β is a real number. In other words, for a given transfer matrix $\mathbf{T}_l$ there must exist a matrix $\mathbf{S}_l$ which transforms $\mathbf{T}_l$ to diagonal form, i.e.

$$\mathbf{S}_l\mathbf{T}_l\mathbf{S}_l^{-1}=\mathbf{Q}_l=\begin{pmatrix} e^{2i\beta_l} & \cdot \\ \cdot & e^{-2i\beta_l} \end{pmatrix}. \tag{8.44}$$

But the matrix $\mathbf{S}_l$ has complex coefficients. When applied to the excitation $\mathbf{u}_l$, this yields a vector

$$\mathbf{U}_l'=\mathbf{S}_l\mathbf{U}_l \tag{8.45}$$

with complex coefficients. It turns out (Hori 1968a) that the ratio of the components of $\mathbf{U}_l'$ is a complex number of unit modulus – the *state ratio*

$$z_l=e^{i\theta_l'}. \tag{8.46}$$

Formally speaking, z_l is a *Cayley transform* which maps the real axis of the excitation ratio ζ_l on to the unit circle. But the new phase angle θ_l' has properties equivalent to those of the angle defined by (8.28); for example, the condition for a stationary state is that (8.31) should apply to θ_N'.

In the representation (8.44), the transformed transfer matrix has the effect of multiplying the state ratio by $\exp(4i\beta_l)$. In a regular lattice we may use the same transformation $\mathbf{S}_0$ for every transfer matrix $\mathbf{T}_l=\mathbf{T}_0$ of the chain, so that $4\beta_0$ is added to the phase angle at each step. The condition (8.31) now reads

$$4N\beta_0=2k\pi \tag{8.47}$$

where k is an integer. Since $2\beta_0$ is simply the phase of an eigenvalue (8.43) of the transfer matrix, it is a function of the spectral variable λ: the condition for a spectral level λ_k is

$$2\beta_0(\lambda_k)=\frac{k}{N}\pi. \tag{8.48}$$

For the simplest model (8.32), for example, we have

$$\mathrm{Tr}\{T_0\} = -(\mathscr{E}_0 - \lambda)/V = \mathrm{Tr}\{Q_0\} = 2\cos 2\beta_0, \tag{8.49}$$

from which we get a spectrum

$$\lambda_k = \mathscr{E}_0 + 2V\cos(k\pi/N). \tag{8.50}$$

Interpreting $(k\pi/N)$ as the wave number, q, of the excitation, we see that this is exactly the spectrum (8.15) for a regular TB chain with appropriate end conditions. From the form of (8.48), where β_0 is a continuous slowly varying function of λ, it is obvious that we thus generate a band of stationary states whose spectral distribution becomes continuous as $N \to \infty$. The corresponding analysis of the spectrum of the regular Kronig–Penney model introduces no new principles.

Notice, however, that if T_l is not the same for every step along the chain, we cannot, in general, find a single transformation S that will reduce all transfer matrices simultaneously to the diagonal form (8.44). Nor may we introduce a different transformation for each different transfer matrix, since this would change the definitions of z_l and of θ'_l at each step. The best we can then do is to introduce a standard transformation S_0 which perhaps diagonalizes some of the transfer matrices, but which leaves others in the *Cayley form*

$$Q_l = S_0 T_l S_0^{-1} = \begin{pmatrix} A_l & B_l \\ B_l^* & A_l^* \end{pmatrix}. \tag{8.51}$$

In the case of a binary alloy, for example, the two different types of atom would be represented by two different transfer matrices, T_0 and T_1, which might both be of the form (8.22) but with different interaction parameters V_0 and V_1 corresponding to differences in the electronic-bound states or to different 'isotopic' masses M_0 and M_1 in the dynamical equation (8.3). It is then discovered that the transformation S_0 that reduces T_0 to the diagonal form (8.44) with phase shift $2\beta_0$ transforms T_1 into

$$Q_1 \equiv S_0 T_1 S_0^{-1} = \begin{pmatrix} (1 + i\tan\gamma)e^{2i\beta_0} & i\tan\gamma\, e^{-2i\beta_0} \\ -i\tan\gamma\, 006\, e^{2i\beta_0} & (1 - i\tan\gamma)e^{-2i\beta_0} \end{pmatrix} \tag{8.52}$$

where the off-diagonal terms are proportional to the *mass difference* parameter μ, i.e.

$$\frac{\tan\gamma}{\tan\beta_0} = \left(\frac{V_0}{V_1} - 1\right) = \left(\frac{M_1}{M_0} - 1\right) \equiv \mu. \tag{8.53}$$

It is obvious by inspection that (8.52) is indeed of the Cayley form (8.51).

To find the spectral gaps in an *ordered alloy* where these two atomic species might, for example, alternate along the chain, we merely note that

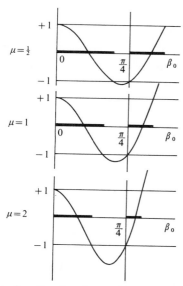

Fig. 8.4. Band gap function for binary alloy, for different mass ratios.

the lth unit cell of this *compound lattice* would be characterized by a matrix corresponding to the effects of the two constituents, i.e.

$$\mathbf{T}_l = \mathbf{T}_0 \mathbf{T}_1. \qquad (8.54)$$

The condition (8.40) that this matrix should have real eigenvalues is easily evaluated in the Cayley representations (8.44) and (8.52); the compound lattice must have spectral gaps for all values of λ where

$$|\cos 4\beta_0 - \tan \gamma \sin 4\beta_0| > 1. \qquad (8.55)$$

Inspection of this condition shows that the spectrum now consists of two bands whose width and separation depends on the parameter (8.53) (fig. 8.4).

It must be emphasized, however, that the problem of a *disordered* alloy with these two constituents cannot be solved simply by a transformation of this kind. Thus, the matrix (8.51) operating on the complex excitation vector $\mathbf{U}_l'$ generates a Cayley transform

$$z_{l+1} = \frac{A_l z_l + B_l}{B_l^* z_l + A_l^*} \qquad (8.56)$$

on the state ratio z_l, which is not simply a rotation of the phase angle θ_l' by a standard phase shift. For this reason, the spectral properties of disordered chains are not simply averaged versions of (8.48).

8.4 Spectral gaps in disordered chains

A disordered chain is characterized mathematically by a random succession of non-identical transfer matrices T_l whose overall product (8.20) describes the propagation of an excitation along the chain. For every such chain, stationary states satisfying standard end conditions such as (8.29) can be found by numerical computation. But the general problem is to find the spectral distribution for an ensemble of chains where the different types of transfer matrix are distributed statistically according to some initial prescription.

The most elementary case would be a *random binary alloy*, where transfer matrices T^A and T^B are distributed at random along the chain with relative concentrations c_A and c_B. As we saw in § 8.2, all types of one-dimensional 'spatial' disorder, such as the 'linear glass' and the 'linear liquid', are equivalent to the postulate that the successive transfer matrices are drawn from a set $\{T^{(\xi)}\}$ where the interatomic spacing ξ has a prescribed probability distribution $P(\xi)$. Thus, for example, the theory of electron states in a 'one-dimensional liquid metal' depends on the properties of a chain whose transfer matrices are of type (8.24) or (8.27), with components depending on the choice of ξ_l at each step, as discussed in § 2.2.

As we saw in § 8.3, the spectrum of a regular chain has a gap wherever the eigenvalues of the transfer matrix are real. Suppose that the gap condition (8.40) for a transfer matrix $T^{(\xi)}(\lambda)$ is satisfied for values of the spectral variable in a range $\Lambda^{(\xi)}(\lambda)$. For different values of the parameter ξ (or, it might be, for atoms of a different kind), the range will be different. But suppose that all such gap regions overlap to some extent – i.e. that there exists a range $\hat{\Lambda}(\lambda)$ which is common to all $\Lambda^{(\xi)}(\lambda)$ for all physically permitted values of ξ (or for all the actual atomic species in the alloy). An excitation whose spectral variable λ lies in this range will encounter transfer matrices with *real* eigenvalues $\{u^{(\pm)(\xi)}\}$ at every step along any disordered chain constructed from the set $\{T^{(\xi)}\}$. If the corresponding eigenvectors $\{W^{(+)(\xi)}\}$ and $\{W^{(-)(\xi)}\}$ form disjoint sets for the permitted values of ξ, the argument of (8.34)–(8.35) remains valid; the excitation vector gets 'stuck' in the directions spanned by $\{W^{(+)(\xi)}\}$ and a small increase in λ cannot add a multiple of 2π to the overall phase angle θ_N. The range $\hat{\Lambda}(\lambda)$ is thus a spectral gap for any chain satisfying these conditions.

This is the general proof (see Hori 1968a; Tong 1968) of the *Saxon–Hutner theorem*, which was originally conjectured (Saxon & Hutner 1949) and proved (Luttinger 1951) in the form: *any spectral region that is a spectral gap for both a pure A-type chain and a pure B-type chain is also a gap for any mixed lattice of A- and B-type atoms.* But the applications of the general theorem cover many other situations. Consider, for example, the electron states of a *Kronig–Penney liquid*, with identical delta functions $\delta_l = \delta_0$ spaced at variable intervals ξ_l. The condition (8.41) for a gap in the energy, κ^2, contains the chain spacing ξ_0, which must now be considered a statistical variable drawn from a distribution $P(\xi)$. But the left-hand side of this condition is a continuous function of ξ_0, so that the gap edges cannot move discontinuously as ξ_0 varies. Thus, in the Borland model of a one-dimensional liquid, where ξ_l lies in a well-defined range (2.7), there could still remain a common gap region for all values of ξ_l (Borland 1961; Makinson & Roberts 1962). In other words, provided that the atomic potentials (i.e. the delta functions δ_0) are sufficiently strong, the effect of 'liquid' disorder is to narrow, but not necessarily to destroy, the energy gaps of the regular one-dimensional model.

At first sight one would not imagine that this theorem would tell us very much about the spectrum of a disordered binary alloy. Consider, for example, the dynamical frequencies of an isotopically disordered alloy, containing a mixture of light and heavy atoms of masses M_0 and M_1 respectively. From (8.3), (8.13) and (8.15) and (8.49), we find a pure chain would have a *phonon band* of the form

$$\omega^2 \equiv \lambda = 2(\Phi/M_0)(1 - \cos 2\beta_0) \tag{8.57}$$

where Φ is a measure of the interatomic force constants and $2\beta_0$ may be identified with the wave number, q, of the excitation. The maximum frequency in this band is $2(\Phi/M_0)^{\frac{1}{2}}$, which is higher than the corresponding limit in a pure M_1 chain. The theorem merely tells us that a mixed chain has a spectral gap for any frequency above *both* these limits – i.e. it cannot have normal modes with a frequency greater than the maximum frequency of a regular chain of the lightest atoms in the mixture.

It turns out, however (Matsuda 1964; see Hori 1968*a*), that if the mass ratio (8.53) is large enough there must also be narrow gaps in the spectrum at certain special wave numbers. Suppose, first of all, that the concentration of heavy atoms is fairly high, so that the *a priori* probability of encountering a long unbroken succession of light atoms is very small. To be precise, let us arbitrarily exclude from the statistical ensemble all chains where more than $(p-1)$ atoms of mass M_0 occur in succession. In applying the theorem we need no longer make reference to the pure chain. Instead we may think of any disordered chain as a random assembly of elements drawn from a set of segments $\{A(s)\}$ of varying length, each containing one M_1 atom and $s-1=0,\ 1,\ 2,\ldots,\ p-1$ atoms of mass M_0. Suppose that $\mathbf{T}^{(s)}(\lambda)$ is the transfer matrix of such a segment, so that a regular chain of such elements would have a spectral gap $\Lambda^{(s)}(\lambda)$: the theorem tells us that the disordered chain should have a gap in the common range $\hat{\Lambda}(\lambda)$ of all these gaps, for $1 \leqslant s \leqslant p$.

The gap region $\Lambda^{(s)}(\lambda)$ is easy to find for any regular chain of segments of type $A(s)$. As in (8.54), the transfer matrix $\mathbf{T}^{(s)}(\lambda)$ is simply the product of a matrix of type $\mathbf{T}_1$ with the product of $(s-1)$ matrices of type $\mathbf{T}_0$. The transformation S_0 that reduces each $\mathbf{T}_0$ to the same diagonal form (8.44) transforms $\mathbf{T}_1$ to the Cayley form (8.52). The trace condition (8.40) for

$$\mathbf{T}^{(s)}(\lambda) \to \mathbf{Q}^{(s)} = (\mathbf{Q}_0)^{s-1}\mathbf{Q}_1 \qquad (8.58)$$

is a generalization of (8.55), i.e.

$$|\tan\gamma \sin 2\alpha - \cos 2\alpha| > 1, \qquad (8.59)$$

where

$$\alpha = s\beta_0. \qquad (8.60)$$

Now suppose that the mass difference μ is quite large, i.e. that $\tan \gamma$ happens to be large enough for (8.59) to be satisfied for

$$\alpha = \pi/2p. \qquad (8.61)$$

By inspection of fig. 8.5 we see that this condition will also be satisfied for all larger values of α up to $\pi/2$. In other words, for the *special frequency* λ_p, where

$$\beta_0(\lambda_p) = \pi/2p, \qquad (8.62)$$

the condition (8.59) is satisfied for each value of s between 1 and p. This frequency thus lies in a common gap $\hat{\Lambda}(\lambda)$ of all the $\Lambda^{(s)}(\lambda)$ and must, therefore, lie in a gap in the spectrum of this restricted ensemble of disordered chains (fig. 8.6).

What is more surprising is that a gap would be found at just this frequency even for chains with values of $s > p$. This is because the condition (8.59) has the periodicity of $\tan 2\alpha$; as we run through the set (8.60) we merely reproduce the same values of 2α as those already considered, together with an integral multiple of π. The gap condition (8.59) is thus satisfied for all values of s, and the spectral value

$$\lambda_p = 2(\Phi/M_0)\{1 - \cos(\pi/p)\} \qquad (8.63)$$

must be in a gap of the perfectly randomized chain, without restriction on the lengths of run of identical atoms.

This argument goes even further. It is easy enough to show that if there is

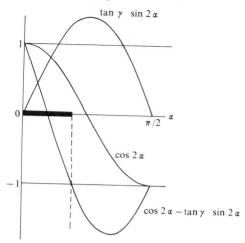

Fig. 8.5. Band gap function for large mass ratio.

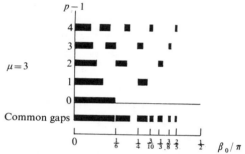

Fig. 8.6. Common gaps for alloys with not more than four light atoms in succession.

a gap in the neighbourhood of (8.62), there will also be gaps of integral multiples of this angle and that these will be common to all regular lattices with segments of type $A(s)$. Indeed, it turns out (Hori 1968a), that we can find an infinite number of special frequencies defined by the spectral values for which

$$2\beta_0(\lambda) = \frac{p-q}{p}\pi, \qquad (8.64)$$

where p and q are integers. The only restriction on these integers is that their ratio should be irreducible and that the mass difference parameter (8.53) should lie in the range

$$q \leqslant \mu \leqslant \cot^2\{\pi/2(q+1)\}. \qquad (8.65)$$

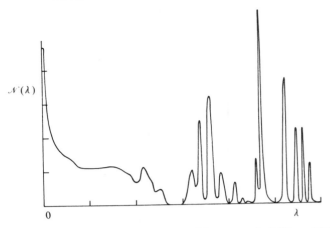

Fig. 8.7. Spectrum of disordered chain with mass ratio 3 (Dean 1960).

On a perfectly disordered chain, where the possibility of finding an infinite run of pure M_0 material cannot be entirely excluded, each of these gaps could, in principle, only be of infinitesimal width. More detailed analysis (Lifshitz 1964) shows, however, that above each excluded special frequency there is a region of genuine spectral levels corresponding to a band of coupled *impurity modes*. These modes are associated with 'islands' of light atoms separated from one another by a 'sea' of heavy atoms. Thus, an island of p light atoms has p distinct modes, spaced as in (8.64), each of which is broadened into a narrow band by interaction (through the heavy atoms) with similar islands along the chain. The overall spectrum is thus just the sum of all such contributions. But since the probability of finding a chain with a very long run of light atoms is very small, there are very few observable spectral levels just below a special frequency. The spectral density thus jumps almost discontinuously at this point. This extraordinary 'structure' in the spectrum of a disordered alloy chain of large mass difference was actually predicted theoretically by Domb, Maradudin, Montroll & Weiss (1959) before it was detected in numerical computations by Dean (1960) (fig. 8.7).

The physical interpretation of this phenomenon in terms of impurity 'islands' suggests, and detailed mathematical analysis confirms, that similar effects are to be observed (Agacy & Borland 1964) in the electron energy spectrum of disordered one-dimensional alloys with sharply differentiated components – e.g. in a Kronig–Penney model where the strengths of the delta functions δ_l are large enough and sufficiently different. In all such cases it is necessary of course (Tong 1968; Roberts 1971) not to rely upon the verbalized statement of the Saxon–Hutner theorem but to study carefully the behaviour of the limit points of the transfer matrices, as explained in detail by Hori (1968a).

Notice, however, that all such structure would be smoothed out in any model, such as a 'linear liquid', where the transfer matrix for each atomic cell is drawn from a statistical ensemble that is distributed over the range of a continuous parameter such as the interatomic spacing ξ. A special frequency structure is not to be expected in a disordered system unless the various components are drawn from a few standard types, so that one can say, for example, that a definite fraction of the sites of an (infinite!) specimen are occupied by *identical* atoms.

In the limit $\mu \to \infty$ the spectrum of special frequencies (8.64) occupies all rational multiples of π along the wave number axis, so that the spectral density becomes one of those pathological functions, discontinuous everywhere, that pure mathematicians delight to invent and applied mathematicians tend to scorn. This particular system is not physically realistic, but the formal limit warns us of unsuspected subtleties in the theory of the excitations of disordered systems. As seems first to have been pointed out by Schmidt (1957), the spectral density and similar physical properties may not always be adequately represented by the simple, smooth functions that our intuition plausibly suggests.

8.5 The spectral density

The phase-angle representation provides a basis for an exact mathematical theory of the *spectral density* $\mathcal{N}(\lambda)$ of excitations on a chain. We define $\mathcal{N}(\lambda)d\lambda$ as the number of spectral levels in the range $\lambda \to \lambda + d\lambda$ per unit cell of the chain. This function is properly defined as the limit of the spectral distribution for a single chain as $N \to \infty$ or, in awkward cases, as the average over an ensemble of such chains.

As we saw in § 8.3, a new stationary state appears each time the final phase angle $\theta_N(\lambda)$ satisfies (8.31). In other words, by definition,

$$\mathcal{N}(\lambda)d\lambda = \lim_{N \to \infty} \frac{1}{2\pi N} [\theta_N(\lambda + d\lambda) - \theta_N(\lambda)], \qquad (8.66)$$

i.e.

$$\mathcal{N}(\lambda)d\lambda = \lim_{N \to \infty} \frac{1}{2\pi N} \frac{\partial \theta_N(\lambda)}{\partial \lambda}. \qquad (8.67)$$

To avoid taking the derivative of a function that is not necessarily continuous, it is convenient to introduce a *cumulated spectral density*

$$\mathcal{D}(\lambda) \equiv \int_{-\infty}^{\lambda} \mathcal{N}(\lambda')\mathrm{d}\lambda' = \lim_{N \to \infty} \frac{1}{2\pi N} \theta_N(\lambda) \tag{8.68}$$

(assuming that $\theta_N(-\infty) = 0$, by definition).

But, of course, we never really want to go to the end of an endless chain. Assuming that the interactions along the chain are of finite range we may treat any stretch of a single chain as a sample from the whole ensemble and use an ergodic hypothesis to transform (8.68) into the average phase change per cell along any such sample. This is just the phase shift (8.30), i.e.

$$\mathcal{D}(\lambda) = \frac{1}{2\pi} \langle \theta_{l+1}(\lambda) - \theta_l(\lambda) \rangle$$

$$= \frac{1}{2\pi} \langle \eta_l(\theta_{l+1}, \lambda) \rangle. \tag{8.69}$$

It is no accident that this formula closely resembles the *Friedel sum rule* for the number of states associated with a scattering centre in a free electron gas, in three dimensions.

To give substance to such an expression, let us recall that the phase shift in the *l*th cell is produced by a transfer matrix

$$\mathbf{T}_l = \mathbf{T}^{(\xi_l)} \tag{8.70}$$

drawn at random from a set $\{\mathbf{T}^{(\xi)}\}$ distributed with probability $P(\xi)$ over the range of a statistical parameter ξ. In § 8.2 this symbol stood for the interatomic spacing on a 'liquid' chain, but here it could also be a selector variable (cf. § 1.5) whose discrete values ξ_α would refer to the different types of atom in an alloy model. In either case, the transfer matrices in the set $\{\mathbf{T}^{(\xi)}\}$ would define a function $\eta(\theta, \xi; \lambda)$ whose value for the *l*th cell would be the phase shift

$$\eta(\theta_{l+1}, \xi_l; \lambda) \equiv \eta_l(\theta_{l+1}; \lambda). \tag{8.71}$$

To evaluate (8.69) we must take an ensemble average of this function over the distribution $P(\xi_l)$. But this average would not be completely defined, since (8.71) also depends on the 'exit' phase angle θ_{l+1}. We must assign to this variable a probability distribution $w_{l+1}(\theta_{l+1})$ so that

$$\mathcal{D}(\lambda) = \frac{1}{2\pi} \iint P(\xi_l)w_{l+1}(\theta_{l+1})\eta(\theta_{l+1}, \xi_l; \lambda)\mathrm{d}\xi_l \mathrm{d}\theta_{l+1}. \tag{8.72}$$

It would not be difficult – and in some cases, not entirely misleading – to evaluate (8.71) approximately in a *random phase approximation* where $w_{l+1}(\theta_{l+1})$ would be assumed independent of θ_{l+1}. However, in a linear chain with statistically independent links any segment is representative of the whole ensemble, so that the distribution of the phase angle at any cell boundary is a stationary function that could depend only on the local parameters. But, because of the topological separability of the chain at each link, this function must be independent of, say, the parameter ξ_l in the previous cell. We can thus assume the existence of a function $w(\theta; \lambda)$ to which both $w_{l+1}(\theta_{l+1})$ and $w_l(\theta_l)$ are identical for a given value of λ. In other words, (8.72) should read

$$\mathscr{D}(\lambda) = \frac{1}{2\pi} \iint P(\xi)w(\theta; \lambda)\eta(\theta, \xi; \lambda)\mathrm{d}\xi\mathrm{d}\theta. \qquad (8.73)$$

But according to (8.30), successive phase angles differ by the phase shift; i.e.

$$\theta_l = \theta_{l+1} - \eta(\theta_{l+1}, \xi_l; \lambda). \qquad (8.74)$$

This imposes a simple probabilistic relation on the corresponding distribution functions; considering all possible ways of occupying the lth cell, we get

$$w_{l+1}(\theta_{l+1}) = \int P(\xi_l)w_l\{\theta_{l+1} - \eta(\theta_{l+1}, \xi_l; \lambda)\}\mathrm{d}\xi_l. \qquad (8.75)$$

Identifying w_{l+1} and w_l with the stationary distribution function w, we get the *self-consistency condition* (Dyson 1953; Schmidt 1957)

$$w(\theta; \lambda) = \int P(\xi)w\{\theta - \eta(\theta, \xi; \lambda)\}\mathrm{d}\xi. \qquad (8.76)$$

This is an *exact* functional equation, whose unique solution for the unknown function $w(\theta; \lambda)$ can be put into (8.73) to provide the complete excitation spectrum of the linear chain.

The existence of the exact *Dyson–Schmidt procedure* is a very important theoretical feature of the theory of one-dimensional disordered systems. The actual solutions of (8.76) may not be easy to calculate, but they provide a standard against which any approximate analytical formula or Monte Carlo computation is to be measured. The method can be applied to excitations on a tree lattice (§ 11.4); it is unfortunate (to say the least of it) that the derivation of (8.76) breaks down for any lattice containing cycles – i.e. for any genuine lattice in two or three dimensions.

If, as in the 'liquid'/'glass' models, $P(\xi)$ is a continuous function of ξ, we may take it for granted that $w(\theta; \lambda)$ is a continuous function of θ, and the solutions of (8.76) can be generated by successive approximations, using, of course, the properties of the transfer matrices, for example (8.24) or (8.26), to define the phase-shift function $\eta(\theta, \xi; \lambda)$. In the very artificial model (Dyson 1953) of the dynamical excitations (8.3) of a chain whose force constants $\Phi_{l,l+1}$ vary according to an exponential or Gaussian distribution,

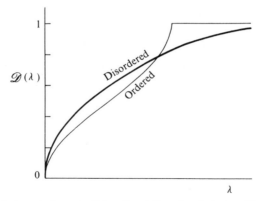

Fig. 8.8. Integrated spectral density of disordered chain with exponentially distributed force constants (Dyson 1953).

it is even possible to solve the integral equation (8.76) analytically and to show that the spectrum acquires a 'tail' going into the gap regions of the unperturbed system (fig. 8.8). This sort of behaviour will be interpreted 'physically' in § 8.6.

In the alloy models, where we write

$$P(\xi) = \sum_\alpha c_\alpha \, \delta(\xi - \xi_\alpha) \tag{8.77}$$

for the probability distribution of ξ over the various alloy components of relative concentrations c_α, the integral equation (8.76) simplifies to

$$w(\theta; \lambda) = \sum_\alpha c_\alpha w\{\theta - \eta_\alpha(\theta; \lambda); \lambda\}, \tag{8.78}$$

where η_α is the phase-shift function of component α. As pointed out by Schmidt (1957), this functional relation is more complicated than it looks. The discussion of § 8.4 suggests, for example, that the function $w(\theta; \lambda)$ cannot be a continuous function of θ in a real spectral gap, where as in (8.35) the phase angle 'sticks' near a real eigenvector of the transfer matrix. It is then much more convenient mathematically to rewrite (8.76) or (8.78) in terms of the integrated distribution of the phase angles

$$W(\theta) = \int_0^\theta w(\theta') \, d\theta', \tag{8.79}$$

which is necessarily continuous and monotone increasing. The Schmidt formula has been used by Agacy & Borland (1964) and by Gubernatis & Taylor (1971, 1973) to study the dynamical spectrum of a disordered binary alloy in the neighbourhood of special frequencies (8.64), showing the extraordinary fine structure that can arise in the density of states (fig. 8.9).

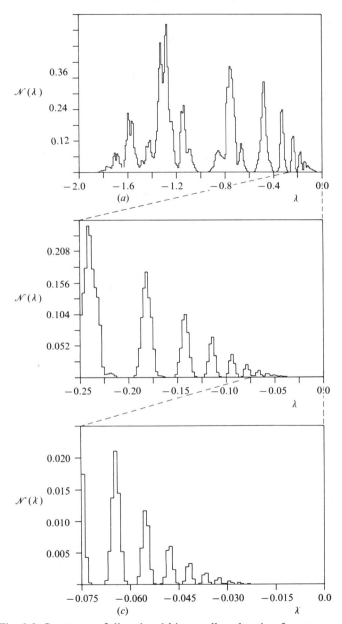

Fig. 8.9. Spectrum of disordered binary alloy showing fine structure near special frequencies. Note successive enlargements of scale in (*a*), (*b*), (*c*) as tail is entered (Gubernatis & Taylor 1973).

It can be shown, for example (Gubernatis, unpublished), that the spectral density in the neighbourhood of a special point λ_s behaves like

$$\mathcal{N}(\lambda) \sim A_i \exp(\alpha_i |\lambda - \lambda_s|^{-\frac{1}{2}}), \qquad (8.80)$$

where A_i and α_i differ to right and left of the special energy, and that the envelope of the successive peaks has the same general form. The inapplicability of perturbation methods in systems like this is obvious!

8.6 Local density approximation

When the probability distribution of the disorder parameter ξ is a continuous function $P(\xi)$ it is not really necessary to solve the Dyson–Schmidt equation (8.76) to discover the form of the excitation spectrum. Provided that the range of disorder is not too extreme it is plausible to treat the spectrum of an infinite chain as the sum of independent contributions from various short sections whose average composition fluctuates from section to section. The 'island' concept that provided a qualitative interpretation of the special frequencies and gaps of a binary alloy (§ 8.4) is thus generalized to a semiquantitative approximation for the overall spectrum.

Suppose that we are trying to calculate the (cumulated) spectral density $\mathcal{D}(\lambda)$ of a disordered chain whose *average* spacing (or equivalent disorder parameter) is ξ_∞. Our first rough estimate might be the spectral density of a *regular* chain with this spacing, i.e.

$$\mathcal{D}(\lambda) \approx \mathcal{D}_0(\lambda; \bar{\xi}_\infty). \qquad (8.81)$$

This approximation is by no means useless; in the case of a Kronig–Penney liquid (§ 8.2) for example, it would tell us where to look for the main 'allowed' bands and for possible gap regions. Notice, however, that the uniqueness of (8.81) depends upon the fact that the one-dimensional system is always topologically ordered (§ 2.2), so that the regular lattice from which the disordered system has been perturbed is well defined.

Now let us choose, at random, a finite segment of the chain encompassing, say, L cells. If the average spacing in this segment were $\bar{\xi}_L$, then $\mathcal{D}_0(\lambda; \bar{\xi}_L)$ would be the spectral density of a perfect chain of this spacing. Suppose we think of the actual chain as a succession of samples of perfect chain, each of length L, but with spacings $\bar{\xi}_L$ drawn from a probability distribution $\mathcal{P}_L(\bar{\xi}_L)$. The spectral density for the whole ensemble would then be given by the *local density approximation* (Blair 1967)

$$\mathcal{D}_L(\lambda) = \int \mathcal{P}_L(\bar{\xi}_L) \mathcal{D}_0(\lambda; \bar{\xi}_L) \mathrm{d}\bar{\xi}_L, \qquad (8.82)$$

which is much more accurate than the simple average (8.81).

If $P(\xi)$ is reasonably well behaved, and if L is not too small, the distribution of $\bar{\xi}_L$ satisfies the Central Limit theorem, and tends to a Gaussian (3.12), i.e.

$$\mathscr{P}_L(\bar{\xi}_L) = (2\pi)^{-\frac{1}{2}} L^{-\frac{1}{2}} \sigma^{-1} \exp\{-L(\bar{\xi}_L - \bar{\xi}_\infty)^2/2\sigma^2\} \tag{8.83}$$

where σ^2 is the variance of individual cell spacings (i.e. of $P(\xi)$). The precise form of $\mathscr{D}_0(\lambda; \xi)$ depends upon the details of the physical system, but the interesting spectral regions are those lying near what would have been band edges of the ideal average chain (8.81). Function theory suggests that such a band edge coincides with a *van Hove singularity* of the spectrum. In a one-dimensional regular lattice, for example, the spectral density should approach unity near the top of the first band like

$$1 - \mathscr{D}_0(\lambda) \sim (\lambda_c - \lambda)^{\frac{1}{2}} \quad \text{for } \lambda < \lambda_c$$
$$\sim 0 \qquad \text{for } \lambda > \lambda_c. \tag{8.84}$$

However the position λ_c of the top of this band must vary with the spacing $\bar{\xi}_L$ according to some scaling law, for example

$$\lambda_c(\bar{\xi}_L) = \lambda_\infty + \alpha(\bar{\xi}_L - \bar{\xi}_\infty) + \ldots \tag{8.85}$$

where λ_∞ is the top of the band for the overall average spacing $\bar{\xi}_\infty$ and where the coefficient α depends on the actual parameters of the model. Putting (8.83)–(8.85) into (8.82), we thus get an expression for the spectral density of the disordered chain. Going well into the gap region above λ_∞, for example, we expect to find an exponentially decaying *band tail* of the form

$$1 - \mathscr{D}(\lambda) \sim \exp\{-L(\lambda - \lambda_\infty)^2/2\alpha^2\sigma^2\}. \tag{8.86}$$

This argument is not, however, complete, since (8.82) and (8.86) depend upon the sampling length L. The choice of this parameter is evidently somewhat arbitrary and cannot be settled by some heuristic device such as varying L to maximize $\mathscr{D}_L(\lambda)$ (Khor 1971). It is plausibly argued, however, and confirmed by comparison with the results of Monte Carlo computations (Roberts, Jones, Khor & Smith 1969), that the best choice for L is the *beat length* in the perfect chain of spacing $\bar{\xi}_L$ that we are supposedly sampling. This is defined as the distance required for the phase angle (8.28) to advance (or retard) by π. At the bottom of a band such as (8.50), this obviously corresponds to a half wavelength π/q of the Bloch function; near the top of a band the phase change in each cell should be measured relative to π, so that the beat length

$$L = \frac{\pi}{\pi - q\xi_L} \tag{8.87}$$

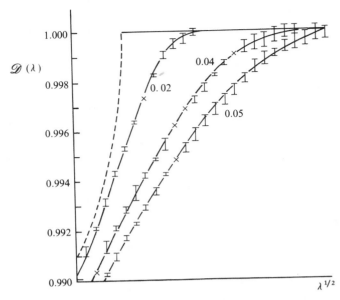

Fig. 8.10. Local density approximation for integrated spectra of Kronig–Penney 'liquid' models compared with Monte Carlo calculations (Jones, Khor, Roberts & Smith 1970).

would again tend to infinity at the band edge. To find q in these expressions we should need to solve a dispersion relation such as (8.15); in other words, the sampling length in (8.82) or (8.86) is taken to vary with both ξ_L and λ, in a 'self-consistent' manner. The resulting formulae for the spectral density are thus somewhat complicated.

To improve the LDA results still further we need to take account of the short-range disorder within each sample L. If the range of disorder is not too great then perturbation theory can be used to estimate the effects on the spectral variable λ (Roberts, Jones & Smith 1969) and thence to modify the function $\mathscr{D}_0(\lambda; \bar{\xi}_L)$ to be included in the integral (8.82). Thus calculation is more elaborate than any 'linear liquid' model really deserves, but it is gratifying (fig. 8.10) that the results are entirely within the range of numerical uncertainty of the Monte Carlo computations with which they have been compared (Jones, Khor, Roberts & Smith 1970). It should be emphasized, however, that there is no analytical justification for choosing the actual beat length (8.87) as a sampling length. The LDA method is thus very instructive as a heuristic, semiquantitative approach to the calculation of the spectrum of a disordered system, but is not, apparently, the starting point for a well-defined and convergent sequence of successive approximations to the exact spectral density and it fails completely to show up such 'pathological' features of the spectrum as the special frequency gaps of the binary alloy model.

8.7 Localization of eigenfunctions

All the normal modes and eigenfunctions of a disordered linear chain are *localized*. This fundamental property was first derived for electrons in a 'one-dimensional liquid' by Mott & Twose (1961) and by Makinson & Roberts (1962) and the same effect was observed by Dean & Bacon (1963) in computed normal modes of an 'alloy' chain. This property was eventually proved to hold for all types of excitation in all the standard models of one-dimensional disorder (Matsuda & Ishii 1970). The implication is – although the point has not yet been proven rigorously (see Ishii 1973) – that the static electrical or thermal conductivity of such a system must be zero (see § 10.10).

The basic idea is not complicated, although it is mathematically quite subtle. Let us consider, for simplicity, an 'alloy' model whose two constituents correspond to two different transfer matrices, $\mathbf{T}$ and $\mathbf{T}'$. Suppose, first, that the spectral variable λ lies in what would be a gap for a perfect lattice of type $\mathbf{T}$, but is in an allowed band for $\mathbf{T}'$. The Saxon–Hutner theorem (§ 8.4) does not apply, and the spectrum may be assumed continuous around λ.

According to (8.32)–(8.40), the eigenvalues of $\mathbf{T}$ must be real for this value of λ. Since this matrix may always be taken to be unimodular, we may choose these so that

$$\mu^{(+)} = 1/\mu^{(-)} > 1 > \mu^{(-)}. \tag{8.88}$$

Now suppose that we start off at the left-hand end of the chain with an excitation vector $\mathbf{U}_0$ of the form (8.33) and pass through N successive cells of this type; according to (8.34) we arrive at an excitation

$$\mathbf{U}_N = (\mathbf{T})^N \mathbf{U}_0 = u_0^{(+)} (\mu^{(+)})^N \mathbf{W}^{(+)} + u_0^{(-)} (\mu^{(-)})^N \mathbf{W}^{(-)}, \tag{8.89}$$

where $\mathbf{W}^{(\pm)}$ are the eigenvectors of $\mathbf{T}$.

In general, the major eigenvalue quickly dominates this expression; $\mathbf{U}_N$ tends to the 'direction' of $\mathbf{W}^{(+)}$ and grows in amplitude exponentially along the chain, with coefficient

$$\gamma = \ln|\mu^{(+)}|. \tag{8.90}$$

But if $u_0^{(+)}$ happened to be zero – that is, if the phase angle θ_0 of the starting excitation $\mathbf{U}_0$ happened to coincide exactly with the phase $\theta^{(-)}$ of the eigenvector $\mathbf{W}^{(-)}$ – then the effect of multiplying by $\mathbf{T}$ would appear quite different: the excitation would *decay* exponentially along the chain, like

$$\mathbf{U}_N = \exp(-\gamma N)\mathbf{W}^{(-)}. \tag{8.91}$$

These results are indeed precisely what we should expect in the gap of a perfect T-type chain. In a 'forbidden energy region', for example, the solution of the Schrödinger equation is a linear combination of exponentially growing and decaying functions of a form precisely analogous to (8.89), i.e.

$$\psi(x) = A e^{\gamma x} + B e^{-\gamma x}. \tag{8.92}$$

If we try to start this function from $x = 0$ with an arbitrary value of $\psi'(0)/\psi(0)$, then the exponentially increasing function will take over; but for one special value of this ratio, only the decaying function is needed, as in (8.91).

The question is: what is the effect of diluting this perfect chain with cells of type $\mathbf{T}'$? Since λ is not in the gap of this transfer matrix, each such cell has the effect of changing the phase angle by a substantial quantity (of the order of $4\beta'$, say, as in (8.44)) on each occasion that it is encountered. Thus, the phase angle of the excitation will no longer tend uniformly to the phase $\theta^{(+)}$ of $\mathbf{W}^{(+)}$, as in (8.35), but varies statistically from cell to cell. All that we can assert of the 'phase-pulling' effect of T-type cells is that they will produce a probability distribution $w(\theta)$ of phase angles, peaked about $\theta^{(+)}$. This distribution must, however, satisfy the Dyson–Schmidt condition (8.76) for our chosen value of λ. We may conjecture that the amplitude still grows exponentially, 'on the average', with a coefficient

$$\gamma = \lim_{N \to \infty} \frac{1}{N} \ln \|\mathbf{T}_N \ldots \mathbf{T}_1 \mathbf{U}_0\|$$

$$= \iint \ln\left\{ \frac{\|\mathbf{T}(\xi)\mathbf{U}(\theta)\|}{\|\mathbf{U}(\theta)\|} \right\} w(\theta) P(\xi) \,\mathrm{d}\xi \,\mathrm{d}\theta. \tag{8.93}$$

In other words, we calculate the logarithmic amplification factor for the effect of a typical transfer matrix $\mathbf{T}(\xi)$ on the norm $\|\mathbf{U}(\theta)\|$ of a typical excitation of phase angle θ, and average over the stationary distribution of phase angles and over the probability distribution $P(\xi)$ of transfer matrices in the chain. This is certainly the result that we would get by naively asserting the statistical independence of the successive matrices in the product $\mathbf{T}_N \ldots \mathbf{T}_1$, as if these were scalar random variables.

Quite generally (8.93) predicts a positive definite value of γ, so that the excitation grows in amplitude as we go to the right along the chain. How, then, are we to construct an eigenfunction that satisfies the boundary conditions (8.29) at both ends of a very long chain? If we start with the prescribed value of phase angle θ_0 on the left, it seems impossible to arrive at the right-hand end with the small or zero amplitude required at this boundary.

Suppose, however, that we were to traverse the chain in the *opposite* direction, starting with the prescribed phase angle θ_N at the *right-hand* end and going to the left. We encounter transfer matrices that are simply transposed versions of the set $\{\mathbf{T}^{(\xi)}\}$ and which have the same general properties. The excitation will, therefore, grow exponentially as we go back along the chain, with just the same coefficient γ.

At first sight, this seems quite paradoxical: how can we say at one moment that the excitation amplitude must, with probability unity, *increase* exponentially as we go to the right, and at the next moment find one that *decreases* exponentially in the same direction? The answer is (Borland 1963) that the latter function is precisely analogous to the decreasing exponential function (8.91) that we found as a special solution in the gap of a perfect lattice. To put it another way: by constructing this function 'backwards' (i.e. from right to left) we ensure that it arrives at the left-hand end with exactly the right phase angle (i.e. the analogue of $\theta^{(-)}$). If we had, in fact, started out from the left with just this phase angle then we should have generated an excitation of decreasing amplitude. The *a priori* probability of hitting upon this special angle is of measure zero, so that we see nothing of this phenomenon in the stationary distribution $w(\theta)$ and in (8.93): nevertheless, for the ensemble as a whole, exponentially increasing and decreasing excitation functions are of equal probability.

This argument can be applied to the calculation of the effect of an 'external' excitation incident, say, on the left-hand end of the chain. Just as in elementary calculations of quantum-mechanical tunnelling, using a wave function like (8.92) in the barrier region, we exclude the exponentially increasing function (which would violate causality conditions). A reflected wave is introduced, so that we can match exactly the boundary condition for an excitation that decays into the forbidden region. The coefficient γ given by (8.93) must be the *transmission constant* into the chain at the spectral value λ.

It is now extremely plausible – although the point does not seem to have been proved rigorously – that the eigenfunctions of this system cannot be *extended* like the Bloch functions of a perfect lattice, but must each be localized to a particular segment of chain, from which the amplitude decays exponentially outward in either direction. As shown in detail by Roberts & Makinson (1962) and by Borland (1963), it is not difficult to show that two functions satisfying the boundary condition (8.29) and growing exponentially inward can turn out to be matched smoothly at some intermediate point of the chain – for example at an 'island' of $\mathbf{T}'$ cells where λ is in an allowed band so that a large overall change of place can easily be produced. The analogue is with a localized impurity state in an energy gap of a regular lattice; the large change of slope of the wave function as it goes through the impurity potential allows it to be matched to the exponentially decaying parts of (8.92) in each direction.

In general, however, it is better to rely on a theorem such as (8.73) asserting that eigenfunctions exist in the spectral region λ and to note that these could not be normalized unless the conditions for exponential growth away from some intermediate point on the chain were rigorously nullified. As in (8.91), this leaves us only with the special exponentially decaying functions; there is no, more or less uniform, 'extended' function in between.

The rigorous proof (Matsuda & Ishii 1970) of the conjecture that the excitation amplitude grows exponentially, with probability unity, for all typical models of one-dimensional disorder, depends on a general theorem concerning Markov processes (Furstenberg 1963). In principle, the *localization constant* $\gamma(\lambda)$ (or its inverse, the *range of localization*) can be evaluated from the integral (8.93), where, of course, the transfer matrices $\{\mathbf{T}^{(\xi)}\}$ and the stationary distribution of phase angle, $w(\theta)$, are both functions of the spectral variable λ. This method has been used by Hirota & Ishii (1971) and by Hirota (1973) to determine the degree of localization in various parts of the spectrum in various standard models. The results are in good agreement with Monte Carlo computations of actual eigenfunctions on sample disordered chains.

The surprising fact is that *all the eigenfunctions of a disordered chain are localized*. In the above discussion we assumed that the spectral variable λ lay outside the allowed bands of at least one type of constituent cell of the chain – in what is sometimes called a *pseudo-gap* region of the spectrum – where the density of states is relatively low, but not zero, because of 'band tailing' (8.86). Intuition tells us that the localization of eigenfunctions is due to the exponential decay of the excitation as it tries to 'tunnel' through stretches of 'forbidden' lattice produced by random fluctuations along the chain.

It turns out, however (Matsuda & Ishii 1970), that the conditions of

Furstenberg's theorem can be satisfied for any value of λ where the transfer matrices $\{T^{(\xi)}\}$ do not all have the same eigenvectors. That is to say, *the eigenfunctions are localized even in a spectral region where the Bloch functions of a regular chain of each of the constituents would be extended.*

To show how this can occur, let us suppose that λ lies in the 'allowed' regions of two types of transfer matrix, T and T'. These must both have complex eigenvalues, but if the corresponding eigenvectors are not the same there is no single transformation S that reduces them simultaneously to diagonal form. Suppose, therefore, that S is chosen, as in (8.44), to diagonalize T whilst T' is transformed into a Cayley matrix (8.51). Since every transfer matrix is unimodular, the matrix elements satisfy the condition $|A|^2 - |B|^2 = 1$ and can be represented parametrically in terms of three real variables by, say

$$A = \cos \alpha + i \sin \alpha \cosh 2\chi; \quad B = e^{i\delta} \sin \alpha \sinh 2\chi. \quad (8.94)$$

In this representation it is very easy to construct the product matrix $[T^n T']$ (where n is an integer), and to calculate its trace:

$$\mathrm{Tr}[T^n T'] = 2(\cos \alpha \cos 2n\beta - \sin \alpha \cosh 2\chi \sin 2n\beta)$$
$$= 2|a'| \cos(2n\beta + \phi') \quad (8.95)$$

where

$$a' \equiv \cos \alpha + i \sin \alpha \cosh 2\chi \equiv |a'| \exp(i\phi'). \quad (8.96)$$

But $|a'| > 1$, so that by an appropriate choice of n we can make (8.95) exceed 2; in other words, the product matrix $[T^n T']$ satisfies the condition (8.40) for real eigenvalues. This product can thus be considered as the transfer matrix of a 'constituent' of the chain, which would, if repeated regularly, produce a spectral gap around λ. The occurrence of the succession $TT \ldots (n$ times $) \ldots T'$ from time to time along the disordered chain is thus sufficient to cause exponential growth of any excitation, in accordance with the general theorem. The argument is reminiscent of the proof of the existence of special frequencies (§ 8.4) in the spectrum of a disordered alloy chain; indeed, the cases where the phase angle β of the matrix T is a rational multiple of π, as in (8.64), must be discussed separately in the complete proof of the Matsuda–Ishii theorem.

In practice, it is found that the eigenfunctions are strongly localized in the pseudo-gap regions; but, within the 'allowed bands' of the 'average chain' (e.g. as in (8.81)), the range of localization becomes very long and is not always apparent in a Monte Carlo computation on a finite model. It can be shown, indeed (Blair 1972; Thouless 1973) that $1/\gamma(\lambda)$ then tends to the *mean free path* that one would calculate for an excitation propagating in

Excitations on a disordered linear chain

the average regular chain and being scattered by deviations from perfect order (§ 10.1). This seems a more natural intuitive explanation of the decay of a localized eigenfunction in this spectral region than the concept of tunnelling through 'forbidden' segments of chain.

It must be emphasized, however, that there is no evidence from analytical theory or from computational experience that these two regimes are sharply separated. The localization constant $\gamma(\lambda)$ seems to vary smoothly as λ goes through the spectrum from a 'fully allowed' band into a 'pseudo-region. This is an important point in principle. There is good reason to believe (§ 9.8) that the excitation spectrum of a two- or three-dimensional disordered system may be divided by *mobility edges* into localized and extended states, whose transport properties are entirely different. In one dimension, however, the topology does not admit of a formal mathematical distinction between a 'localized' and a 'randomly scattered' function.

The intimate connection between the spectral density and the range of localization on a disordered chain is beautifully illustrated by a dispersion relation due originally to Herbert & Jones (1971) and generalized by Thouless (1972). By taking the logarithm of the Green function $G_{1N}(\lambda)$ for a tight-binding Hamiltonian such as (8.12), with only nearest neighbour matrix elements $V_{l,l+1}$, it is easy to show that the localization constant (8.93) must satisfy

$$\gamma(\lambda) \equiv \lim_{N \to \infty} \frac{1}{N} \ln|G_{1N}(\lambda)|$$

$$= \lim_{N \to \infty} \frac{1}{N} \left\{ \sum_{\alpha} \ln|\lambda - \lambda_{\alpha}| - \sum_{l} \ln|V_{l,l+1}| \right\}$$

$$= \int \ln|\lambda - \lambda'| \mathcal{N}(\lambda') d\lambda' - \langle \ln|V| \rangle, \tag{8.97}$$

where $\mathcal{N}(\lambda)$ is the spectral density in the ensemble limit. The essence of the proof (Ishii 1973) is that an excitation starting out correctly from $l = 1$ and vanishing at $l = N$, must give a singularity of $G_{1N}(\lambda)$ at each eigenvalue $\lambda = \lambda_{\alpha}$ (see § 9.2).

This sort of result suggests that other properties of excitations on a linear chain are best deduced in the language of *Green functions,* which will be explained and applied in later chapters (see § 9.2). We shall there use the known exact properties of one-dimensional systems to illustrate various general mathematical principles and to test the validity of various types of approximation.

9

Excitations on a disordered lattice

'The genuine spirit of localism'
Borrow

9.1 The TBA model

The simplest type of three-dimensional disordered system is a regular
lattice on which different atoms or spins have been distributed at random
(chapter 1). Like every one-dimensional system, such a system is topologically ordered. We are thus concerned with precisely the same types of
dynamic, magnetic and electronic excitations as those considered in § 8.1 in
the simpler context of a disordered linear chain. We concentrate our
attention on the set of equations

$$(\mathscr{E}_l - \lambda)u_l + \sum_{l' \neq l} V_{ll'}u_{l'} = 0 \qquad (9.1)$$

which generalize (8.12) to the case where the site labels l and l' run over a
regular lattice in real, three-dimensional space. The effects of substitutional
disorder on nearly free electrons or in resonance bands will be discussed in a
broader context in § 10.9.

It is often convenient to think of these equations as representing the
electronic properties of a *tight-binding alloy* (TBA), whose atomic orbitals
have different bound state energies $\mathscr{E}_l$ on the different constituents of the
alloy, and different overlap integrals $V_{ll'}$ between different cells. Mathematically speaking, however, it is scarcely necessary to concern ourselves with
the physical interpretation of the symbols in (9.1). Thus, the *amplitude
variable u_l*, shown as if a scalar quantity, might have many components,
representing, say, the Cartesian components of the displacement of the
atom on the lth site (as in (8.3)), or the relative contributions of the
atomic orbitals to an LCAO wave function (as in (8.11)). Again, the
spectral variable λ is not necessarily an energy; it may stand for the square,
ω^2, of a dynamical excitation frequency. The physical disorder introduced
through the *diagonal elements $\mathscr{E}_l$* and/or the *off-diagonal elements V_{ll}* needs

to be defined only through the statistical properties of these parameters of the model, regardless of whether these arise from isotopic, atomic, magnetic or other forms of cellular disorder (cf. (8.13)). In the present chapter, therefore, we may appear to be studying the mathematical properties of a very simple and specialized model; in reality, this theory applies to a wide range of physical systems and phenomena. In practice, the TBA model is often simplified further by ignoring disorder in the off-diagonal elements (see § 9.3).

Algebraically speaking, the most serious consequence of going from a one-dimensional to a three-dimensional model of this type is that we lose all the advantages of the transfer matrix representation (8.19). The same fundamental difficulty arises as in the statistical mechanics of the Ising model (cf. § 5.7); because a two-dimensional or three-dimensional lattice is multiply connected, the propagation of order, or of an excitation, can no longer be reduced to a simple product of independent matrices as in (8.20). In the case of the linear chain, this was the key to the statistical self-consistency of the Dyson–Schmidt equation (8.76) from which the exact spectrum can be deduced. In higher dimensions, no rigorous analogue of this theorem seems to exist.

Another special property of one-dimensional disorder is the possibility of representing *all* excitation problems in transfer matrix language, including, for example, the theory of electron states in the Kronig–Penney liquid and other random potentials (§ 8.2). In general, however, it does not always seem feasible to reduce the problem of electron propagation in a substitutionally disordered array of scattering centres to a convergent generalization of the equations (9.1). The theory of the electron states of a disordered system near the 'free band' limit will be discussed in chapter 10.

9.2 The Green function formalism

Most of the approximate methods that have been applied to real disordered systems have been set up in the *Green function* formalism which is, of course, the accepted canonical language of advanced quantum field theory. Since the basic properties of Green functions, *resolvents, propagators* and similar mathematical subjects are fully discussed in numerous standard texts, the present section consists mainly of definitions and theorems without explicit proof.

The standard TBA equations (9.1) are written schematically in the form

$$(\mathscr{H} - \lambda)u = 0 \qquad (9.2)$$

where the *Hamiltonian* is the matrix operator

$$\mathcal{H}_{ll'} = \mathcal{E}_l \delta_{ll'} + V_{ll'}. \tag{9.3}$$

(It is a matter of taste whether additional symbols for site *annihilation* and *creation* operators, a_l, a_l^*, or for *site projection operators* $|l\rangle$, $\langle l'|$, etc. are included in this formula to aid later manipulations.) By definition, the *Green function* is the inverse operator

$$G(\lambda) = (\lambda - \mathcal{H})^{-1}. \tag{9.4}$$

In other words, we must solve the *inhomogeneous* equations

$$\sum_{l'} \{\mathcal{H}_{ll'} - \lambda \delta_{ll'}\} G_{l'l''}(\lambda) = -\delta_{ll''}. \tag{9.5}$$

Evidently $G_{ll'}(\lambda)$ measures the amplitude propagated to site l' by an excitation of unit amplitude and 'frequency' λ applied to the site l.

The key mathematical property of the Green function is that it is diagonal in a matrix representation in which the Hamiltonian is diagonal and that it has poles at the real values of λ corresponding to the eigenvalues $\lambda^{(\alpha)}$ of (9.2). If $u_l^{(\alpha)}$ is the corresponding eigenfunction, then

$$G_{ll'}(\lambda) = \sum_\alpha u_l^{(\alpha)} \frac{1}{\lambda - \lambda^{(\alpha)}} u_{l'}^{(\alpha)}. \tag{9.6}$$

The total number of eigenvalues in a small range of the spectral variable λ can be calculated by treating λ as a complex variable with infinitesimal 'positive' imaginary part $+i\varepsilon$ and carrying out a contour integration. This yields the general theorem for the *spectral density*

$$\mathcal{N}(\lambda) = -\frac{1}{\pi} \text{Im}\{\text{Tr}[G^+(\lambda)]\}$$

$$= -\frac{1}{2\pi i} \text{Tr}[G^+(\lambda) - G^-(\lambda)]. \tag{9.7}$$

In the *site representation* (9.6) we evaluate the trace over a lattice of N sites, so that the *normalized spectral density* (cf. (8.66)) for a large system is given by

$$\mathcal{N}(\lambda) = \lim_{\varepsilon \to 0^+} \left\{ \lim_{N \to \infty} -\frac{1}{2\pi N i} \sum_l [G_{ll}(\lambda + i\varepsilon) - G_{ll}(\lambda - i\varepsilon)] \right\}. \tag{9.8}$$

But the more abstract formulation (9.7) shows that we may evaluate the trace in any convenient representation of the operators $G^+(\lambda)$ and $G^-(\lambda)$.

Indeed, for a perfectly *ordered* system, where $\mathcal{H}_{ll'}$ would have the fundamental translational symmetry (8.2), we know that the eigenfunctions of the Hamiltonian are Bloch functions of the form (1.43). In this case, therefore, the Green function must be diagonal in a *reciprocal space* or *momentum representation*:

$$G_{qq'}(\lambda) \equiv \sum_{ll'} e^{iq \cdot l} G_{ll'}(\lambda) e^{-iq' \cdot l'}$$

$$= \frac{1}{\lambda - \lambda(q)} \delta_{qq'} \tag{9.9}$$

where $\lambda(q)$ might be given, for example, by the dispersion formula (8.15) for a single tight-binding band. We may verify that (9.7) then reproduces the usual expression for the density of states in such a band as an integral over a thin energy shell in q-space.

Formulae such as (9.6) and (9.7) are *exact* and might, therefore, be applied to any system such as a disordered chain or lattice. But we are interested eventually only in the *average* properties of a very large system or of a whole ensemble of such systems with appropriate statistical parameters. The spectral density thus comes to be defined as the ensemble average of (9.7):

$$\mathcal{N}(\lambda) = -\left\langle \frac{1}{2\pi i} \mathrm{Tr}[G^+(\lambda) - G^-(\lambda)] \right\rangle$$

$$= -\frac{1}{2\pi i} \mathrm{Tr}[\langle G^+(\lambda) \rangle - \langle G^-(\lambda) \rangle], \tag{9.10}$$

since the operations $\langle\ \rangle$ and Tr obviously commute. This has the advantage that the Green function, which is as statistically homogeneous as the system itself, is reduced by ensemble averaging to a function with perfect lattice translational symmetry, i.e.

$$\langle G_{ll'}(\lambda) \rangle \equiv \mathcal{G}(l - l'; \lambda). \tag{9.11}$$

From (9.10) we immediately deduce that the spectral density is just the imaginary part of this function at $l' = l$:

$$\mathcal{N}(\lambda) = \frac{1}{\pi} \mathrm{Im}\ \mathcal{G}^+(0; \lambda). \tag{9.12}$$

Taking a Fourier transform of (9.11), as in (9.9), we get

$$\langle G_{qq'}(\lambda) \rangle = \mathcal{G}(q; \lambda) \delta_{qq'}, \tag{9.13}$$

which is again diagonal in the reciprocal space representation. But the excitations of a disordered system are not themselves Bloch waves; in place of (9.9) we must interpret the imaginary part of $\mathcal{G}^+(q; \lambda)$ as the probability density of finding a Fourier component of wave vector q amongst the excitations of 'energy', or 'frequency', λ.

9.3 Propagator and locator expansions

It is natural to seek a solution of the Green function equations (9.4) or (9.5) in the form of a series expansion in powers of some supposedly small 'perturbation' associated with the disorder. We therefore write the Hamiltonian (9.3) in the form

$$\mathscr{H} = \bar{\mathscr{H}} + \mathscr{W} \qquad (9.14)$$

where the 'unperturbed' Hamiltonian $\bar{\mathscr{H}}$ has the perfect crystalline symmetry, whilst $\mathscr{W}$ contains all the effects of disorder. These effects are minimized by defining $\bar{\mathscr{H}}$ as the Hamiltonian of a *virtual crystal* with the average properties of the material, i.e.

$$\bar{\mathscr{H}}_{ll'} = \langle \mathscr{H}_{ll'} \rangle = \bar{\mathscr{E}} \delta_{ll'} + \bar{V}_{ll'}. \qquad (9.15)$$

In other words, the ensemble average of the perturbation vanishes:

$$\langle \mathscr{W} \rangle = 0. \qquad (9.16)$$

From (8.15) and (9.9) we may deduce that there corresponds to $\bar{\mathscr{H}}$ a Green function or *virtual crystal propagator*

$$\bar{G}_{qq'}(\lambda) = \{\lambda - \bar{\lambda}(\mathbf{q})\}^{-1} \delta_{qq'}, \qquad (9.17)$$

with a dispersion law

$$\bar{\lambda}(\mathbf{q}) = \mathscr{E} + \sum_{\mathbf{h} \neq 0} \bar{V}(\mathbf{h}) e^{i\mathbf{q} \cdot \mathbf{h}}. \qquad (9.18)$$

The Dyson equation for the Green function of the disordered system (9.4) can be expanded in a series in $\mathscr{W}$, i.e.

$$G = \bar{G} + \bar{G}\mathscr{W}G$$
$$= \bar{G} + \bar{G}\mathscr{W}\bar{G} + \bar{G}\mathscr{W}\bar{G}\mathscr{W}\bar{G} + \dots \qquad (9.19)$$

In the site representation (9.5), this *propagator expansion* reads

$$G_{ll'} = \bar{G}_{ll'} + \sum_{m,m'} \bar{G}_{lm}\mathscr{W}_{mm'}\bar{G}_{m'l'} + \sum_{m,m',n,n'} \bar{G}_{lm}\mathscr{W}_{mn}\bar{G}_{nn'}\mathscr{W}_{n'm'}\bar{G}_{m'l'} + \dots \qquad (9.20)$$

In this expression, each virtual crystal propagator $\bar{G}_{lm}$ is a function of λ, obtained from (9.17) by a Fourier transformation as in (9.9).

To determine the spectral density we should now follow (9.10) or (9.22) by taking an ensemble average of (9.20), term by term. But the resulting series cannot be resummed exactly. The crudest approximation is to replace each factor $\mathscr{W}$ by its ensemble average (9.16). In the *virtual crystal approximation* (VCA), where

$$\langle G_{ll} \rangle \approx \bar{G}_{ll}, \qquad (9.21)$$

the statistical fluctuations of the disorder are ignored; the spectrum of the disordered material is taken to be that of a regular crystal with the average parameters (9.15). Like the corresponding approximation (8.81) for the spectrum of a linear chain, this is useful in identifying the spectral regions in which excitations may be expected to occur, but completely fails to describe the more interesting features of the spectrum.

Indeed, it scarcely needs to be emphasized that the average Green function of a randomized Hamiltonian must be quite different from the Green function of the average Hamiltonian. In the series (9.20) these differences arise directly from the fact that no restrictions are placed on the site labels over which each term is summed. In the diagrammatic representation (fig. 9.1) of this series, any site may be revisited a number of times on

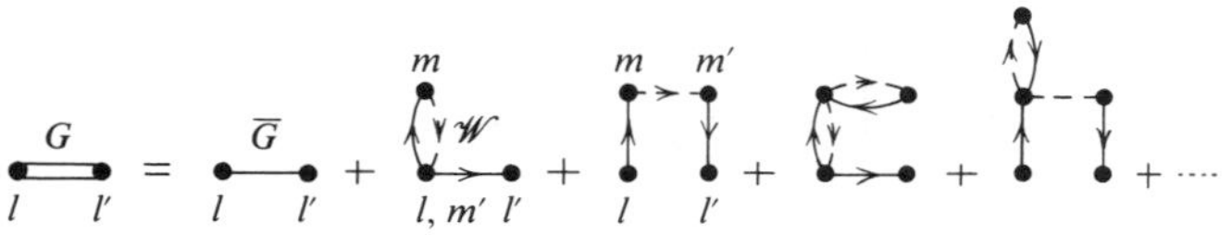

Fig. 9.1

the same graph, showing that a perturbation matrix element $\mathcal{W}_{mn}$ may occur several times in the same product. In spite of (9.16), the ensemble average of such a factor need not vanish; indeed, if our system is not to be perfectly ordered, the *variance* of the disorder perturbation must not be zero:

$$\langle \mathcal{W}^2 \rangle \neq 0. \tag{9.22}$$

Thus, when we try to evaluate the ensemble average of (9.20), we get successively more complex contributions from successive terms, for whose sum there is no exact algorithm.

To obtain a decisive improvement on the VCA formula (9.21) it is not sufficient to evaluate the first few terms in the average of the propagator series (9.20). For a reasonably simple analytical formula which sums at least a part of this series to infinite order, it is convenient to omit from the Hamiltonian all terms referring to off-diagonal disorder (cf. § 9.8). Except in the case of lattice dynamics with variations of isotopic mass (§ 8.1), it is seldom physically justifiable to assume that the interaction matrix elements $V_{ll'}$, are the same between all different types of atom in neighbouring cells. But a model in which

$$\mathcal{W}_{ll'} = (\mathcal{E}_l - \bar{\mathcal{E}})\delta_{ll'} = w_l \delta_{ll'}, \tag{9.23}$$

i.e. where the effects of disorder are to be seen primarily as variations of the local atomic energy levels $\mathcal{E}_l$, has many interesting properties. Like the

Ising model in the theory of order–disorder phenomena, it exemplifies many of the characteristic spectral features of disordered systems at minimum cost in mathematical complexity.

To specify such a model completely, we need to know the probability distribution $P(w)$ of the random variable w_l, which measures the deviation of the atomic level $\mathscr{E}_l$ from the average $\bar{\mathscr{E}}$ in the virtual crystal (which we can now, by convention, take as the zero of the spectral variable λ). For historical reasons, the case where $P(w)$ is a *continuous* function is called the *Anderson model* of disorder (§ 9.9). The elementary theory of binary alloys (§ 1.2) is based on what might be called the binary random alloy (BRA) model, where $\mathscr{E}_l$ takes only two *discrete* values $\mathscr{E}_A$ and $\mathscr{E}_B$, with relative probabilities (i.e. atomic concentrations) c_A and c_B. We naturally assume, for simplicity, that there is no correlation between the values of w on neighbouring sites, although the effects of short-range order (§ 1.5) can be included approximately, if necessary (see e.g. Elliott, Krumhansl & Leath 1974). For much of the present chapter the analysis applies to both types of model, although there are also some subtle differences between the continuous and discrete cases.

Under these conditions, the propagator series (9.20) takes the form

$$G_{ll'} = \bar{G}_{ll'} + \sum_{l'} \bar{G}_{ll''} w_{l'} \bar{G}^{l''l'} + \sum_{l'',l'''} \bar{G}_{ll''} w_{l'} \bar{G}_{l''l'''} w_{l'} \bar{G}_{l'''l'} + \ldots \quad (9.24)$$

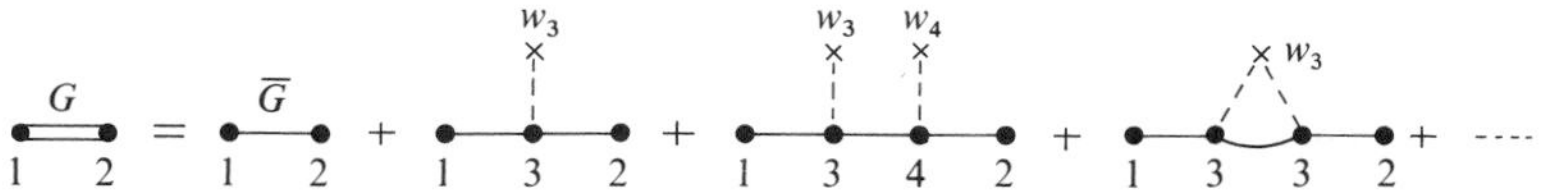

Fig. 9.2

represented diagrammatically in fig. 9.2. It is easy now, however, to remove all unbroken strings of identical factors. Diagrammatically, (fig. 9.3) these

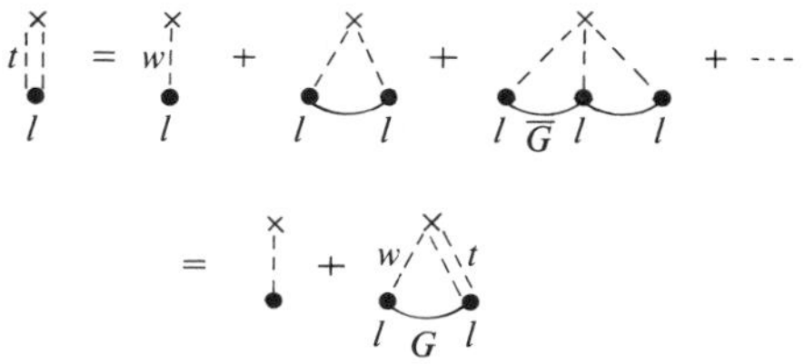

Fig. 9.3

correspond to multiple scattering of the excitation from the same site without excursions to other sites. This is the prescription for the introduction of a *site t-matrix*, t_l, in place of the perturbing 'potential' w_l. The series can be summed by a Dyson equation,

$$t_l \equiv w_l + w_l \bar{G}_{ll} + w_l \bar{G}_{ll} w_l \bar{G}_{ll} w_l + \dots$$

$$= w_l + w_l \bar{G}_{ll} t_l$$

$$= \frac{w_l}{1 - w_l \bar{G}_{00}} \tag{9.25}$$

since $\bar{G}_{ll}$ has the same value, $\bar{G}_{00}$ say, at every lattice site. Putting this expansion into the propagator series (9.24), we get the series

$$G_{ll'} = \bar{G}_{ll'} + \sum_{l' \neq l,l'} \bar{G}_{ll'} t_{l'} \bar{G}_{l'l'} + \sum_{\substack{l' \neq l,l'' \\ l'' \neq l',l'}} \bar{G}_{ll'} t_{l'} \bar{G}_{l'l''} t_{l''} \bar{G}_{l''l'} + \dots \tag{9.26}$$

where (fig. 9.4) successive labels may not immediately repeat.

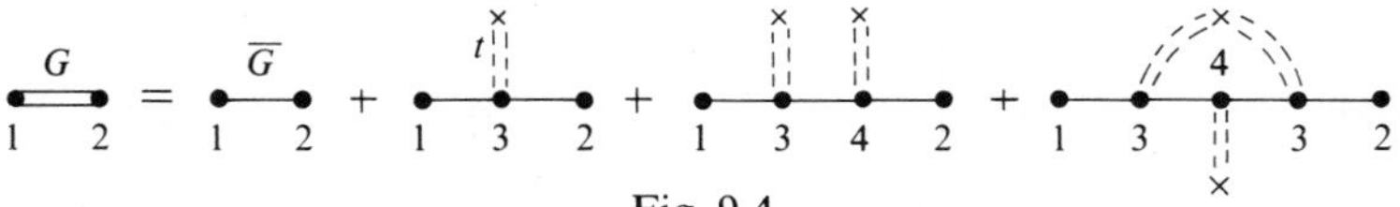

Fig. 9.4

In the series (9.26) each factor t_l is a random variable related by (9.25) to the perturbation w_l on the lth site. The problem of correlations induced by eventual repetitions of the same factor in a given product still remains. But when we take an ensemble average, these correlations have much less weight in (9.26) than they did in (9.24). We thus expect a much better answer from the *average t-matrix approximation* (ATA) where these factors are *decoupled* by replacing each t_l by the ensemble average

$$\bar{t} \equiv \left\langle \frac{w_l}{1 - w_l \bar{G}_{00}} \right\rangle = \int \frac{w}{1 - w \bar{G}_{00}} P(w) \mathrm{d}w. \tag{9.27}$$

The series thus obtained from (9.26), i.e.

$$G_{ll'}^{\mathrm{ATA}} = \bar{G}_{ll'} + \sum_{l' \neq l,l'} \bar{G}_{ll'} \bar{t} \bar{G}_{l'l'} + \sum_{\substack{l' \neq l,l'' \\ l'' \neq l',l'}} \bar{G}_{ll'} \bar{t} \bar{G}_{l'l''} \bar{t} \bar{G}_{l''l'} + \dots \tag{9.28}$$

can be summed. Comparing (9.24) with (9.26), we see that G^{ATA} is just what we should have obtained from a propagator series if the perturbation on every site were the same quantity Σ corresponding to the same t-matrix t. Inverting (9.25), this is given by the simple algebraic relation

$$\Sigma = \frac{\bar{t}}{1 + \bar{t} G_{00}}. \tag{9.29}$$

But the apparent effect of such a 'perturbation' can be no more than to shift the zero of the spectral variable λ by the self-energy Σ.

But the terms in the general propagator expansion (9.20) are all, implicitly, functions of λ. In constructing the site t-matrices (9.25), for example, we used $\bar{G}_{00}$, the 'same site' virtual crystal propagator in the site representation. This element is derived from (9.9) and (9.17) by a Fourier inversion

$$\bar{G}_{00} \equiv \bar{G}_{ll}(\lambda) = N^{-1}\sum_{\mathbf{q}}\{\lambda - \lambda(\mathbf{q})\}^{-1}. \tag{9.30}$$

In (9.27) and (9.29), we should have made it clear that the average t-matrix $\bar{t}(\lambda)$ and the self-energy $\Sigma(\lambda)$ must both be functions of λ. When, therefore, we sum the series (9.28), the singularity of (9.17) at the pole $\bar{\lambda}(\mathbf{q})$ will appear to be shifted by an amount $\Sigma(\lambda)$ which itself depends on λ. In other words, the ATA expression for the average Green function of the medium must, like (9.13), be diagonal in the momentum representation, with amplitude

$$\mathscr{G}^{\mathrm{ATA}}(\mathbf{q};\lambda) = \{\lambda - \bar{\lambda}(\mathbf{q}) - \Sigma(\lambda)\}^{-1}. \tag{9.31}$$

According to the prescription (9.12), the spectral density can be obtained from this function by giving λ a positive infinitesimal imaginary part and integrating over $\mathbf{q}$. It turns out, however, that the defining equations (9.27), (9.29) and (9.30) make $\Sigma(\lambda)$ into a complex function whose imaginary part is by no means infinitesimal in this limit. The prescription (9.12) is still correct, but the physical interpretation of (9.31) is no longer trivial. In simple terms Re $\Sigma(\lambda)$ measures the shift in the spectral levels produced by the disorder whilst Im $\Sigma(\lambda)$ measures their apparent broadening. Of course the true spectral levels of the disordered system are all perfectly sharp. But these excitations are not individually invariant under translations of the virtual crystal lattice and are not, therefore, eigenstates of the crystal momentum. In the reciprocal lattice representation, the Green function of the actual disordered system must have off-diagonal elements which average to zero in the ensemble average $\langle G \rangle$. The physical effects of these off-diagonal elements – finite lifetime and scattering of Bloch-like excitations of the virtual crystal – are significant in transport phenomena (§ 10.10) but are represented here only by the finite imaginary part of the self-energy in the average one-particle Green function.

The ATA equations (9.27)–(9.31), although somewhat more complicated to solve, are obviously a considerable improvement on the VCA formula (9.17). The statistical fluctuations associated with the disorder are taken into account over the whole range of relative concentrations of the constituents and some physical features of the model are correctly described (§ 9.4). But the method also yields some 'unphysical' results; for example, the spectrum of the BRA model is always split into two bands, even when the atomic levels λ_A and λ_B are very close together.

It is possible to improve on the approximation (9.26) by the usual techniques of graphical analysis of perturbation series (see e.g. Yonezawa & Morigaki 1973; Elliott, Krumhansl & Leath 1974). By the method of cumulants (§ 5.10), for example, Yonezawa & Matsubara (1966) were able to make further partial summations of infinite sequences of terms, yielding closer approximations to the average Green function. In practice, however, these and other improvements on the ATA formula are obtainable much more directly by making appropriate approximations in the 'compact' representations of the Green function. This approach will be illustrated when we derive the *coherent potential approximation* (CPA) in § 9.4.

An alternative, 'expanded' representation of the total Green function of a disordered system can be generated from equation (9.5) written in the form

$$(\lambda - w_l)G_{ll'}(\lambda) = \delta_{ll'} + \sum_{l''} V_{ll''}G_{l''l'}(\lambda). \tag{9.32}$$

Dividing through by $\lambda - w_l$, we obtain a Dyson equation symbolized by

$$G = g + gVG, \tag{9.33}$$

where the *locator*

$$g_{ll'} \equiv g_l\delta_{ll'} \equiv \frac{1}{\lambda - w_l}\delta_{ll'} \tag{9.34}$$

is simply a Green function that would 'locate' the excitation firmly on the lth site in the local spectral level at $\lambda = w_l$. But each of these bound states is linked to its neighbours by an *interactor* matrix element $V_{ll'}$. Expanding (9.33) symbolically in powers of gV, we obtain the *locator expansion* (Matsubara & Toyozawa 1961)

$$G = g + gVg + gVgVg + \ldots \tag{9.35}$$

In the usual site representation, this means

$$G_{ll'} = g_l\delta_{ll'} + g_lV_{ll'}g_{l'} + \sum_{l''} g_lV_{ll''}g_{l''}V_{l''l'}g_{l'} + \ldots \tag{9.36}$$

Since, by definition, the interactor matrix elements $V_{ll'}$ do not exist for $l = l'$, this series is similar to the t-matrix expansion (9.26) in that successive summation indices cannot immediately repeat. But the usual difficulty of dealing with correlations induced by taking ensemble averages of terms containing powers of g_l cannot be avoided. Here again, partial diagrammatic summations by a cumulant method (Matsubara & Kaneyoshi 1966) does not, in the end, yield better results than can be obtained more directly by the CPA method (§ 9.4).

Notice, however, the ease with which the series can be summed if the stochastic factors g_l are all replaced by their ensemble average

$$\bar{g}(\lambda) \equiv \langle g_l \rangle = \int \frac{1}{(\lambda - w)} P(w) dw. \tag{9.37}$$

Let us think of this as defining a modified spectral variable

$$\lambda^*(\lambda) = \{\bar{g}(\lambda)\}^{-1}: \tag{9.38}$$

our approximation for the average Green functions reads like

$$\langle G_{ll'} \rangle \approx \frac{1}{\lambda^*} \delta_{ll'} + \frac{1}{\lambda^{*2}} V_{ll'} + \frac{1}{\lambda^{*3}} \sum_{l''} V_{ll''} V_{l''l'}. \tag{9.39}$$

Now make a Fourier transformation of the interactors to a momentum representation

$$V_{ll'} = N^{-1} \sum_{q} V_q e^{iq \cdot (l - l')}, \tag{9.40}$$

whence (9.39) takes the form

$$\langle G_{ll'} \rangle \approx \frac{1}{\lambda^*} \delta_{ll'} + \frac{1}{\lambda^{*2}} \sum_{q} (V_q)^2 e^{iq \cdot (l-l')} + \frac{1}{\lambda^{*3}} \sum_{q} (V_q)^3 e^{iq \cdot (l-l')} + \dots \tag{9.41}$$

which contains a geometric series in powers of (V_q/λ^*) that can be summed immediately. The final result is evidently diagonal in the momentum representation (9.13), i.e.

$$\mathcal{G}(q; \lambda) \approx \{\lambda^*(\lambda) - V_q\}^{-1}. \tag{9.42}$$

But this is only another way of saying that (9.39) is the locator expansion for the Green function (9.17) of the virtual crystal lattice (where, by (9.18) and (9.40) the dispersion formula is $\bar{\lambda}(q) \equiv V_q$ at the value λ^* of the spectral variable). In other words, this approximation gives a result similar to the ATA formula (9.31) with an apparent self-energy

$$\Sigma(\lambda) = \lambda - \lambda^*(\lambda). \tag{9.43}$$

This approximation has no special merit in itself, but shows clearly, by its derivation, the use of a Fourier transformation in reducing a graphical series in the site representation to a form that can be summed.

On the other hand, it is possible by straightforward algebraic manipulation of (9.32), to write the site-diagonal elements of the total Green function in the form

$$G_{ll}(\lambda) = \{\lambda - \Sigma(l; \lambda)\}^{-1} \tag{9.44}$$

where the self-energy can be expanded as a locator series like (9.36), i.e.

$$\Sigma(l; \lambda) = w_l + \sum_{l'} V_{ll'} g_{l'} V_{l'l} + \sum_{l', l''} V_{ll'} g_{l'} V_{l'l''} g_{l''} V_{l''l} + \dots, \tag{9.45}$$

with the restriction that the site label should not return to l in any intermediate factor of any term. If this series can be summed (but see § 9.8) then we should be able to deduce the density of states (9.12) from the analytical properties of $\Sigma(l; \lambda)$ near the real λ axis.

9.4 The coherent potential approximation

In the statistical mechanics of order–disorder phenomena, we soon come to appreciate (§§ 5.2, 6.2) the power of *mean field approximations*. These are approximations derived from the 'compact' representation of the equations of the system by applying the phenomenological principle that the 'microphysics' of local objects – single atoms or clusters of atoms on adjacent sites (§ 5.4) – should be consistent with the 'macrophysical' properties imputed to the assembly in which they are immersed and of which they are part. Thus, for example, the key MFA equations (5.4) and (5.5) for the spin disorder model tell us that the spin polarization at any lattice site should be just what is needed, on the average, to generate the macroscopic 'effective field' that causes this polarization.

In the theory of the excitation spectrum of a disordered system, the same general principle leads directly to the *coherent potential approximation* (CPA). Following Soven (1967) and Taylor (1967), we suppose that the assembly has a spectrum described 'macroscopically' by a *medium propagator* $\tilde{G}(\lambda)$. For the moment, this function is not fully defined, but we assume that it has similar analytical properties to the ATA Green function (9.31). Thus, we assume that the medium propagator differs from the virtual crystal propagator $\bar{G}(\lambda)$ by a self-energy correction $\tilde{\Sigma}(\lambda)$, i.e.

$$\tilde{G}(\lambda) = \bar{G}\{\lambda - \tilde{\Sigma}(\lambda)\}. \qquad (9.46)$$

Physically speaking, this means that the system behaves as if a *coherent potential* $\tilde{\Sigma}(\lambda)$ had been assigned to each lattice site. Thus, the random potential w_l at the lth site now looks like an object with effective potential

$$\tilde{w}_l = w_l - \tilde{\Sigma}(\lambda). \qquad (9.47)$$

'Microphysically speaking', this is a perturbation of the uniform medium and hence, as in (9.25), scatters with an apparent t-matrix

$$\tilde{t}_l = \frac{\tilde{w}_l}{1 - \tilde{w}_l \tilde{G}_{00}}. \qquad (9.48)$$

This scattering is, of course, different from site to site, because w_l is a stochastic variable. But the overall effect of the site potentials is already supposed to have been accounted for in the medium propagator $\tilde{G}$. For self-consistency, therefore, the ensemble average of all single-site t-matrices (9.48) must vanish; we must have

$$\langle \tilde{t}_l \rangle \equiv \left\langle \frac{\tilde{w}_l}{1 - \tilde{w}_l \tilde{G}_{00}} \right\rangle$$

$$\equiv \int \frac{\{w - \tilde{\Sigma}(\lambda)\}P(w)\mathrm{d}w}{1 - \{w - \tilde{\Sigma}(\lambda)\}\tilde{G}_{00}\{\lambda - \tilde{\Sigma}(\lambda)\}} = 0. \quad (9.49)$$

The numerical solution of this implicit equation for the variable $\tilde{\Sigma}(\lambda)$ gives the coherent potential as a function of λ. Putting this into (9.46) we get the medium propagator and hence obtain the excitation spectrum of the system.

The CPA formula is now recognized to be the best *single-site* approximation for the spectral properties of a disordered system (see e.g. Yonezawa & Morigaki 1973; Elliott, Krumhansl & Leath 1974). Although attributed to Soven (1967) and Taylor (1967), it is noteworthy that the same mathematical approximation was used by Hubbard (1963, 1964) in his theory of electron correlation effects in narrow bands. The same result can be obtained in several apparently different ways. Thus, for example, the t-matrix expansion (9.26) can be partially summed by replacing the virtual crystal propagator by a renormalized medium propagator $\tilde{G}$ and correcting for multiple counting of diagrams (Onodera & Toyozawa 1968; Leath 1968; Aiyer, Elliott, Krumhansl & Leath 1969). The CPA condition (9.49) then arises from the highly plausible (but not exact) identification of this function with the average Green function of the system, i.e.

$$\tilde{G} \approx \langle G \rangle. \quad (9.50)$$

Matsubara & Yonezawa (1967) also obtained the CPA formulae by approximating to the value of an infinite continued fraction (§ 9.7) that was deduced from cumulant expansion of the propagator series (9.19). In fact (Yonezawa 1968) the supposedly 'exact' cumulant series tends to over-emphasize multiple-occupancy effects, giving rise to spurious poles, which are eliminated by the subsequent approximation.

The most instructive alternative derivation (Leath 1970; Ducastelle 1971) is from the locator expansion (9.36). This is a straighforward piece of algebra, whose success depends upon the same topological and stochastic principles as the transformation of the propagator series (9.24) into the non-repeating t-matrix expansion (9.26) and the decoupling approximation implicit in going from (9.27)–(9.29) to (9.49). Starting from the Dyson equation (9.31) we relate the exact Green function to a *renormalized interactor*

$$U \equiv V + VGV, \tag{9.51}$$

which satisfies its own self-consistency condition

$$U = V + VgU, \tag{9.52}$$

in terms of the locator g. In an effective medium with propagator $\tilde{G}$, this equation would look like

$$\tilde{U} = V + V\tilde{g}\tilde{U}, \tag{9.53}$$

where the function $\tilde{g}$ is a *medium locator* satisfying the analogue of (9.31), i.e.

$$\tilde{G} = \tilde{g} + \tilde{g}V\tilde{G} = (\tilde{g}^{-1} - V)^{-1}. \tag{9.54}$$

Thus, eliminating V in favour of the medium interactor U, we get an exact series expansion for the ensemble average of (9.51)

$$\langle U \rangle = \tilde{U} + \tilde{U}\langle (g - \tilde{g})U \rangle$$
$$= \tilde{U} + \tilde{U}\langle (g - \tilde{g}) \rangle \tilde{U} + \tilde{U}\langle (g - \tilde{g})\tilde{U}(g - \tilde{g}) \rangle \tilde{U} + \dots \tag{9.55}$$

This, in turn, can be rearranged into a series like (9.28) without immediate repetitions. The equivalence of $\langle U \rangle$ with $\tilde{U}$ demanded by (9.50) is then approximately achieved by statistically decoupling the terms and making them vanish on the average, i.e.

$$\left\langle \frac{g - \tilde{g}}{1 - (g - \tilde{g})\tilde{U}} \right\rangle = 0. \tag{9.56}$$

This is the analogue, in the locator representation, of the CPA 'non-scattering' conditions (9.49). But further algebra shows that the result is just the same as if we had put the coherent potential into the definition of the locator, i.e. by analogy with (9.34)

$$\tilde{g}_{ll'} = \frac{1}{\lambda - w_l - \tilde{\Sigma}(\lambda)}\delta_{ll'}. \tag{9.57}$$

In other words, the single-site CPA formula is apparently valid in the limit where the coupling constant V is small, so that the locator expansion (9.35) converges, as well as in the obvious case where the disorder itself is a weak perturbation.

Another valuable characteristic of the CPA propagator $\tilde{G}(\lambda)$ is that it conforms to all the necessary analytic conditions imposed by the interpretative equations (9.10)–(9.13) on the average Green function $\langle G(\lambda) \rangle$. Considered as a function of the complex variable λ, this must be a *herglotz function* (Shohat & Tamarkin 1943; Akhiezer 1965) satisfying

$$\text{Im}\langle G(\lambda) \rangle \gtrless 0 \quad \text{for Im } \lambda \gtrless 0, \tag{9.58}$$

and having no poles or cuts except on the real axis. This ensures that the spectral density (9.12) is always positive definite and excludes the appearance of 'unphysical' effects such as states growing exponentially with time. The proof that the CPA spectrum satisfies the *sum rule* for the total number of states in the bands (Hasegawa & Nakamura 1969; Yonezawa & Morigaki 1973) depends ultimately on the general theorem (Muller-Hartmann 1973; Ducastelle 1974a) that the medium propagator $\tilde{G}(\lambda)$ and the coherent potential $\tilde{\Sigma}(\lambda)$ are both herglotz functions. Even when applied to the Anderson model of disorder (§ 9.9) the CPA condition (9.49) leads to a complex self-energy $\tilde{\Sigma}(\lambda)$ which is an analytic function of λ everywhere off the real axis (Brouers 1971).

To understand the main properties of the CPA spectrum, we consider the standard BRA model, where (9.49) or (9.56) reduces to the algebraic equation

$$\frac{c_A(w_A - \tilde{\Sigma})}{1 - (w_A - \tilde{\Sigma})\tilde{G}_{00}} + \frac{c_B(w_B - \tilde{\Sigma})}{1 - (w_B - \tilde{\Sigma})\tilde{G}_{00}} = 0. \tag{9.59}$$

This can be solved immediately for $\tilde{G}_{00}$ as an algebraic function of $\tilde{\Sigma}$; hence, using (9.46), a value of λ can be associated with any chosen value of $\tilde{\Sigma}$, thus defining the function $\tilde{\Sigma}(\lambda)$; the spectrum then follows from, say, (9.12).

In limiting cases, however, the CPA equations can be solved explicitly. Thus, (9.59) can be rewritten in the form

$$\tilde{\Sigma} = \frac{\bar{w} - w_A w_B \tilde{G}_{00}}{1 - (w_A + w_B - \tilde{\Sigma})\tilde{G}_{00}}, \tag{9.60}$$

which is well adapted to evaluation by iterative substitution in the denominator. In the lowest approximation

$$\tilde{\Sigma} \approx \bar{w} \equiv c_A w_A + c_B w_B, \tag{9.61}$$

showing that the coherent potential reduces to the virtual crystal potential in the limit of weak disorder. Although we have chosen a spectral scale for which $\bar{w}$ vanishes, this result shows that the CPA equations transform correctly under arbitrary changes of the spectral zero.

The next approximation to (9.60) yields

$$\tilde{\Sigma}(\lambda) \approx \bar{w} + c_A c_B (w_A - w_B)^2 \tilde{G}_{00}(\lambda). \tag{9.62}$$

This is just what would have been obtained if we had treated w_A and w_B as perturbations from the average potential $\bar{w}$, giving rise to scattering between the VCA Bloch states (cf. § 10.1). Near the bottom of the unperturbed band $\bar{\lambda}(\mathbf{q})$, the VCA propagator (9.30) is essentially negative, so that the band edge is lowered by an amount proportional to the disorder variance (9.22), i.e.

$$c_A c_B (w_A - w_B)^2 = \tfrac{1}{2} \langle w^2 \rangle. \tag{9.63}$$

But the familiar 'tail' in the density of states produced by fluctuations of concentration (e.g. fig. 8.10) does not show up in the CPA spectrum.

On the other hand, when $w_A - w_B$ is large compared with the virtual crystal band width B, we may approximate to the Green function itself by writing (9.30) in the form

$$\bar{G}_{00}(\lambda) - \tilde{\Sigma} = N^{-1} \sum_q \{\lambda - \tilde{\Sigma} - \bar{\lambda}(\mathbf{q})\}^{-1}$$

$$\approx (\lambda - \tilde{\Sigma})^{-1} \tag{9.64}$$

when the denominator is large. Putting this into (9.60), we get a formula for $\tilde{\Sigma}$, from which we derive the medium propagator

$$\tilde{G}(\lambda) \approx c_A \bar{G}(\lambda - \lambda_A) + c_B \bar{G}(\lambda - \lambda_B). \tag{9.65}$$

In other words, the system behaves as if there were two independent Bloch type bands, each of the normal width B, centred respectively on the atomic levels λ_A and λ_B. This is the regime where the locator representation (9.57) would be appropriate.

It is convenient to indicate the 'strength' of the disorder by the parameter

$$\delta = |\lambda_A - \lambda_B|/B, \tag{9.66}$$

which measures the distance between the atom levels in units of the virtual crystal band width B. In phonon models, this parameter is proportional to the mass difference (8.53) of the constituent atoms. The spectrum of the BRA model depends only on the underlying lattice structure, the relative concentrations c_A, c_B of the two constituents, and on the parameter δ.

A major virtue of the CPA formula is that it gives a well-defined answer throughout the whole range of these parameters. On the borders of the parametric rectangle mapped out in fig. 9.5 (Leath 1973), the results are particularly simple. Thus, for small values of δ, the weak scattering limit (9.62) is valid at all concentrations. Then, keeping c_A or c_B small, we can go to large values of δ, where the disorder shows up as a side band due to a low concentration of more-or-less isolated impurities. In these circumstances, however, we pass across the range of concentrations using the *split band formula* (9.65). In the interior of this region the CPA spectrum behaves as might be expected, interpolating smoothly between the various limiting regimes. Thus (fig. 9.6), the single band of the weak scattering limit broadens and splits into two bands as δ passes through a value near 1, and these bands move apart, following the movement of the atomic levels

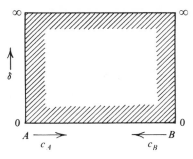

Fig. 9.5. CPA formula gives simple results near the boundaries of this parametric region.

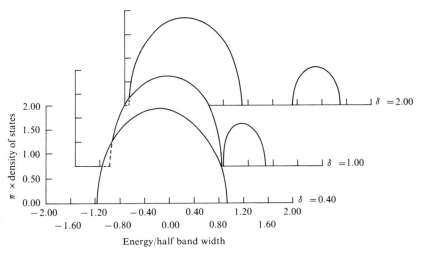

Fig. 9.6. CPA spectra for concentration $c_B = 0.15$, showing band splitting as δ increases (Velicky, Kirkpatrick & Ehrenreich 1968).

themselves, until we are at the split band limit. We also note that any van Hove singularities in the spectrum of the perfect crystal are blunted by the disorder.

All this is highly satisfactory. But the true test of a mathematical approximation is whether it reproduces closely the exact answer to the problem. Comparison with machine computations shows that the CPA spectrum is often quite near the correct results. Thus (fig. 9.7), in a three-dimensional lattice, with δ not too large, and neither c_A or c_B below the percolation threshold (§ 9.10), the agreement is excellent. But where one or other

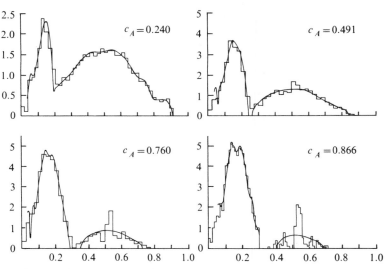

Fig. 9.7. Three-dimensional CPA spectra (phonon model), for various concentrations, compared with computer simulations (Taylor 1967).

constituent is in low concentration, with the possibility of localized 'impurity modes' appearing outside the main band (§ 9.6), the complicated details of the spectral density are entirely lost. This smearing out of significant 'structure' in the CPA spectrum shows up very strongly in the one-dimensional case (fig. 9.8) where all the clutter of singularities at 'special frequencies' (§ 8.4) is replaced by a smooth tail. We know, of course, that these 'pathological' effects are sharpened by the strong assumption of the standard BRA model that each of the two components has only a single, precisely defined atomic level (or mass). For actual physical systems, it may be more realistic to introduce a more continuous probability distribution of disorder, as in the Anderson model, or to study the CPA spectrum of a *multi-component alloy* (Yonezawa 1973) where a great many interesting effects of coalescing and splitting bands can be observed.

Perhaps the most serious defect of the CPA formula is that the split-band limit (9.65) is not, in fact, a correct representation of the true spectrum in the limit as $\delta \to \infty$. In this case, the atomic levels λ_A and λ_B are so separated that excitations on the A sites cannot penetrate B sites and vice versa. Consideration of fluctuations producing relatively extended regions of, say, pure A-type material supports the conclusion that there will be a band of normal width centred on λ_A. But computer simulations (Kirkpatrick &

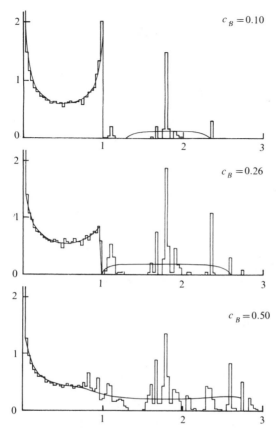

Fig. 9.8. One-dimensional CPA spectra (one-dimensional phonon model), for various concentrations, compared with computer simulations (Taylor 1967).

Eggarter 1972) have shown that this band is not smooth, as predicted by (9.65), but has considerable 'structure', including a remarkable dip, near the centre, with a sharp singularity precisely at $\lambda = \lambda_A$ (fig. 9.9). The appearance of such spectral structure, even in a three-dimensional model at relatively high concentration, can be explained by reference to the relatively frequent occurrence of particular configurations of atoms (fig. 9.10) which can sustain a localized excitation (see § 9.9), even when not completely isolated by atoms of the other type. But this explanation only emphasizes the inability of the single site CPA to include such effects in its description of the spectrum.

Excitations on a disordered lattice

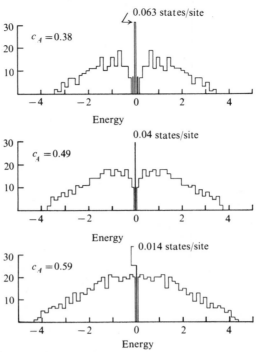

Fig. 9.9. Computed spectra for three-dimensional TBA model for various concentrations in the split band limit (Kirkpatrick & Eggarter 1972).

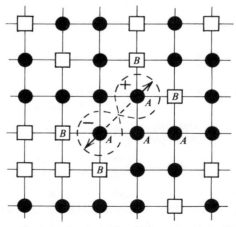

Fig. 9.10. Localized mode on a pair of 'A' atoms not entirely isolated by 'B' atoms.

9.5 Local environment corrections to CPA

A *single site* or *mean field* approximation such as CPA fails to take into account the variety of *local environments* of each type of atom. In the elementary theory of order–disorder phenomena (§ 5.2), the effects of statistical fluctuations are allowed for by systematic analysis (§§ 5.3, 5.4) of the behaviour of larger and larger *clusters* of atoms on contiguous lattice sites. Physical intuition, and actual experience with one-dimensional models (§ 8.6), combine to suggest that a similar procedure should generate a succession of approximations converging on the true excitation spectrum of any substitutionally disordered system.

It is not difficult to think of a prescription that would give nearly perfect results if it could be carried out for a very large cluster. In practice, however, we are severely constrained by the tedium and expense of the necessary computations. What we need is a scheme that gives the best results for the smallest size of cluster. In particular, we should attempt to preserve the significant virtues of the standard CPA formula, such as herglotz analyticity (9.58) and propagator/locator duality. Many of the phenomenological schemes for 'going beyond single-site CPA' (Ducastelle 1972) do not satisfy these conditions (Capek 1971; Nickel & Butler 1973) and must therefore be rejected.

Formally speaking, it is very easy to show (Leath 1973, 1974) that the algebraic steps leading up to the standard CPA equation (9.49) can be followed when each symbol, $\tilde{t}$, $\tilde{w}$, $\tilde{\Sigma}$, $\tilde{G}_{00}$, is interpreted as an $n \times n$ matrix. We may further verify that this equation has an algebraic solution for the medium propagator (9.50) and that this is identical with the result arrived at from the locator equation (9.56) assuming a similar generalization to matrix language. It turns out, moreover (Ducastelle 1974a), that this estimate of the average Green function is a herglotz function.

What we really mean by this algebraic generalization of CPA is that we have divided the crystal lattice into a regular superlattice of distinct clusters (fig. 9.11), each of which is treated as a quasi-independent 'molecule' with its own internal excitations. But this superlattice is itself substitutionally disordered, since many different types of 'molecule' are generated by the random distribution of A- and B-type atoms on the n sites of a cluster. In this representation, for example, the matrix $\tilde{w}$ would be related, as in (9.47) to a diagonal matrix w, whose elements

$$w_{pq} = w_p \delta_{pq} \quad (1 \leqslant p, q \leqslant n) \tag{9.67}$$

would depend on the type of atom at the pth site of the cluster. The

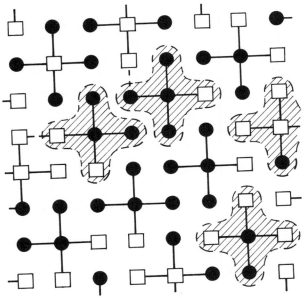

Fig. 9.11. Dissection of alloy model into superlattice of 'clusters' of various types.

ensemble average $\langle \; \rangle$ would then run over all possible distributions of atoms with appropriate statistical weights. To get the *molecular coherent potential approximation* (MCPA) we apply 'single site' CPA to this assembly of 'molecules'. Thus, the self-consistency condition (9.49) could be written more explicitly

$$\langle \tilde{t}_{pq} \rangle = 0, \tag{9.68}$$

showing that we want to construct a medium in which, on the average, every element of the effective scattering matrix of the cluster vanishes.

In principle, by constructing larger and larger clusters we approach closer and closer to the ideal situation in which we are calculating the total T-matrix of a substantial volume of specimen, modified by the properties of the smoothed medium in which it is embedded. But now the 'self-energy' symbol $\tilde{\Sigma}$ represents an $n \times n$ matrix, with diagonal and off-diagonal elements $\tilde{\Sigma}_{pq}$ between the various sites of the cluster. Not only does this add extravagantly to the computational labour; it also has unphysical features. In the definition of $\tilde{\Sigma}$ we tacitly assume that it is *cluster diagonal* – that it has no elements linking sites in different clusters. This is the necessary and sufficient condition that the MCPA equations should yield a herglotz Green function and that the propagator and locator definitions should be equivalent. But it is obviously inconsistent with the statistical translational invariance of the whole system; on the average, a pair of neighbouring lattice sites should be linked by the same operators, whether the atoms belong to the same or different 'clusters'. Thus, by carving up the original lattice into clusters of sites we have introduced artificial boundaries which are of no physical significance. For this reason there is no guarantee that a complete MCPA calculation yields the best estimate of the spectrum for a given size of cluster.

We are faced, therefore, with a choice between several approximation schemes that are not simply formal generalizations of single site CPA and do not necessarily share all its mathematical virtues. We might, for example (Berk & Tahir-Kheli 1973; Berk, Shazeer & Tahir-Kheli 1973; Zittartz 1974), replace (9.68) by the less stringent condition

$$\langle \tilde{t}_{1q} \rangle = 0 \quad (1 \leqslant q \leqslant n). \tag{9.69}$$

In other words, we no longer insist that all matrix elements of the average T-matrix of a cluster should vanish, but only those linking the central atom $(p=1)$ with itself or other atoms. This reduces the number of equations from n^2 to n, and also has the very desirable effect of generating a self energy matrix $\tilde{\Sigma}$ that is translationally invariant: the Hamiltonian of the system is modified by the introduction of a *non-local coherent potential*,

$$\tilde{\Sigma}_{ll'} = \tilde{\Sigma}(l - l'), \tag{9.70}$$

which depends only on the lattice distance $l - l'$ between the two sites l and l', regardless of whether they belong to the same or different clusters. The self-energy (9.70) is thus *momentum-dependent* in its Fourier transform, requiring the interpretation (9.13) of its real and imaginary parts. It is not known, however, whether these equations (which obviously reduce to the single site CPA formula for $n = 1$) have satisfactory analyticity and propagator/locator duality properties.

To reduce the computational labour further, the only option is to return to a local, scalar, coherent potential, as in the single-site theory. By analogy with the Bethe–Peierls approximation for order–disorder phenomena (§ 5.4) we may, for example, replace the CPA condition (9.49) by a generalization of (9.50), i.e.

$$\tilde{G} = \langle G_{11} \rangle, \tag{9.71}$$

whereby the renormalized medium propagator is equated to the average diagonal element of the Green function for the central atom of the cluster

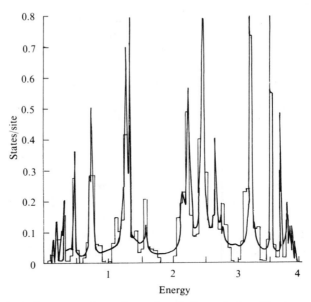

Fig. 9.12. Spectrum of one-dimensional alloy calculated by seven-site cluster approximation, compared with exact spectrum (histogram) (Bulter 1973).

(Tsukada 1972; Capek 1972; Brouers, Cyrot & Cyrot-Lackman 1973). This certainly reproduces CPA in the single site case, but for $n > 1$ the results are not satisfactory (Nickel & Butler 1973; Butler 1973). The reason seems to be that the density of states at the centre of a cluster is relatively insensitive to the properties of the surrounding medium, so that $\tilde{\Sigma}$ is not adequately controlled by this condition. Much better results are obtained (Butler 1973; Miwa 1974) if the self-consistency condition is applied to a *boundary* site of the cluster, which is, of course, in immediate contact with the surrounding medium. Indeed, for a one-dimensional lattice, where the 'cluster' makes contact with the 'medium' at disconnected points, this approximation is mathematically equivalent to the much more elaborate MCPA formalism and reproduces many of the more spectacular features of the excitation spectrum (fig. 9.12). Equivalent propagator and locator forms of this theory can be constructed (Brouers & Ducastelle 1975) but the analyticity of the Green function has not been proved.

As pointed out in § 5.4, a *compact* representation such as CPA and its generalizations has the great advantage of exhibiting a variety of complex analytical behaviour through the solution of a small number of algebraic or integral equations. This sort of behaviour cannot appear in any truncated series generated by an *expanded* representation of the equations, except as a result of partial summation of an infinite subset of terms, as in a Dyson equation. Although diagrammatic techniques can be used to derive many such summations, they must always be guided by the same 'phenomenological' considerations as are used directly in the compact equations. In generalizing CPA, for example, we see immediately that the important interactions are those between adjacent sites in the lattice – i.e. sites within the same 'cluster' – and that formal diagrammatic 'pair' terms for more distant sites are negligible. For this reason the use of 'Bethe' boundary conditions (§ 11.4) to continue the cluster outwards without sharp truncation gives very good results (Sen & Yndurain 1976).

9.6 Spectral bounds and band tails

In the one-dimensional case, considerable information about the spectra of alloy models could be deduced from the Saxon–Hutner theorem (§ 8.4). This told us that any spectral region which is a spectral gap for both a pure *A*-type chain and a pure *B*-type chain is necessarily a gap for any mixed lattice of *A*- and *B*-type atoms. The general theorem for the bounds of the spectrum of any tight-binding model in any number of dimensions (Cyrot-Lackman 1972; Carroll 1975; Ducastelle 1974*b*) has a very simple algebraic proof.

Consider the equations (9.1) for an eigenvalue of the TBA model:

$$(\lambda - \mathscr{E}_l)u_l = \sum_{l' \neq l} V_{ll'}u_{l'}; \quad l = 1 \dots N. \tag{9.72}$$

Dividing each equation by the amplitude $|u_l|$ of the eigenfunction, we get the inequality

$$|\lambda - \mathscr{E}_l| \leqslant \sum_{l' \neq l} |V_{ll'}|\{|u_{l'}|/|u_l|\}. \tag{9.73}$$

Now suppose that l labels the site where this eigenfunction has maximum amplitude; the ratio $\{|u_{l'}|/|u_l|\}$ is always $\leqslant 1$, so that

$$|\lambda - \mathscr{E}_l| \leqslant \sum_{l' \neq l} |V_{ll'}| \equiv B_l. \tag{9.74}$$

For every eigenvalue λ of (9.72), there exists at least one site, l, for which this inequality is satisfied. We have thus proved the *Hadamard–Gerschgorin theorem* of matrix algebra: λ is contained in the union of the 'discs' centred on $\mathscr{E}_l$ with radii B_l.

The consequences of this theorem are obvious. In a perfect crystal, where

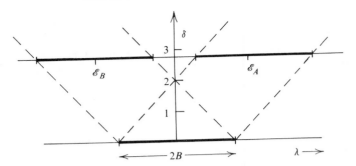

Fig. 9.13. Spectral limits for binary alloy model.

$\mathscr{E}_l$ is the same for all sites and where $V_{ll'} = V$ for each of z nearest neighbours, we find, as in (8.16), that the total width of the band cannot exceed

$$B = 2zV. \tag{9.75}$$

For the BRA model, the spectrum must lie wholly within the region covered by two bands, each of the 'perfect' width (9.75), centred on the local levels λ_A, λ_B of the two constituent atoms (fig. 9.13). This shows that a gap must open up when the parameter δ, defined in (9.66), exceeds 2.

The Saxon–Hutner theorem is obviously included in (9.74). Unfortunately, the subtle use of this theorem to demonstrate the existence of gaps and special frequencies in the alloy chain cannot, apparently, be extended to lattices of more than one dimension. Suppose, for example, that we consider the spectral gaps of a superlattice of identical 'objects' analogous to segments of type $\{A(s)\}$ (§ 8.4) – i.e. one atom of type B embedded in an arrangement of s atoms of type A. There would be many different, non-equivalent 'objects' of this type, so that it is most unlikely that the same gap would be found for all arrangements and for many different values of s. As we have seen (e.g. figs. 9.9, 9.10) there is evidence of special spectral 'features' in BRA models in two and three dimensions, but these do not seem to be subject to the same simple arithmetical principles as in the one-dimensional case.

Many applications of the TBA model tacitly assume that all the 'overlap integrals' have the same sign. Let us suppose, for example, that these are all positive or zero, and that the 'zero of energy' has also been shifted so that $\mathscr{E}_l \geqslant 0$ for all l. From (9.74) we deduce that there exists a site l such that

$$|\lambda| \leqslant B_l + \mathscr{E}_l. \tag{9.76}$$

In other words, if the spectrum has positive eigenvalues, these must be bounded above by a maximum eigenvalue

$$\lambda_{max} \leqslant \max_l \{B_l + \mathscr{E}_l\}. \tag{9.77}$$

This inequality, however, is only half the statement of the *Perron–Frobenius theorem* concerning the maximum eigenvalue of a matrix with non-negative elements. Making the formal identification $V_{ll} \equiv \mathscr{E}_l$, this theorem tells us (Cyrot-Lackmann 1972)

$$\min_l \left\{ \sum_{l'} V_{ll'} \right\} < \lambda_{max} < \max_l \left\{ \sum_{l'} V_{ll'} \right\}, \tag{9.78}$$

which goes beyond (9.77) by putting an explicit lower bound on the position of the top of the spectrum. In the BRA case, where the two bounds differ by $|\lambda_A - \lambda_B|$, this tells us nothing new, but in some models of topological disorder (§ 11.3) the two ends of (9.78) may come close enough together to fix unequivocally the edge of the band.

The behaviour of the BRA spectrum near each band edge can be deduced heuristically (Lifshitz 1964) from the *local density* principle (§ 8.6). Suppose, for example, that λ_0^A is the lower edge of the lower band in the split band case and that according to (9.74) this bound would be reached in an infinite crystal of type A atoms. The addition of an average concentration c_B of B-type atoms can only have the effect of increasing each eigenvalue. To get an eigenvalue λ near to λ_0^A, we must look for a sufficiently large region that has been left free of B atoms by statistical fluctuations of local concentration. Within this region, the excitation would have the wave number q given by a dispersion relation of the type (8.15), i.e.

$$\begin{aligned} \lambda &\approx \lambda_0^A + \tfrac{1}{2}B(1 - \cos aq) \\ &\approx \lambda_0^A + \tfrac{1}{4}Ba^2q^2, \end{aligned} \tag{9.79}$$

where B is the band width parameter and a is a length of the order of the lattice spacing.

But such a region would not be adequately sampled by the excitation unless it were larger than a half wavelength,

$$L \sim \pi/q, \tag{9.80}$$

or, near an upper band edge, the corresponding beat length (8.87). In a lattice of d spatial dimensions, such a region would contain

$$N \sim (L/a)^d \tag{9.81}$$

sites. The *a priori* probability that this volume would be empty of B atoms is of the form

$$(c_B)^N \sim \exp[(L/a)^d \ln c_B]. \tag{9.82}$$

This we interpret as the probability of finding a state of the chosen energy λ;

from (9.79)–(9.82) we get an estimate of the functional form of the spectral density

$$\mathcal{N}(\lambda) \propto \exp[(\pi/qa)^d \ln c_B]$$

$$\sim \exp\left[\alpha\left\{\frac{\lambda - \lambda_0^A}{B}\right\}^{-\frac{1}{2}d} \ln c_B\right], \tag{9.83}$$

where α is a numerical factor, of the order of unity, that cannot be precisely determined by this sort of argument.

Although the *Lifshitz formula* (9.83) has not, apparently, been tested against computer simulated spectra for a general three-dimensional BRA model, it looks consistent with known observations. Exponential tails of this kind are, of course, quite beyond the capacities of CPA and its generalizations.

In the one-dimensional case, where the exponent behaves as $(\lambda - \lambda_0^A)^{-\frac{1}{2}}$, this formula agrees with the characteristic tail function (8.80) on either side of a special frequency. Note, however, that (9.83) is not equivalent to the tail function (8.86) for a 'linear liquid' model whose statistical definition and spectral properties are entirely different from those of a binary alloy. Nor is there any reason why the Lifshitz formula should apply, for example, to the spectral tails of an Anderson TBA model where the disorder variable w_l has a continuous range. Indeed, the application of exact algebraic inequalities to the spectrum of this model (Harris 1973) shows a possible technique by which the Lifshitz *Ansatz* could be made mathematically rigorous.

9.7 Spectral moments and continued fractions

The spectral density $\mathcal{N}(\lambda)$ of the excitations governed by a Hamiltonian $\mathcal{H}$ must be consistent with the *spectral moments*

$$\mu_p \equiv \int \lambda^p \mathcal{N}(\lambda) d\lambda = N^{-1} \langle \mathcal{H}^p \rangle. \tag{9.84}$$

As in the moment expansion (5.163) for the thermodynamic properties of a cooperative system (§§ 5.10, 6.5), the brackets $\langle \ \rangle$ encompass the trace of quantum operators and ensemble averaging. Since the first few of these characteristic invariants of the spectrum can often be calculated very easily, one naturally hopes to obtain useful information about the unknown function $\mathcal{N}(\lambda)$ by a direct attack on the Hamiltonian.

For the TBA model (Cyrot-Lackmann 1972), the Hamiltonian is most simply represented as the matrix (9.3) acting on the local site orbitals $|l\rangle$. With the convention (cf. (9.23), (9.78))

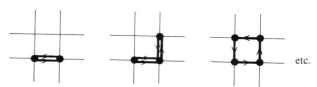

Fig. 9.14. Graphs for moments.

$$V_{ll} \equiv \mathscr{E}_l \equiv w_l \qquad (9.85)$$

for the diagonal elements, we can write (9.84) in the simple-seeming form

$$\mu_p = N^{-1} \Big\langle \sum_{0,\,1,\,2,\,\ldots,\,(p-1)} V_{01} V_{12} V_{23} \ldots V_{(p-1)0} \Big\rangle \qquad (9.86)$$

where there are no restrictions on each summation over each site label **0, 1, 2, . . ., (p − 1)**. It is very easy to write down explicit algebraic formulae for the first three or four moments and to work out the ensemble averages over statistical distributions of the matrix elements.

But this procedure becomes unacceptably laborious as the order p increases. Consider, for example, the simplest possible case – an ordered simple cubic lattice, where the diagonal elements (9.85) all vanish and only the nearest-neighbour interaction V survives. It is obvious that each non-zero term in (9.86) corresponds to a connected graph of $p = 2q$ lattice steps, which may retraverse any segment any number of times in either direction, but which eventually returns to the origin (fig. 9.14). Elementary combinatorial arguments show that the number of such terms is given by

$$\mu_{2q}/(V)^{2q} = \sum_{r,s,t} \frac{(2q)!}{(r!)^2(s!)^2(t!)^2} \qquad (q = r+s+t). \qquad (9.87)$$

It is clear that the magnitude of this expression, and the number of distinct graphs that contribute to it, both increase very rapidly with q. Introduce some disorder requiring, say, statistical averages of powers of the diagonal matrix elements at each vertex and the computational work soon becomes unmanageable. As with most efforts to grind out the successive terms of a graphical sequence, the process slows to a halt after a dozen or so terms.

We might have available, therefore, a list of, say, the first twenty moments of the spectral distribution: how much can we learn about the spectrum itself? Experience has shown that direct attempts to solve the *moment problem* (Shohat & Tamarkin 1963) yield rather disappointing results. The exact formulae (Gaspard & Cyrot-Lackmann 1973; Yndurain & Yndurain 1975) are generalizations of the Tchebytcheff inequalities and

Excitations on a disordered lattice

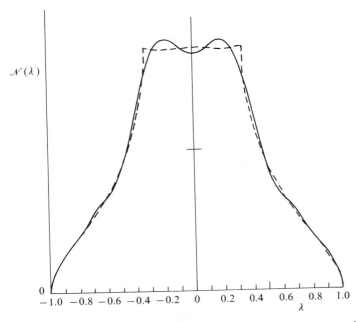

Fig. 9.15. Spectrum approximated with 13 moments ———, compared with exact result – – – (Fisher & Camp 1972).

merely provide upper and lower bounding curves for integrated functions such as the cumulated spectral density (8.68). We thus get very little information about interesting spectral 'features' such as localized mode peaks, sharp edges or band tails. It is instructive to observe, for example, that knowledge of the exact moments (9.87) up to a high order gives only a very imperfect representation of the van Hove singularities (fig. 9.15) of the exact spectrum of a simple, narrow, tight-bound band (Fisher & Camp 1972).

The source of the mathematical difficulty is that the formal Fourier inversion of (9.84), i.e.

$$\mathcal{N}(\lambda) = \frac{1}{2\pi} \int_{-\infty}^{\infty} dx \, e^{ix\lambda} \sum_{p=0}^{\infty} \frac{(-ix)^p}{p!} \mu_p \qquad (9.88)$$

is valid only in the limit of an infinite sum and produces meaningless results if the series is truncated at some finite value of p. To avoid this difficulty, and thus to construct an optimal estimate of the spectrum, it is almost essential to represent $\mathcal{N}(\lambda)$ through its *Hilbert transform*

$$\mathscr{G}(\lambda) = \int_{-\infty}^{\infty} \frac{\mathscr{N}(\lambda')}{\lambda - \lambda'} \, d\lambda'. \tag{9.89}$$

This, in its turn, may be expanded as a *continued fraction* (Wall 1948) of the form

$$\mathscr{G}(\lambda) = \cfrac{a_0}{\lambda - a_1 - \cfrac{b_1}{\lambda - a_2 - \cfrac{b_2}{\lambda - \dots}}}. \tag{9.90}$$

The coefficients a_0, a_1, b_1, etc. are functions of the moments up to some definite order in each case. The successive approximants converge rapidly towards a function that is an asymptotic representation of $\mathscr{N}(\lambda)$. Most significantly, truncation of this sequence at any finite coefficients a_n, b_n yields a herglotz function (9.58) of the complex variable λ. Indeed, such approximants correspond to the *Padé approximants* to $\mathscr{G}(\lambda)$, whose limiting form can often be deduced by extrapolation from computed values of the coefficients a_0, a_1, etc.

What must be emphasized, however, is that this approach to the problem of the excitation spectrum is not really independent of the methods discussed in previous sections of this chapter. It is obvious, for example, that the Hilbert transform (9.89) of the spectral density is simply the site diagonal element (9.30) of the ensemble-averaged Green function. There seems little merit in constructing this function by algebraic manipulation of the spectral moments when we already have an elaborate theory of the Green function itself. It is easy to see, for example, that the term of order p in the locator expansion (9.36) of the ensemble-averaged Green function $\langle G_{ll} \rangle$ is closely related to the expression (9.86) for the pth moment. In the case of an ordered system, where the locator factor at each vertex reduces to

$$g_l(\lambda) = \frac{1}{\lambda}, \tag{9.91}$$

we could, in fact, reduce (9.37) to

$$G_{00}(\lambda) = \lambda^{-1} (1 + \sum_p \mu_p / \lambda^p), \tag{9.92}$$

with the moments to be evaluated precisely by (9.87). The connection between the approximation (9.39) and the Fourier inversion (9.88) is very instructive. Notice also that, rewriting the Green function (9.92) in the form (9.44), we may express the self-energy as a locator series (9.45) where the graphs are irreducible. The coefficients in the continued fraction (9.90) can thus be determined from modified moments which differ from (9.86) and (9.87) by omitting graphs that revisit the origin before the final return.

From a practical point of view, however, the most direct procedure for generating the coefficients $a_0, a_1, \ldots, b_1, b_2, \ldots$ in the continued fraction representation (9.90) is the *recursion method* of Haydock, Heine & Kelly (1972, 1975). This is a perfectly general algebraic technique, in which a basis set of states $|s\}$ is generated from some starting state $|1\}$ by multiplying again and again by the Hamiltonian. Each successive state is defined by the algorithm

$$|s+1\} = \mathscr{H}|s\} - a_s|s\} - b_{s-1}|s-1\}, \tag{9.93}$$

together with conditions on the coefficients a_s and b_s ensuring that $|s+1\}$ is orthogonal to its immediate predecessors $|s\}$ and $|s-1\}$. It can then easily be shown that $|s+1\}$ is orthogonal to all lower states, $|s-2\}$, $|s-3\}$, etc., so that the set of all states generated in this way forms an orthogonal basis for the representation of the Hamiltonian. In this representation (duly normalized) the only non-vanishing matrix elements of $\mathscr{H}$ are those of the form

$$\mathscr{H}_{ss} \equiv \frac{\{s|\mathscr{H}|s\}}{\{s|s\}} = a_s, \tag{9.94}$$

and

$$\mathscr{H}_{s,s-1} = \mathscr{H}_{s-1,s} \equiv \frac{\{s-1|\mathscr{H}|s\}}{\{s-1|s-1\}^{\frac{1}{2}}\{s|s\}^{\frac{1}{2}}} = (b_{s-1})^{\frac{1}{2}}. \tag{9.95}$$

The Hamiltonian matrix is thus *tridiagonal* and it follows quite simply that the corresponding Green function (9.4) has precisely the form of a continued fraction (9.90), with the coefficients defined by the recursion algorithm (9.93).

Physical intuition and computational simplicity combine to suggest that this method should be applied in the local site representation, where the initial function $|1\}$ would be a chosen site orbital $|l\rangle$, and where the Hamiltonian (9.3) would introduce new neighbours at each successive stage of calculation. It is obvious that contributions to the coefficients a_q and b_{q-1} come from closed lattice paths of up to $2q$ steps and thus include all the moments (9.86) up to this order. In general, these coefficients can only be computed numerically, stage by stage, but the algorithm (9.93) is highly efficient and can be carried through to a high order. In some respects, this method resembles the Kikuchi method (§ 5.4) for the systematic improvement of 'cluster' approximations in the theory of order–disorder phenomena by noting the effects of adding sites in succession.

In the same spirit, we make the most of the *recursion relations*

$$g_{s-1}(\lambda) = \frac{1}{\lambda - a_s - b_s g_s(\lambda)} \qquad (9.96)$$

between successive approximants to the continued fraction (9.90). At the base of this hierarchy we have $g_0(\lambda)$, which is precisely the Green function $\mathscr{G}(\lambda)$. Suppose that the coefficients a_s and b_s are effectively constant for all values of s beyond the qth stage. Then the approximant g_q must satisfy a self-consistency equation,

$$g(\lambda) = \frac{1}{\lambda - a - bg(\lambda)}, \qquad (9.97)$$

which has the elementary solution

$$g(\lambda) = \frac{\lambda - a}{2b} \left\{ 1 - \left(1 - \frac{4b}{(\lambda - a)^2} \right)^{\frac{1}{2}} \right\}. \qquad (9.98)$$

This formula would thus provide an exact termination for the sequence. All this is really an optimal computational procedure for solving, to an appropriate degree of accuracy, the Dyson equation for the self-energy (9.45) as a graphical expansion in the locator representation.

How far can we go by this method? The only system for which the coefficient a_s, b_s become exactly independent of s (Jacobs 1973) is an ordered tree of coordination number z (§ 5.4). We then have $a_1 = a_2 = \ldots = 0$, and $b_1 = zV^2$, $b_2 = b_3 = \ldots = (z-1)V^2$. The Green function can then be written down at once from (9.90), (9.96) and (9.98). For a linear chain this gives a band of width $2B = 4|V|$ with density of states

$$\mathscr{N}(\lambda) = \frac{1}{\pi}(B^2 - \lambda^2)^{-\frac{1}{2}}, \qquad (9.99)$$

exactly as one would deduce from, say, (8.37). But for a genuine, ordered, two- or three-dimensional lattice, multiple connectivity leads to closed chains which do not allow the coefficients to settle down to constant values. The method, therefore, is useful mainly for constructing an approximation to the density of states without going to the labour of solving the eigenvalue equations and summing over the Brillouin zone. Once again, van Hove singularities associated with long-range lattice order will not be correctly reproduced.

It must be remembered, however, that the continued fraction represents the *local* Green function. For a disordered system, the coefficients a_s, b_s depend on the local circumstances. Thus, the very first coefficient, a_1, would normally depend on whether an A atom or a B atom were on the starting site l, whilst b_1 would be immediately affected by off-diagonal disorder (§ 9.8). There is no exact algebraic procedure for expressing the ensemble average Green function as a continued fraction with 'averaged' coefficients. Strictly speaking, to evaluate the density of states of the whole system, it is necessary to compute the locator for each possible atomic configuration of a finite cluster of sites and then to take an ensemble average.

To avoid this labour we may insist that the self-consistency condition (9.97) need only be satisfied 'in the mean' (Jacobs 1973). Suppose, for example, that we have calculated the coefficients a_1 and b_1 for all distinct occupations of the starting site l. In the BRA model, where a_1 turns out to be just the random potential w_l, and where b_1 is the same for all sites, this is a trivial calculation and we can easily solve the approximate self-consistency equation

$$\tilde{g}(\lambda) = \left\langle \frac{1}{\lambda - a_1 - b_1 \tilde{g}(\lambda)} \right\rangle \qquad (9.100)$$

for the medium locator $\tilde{g}$. When written out at length, this condition looks much the same as, say (9.54), showing that this is a single-site approximation in the same spirit as CPA. Notice, however, that (9.100) does not give the correct answer in the single-band limit because (9.97) does not yield the exact band structure of the regular crystal when the disorder vanishes. On the other hand, the analytic properties of a continued fraction representation are very satisfactory. The approximation can be improved by introducing several stages of the recursion relation (9.96) into the expression for the locator, before terminating the hierarchy with a medium locator in the final denominator (Jacobs 1974). We thus construct a cluster approximation that is reasonably computable and goes beyond single-site CPA.

The algebraic economy of the continued fraction representation can be further exploited by the analytical device of introducing an extended Hilbert space where the equivalent Hamiltonian automatically represents all the different disordered configurations to be found in the statistical ensemble (Mookerjee 1973; Bishop & Mookerjee 1974; Mookerjee 1975; Bishop 1975). This device provides an efficient machinery for calculating the various graphs that contribute to a typical 'cluster generalization' of CPA whilst keeping careful control of the analytical properties of the Green function.

At first sight the recursion algorithm (9.93) appears to map the spectrum of our system on to the spectrum of a fictitious one-dimensional system, whose equivalent Hamiltonian has diagonal matrix elements a_s, and nearest-neighbour transfer matrix elements $(b_s)^{\frac{1}{2}}$. The qth site along this chain represents, so to speak, the qth 'shell' of sites in the original lattice – i.e. the new sites reached when the number of steps on a closed graph from the origin is increased from $2(q-1)$ to $2q$. But this apparent reduction of dimensionality is not topologically significant and for a disordered system the algebraic mapping is not immediately useful because the coefficients themselves depend on the starting site, l, in the original lattice. We cannot make free use of our extensive understanding of the properties of one-dimensional systems, because there is no guarantee that the spectrum of a particular equivalent chain, computed to some approximation of order q, converges uniformly on the desired density of states.

It is obvious, however, that in a disordered system the successive coefficients a_s, b_s will seem to behave like random variables drawn from some probability distribution $P_s(a_s, b_s)$. The successive approximants, $g_{s-1}(\lambda)$, $g_s(\lambda)$ could likewise be considered as stochastic variables drawn from the distributions F_{s-1}, F_s, respectively. Because these variables are linked algebraically by the recursion relation (9.96), there must be a functional relation between their probability densities; i.e., by elementary arguments

$$F_s\{g_a(\lambda)\} = \iint P_s(a_s, b_s) F_{s-1}\left\{\frac{1}{\lambda - a_s - b_s\, g_s(\lambda)}\right\} da_s db_s. \qquad (9.101)$$

Now suppose, further, that the equivalent chain turns out to be statistically homogeneous in the sense that the coefficients a_s, b_s seem all to be drawn from the *same* distribution $P(a, b)$, regardless of the index s. It would then be reasonable to assume (although this demands careful proof in any particular case) that the approximants $g_s(\lambda)$ would similarly tend to a stationary distribution $F\{g(\lambda)\}$ independent of s. In this case (9.101) becomes an integral equation,

$$F\{g(\lambda)\} = \iint P(a, b)\, F\left\{\frac{1}{\lambda - a - bg(\lambda)}\right\} da\, db, \qquad (9.102)$$

from whose solution we could calculate the medium locator $\bar{g}(\lambda)$, etc.

This is not, in general, a practical procedure for finding the spectrum of a disordered lattice. It is easy to see, however, that the condition of statistical homogeneity of the diagonal and off-diagonal elements of the Hamiltonian is satisfied for an actual linear chain model. The integral equation (9.102) (or the corresponding equation for the probability distribution of the 'local' self-energy (9.45)) thus provides an exact solution of the spectral problem in one dimension (Economou & Cohen 1971). The above derivation strongly suggests, however, that this is none other than the Dyson–Schmidt equation (8.76) clothed in the Green function symbolism (Eggarter 1973).

Comparing (9.102) with the 'mean field' approximation (9.100), we realize that the standard problem of allowing for statistical fluctuations of local atomic configurations cannot be avoided; a stochastic variable such as the locator cannot be replaced by its average value. Notice, indeed, that the CPA *ansatz,* however it may be generalized to clusters of finite size, cannot reproduce the exact spectrum even for a one-dimensional system. In this respect, the theory of excitations on a substitutionally disordered lattice is more difficult than the theory of order–disorder phenomena (§ 5.4), where the Bethe–Peierls 'cluster' method gives exact answers on a linear chain, or on any regular lattice of higher coordination number.

9.8 Off-diagonal disorder

It is physically unrealistic to confine the effects of alloy disorder to the site-diagonal elements of the Hamiltonian. In the theory of lattice vibrations, for example, each of the force constant (8.2) must surely depend on the types of atom being linked. In the theory of electronic states in alloys we cannot assume that atoms of the different constituent species all have the same transfer matrix elements – i.e. all have resonances of the same intrinsic width (§ 10.3).

Off-diagonal disorder is introduced into the TBA model by treating the transfer matrix elements $V_{ll'}$ as random variables. Thus, the nearest neighbour matrix elements in a binary alloy are given distinct values, V^{AA}, V^{BB}, V^{AB}, according as they refer to AA-, BB- or AB-type pairs of atoms. In general, this type of disorder is additional to the usual diagonal disorder associated with, say, two distinct atomic levels $\mathscr{E}_A$, $\mathscr{E}_B$.

In a model representing a genuine physical system, the values of these parameters would not necessarily be arithmetically related. It turns out, however (Shiba 1971), that the spectral theory is greatly simplified if V^{AB} is taken to be the geometric mean of the other two parameters. This is equivalent to writing

$$V^{\alpha\beta} = \gamma^\alpha V * \gamma^\beta, \tag{9.103}$$

where each of the two factors γ^α, γ^β can take either of two values γ^A, γ^B. Although this relation has no *a priori* physical justification, it facilitates the study of simple theoretical models where $(\gamma^A/\gamma^B)^2$, the ratio of A-type to B-type band width, is far from unity.

To bring this effect into the formalism of the TBA model, let us define a local operator (Brouers, Ducastelle & van der Rest 1973)

$$\gamma \equiv \gamma_l |l\rangle\langle l|, \tag{9.104}$$

which introduces a factor γ_l for each site label l of any matrix in the site representation. We define γ_l to have value γ^A or γ^B according as this site is occupied by an atom of type A or B. The interactor in the disordered system can therefore be written as a matrix

$$V \equiv V_{ll'} = \gamma_l V_{ll'}^* \gamma_{l'} \equiv \gamma V^* \gamma, \tag{9.105}$$

where now V^* represents a configurationally independent nearest-neighbour interaction, as in (9.103).

The total Green function, G, of our system may be expressed as a locator expansion (9.35) generated by iteration of the Dyson equation (9.33). But this equation remains formally invariant under a transformation by the operator γ: in terms of

$$G^* = \gamma G \gamma \quad \text{and} \quad g^* = \gamma g \gamma, \tag{9.106}$$

we obtain

$$G^* = g^* + g^* V^* G^*. \tag{9.107}$$

Since V^* is a translationally invariant interactor, we can treat (9.107) as the defining equation for a system whose disorder is entirely site-diagonal. Each 'locator' g_l^* differs from the standard TBA form (9.34) by a factor γ_l^2, but this can be carried through the algebra into a straightforward generalization of the single-site CPA condition (9.56). Once we have determined the average Green function $\langle G_{00}^* \rangle$ in this approximation, we can go back through (9.106) to the medium propagator $\langle G_{00} \rangle$, and hence determine the spectrum of our original system.

Since the TBA model itself is not very realistic, and the numerical results obtained by this method do not seem to exhibit any unexpected features, there is little incentive to study more complex models of off-diagonal disorder where, for example, the transfer matrix elements are not assumed to be related by the Shiba condition (9.103). This condition, indeed, provides a unique simplification of the problem. Even the assumption that V^{AB} is the *arithmetic* mean of V^{AA} and V^{BB} (Schwartz, Krakauer & Fukuyama 1973) does not eliminate the fundamental characteristic of off-diagonal disorder – that it refers to *pairs* of lattice sites and cannot, in general, be reduced to a single-site effective Hamiltonian with a purely local coherent potential. To deal with this, therefore, one is forced to the much greater mathematical complexity of a 'cluster' generalization of CPA, as discussed in § 9.5 (Bergmann & Halpern 1974; Moorjani, Tomoyasu, Sokolski & Bose 1974; Cubioti, Donato & Jacobs 1975).

9.9 Anderson localization

In a one-dimensional system, *all* the excitations are localized by disorder (§ 8.7). But the method used to prove this theorem cannot be generalized to any two- or three-dimensional model. The question naturally arises whether the eigenstates of any disordered system must be similarly localized, or whether we may sometimes find *extended* states analogous to the Bloch functions of an ordered crystal. The properties of the eigenstates of the TBA Hamiltonian (9.1) provide important evidence on this fundamental question.

Strictly localized modes can always be found in the BRA model (§ 9.4). These occur when the atomic species are differentiated into a few sharply distinct types, so that, for example, an excitation on the 'A' atoms cannot pass through the 'B' atoms and vice versa. Such a mode is thus confined to a more or less isolated cluster of one or the other species and shows up as a delta-function singularity in the excitation spectrum (fig. 9.9). But spectral 'features' of this kind do not occur when the disorder parameter has a continuous distribution of values – as, for example, in almost all models of topological disorder. The fact that some special modes *can* be localized for a rather special class of TBA models proves nothing concerning the general properties of eigenstates in disordered systems of more than one dimension.

The fundamental theorem concerning the effects of substitutional disorder on the excitations of a lattice of more than one dimension is due to Anderson (1958): *the eigenfunctions are localized if the 'strength' of the disorder exceeds some definite value.* Later studies have shown, moreover (Ziman 1969a), that the point at which the *Anderson transition* occurs also depends on the value of the spectral variable, λ, so that under some circumstances the spectrum may be divided by so-called *mobility edges* into ranges where either all the states are localized or all are extended (fig. 9.16).

Unfortunately there is no rigorous proof of the *Anderson theorem* identifying the precise point at which the transition must occur. As in the theory of the three-dimensional Ising model (chapter 5), where the existence of a phase transition is not in doubt but where the critical temperature and other properties in that neighbourhood are very difficult to calculate, we learn more from a simplified schematic mathematical analysis (e.g. the

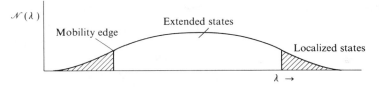

Fig. 9.16. Localized states and mobility edges.

mean field approximation, § 5.2) than from the attempt to arrive at an exact answer. For the moment, therefore, we follow a simplified version of Anderson's original proof of 'the absence of diffusion in certain random lattices'.

The essence of the *Anderson model* is that it is a TBA model where the diagonal disorder variable w_l, defined in (9.23), is drawn from a *continuous* probability distribution $P(w)$. Since the conditions under which the transition occurs are not thought to be very sensitive to the form of this distribution, it is mathematically convenient to assume that this is a *uniform distribution*

$$P(w) = 1/W \quad \text{for } -\tfrac{1}{2}W \leqslant w \leqslant \tfrac{1}{2}W, \tag{9.108}$$

of 'width' W. As in (9.66), the 'strength' of the disorder is best measured by the ratio

$$\delta = W/B, \tag{9.109}$$

where the tight-binding band width B, due to interatomic overlap (8.16), also provides an intrinsic scale factor for the spectral variable λ.

We may start from the formal expression (9.44) for the site-diagonal elements of the total Green function

$$G_{ll}(\lambda) = \{\lambda - \Sigma(l; \lambda)\}^{-1}. \tag{9.110}$$

In this representation, the self-energy function $\Sigma(l; \lambda)$ depends on the chosen site, l, and on the complex variable λ. Now suppose we 'create' a displacement (or an electron state) entirely on this site. The subsequent history of the amplitude $u_l(t)$ on this site is given by time-dependent equations of motion derived from the Hamiltonian (9.2). But, by the definition (9.5) of the Green function, this is equivalent to a Fourier transformation of (9.110) from the 'frequency' variable λ to the time variable t.

The question whether the excitation remains localized to some extent on the lth site depends on whether or not the amplitude $u_l(t) \equiv G_{ll}(t)$ remains non-zero in the limit of large t. But this, in turn, depends on the analytical properties of the function $\Sigma(l; \lambda)$. The Fourier transformation of $G_{ll}(\lambda)$ to $G_{ll}(t)$ can be represented by an integration along a contour 'just above the real axis' in the λ-plane. As is well known from the general theory of Green functions, if $G_{ll}(\lambda)$ has a pole where the self-energy $\Sigma(l; \lambda)$ has a non-zero imaginary part, then $G_{ll}(t)$ acquires a factor that decays exponentially with time and we cannot be dealing with a stationary state that is localized on this site. In other words *a necessary condition that there should be an eigenstate of 'frequency' λ_0 localized on or near the site l is that the imaginary part of $\Sigma(l; \lambda)$ should vanish as $\lambda \to \lambda_0 + i\varepsilon$*. This is really no more than a formal *description* of a localized state in the Green function language, and does not depend on the model.

The self-energy has an exact locator expansion (9.45), i.e.

$$\Sigma(l; \lambda) = w_l + \sum_{l'} V_{ll'} g_{l'} V_{l'l} + \sum_{l',l''} V_{ll'} g_{l'} V_{l'l''} g_{l''} V_{l''l} + \ldots \quad (9.111)$$

When λ is real, each term in this series is also real. If the sequence of partial sums of such terms has a limit this too must be real. In other words, a *sufficient* condition that states in the spectral region round λ should be localized is that the series (9.111) should converge.

Let us consider the term of order V^{L+1}, where L is a large integer. From (9.45) we know that each contribution,

$$V T_j^{(L)} = V g_{l'} V g_{l''} \ldots V, \quad (9.112)$$

to such a term corresponds graphically to a path which wanders through the lattice, with the sole restriction that it does not return to the starting point l before the final step. Since every site has z neighbours, there must be something like z^L contributions to this term, each of which contains a product of L successive factors of the type

$$V g_{l'} \equiv \frac{V}{\lambda - w_{l'}}. \quad (9.113)$$

Unfortunately, we know little about such factors except the distribution of the stochastic variable $w_{l'}$. But in the spirit of the approximations of §§ 9.3 and 9.4, let us make the simplifying assumption that the *successive each such product are statistically independent*. The probability distribution of the product (9.112) can then be deduced at once from the distribution of the logarithm of each factor (9.113) – i.e.

$$\langle \ln |T_j^{(L)}| \rangle = L \langle \ln |Vg| \rangle. \quad (9.114)$$

It is then easy to show (Economou & Cohen 1972; Athreya, Subramanian & Kumar 1973) that a series of which $T_j^{(L)}$ is the Lth term converges with probability 1 if $\langle \ln|Vg| \rangle < 0$, and diverges with probability 1 if this quantity is ≥ 0.

In the present case, however, the Lth term in (9.111) is the sum of z^L terms, of the type (9.112), which are not all of the same sign. The only certainty we have is that our series must converge at least as well as the related geometric series with Lth order term

$$\sum_{j=1} z^L |T_j^{(L)}| \sim z^L V^L \exp\{L\langle \ln|g| \rangle\}. \tag{9.115}$$

This series converges absolutely if

$$zV \exp\langle \ln|g| \rangle < 1, \tag{9.116}$$

which we take as our first estimate of the condition for localization (Ziman 1969).

For the Anderson distribution (9.108), we easily evaluate

$$\langle \ln|g| \rangle \equiv -\frac{1}{W} \int_{-\frac{1}{2}W}^{\frac{1}{2}W} \ln|\lambda - w| dw. \tag{9.117}$$

For $\lambda = 0$, the localization condition (9.116) then simplifies to

$$2zVe/W < 1 \tag{9.118}$$

– i.e. localization occurs at the centre of the band when the disorder strength (9.109) exceeds the critical ratio

$$\delta_c = e. \tag{9.119}$$

In other words *states at the centre of the band must be localized when the energy of the individual atomic states varies at random over a range somewhat greater than the width of the band produced by overlap between adjacent atomic orbitals*. This is the basic result obtained by Anderson (1958).

Inspection of (9.116) and (9.117) shows, however, that the whole spectrum does not become localized, all at once, as δ passes through δ_c. For smaller values of δ, states near the edges of the spectrum may already be localized whilst those in the centre of the band are still extended (Ziman 1969). Thus (fig. 9.17) the first effect of disorder is to produce 'tails' of localized states at the edges of the original tight-bound band: as the disorder increases, these tails become longer, until eventually, as we approach the condition (9.118), the mobility edges move inward and coalesce at the band centre.

Although the localization criterion (9.116) gives a good qualitative description of the circumstances under which, in theory, an *Anderson transition* would be observed, a result such as (9.119) is obviously only a crude estimate of the critical disorder strength. The whole subject seems,

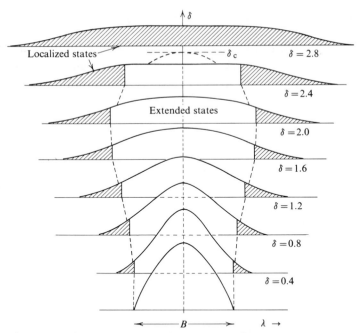

Fig. 9.17. In the Anderson model, as disorder (δ) increases more states are localized, until the mobility edges coalesce at the centre of the band at the Anderson transition (δ_c).

indeed, to be of considerable mathematical subtlety, with conceptual and numerical uncertainties that have not been resolved (see Thouless 1974). It is argued, for example (Anderson 1958; Thouless 1970) that a condition such as (9.116) is much too lax, since the convergence of the locator expansion is, in reality, dominated by the largest values of $T_j^{(L)}$, far out in the tail of the probability distribution of this variable. But provided that the series satisfies the necessary conditions for 'convergence with probability 1' (see e.g. Doob 1953) divergences of this kind belong to a set of measure zero in the statistical ensemble generated by the random variables w_l and can, therefore, be ignored. Indeed, the locator expansion (9.111) cannot be *uniformly* convergent as a function of the real variable λ, since the Green function (9.36) must have a singularity at any actual eigenvalue of the Hamiltonian.

Another objection to this simple treatment is that the series (9.111) can *never* be absolutely convergent near the band centre, since we must eventually encounter a factor g_l with an infinitesimal denominator. Anderson (1958) therefore deliberately replaced the simple locator expansion by a *renormalized perturbation expansion*. By well-known techniques of graphical analysis, the term of order V^{L+1} is expressed as the sum of products like

$$V\mathcal{T}_j^{(L)} = V_{ll'}g_l'V_{l'l''}g_{l'}' \ldots V_{l^{(L)}l},\qquad(9.120)$$

which differs from (9.112) in that the site labels follow a *self-avoiding walk* (§§ 5.10, 7.8). The fact that a vanishing denominator on a given site cannot occur more than once thus reduces the order of divergence of such a product. Moreover, it can be argued that possible strong singularities due to the accidental proximity of other sites of nearly the same energy are removed in the renormalized locators

$$g_l' = \{\lambda - w_l - \Sigma'(l)\},\qquad(9.121)$$

where the strong interaction would automatically produce a large energy shift $\Sigma'(l)$. Unfortunately, this self-energy correction is not simply a function of λ and of the site label l, but really depends on the order L of the term, and on the disorder potentials on other sites, in a very complicated, essentially incalculable, manner.

If now we assume, as in (9.114), that successive factors in (9.120) are statistically independent, we arrive quite easily at Anderson's original minimum estimate of the critical disorder ratio δ_c. According to (5.187), the total number of distinct self-avoiding walks of length L is of the order of ζ^L, where ζ is the *connective constant* of the lattice. In (9.116) and (9.118), therefore, we should replace z by ζ, and hence reduce (9.119) by a factor ζ/z. Unfortunately, this 'correction' to the general localization criterion (9.116) would produce mobility edges well inside the edges of the standard band of Bloch states of a nearly perfectly ordered system, which is quite unphysical.

The correlations induced by renormalization cannot be ignored. Taking a pessimistic view of these and of the influence of large factors in the long tail of the probability distribution of $\mathcal{T}_j^{(L)}$, Anderson arrived at a much larger estimate of δ_c than is suggested by the above simple arguments. On the other hand, rough approximations to the locators in (9.120) derived from effective-medium theory (§§ 9.4, 9.5) yields localization criteria that are not so very far from the results calculated from (9.116) (Economou & Cohen 1972; Licciardello & Economou 1975; but see Mookerjee 1974).

The advantages to be gained from renormalizing the locator expansion may well prove illusory. Indeed, it is worth noting that a singular factor $Vg_{l'}$ with vanishing denominator can occur in the terms of the series (9.111) only if there exist other sites, l'', l''', say, for which

$$w_{l''} < \lambda < w_{l'''}.\qquad(9.122)$$

Under these circumstances, $g_{l'}$ and $g_{l''}$ will be of opposite signs. Let us assume, as in (9.114), that the successive factors in (9.112) behave as if statistically uncorrelated. Under these circumstances, a product $T_j^{(L)}$ may turn out to be positive or negative. For large values of L, the signs of the contributions to a term

$$S^{(L)} = \sum_j T_j^{(L)} \tag{9.123}$$

of (9.111) will appear to be random, with equal *a priori* probability of being positive or negative (Kikuchi 1970). A divergent factor in any one product $T_j^{(L)}$ may thus be compensated algebraically in $S^{(L)}$, without recourse to graphical renormalization. Far from behaving like (9.115), we may expect

$$S^{(L)} \sim 0. \tag{9.124}$$

This does not imply, however, that the locator expansion (9.111) always converges. Fluctuations can only be suppressed – i.e. the series converges with probability 1 – if (Doob 1953) the *variance series*, with Lth term

$$\{S^{(L)}\}^2 = |\sum_{j=1}^{z^L} T_j^{(L)}|^2$$
$$\sim z^L \{V^L \exp(L\langle\ln|g|\rangle)\}^2 \tag{9.125}$$

also convergences. In other words, for values of λ where (9.122) can be satisfied, we should replace the localization criterion (9.116) by the less stringent condition

$$zV \exp\langle\ln|g|\rangle < z^{\frac{1}{2}}. \tag{9.126}$$

For the Anderson model, the condition (9.122) reads

$$-\tfrac{1}{2}W < \lambda < \tfrac{1}{2}W. \tag{9.127}$$

The critical ratio (9.119) for the Anderson transition at the band centre would now become

$$\delta_c = e/z^{\frac{1}{2}}. \tag{9.128}$$

At the band edges, however, where all the factors in (9.113) would have the same sign, without singularities, the physically reasonable results given by (9.116) are still valid.

To avoid such mathematical subtleties as the convergence of a series with random terms, we may try to calculate the site-diagonal Green function (9.110) by some other means. Unfortunately, the approximations inherent in the CPA method and its generalizations (§§ 9.4, 9.5) always imply that we are dealing with extended states (Haydock & Mookerjee 1974). But suppose (Abou-Chacra, Anderson & Thouless 1973; Beeby 1973; Abou-Chacra & Thouless 1974) that we write down explicitly the hierarchy of equations that define the terms (9.120) of the renormalized perturbation expansion. Ignoring terms of order V^3, the first of these takes the form

$$\Sigma(l; i\lambda) - w_l = \sum_{l'} V_{ll'}\{\lambda - w_{l'} - \Sigma'(l'; \lambda)\}^{-1} V_{l'l}. \tag{9.129}$$

Strictly speaking, the function $\Sigma'(l'; \lambda)$ is not the same as the self-energy $\Sigma(l; \lambda)$ in the true site-diagonal Green function. But, if we conflate these two quantities, we have a relatively simple *self-consistency condition* by which we might generate the function $\Sigma(l; \lambda)$; indeed, iteration of this equation would produce many of the terms of the locator expansion (9.111). But because w_l is a random variable, we may suppose $\Sigma(l; \lambda)$ to be a stochastic variable with a stationary distribution function $F(\Sigma)$. Since (9.129) links the site l only with its z neighbours, it is relatively easy to write down the integral equation that makes $F(\Sigma)$ consistent with this connection. The criterion for localization is now, simply, that this integral equation should have a well-behaved solution for real values of $\Sigma(\lambda)$, as λ itself tends to real values. The algebra is rather complicated, but yields well-defined localization limits similar to those estimated by Anderson from the convergence of the renormalized series. The localization of *all* states in one-dimensional systems (§ 8.7) is also satisfactorily confirmed.

Because of its inherent approximations, this method is not necessarily more reliable than those based on the convergence of a random series. It is worth remarking, however, that the self-consistency condition (9.129) is almost identical with the self-consistency condition (9.97) for an approximant to the continued fraction (9.90) for the Green function; the integral equation for the probability distribution of $\Sigma(l; \lambda)$ must be equivalent to the integral equation (9.102). In other words, the self-consistency approach to Anderson localization follows essentially the same path as the *recursion method* (§ 9.7) for finding the spectrum of a random alloy – a method equivalent to the exact *Dyson–Schmidt* solution (§ 8.5) for the spectrum of a disordered linear chain. It would clearly be an economy of intellectual effort to exploit these equivalences, which are ignored in the published literature.

In the fog of speculations and uncertainties surrounding the Anderson theorem, it is reassuring to look at the empirical evidence concerning the transition. Exact eigenstates for a simulated Anderson model of a few hundred sites can readily be computed. In a finite system, the concept of localization is not rigorously defined, but various practical indicators of the effect have been suggested.

366 *Excitations on a disordered lattice*

For example, the *participation ratio* (Bell & Dean 1970)

$$\alpha = \{\langle |u_l|^2 \rangle\}^2 / \langle |u_l|^4 \rangle \qquad (9.130)$$

measures the fraction of lattice sites that are actively excited in an eigenstate of amplitude u_l. For extended states, we must have $\alpha \sim 1$, whilst for a localized state the value of α should decrease as N^{-1} when we increase the size of the model. Again (Edwards & Thouless 1972) the eigenvalue λ of a localized excitation should be insensitive to conditions imposed on the wave functions on distant surfaces of the model. It is easy to compute the shift $\Delta\lambda$ when we change from 'periodic' to 'antiperiodic' boundary conditions. For extended states we expect $N\Delta\lambda$ to be of the order of the band width B; a criterion for localization is that $N\Delta\lambda$ decreases as N increases. Another indicator (Schönhammer & Brenig 1973) can be deduced from the standard formulae for mobility of electrons in such a system (see § 10.10).

These criteria are not necessarily equivalent mathematically, and have not been standardized; the numbers given in the literature are not sufficiently reliable to justify quotation or detailed comparison in terms of lattice type. But the computer simulations all agree in the observation of a relatively sharp transition from extended to localized states, in qualitative conformity with the theorem. There seems little doubt, however, that the true value of the disorder ratio (9.109) for the Anderson transition at the centre of the band is much less than originally estimated by Anderson (1958), and lies in the neighbourhood of

$$\delta_c \approx 1 \qquad (9.131)$$

for simple lattices in two and three dimensions. This fact disconfirms some of the more conjectural analyses of what is, after all, a well-posed mathematical question about a very simple theoretical model!

Although it is now well established that the spectrum of an Anderson model has regimes of extended and localized states, depending on the degree of disorder, the nature of the transition between these regimes is not well understood. A simple argument suggests that these regimes are essentially distinct (Mott 1967; Halperin 1973). Suppose that an extended state and a localized state were to coexist at the same point in the spectrum; any infinitesimal perturbation of the disorder would mix these states, producing two extended states. Thus, the existence of a 'mobility edge' separating one regime from the other at a well-defined spectral point λ_c is assured. The observation by Kirkpatrick & Eggarter (1972) of an infinitesimal gap in the continuous spectrum around a special localized frequency in an alloy model (§ 9.4; fig. 9.9) vividly illustrates this principle. On the other hand this

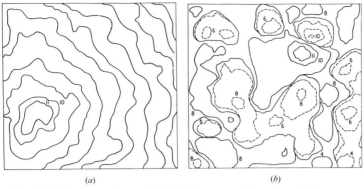

Fig. 9.18. Contour plots of ln $|\psi(r)|$ in a two-dimensional Anderson model: (a) ground state; (b) 'localized' state (Khor & Smith 1971a).

argument for a sharp transition is not rigorous and the possibility of 'percolating' states (§ 9.10) coexisting with localized states cannot be altogether discounted (Kumar & Subramanian 1974).

The main characteristics of eigenstates within the localization regime can be determined from off-diagonal elements $G_{ll'}(\lambda)$ of the Green function (9.36). An approximate summation of the renormalized perturbation expansion (Anderson 1972) shows that this falls off exponentially as a function of the distance $R = |l - l'|$ with a range that increases as $(\lambda - \lambda_c)^{-\frac{1}{3}}$ as we approach the mobility edge (see also Fujita & Hori 1972, 1973). As in the one-dimensional case, however (§ 8.7), this general trend does not exclude the possibility that such a function may have subsidiary peaks by accidental resonance with a favourable site at some distance from the main centre. The fact that the states in band tails of the Anderson model are almost certainly exponentially localized can be exploited to estimate the spectral density by a direct generalization of the *local density approximation* (§ 8.6) that works so well in one-dimensional models (Khor & Smith 1971a). These states are simply localized by fitting into regions where there are favourable fluctuations of the disorder potential. It can also be shown (see § 10.10) that electrons in such states have zero d.c. mobility unless they are given energy to 'hop' (§ 13.3).

But what happens as we cross a mobility edge, by changing λ or by decreasing the disorder? Exponentially localized functions spread and extend through the system – perhaps with an intermediate regime of *power law states* (Last & Thouless 1974) and certainly into a regime of highly irregular wave functions (fig. 9.18) with a random distribution of local

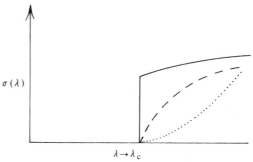

Fig. 9.19. Various ways in which conductivity might behave at a 'mobility edge'.

maxima (Khor & Smith 1971b). It is not clear, however, whether this transition is accompanied by a jump discontinuity of the mobility as a function of λ, as suggested by Mott (1967, 1974) and Mott & Davis (1971) or whether (Kumar & Subramanian 1974) the conductivity in the extended states appears to rise steadily from zero (fig. 9.19). We shall return to this question, which is more general than the Anderson model, in § 9.11.

Although the mathematical theory of Anderson localization has been applied almost exclusively to the Anderson model, where the diagonal disorder distribution function $P(w)$ has the simple form (9.108), the same arguments are valid for any similar model system. By a mathematical curiosity (Lloyd 1969), the *exact* spectrum can be calculated for a system where the disorder variables $\{w_l\}$ are drawn independently from a *Lorentz distribution*

$$P(w) = \frac{1}{\pi} \frac{\Gamma}{w^2 + \Gamma^2}. \tag{9.132}$$

This distribution is automatically centred at $\lambda = 0$ in the band of the virtual crystal propagator $\bar{G}(\lambda)$, as in (9.17), and is thus characterized by the *width* Γ.

The key to the theory of the *Lloyd model* is that the Green function $G(\lambda; \ldots w_l \ldots)$ of the disordered system, considered as a function of the many (complex) variables $\{w_l\}$, retains its functional form under ensemble averaging. Let us suppose that we are dealing with the 'retarded' Green function (cf. (9.7)) where λ has an infinitesimal 'positive' imaginary part. The effect of the disorder at the site l is expressed through the Dyson equation (9.19), i.e.

$$G_{ll'}(\lambda; \ldots w_l \ldots) = G_{ll'}(\lambda; \ldots 0 \ldots) +$$
$$+ G_{ll}(\lambda; \ldots 0 \ldots) w_l G_{ll'}(\lambda; \ldots w_l \ldots). \tag{9.133}$$

This has the algebraic solution

$$G_{ll'}(\lambda; \ldots w_l \ldots) = G_{ll'}(\lambda; \ldots 0 \ldots) +$$

$$+ G_{ll}(\lambda; \ldots 0 \ldots) \frac{w_l}{1 - w_l G_{ll}} G_{ll'}(\lambda; \ldots 0 \ldots), \qquad (9.134)$$

where

$$G_{ll} \equiv G_{ll}(\lambda; \ldots 0 \ldots). \qquad (9.135)$$

These equations are valid, regardless of the values of the other disorder variables, which are statistically independent of w_l. Integrating (9.134) over the probability distribution (9.132), and using elementary algebraic identities, we get

$$\int_{-\infty}^{\infty} G(\lambda; \ldots w_l \ldots) P(w_l) \mathrm{d}w_l = G(\lambda; \ldots -\mathrm{i}\Gamma \ldots). \qquad (9.136)$$

In other words, the average effect of the disorder at the lth site is to shift the site-diagonal element of the Hamiltonian by $-\mathrm{i}\Gamma$. But in the ensemble average, the same shift will be found at every site, which corresponds to a uniform self-energy correction to the spectral variable λ:

$$\langle G(\lambda) \rangle = G(\lambda; -\mathrm{i}\Gamma, -\mathrm{i}\Gamma \ldots)$$

$$= G(\lambda + \mathrm{i}\Gamma; 0, 0 \ldots)$$

$$= \bar{G}(\lambda + \mathrm{i}\Gamma). \qquad (9.137)$$

Since the virtual crystal propagator (9.17) is that of a perfect system with Bloch-like excitations (9.18), we have an exact analytical expression for the average Green function – hence the spectral density (9.7) – of the system. This theorem is evidently valid for any regular lattice, in any number of dimensions. The fact that both the ATA (§ 9.3) and CPA (§ 9.4) methods give the same exact spectral function for this model (Kumar & Baskarian 1973) suggests that the CPA spectrum of the standard Anderson model (Brouers 1971) cannot be far from reality, except in the 'tail' regions.

It must be emphasized, however, that the fact that (9.137) has a finite imaginary part does not mean that all the states of the Lloyd model are extended. The condition for localization applies to the site-diagonal Green function (9.110) of a particular sample system, not to the ensemble-averaged function (9.11) (Anderson 1970; Saitoh 1970; Lehmann 1971). As one might expect from (9.131), localization occurs at the centre of the band when the width parameter Γ is about equal to the 'perfect' band width B (Licciardello & Economou 1975). But this model has its value as a test-bed for mathematical speculations concerning the spectral properties of substitutionally disordered systems (see e.g. Lehmann 1973). It is interesting to note, for example, that the spectral density function (9.7) that might be deduced from (9.137) would exhibit no unusual or singular behaviour as a function of λ as one crossed a mobility edge from localized to extended states (cf. Edwards & Thouless 1971; Thouless 1971). This suggests that the 'Anderson transition' is not analogous to the phase transition of a cooperative assembly, where the thermodynamic functions are non-analytic through the critical point (chapter 5).

9.10 Percolation theory

The question of Anderson localization in a simple binary alloy model does not differ in principle from the problems discussed in the previous section. For weak disorder, for example, a criterion such as (9.116) correctly indicates that states in the band tails should be localized (Ziman 1969a). But the predictions of this criterion are clearly false in the split band limit, where the disorder strength δ defined in (9.66) is very large. In this limit, the atomic levels of A atoms and B atoms are so far apart in energy that there can be no electron state that has finite amplitudes simultaneously in both types of atom. As in the derivation of (9.83), we assume that electrons in the 'A band' can be found only on A sites and cannot propagate through B sites – and vice versa. This argument was used to explain the unusual spectral structure (fig. 9.9) found by Kirkpatrick & Eggarter (1972) in such models; highly localized excitations could occur on particular configurations of atoms, partially isolated by atoms of the 'wrong' type (fig. 9.10).

It is evident, however, that an electron state would *not* be localized if it covered the sites of an *infinite cluster* of A atoms – that is, a set of sites, connected by the nearest-neighbour matrix elements $V_{ll'}$ extending to all regions of the system. It is intuitively obvious that such clusters must exist if the concentration c_A of favourable sites is sufficiently high. It is also obvious that an occasional very large cluster may occur by statistical fluctuation, even when c_A is small. From the point of view of the theory of electron localization, the fundamental question is whether extended states of this kind may be found with finite non-zero probability for all values of the concentration or whether there is any analogue of the Anderson transition from localized to extended states in this model.

This is none other than the *site percolation* problem of classical probability theory (Broadbent & Hammersley 1957): '*atoms*' *are distributed at random on the sites of a regular lattice in such a way that any given site has probability p of being occupied; what is the probability $P(p)$ that a given atom belongs to an infinite cluster?* In thus defining the problem, we use the

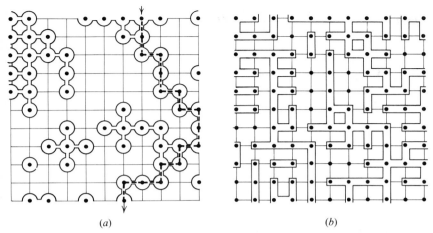

(a)   (b)

Fig. 9.20. Percolation on a lattice: (a) site problem showing a percolation path; (b) bond problem, with no percolation path crossing this
block.

natural nomenclature of crystal physics, but *percolation theory* has much
wider applicability, to problems as diverse as the percolation of a fluid
through a porous medium and the spread of disease through an orchard
(Frisch & Hammersley 1963).

The companion problem of *bond percolation* applies to a lattice where a
fraction p of the 'bonds' are 'favourable' (i.e. open to traversal, not
blocked, etc.). The function $P(p)$ then refers to the probability that a given
favourable bond is part of an infinite cluster linked by such bonds. The
bond percolation problem on a *regular* lattice does not arise so naturally
out of the physics of substitutional disorder, but plays an important part in
the theory of transport in topologically disordered systems (§ 11.4) and in
random continuous media (§ 13.4). Mathematically speaking, the site and
bond problems (fig. 9.20) are so similar that we treat them both together in
this section. For all the physical models in which we are interested it can
also be shown (Broadbent & Hammersley 1957) that the above definition of
the *percolation probability* function $P(p)$ is equivalent to definitions derived
from other concepts such as the existence of connected 'favourable' paths
extending from one boundary to another of an 'infinitely thick' specimen.

The fundamental theorem of percolation theory (Broadbent & Hammersley 1957) proves that *the percolation probability $P(p)$ is of measure
zero for $p < p_c$*, where the critical concentration p_c is characteristic of the
lattice type (but is not usually the same for bond percolation as for site

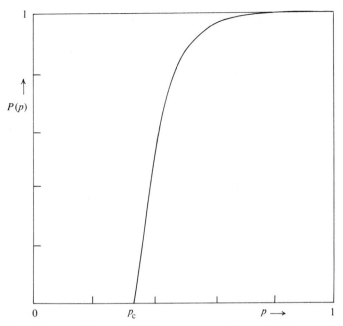

Fig. 9.21. Percolation probability on a tree of coordination number $z = 4$ (cf. fig. 7.10).

percolation on a given lattice). This theorem is illustrated perfectly by the exact solution (Fisher & Essam 1961) for bond percolation on a *tree* (§ 5.4) of coordination number z. This is none other than the problem of gel formation in a polymer solution (§ 7.5). The percolation probability $P(p)$ is precisely equivalent to the *gel fraction* (7.44), which is shown to be zero for

$$p < \frac{1}{z-1} \equiv p_c.$$

(9.138)

Above this critical concentration, the percolation probability is given as a function of p by the equation (7.43), which in our present nomenclature reads

$$\sum_{j=0}^{z-2} \{1 - P(p)\}^{j/z} = 1/p.$$

(9.139)

Solving this equation, we see, as in fig. 9.21, that $P(p)$ rises monotonically from zero at p_c to a value approaching unity, which it must eventually attain at $p = 1$. The corresponding site percolation problem on the same lattice can be solved with equal facility: *a percolation transition* occurs at the same critical concentration (9.138).

For physically realistic lattices, however, there are few exact formulae for $P(p)$ or even for p_c (see e.g. Essam 1972). Mathematically speaking, percolation theory is closely related to other lattice-graphical problems that arise in the theory of disordered systems, such as the properties of the Ising model (§ 5.10) and of 'ferroelectric' models (§ 5.8), and the polymer excluded volume problem (§ 7.8). But these connections (Fortuin & Kasteleyn 1971; Temperley & Lieb 1971) are abstract and formal, without much practical significance.

Recalling, for example, that the *branching number* $(z-1)$ of a tree is analogous to the *connective constant* ζ of a multiply-connected lattice (§ 5.10), we might hope for a simple generalization of (9.138). Unfortunately this only informs us that the site and bond *percolation thresholds* have a lower bound:

$$p_c^S, p_c^B \geqslant 1/\zeta. \qquad (9.140)$$

This is easily proved (Broadbent & Hammersley 1957). In § 5.10 we defined $u(n)$, the total number of distinct n-stepped *self-avoiding walks* from a fixed vertex of a given lattice. For a concentration p of 'favourable' sites or bonds on this lattice, the expected number of 'open' paths of this type must be

$$N_n^{SA}(p) = p^n u(n). \qquad (9.141)$$

Let $\pi_n(j)$ be the probability that the number of such open paths is exactly j. Then the probability $P_n^{SA}(p)$ that at least one such path is open can be compared with the expected number of paths:

$$\begin{aligned} P_n^{SA}(p) &\equiv \sum_{j \geqslant 1} \pi_n(j) \\ &\leqslant \sum_{j \geqslant 1} j\pi_n(j) \equiv N_n^{SA}(p). \end{aligned} \qquad (9.142)$$

By (5.187),

$$u(n) \sim \zeta^n \qquad (9.143)$$

in the limit of large n. Thus, from (9.141)–(9.143) we have

$$P^{SA}(p) \equiv \lim_{n \to \infty} P_n^{SA}(p) \to 0 \quad \text{if } p < 1/\zeta. \qquad (9.144)$$

But *any* infinite cluster contains at least one infinite self-avoiding walk, so that $P(p)$ itself must likewise vanish – whence (9.140) follows.

This theorem is not, however, very useful, since the correct values for p_c for standard lattices (table 9.1) considerably exceed the values of $1/\zeta$ that can be deduced from table 5.1. Equality in (9.140) is found only for a genuine tree, where every walk is self-avoiding, and 'wasteful' cycles of favourable bonds or sites cannot occur.

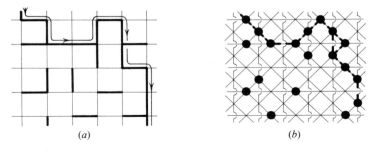

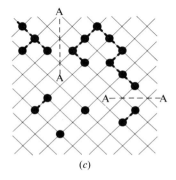

Fig. 9.22. (*a*) Bond percolation path on square net *S*: $p > p_c^B$. (*b*) Site percolation on the covering lattice S^C of the same model. (*c*) Removal of cross-links to turn S^C into a square net cuts percolation path at *A–A*: hence $p < p_c^S$.

The *bond* percolation threshold cannot, in fact, exceed the *site* percolation threshold for the same lattice:

$$p_c^B \leqslant p_c^S. \qquad (9.145)$$

The general proof of this theorem (Hammersley 1961) is quite lengthy, but can be illustrated by reference to a particular case (Fisher 1964). Suppose that *p* is above the *bond* percolation threshold for a planar square lattice *S*, so that fig. 9.22(*a*) shows part of an infinite percolation path. Transform the lattice *S* into its *covering lattice* S^C: each bond of *S* becomes a site of S^C; if two bonds of *S* meet at a vertex, then the corresponding sites of S^C are joined by a direct link of S^C (fig. 9.22(*b*)). It is obvious that the original percolation path remains connected, so that the *site* percolation threshold

Table 9.1 *Thresholds for site and bond percolation on regular two-dimensional and three-dimensional lattices. Figures represented 0.6527 . . ., etc. are exact.*

| Lattice | Coordination number | Packing fraction | Percolation threshold | | Critical bond number | Critical volume fraction |
			Bonds	Sites		
	z	η	p_c^B	p_c^S	zp_c^B	ηp_c^S
Honeycomb	3	0.61	0.6527 . . .	0.70	1.96 . . .	0.427
Square	4	0.79	0.5000 . . .	0.59	2.00 . . .	0.466
Triangular	6	0.91	0.3473 . . .	0.5000 . . .	2.08 . . .	0.455 . . .
Tetrahedral (diamond)	4	0.34	0.39	0.43	1.56	0.143
s.c.	6	0.52	0.25	0.31	1.50	0.161
b.c.c.	8	0.68	0.18	0.24	1.44	0.163
f.c.c. ⎫ h.c.p. ⎭	12	0.74	0.12	0.20	1.44	0.148

on S^C must be identical with p_c^B on the original lattice S. But we can turn S^C into a new version of S by removing all the 'crossed-links' (fig. 9.22(*c*)). This now represents a *site* percolation problem, on a square lattice, with the original concentration p of favourable sites. But the links that were cancelled may have been necessary for the connectivity of the percolation path, so that p may lie below the percolation threshold, p_c^S, for this system. In other words (9.145) is true for this particular lattice type.

For a tree, the two thresholds have the same value (9.138). But for typical two- and three-dimensional lattices, there are substantial differences between the critical concentrations for bond and site percolation (table 9.1). Note, in particular, that these differences increase as we go to more closely packed lattices which are more 'highly connected'.

The covering-lattice transformation of fig. 9.22 is only one of several topological transformations that can be applied to the problem of percolation on a lattice (see Essam 1972). For example, the *duality transformation* (fig. 5.13) leads to simple algebraic relations between the percolation thresholds on the two lattices, analogous to the relation (5.182) between low- and high-temperature formulae for the partition function of the Ising model under the same transformation. Just as Kramers & Wannier (1941) were able to locate the critical temperature for the phase transition on a *self-dual* lattice, Sykes & Essam (1964) were able to find the *exact* bond percolation thresholds for several planar lattices.

Notice also the exact value $p_c^S = \frac{1}{2}$ for the triangular net. This is the

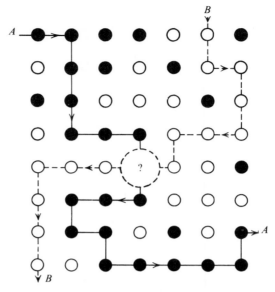

Fig. 9.23. Site percolation in two dimensions: contradiction at site where infinite *A* chain intersects infinite *B* chain.

minimum possible value for site percolation on a two-dimensional lattice. For if this were not so we could have a distribution of *A*-type atoms with concentration c_A above the percolation threshold, in the range $p_c^S < c_A < \frac{1}{2}$. On the remaining sites we could put *B*-type atoms with concentration $c_B = (1 - c_A) > \frac{1}{2}$, which would also be above the percolation threshold (fig. 9.23). This means that infinite clusters of both *A*-type and *B*-type atoms would coexist on the same plane. But if we take an infinite chain of *A* atoms from an *A*-cluster and a similar chain of *B* atoms from a *B*-cluster, they must surely (the proof of this by Harris (1960) is quite lengthy) intersect. The site of intersection must be *both* A and B, which is absurd. This argument is not, of course, valid in three dimensions, where infinite percolation clusters of both *A*- and *B*-types may interpenetrate one another and coexist.

Numerical values of p_c for both site and bond percolation on various simple lattices are quoted in table 9.1 (Essam 1972). Apart from a small number of exact results for two-dimensional lattices, these numbers have been derived by laborious computations along the same lines as the series expansions (§ 5.10) and Monte Carlo calculations (§ 6.6) used to determine the critical parameters of models of cooperative phenomena. In table 9.1

these estimates of p_c^B and p_c^S are quoted only to the second decimal place, which seems to be the current measure of agreement between the results obtained by different methods.

The most interesting feature of this table is that the numbers conform quite closely to a few simple empirical rules. We note, for example, that the *bond* percolation threshold for all three-dimensional lattices is very nearly inversely proportional to the coordination number. Indeed the formula

$$zp_c^B \approx d/(d-1) \tag{9.146}$$

for the *critical bond number* in the lattice of dimensionality *d* is satisfied to within a few per cent in all cases (Vyssotsky, Gordon, Frisch & Hammersley 1961). In other words: *percolation occurs in a regular three- (two-) dimensional network if there are on the average more than 1.5(2) 'favourable' links to any node* (Ziman 1968). This rule of thumb is sufficiently accurate for almost all practical applications of bond percolation theory.

For site percolation, the corresponding *dimensional 'invariant'* is derived by reference to the *packing fraction η* – the standard metallurgical parameter representing the proportion of the total volume of a crystal that is actually occupied by touching spheres centred on the lattice sites (§ 2.11). As we see from table 9.1, the *critical volume fraction $\eta p_c^{S'}$* is very nearly constant for all lattices of the same dimensionality (Scher & Zallen 1970): *site percolation occurs in a regular three- (two-)dimensional assembly when 'favourable' regions occupy about 15 per cent (45 per cent) of the total volume.*

Although these formulae have been scrutinized in the hope that they might be deduced from more general principles (see e.g. Shante & Kirkpatrick 1971; Pike & Seager 1974) they remain tantalizingly approximate and empirical. Elaborations, to percolation along links to more distant neighbours, scarcely illuminate their mathematical status. As we shall see (§ 13.3) generalizations of these rules to *topologically* disordered networks or assemblies are very useful – but highly conjectural.

For more detailed practical applications of percolation theory we need to know the behaviour of the percolation probability function $P(p)$. We know, of course, that this is strictly zero below p_c, and that it tends to unity as $p \to 1$. It must always be remembered, incidentally, that $P(p) < 1$ for $p < 1$. This is because there is always a finite probability of finding isolated small clusters that are not connected to the main percolation region. Thus, allowing only for 1-site clusters,

$$1 - P(p) < P_1 = (1-p)^z. \tag{9.147}$$

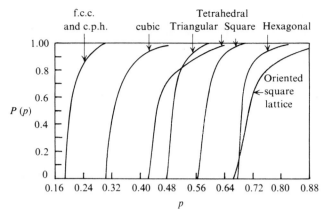

Fig. 9.24. Site percolation probability function for various lattice types
(Frisch, Hammersley & Welch 1962).

Unfortunately, the behaviour of $P(p)$ for real lattices can only be determined by laborious computation. The evidence of Monte Carlo computations (Frisch, Hammersley & Welsh 1962) is that all percolation models
give much the same shape of $P(p)$, apart from some shift of the critical
concentration p_c (fig. 9.24). But the available data do not seem sufficiently
accurate to locate the *critical exponent* (cf. § 5.12) in an expression such as

$$P(p) \sim |p - p_c|^\alpha \tag{9.148}$$

for p near p_c. The only exact result is that, for a tree lattice, where $P(p)$ is
given analytically by (9.139), the exponent α is always unity.

In attempting to understand the deeper mathematical properties of the
percolation phenomenon, the analogy with the theory of the Ising model
can be exploited (Fortuin & Kasteleyn 1971; Fortuin 1972; Essam 1972).
For example, the percolation function $P(p)$ corresponds to the net magnetization – the proportion of the whole crystal that is aligned by connection
to the infinite master cluster. By these arguments, for example, scaling
theory and renormalization group methods (§ 5.12) can be extended to
include discussion of the critical exponents of the percolation system
(Dunn, Essam & Loveluck 1975; Dunn, Essam & Ritchie 1975; Young &
Stinchcombe 1975).

A curious point, noted by Bishop (1973) is that the following formula

$$(\mathscr{k}T_c/zJ) \approx \{(1 + 1/p_c^S) \tanh^{-1}p_c^S\}^{-1} \tag{9.149}$$

gives the critical temperature (5.6) of the Ising model to within a few
percent for all the lattices noted in table 9.1. It is easy to verify that this

formula is exact on a tree lattice, where the Bethe formula (5.17) for T_c and the percolation formula (9.138) are both true. Perhaps this empirical formula says something about the role of the connective constant in both § 5.10 and in (9.140).

Percolation arguments play a varied and significant part in the discussion of many properties of disordered systems. But we cannot end this section without returning briefly to our starting point – the question of Anderson-type localization of electron states on, say, the A atoms of a disordered alloy in the split-band limit. Clearly, if the concentration of such atoms, c_A, is below the site percolation threshold p_c^S, then all clusters of these atoms are of finite extent and may be expected to allow only *localized* states. Again, if $c_A > p_c^S$, there are certainly infinite clusters of A atoms capable of supporting *extended* states. But however much c_A exceeds the percolation threshold, there still exist 'isolated' A atoms or clusters, with finite probability (9.147), on which we might expect to find only localized modes. But this would contradict the principle that extended and localized states cannot coexist at the same energy. Here, once again, the remarkable results of Kirkpatrick & Eggarter's simulation experiment (§ 9.4) suggest the possibility of a much more complicated behaviour where the continuous spectrum of extended states either resonates with and delocalizes these modes, or else actually splits apart to accommodate them in a 'special' gap.

Let us note, finally, a trivial point: 'percolation' is a meaningless concept for a *one-dimensional system*. It is obvious that the smallest density of 'unfavourable' sites, or bonds, distributed at random along a chain, will simply cut it into segments of finite length. There is no getting around the blockages; there are no infinite clusters; the 'percolation threshold' has risen to the limit $p_c = 1$. This is clearly consistent with the theorem of § 8.7 – that all excitations are localized on a disordered linear chain – although it does not provide a proof in the general quantum-mechanical case. It is also related to other pathological properties of one-dimensional models, such as the absence of topological disorder (§ 2.4) and the absence of phase transitions (§§ 5.5, 6.1). Once again we observe the fundamental topological deficiencies of any one-dimensional model as a realistic representation of a genuine three-dimensional physical system.

9.11 Maze conduction

From percolation theory we deduce the *existence* of percolation paths through a disordered system: the question is – what contribution do such paths make to the *transport properties* of the system? Before any attempt to calculate such properties quantum mechanically, we should have some understanding of the corresponding *classical* problem: what is the average conductance of a *maze* – i.e. a network of 'conductors' interconnected irregularly according to some statistical prescription? The percolation models of § 9.10 can be treated as simple *lattice mazes,* where we assign a standard conductance σ_1 to each 'favourable' link, and a much smaller conductance σ_2 to each 'unfavourable' link. Strictly speaking, percolation theory applies only to the extreme case where $\sigma_2 = 0$ (i.e. there is *no* conduction through an unfavourable bond or site), but the generalized case is often more physically realistic. Thus, in an application of the theory of *maze conduction* to the diffusion of a third component, C, in an $A_x B_{1-x}$ alloy (Pike, Camp, Seager & McVay 1974) it is natural to assume that atoms of type C can diffuse very rapidly through an A-type cell of the lattice, but are not absolutely prohibited from passing through a cell containing a B atom.

The archetypal physical model for maze conduction is a regular lattice of electrical conductors from which a proportion $(1-p)$ of the 'bonds', or of the 'sites' (i.e. connection nodes) are missing. From all the arguments of percolation theory, it is obvious that the *bulk conductance* $\sigma(p)$ should be zero for $p < p_c$: every path from any lattice site eventually terminates at a non-conducting link. Above the percolation threshold it is easy to suppose (Ziman 1968) that the system should behave as if it had an average conductance proportional to the percolation function $p(p)$. But this supposition is not correct. Direct measurements on a sheet of colloidal graph paper with holes punched in it (Last & Thouless 1971), or on a wire mesh with junctions snipped out (Watson & Leath 1974), agree with computations for a simulated two- or three-dimensional model (Kirkpatrick 1971). In the 'critical' region the conductivity rises as

$$\sigma(p) \sim |p - p_c|^\beta, \tag{9.150}$$

with an exponent β in the neighbourhood of 1.5. With increasing p, this leads smoothly into a linear segment, right up to $p = 1$ (fig. 9.25). In other words, the apparent 'percolative mobility' $\sigma(p)/P(p)$ is not constant, but decreases as we approach p_c.

The best available explanation of this effect comes from the *exact* solution of the problem of maze conduction on a regular *tree* (Stinchcombe 1974). The method is entirely analogous to the Dyson–Schmidt solution for

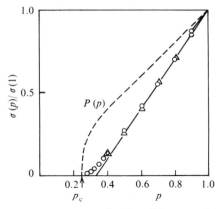

Fig. 9.25. Maze conduction on a simple cubic network: ○, △ computer simulation; ——, effective medium theory; – – –, percolation probability $P(p)$ (Fitzpatrick 1971).

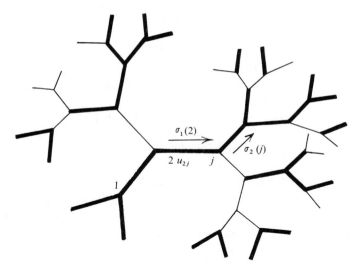

Fig. 9.26. Conduction on a tree with random links.

the spectrum of a disordered linear chain (§§ 8.5, 9.7). Suppose that $\sigma_i(j)$ is the total conductivity that would be measured from a site j along all favourable paths and branches to the surface of the specimen except those paths through a particular neighbouring site i (fig. 9.26). Elementary theory (Kirchhoff's laws!) connects the values of this function on successive branches:

$$\sigma_1(2) = \sum_{j=1}^{z-1} \left\{ \frac{1}{u_{1j}} + \frac{1}{\sigma_2(j)} \right\}^{-1}, \tag{9.151}$$

where u_{2j} is the actual conductance of the link from site 2 to site j. This is the analogue of (8.74) or (9.96) in the Dyson–Schmidt theory.

But the *outward conductivity* $\sigma_i(j)$ is a statistical variable that depends on the detailed distribution of favourable links in the neighbourhood of the sites (i, j). Moreover, the branches 'outwards' from site 2 are never reconnected, so that the variables $\sigma_2(j)$ in (9.151) are statistically independent of one another. In the spirit of (8.76) or (9.102), we assume that each of the variables $\sigma_1(2)$ and $\sigma_2(j)$ is drawn from the same stationary probability distribution, which must be consistent with (9.151).

The situation is complicated by the appearance of $z-1$ terms on the right-hand side. But since these are a sum of statistically independent terms, the equations are simplified by representing the probability distribution of $\sigma_i(j)$ by its *cumulant generating function* (cf. (5.165)), defined by the ensemble average

$$B(x) \equiv \langle \exp\{-x\sigma_i(j)\} \rangle. \tag{9.152}$$

The statistical distribution of branch conductances u_{ij} is given *a priori* by some probability distribution $g(u)$; in the percolation model this would have magnitude p at the standard branch conductance σ_1, and $(1-p)$ at conductance $\sigma_2 = 0$. From (9.151) we then deduce the consistency condition

$$B(x) = \{C(x)\}^{z-1}, \tag{9.153}$$

where

$$C(x) \equiv \int g(u)\langle \exp\{-x[1/u + 1/\sigma_i(j)]^{-1}\} \rangle du. \tag{9.154}$$

The problem now is to express the ensemble average in (9.152) in terms of the corresponding average in (9.154). By a Laplace transformation and some elementary algebraic manipulations, Stinchcombe was able to write down a non-linear integral equation of the form

$$\int_0^\infty e^{-tx} C(x) dx = \int g(u) \left[\frac{1}{t+u} + \frac{u^2}{(t+u)^2} \int_0^\infty e^{-utx/(u+t)} \{C(x)\}^{z-1} dx \right] du. \tag{9.155}$$

The maze conductance problem is solved by finding a function $C(x)$ that satisfies (9.155), subject to suitable boundary conditions that can easily be written down. The definition (9.152) implies that the average total conductivity of the network, from any site along *all* favourable branches, must be

$$\sigma(p) = \{z/(z-1)\}\langle \sigma_i(j) \rangle = -zC'(0). \tag{9.156}$$

Note that this analysis is valid for more complex situations where, for example, the conductance σ_2 of an 'unfavourable' link is not taken to be zero.

An equation such as (9.155) can scarcely be expected to have a solution that can be represented in closed analytical form. But the behaviour of $\sigma(p)$ can be determined by judicious mathematical approximations over various ranges of p. One may confirm (9.138), for example, by showing that $\sigma(p)=0$ for all values of p below the exact percolation threshold $p_c = 1/(z-1)$.

Again, for $p_c \ll p < 1$, the solution is very close to what one can obtain from (9.151) by simply replacing each variable $\sigma_i(j)$ by its ensemble average. Since only those $p(z-1)$ branches contribute where $u_{2j} = \sigma_1 \neq 0$, we get

$$\langle \sigma_i(j) \rangle = p(z-1)\{1/\sigma_1 + 1/\langle \sigma_i(j) \rangle\}^{-1}. \qquad (9.157)$$

Noting, from (9.156), that the outward conductivity $\sigma_i(j)$ is not quite the same as $\sigma(p)$, we obtain the very simple formula

$$\sigma(p) = z\sigma_1(p-p_c). \qquad (9.158)$$

This is precisely the formula (Kirkpatrick 1971) that may be deduced by appeal to the *effective medium* concept (§ 9.4) which we have already used in deriving the CPA formula (9.49). The analogy with other mean field approximations, such as (9.100), is clear.

But although (9.158) gives exactly the right answer at $p-1$, and vanishes at p_c, it is not a good approximation to the solution of the Stinchcombe equation (9.155) in the critical region. The behaviour of $\sigma(p)$ for $p > p_c$ is of the form (9.150), with an exponent $\beta = 2$. This again can be explained by intuitive probabilistic arguments (Stinchcombe 1973; Essam, Place & Sondheimer 1974). In (9.157) we have neglected fluctuations in $\sigma_i(j)$. Near the percolation threshold it is incorrect to suppose that the outward conductivity of every branch $\sigma_2(j)$ is near to the average value $\langle \sigma_i(j) \rangle$. This branch has a finite probability of *not* belonging to an infinite cluster and hence of contributing nothing to the outward conductivity $\sigma_1(2)$. In other words, as p approaches p_c, each percolative cluster loses more and more alternative 'infinite' branches. Instead of providing innumerable parallel paths to the surface, the cluster approximates more and more closely to a single, infinitely long chain, of infinite total resistance. This is why the apparent 'mobility' in a percolative cluster vanishes as

$$\sigma(p)/P(p) \sim |p - p_c| \qquad (9.159)$$

in the tree model.

It is not clear whether the same explanation of the behaviour of maze

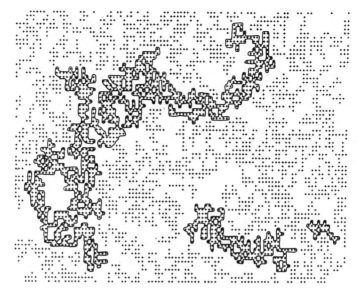

Fig. 9.27. Large computer-generated clusters on a square lattice for $p = 0.53$, below the site percolation threshold ($p_c^S = 0.59$). These clusters are predominantly simply connected (Stanley *et al.* 1976).

conductivity in the critical region is valid for an ordinary two- or three-dimensional lattice. It has been argued, for example (Kirkpatrick 1973), that the conductivity is *not* mainly along a single one-dimensional 'critical path' of favourable links, but that the percolation clusters are already topologically complicated and provide many parallel multiply-connected pathways for the current. This conjecture is not, however, confirmed by computer simulation (Stanley *et al.* 1976) (fig. 9.27) and renormalization group methods (§ 5.12) are needed to get near the correct answer analytically (Stinchcombe & Watson 1976).

The subtleties that arise even in such a 'black and white' problem as the conductivity of a regular percolative maze indicate the difficulty of settling more delicate questions, such as the behaviour of the (quantum) transport properties in an Anderson transition (cf. fig. 9.19). It is not difficult (Thouless 1974) to interpret Anderson localization as a percolation effect. In any TBA model, we may say that an electron of energy $\mathscr{E}$ cannot easily pass through a site whose 'atomic' level $\mathscr{E}_i$ differs from $\mathscr{E}$ by more than, say γV where $\gamma \sim 1$. Out of the whole range W of the disorder potential, only a fraction

$$p \approx 2\gamma V/W \tag{9.160}$$

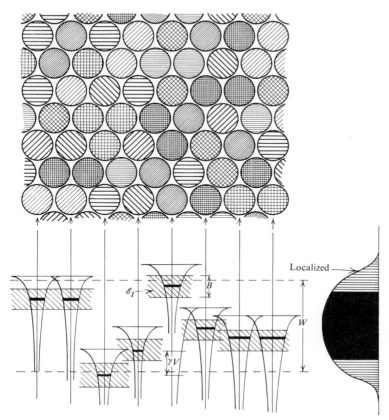

Fig. 9.28. Variations of energy in Anderson model divide the sites into various types which do not communicate if the energy difference exceeds γV. States are extended or localized according as they may percolate or not through sites at the same type.

of the lattice sites may be deemed 'favourable' (fig. 9.28). From the point of view of the electron, therefore, we have a percolation problem; the electron will be localized if

$$p < p_c^S \approx \gamma'/z, \qquad (9.161)$$

where we see from table 9.1 that $\gamma' \approx 2$. From (9.160) and (9.161) we obtain a localization condition which is not so very far from (9.131). It is easy to see, moreover, that the probability of localization increases as $\mathscr{E}$ approaches the edges of the band, where the proportion of 'favourable' sites falls much below (9.160).

This argument is obviously very crude and takes no account of resonant tunnelling through unfavourable sites (Thouless 1974). But the percolation analogy suggests that the behaviour of the conductivity as we go through the mobility edge may be determined as much by the long-range topology of the available paths as by more local characteristics.

10
Electrons in disordered metals

—

'And pictures in our eyes to get
Was all our propagation'
Donne

10.1 The NFE model

The TBA model (§ 9.1), in whose title appears the word 'alloy', is of limited validity as a model for the electronic properties of a *metal*. The *conduction* electrons, which are responsible for the characteristic metallic behaviour of a wide variety of ordered and disordered materials, cannot be satisfactorily described in terms of tight-bound wave functions. It is not correct to suppose, for example, that an energy eigenstate of an electron can be written as a linear combination of a small number of atomic orbitals (as in (8.10)); we must return to the one-electron Schrödinger equation (8.9), i.e.

$$\{-(\hbar^2/2m)\nabla^2 + \mathscr{V}(\mathbf{r})\}\Psi_j(\mathbf{r}) = \mathscr{E}_j\Psi_j(\mathbf{r}) \qquad (10.1)$$

for the eigenfunction $\Psi_j(\mathbf{r})$, of energy $\mathscr{E}_j$, in the potential function $\mathscr{V}(\mathbf{r})$ throughout the volume of the specimen.

The solution of this differential equation for the continuous function $\Psi_j(\mathbf{r})$ is a much more formidable mathematical problem than the solution of the matrix equations (8.11) or (9.1) for the site amplitudes u_l. Even for an ordered crystal, where the analysis is immensely simplified by the Bloch theorem (§ 1.1), the 'band structure problem' has generated a vast literature of specialized mathematical analysis, physical investigation and elaborate computation. But we must not be deterred by the evident difficulties of extending these techniques to systems without lattice translational symmetry: *disordered metals,* in the form of alloys and liquids, are too important in the everyday world to be ignored by theoretical physics!

In the present chapter we shall be particularly concerned with electrons in *free bands* – i.e. where there is a quasi-continuous spectrum of one-electron excited states, with non-zero spectral density $\mathscr{N}(\mathscr{E}_F)$ immediately above a distinct Fermi energy $\mathscr{E}_F$. We postpone until chapter 11 the much more subtle questions that must be asked about systems such as *amorphous* or *liquid semiconductors,* where band gaps near the Fermi energy dominate the electrical properties. But we need not restrict ourselves particularly to 'simple' metals without the d-bands and f-bands of transition and rare earth metals. It turns out, moreover, that much of the analysis of the present chapter applies both to substitutionally disordered materials, such as crystalline alloys, and to topologically disordered systems such as liquid or 'glassy' metals (§ 2.13).

It is well known that many of the electrical properties of such a system can be reproduced qualitatively by the *nearly free electron* (NFE) model. In the first instance, we assign n electrons per unit volume to the *plane wave* states

$$|\mathbf{k}\rangle = \frac{1}{\sqrt{V}} e^{i\mathbf{k}\cdot\mathbf{r}}, \tag{10.2}$$

with 'unperturbed' energies

$$\mathscr{E}_{\mathbf{k}}^{0} = \hbar^{2}k^{2}/2m. \tag{10.3}$$

These states are occupied up to a Fermi energy

$$\mathscr{E}_{F} = \hbar^{2}k_{F}^{2}/2m, \tag{10.4}$$

defining a sphere in momentum space of radius

$$k_{F} = (3\pi^{2}n)^{\frac{1}{3}}. \tag{10.5}$$

The *density of states* at the *Fermi level* is given by

$$\mathscr{N}(\mathscr{E}_{F}) = 3n/2\mathscr{E}_{F}, \tag{10.6}$$

and each electron state on the *Fermi surface* carries an electric current of magnitude

$$j_{F} = e\hbar k_{F}/m. \tag{10.7}$$

These standard equations define symbols that will be much used in the present chapter.

In this context, the potential $\mathscr{V}(\mathbf{r})$ seen by an electron appears always through its matrix elements in a *momentum representation*. These are simply the Fourier components.

$$\langle\mathbf{k}|\mathscr{V}|\mathbf{k}'\rangle = \frac{1}{V}\int \mathscr{V}(\mathbf{r})e^{-i(\mathbf{k}-\mathbf{k}')\cdot\mathbf{r}}\mathrm{d}^{3}\mathbf{r} \equiv \mathscr{V}(\mathbf{k}-\mathbf{k}'). \tag{10.8}$$

For a perfect crystal, where the potential must have the lattice translational symmetry (2.1), these components are non-zero only at the vectors of the reciprocal lattice, where $\mathbf{k}-\mathbf{k}'=\mathbf{g}$. In the NFE formalism for the band structure such delta-function matrix elements of the potential give rise to splitting of the unperturbed energies at the boundaries of the Brillouin zones and hence, eventually, to band gaps and other familiar features of the electronic structure of crystalline metals and semiconductors.

In the absence of long-range order, however, coherent diffraction effects of this kind are not to be expected. Assuming that the zero of energy in (10.3) is the average potential

$$\bar{\mathcal{V}}(\mathbf{r}) = \mathcal{V}(0), \tag{10.9}$$

we are left with the perturbing effect of a continuous, relatively weak function of $(\mathbf{k} - \mathbf{k}')$. After a further shift of the zero of energy, it is reasonable to assume (cf. § 10.4) that changes of order $|\mathcal{V}|^2$ in the density of states (10.6) or in the *Fermi current* (10.7) can be ignored. For the moment we concentrate on the *transitions* produced by the *off-diagonal* matrix elements of $\mathcal{V}(\mathbf{r})$ between the simple plane-wave states (10.2). This *scattering* of free electrons by the atomic disorder of the metal gives rise to such characteristic *transport coefficients* as the electrical *resistivity* ρ.

Elementary quantum theory tells us that the transition rate per unit solid angle between states $|\mathbf{k}\rangle$ and $|\mathbf{k}'\rangle$ on the Fermi surface is given by the 'golden rule' formula

$$\mathcal{Q}(\theta) = (2\pi/\hbar)|\langle \mathbf{k}|\mathcal{V}|\mathbf{k}'\rangle|^2 \tfrac{1}{2}\mathcal{N}(\mathcal{E}_F)/4\pi. \tag{10.10}$$

Because electron spin does not change in potential scattering, only half the total density of states (10.6) is available after the transition.

This scattering phenomenon is precisely analogous to the diffraction of an external beam of X-rays or neutrons (§§ 4.1, 4.2) by the assembly of atoms or ions. Suppose, for simplicity (cf. § 10.2), that the total potential seen by a conduction electron on the Fermi surface can be written, as a *superposition* of N identical 'atomic' potentials,

$$\mathcal{V}(\mathbf{r}) = \Sigma_i v(\mathbf{r} - \mathbf{R}_i), \tag{10.11}$$

each centred on an atomic position $\mathbf{R}_i$. Precisely as in (4.6)–(4.7) we may write

$$|\langle \mathbf{k}|\mathcal{V}|\mathbf{k}'\rangle|^2 = N^{-1}S(\mathbf{q})|v(\mathbf{q})|^2, \tag{10.12}$$

where the atomic form factor $|v(\mathbf{q})|^2$ is deduced from the Fourier transform of the atomic potential $v(\mathbf{r})$, and $S(\mathbf{q})$ is the *structure factor* (4.8) of the assembly. Each of these functions is to be evaluated for the value of the *scattering vector* (4.3), i.e.

$$\mathbf{q} = \mathbf{k}' - \mathbf{k}. \tag{10.13}$$

From the *Boltzmann equation* of elementary transport theory, we deduce the formula

$$\sigma \equiv 1/\rho = \tfrac{1}{3}j_F^2\tau\mathcal{N}(\mathcal{E}_F) \tag{10.14}$$

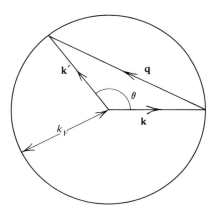

Fig. 10.1. Scattering geometry.

for the *electrical conductivity*. The *relaxation time*, τ, is the inverse of an average of the scattering probability (10.10);

$$\frac{1}{\tau} = \int (1 - \cos\theta)\,\mathcal{Q}(\theta)\mathrm{d}\Omega \tag{10.15}$$

over the whole solid angle. The factor

$$(1 - \cos\theta) = 2(q/2k_F)^2 \tag{10.16}$$

strongly weights the integral towards large angles of scatter, near the upper limit, $2k_F$, of the scattering vector q (fig. 10.1).

Putting together (10.12)–(10.16), we obtain the basic *NFE formula* for the electrical resistivity of a metallic liquid (Bhatia & Krishnan 1948; Gerstenkorn 1952; Ziman 1961):

$$\rho_L = \frac{3\pi}{\hbar} \frac{1}{j_F^2} \frac{1}{N} \int_0^1 |v(q)|^2 S(q)\,4\left(\frac{q}{2k_F}\right)^3 \mathrm{d}\left(\frac{q}{2k_F}\right). \tag{10.17}$$

Within the same theoretical framework we can also treat (10.17) as a function $\rho(\mathscr{E})$ of the position of the Fermi level, and hence calculate the *absolute thermoelectric power*

$$Q = \frac{\pi^2}{3} \frac{k^2 T}{|e|\mathscr{E}_F}\xi, \tag{10.18}$$

where

$$\xi \equiv -\mathscr{E}_F \left[\frac{\partial \ln \rho(\mathscr{E})}{\partial \mathscr{E}}\right]_{\mathscr{E}=\mathscr{E}_F} \tag{10.19}$$

includes the variation of the Fermi radius and Fermi current with a hypothetical change of $\mathscr{E}_F$.

Since these properties can be measured experimentally for a variety of materials under a wide range of pressures, temperatures, etc. the basic NFE formulae (10.17) and (10.18) provide a powerful means of testing the theory of electrons in disordered systems (see e.g. Faber 1972). But to a critical mind the above derivation must seem too naive and bald to be entirely convincing. We now embark upon an extensive discussion of the theoretical significance of the symbols that appear in (10.17) and the conditions under which this direct and apparently unambiguous formula may be assumed to be approximately valid.

10.2 Screened pseudopotentials

To make sense of a formula such as (10.17), we must know the physical magnitudes of the various factors. For k_F and j_F, we refer to (10.5) and (10.7), since the density of conduction electrons is usually known for any metallic liquid from the effective valency of the ions (cf. § 10.10). We also have direct measurements of the structure factor $S(\mathbf{q})$ by X-ray or neutron diffraction (chapter 4), or we may use some standard model formula such as the PY formula (2.46) for the pair correlation function in a hard-sphere liquid. But the proper choice of the 'atomic potential' whose Fourier transform $v(\mathbf{q})$ appears in the atomic form factor $|v(\mathbf{q})|^2$ is a much more subtle question, whose complete answer would take us deep into the general theory of the metallic state. In the present context we can only hint at the subtleties of the problem.

In the first place, a metal is a *many body system,* where the mutual interactions of the conduction electrons are just as important as their individual interactions with the metallic ions. The main effect of the *electron–electron interaction* is to *screen* all the electrostatic forces in the system. Thus, the long-range Coulomb interaction,

$$v_b(\mathbf{r}) \approx -Ze^2/r, \qquad (10.20)$$

between each electron and the charge $Z|e|$ of each ion of valency Z, is reduced to a short-ranged *screened Coulomb interaction*

$$v_s(\mathbf{r}) \approx -\frac{Ze^2}{r} \exp(-\lambda r), \qquad (10.21)$$

where the *screening length,* $1/\lambda$, is easily calculated from

$$\lambda^2 = 4\pi e^2 \mathcal{N}(\mathscr{E}_F). \qquad (10.22)$$

This, of course, is a crude description of the results deduced by the *linear response* method, where the effects of screening are all contained in a *dielectric function,* $\varepsilon(q)$, depending on the wave number of the field to be screened. Strictly speaking, this function should also depend upon the frequency; but all dynamical motions of the ions in a liquid metal are so slow compared with electron response times (e.g. plasma frequencies) that we may use the *static dielectric function* without risk of error.

For small values of q, the dielectric function always takes the form

$$\varepsilon(q) \approx 1 + \lambda^2/q^2. \tag{10.23}$$

In the momentum representation, the coulomb potential (10.20) takes the form

$$v_b(q) \approx -4\pi ZNe^2/q^2. \tag{10.24}$$

In the linear response approximation, we calculate the screened potential

$$v_s(q) = v_b(q)/\varepsilon(q)$$
$$\approx -4\pi ZNe^2/(\lambda^2 + q^2) \tag{10.25}$$

which is indeed the Fourier transform of (10.21).

If each metal ion were simply a point charge we could use (10.25) for the form factor in the NFE formula (Gerstenkorn 1952). This is obviously a poor approximation, but for small values of q it correctly describes the integrand in (10.17). In this limit, the structure factor $S(q)$ is given by the classical formula (4.22), as if the liquid were a continuous medium with compressibility κ_T. The NFE formula is thus consistent with a continuum model, in which electron scattering is analogous to the optical scattering (§ 4.4) by variations of potential associated with density fluctuations in the liquid. For the electron case, the *deformation potential* γ in (4.24) is just the long-wave limit of (10.25), i.e.

$$\gamma = v_s(q) \to ZN/\mathcal{N}'(\mathscr{E}_F) \approx \tfrac{2}{3}\mathscr{E}_F. \tag{10.26}$$

In other words, a part of the electrical resistance of any metallic liquid may be ascribed (qualitatively!) to a *plasma resistance* (Ziman 1961) which depends only on the electron density and which does not depend on the details of the arrangement of the ions or on their 'chemical' characteristics.

For shorter wavelengths, comparable to the interatomic distance, it is quite wrong to replace the 'bare' potential of an ion, $v_b(\mathbf{r})$, by its long range coulomb potential (10.20). But since the core regions of neighbouring ions never really overlap, we may still write the total *bare* potential as a superposition,

$$\mathscr{V}_b(\mathbf{r}) = \Sigma_i v_b(\mathbf{r} - \mathbf{R}_i). \tag{10.27}$$

In the linear response approximation the effective matrix elements of the total *screened* potential are of the form (10.12), with atomic form factor

$$|v_s(q)|^2 = |v_b(q)/\varepsilon(q)|^2. \tag{10.28}$$

The electron–electron interaction simply replaces the bare potentials of the separate ions by the corresponding screened potentials, v_s, with Fourier components defined by (10.28). Indeed the superposition property (10.27) is preserved by these linear transformations so that we can express the *total* screened potential in real space as if it were a sum

$$\mathcal{V}_s(\mathbf{r}) = \Sigma_i v_s(\mathbf{r} - \mathbf{R}_i) \tag{10.29}$$

of independent potentials centred on the various ions.

But the function $v(q)$ that is to appear in the NFE formula (10.17) cannot really be anything like the self-consistent one-electron potential of a screened ion. By definition, such a potential must have as many bound eigenstates as there are electrons in the ion core. Although such core states are already occupied and hence inaccessible to the conduction electrons, they must certainly figure amongst the solutions of the one-electron Schrödinger equation in each atomic sphere. But in (10.10) we are using the Born approximation to describe the scattering of electrons by an assembly of such objects at the Fermi level $\mathscr{E}_F$. The deep core states do not permit this approximation, even for scattering from a *single* atom or ion.

In the theory of electronic band structure, the NFE formalism is often preserved by introducing a *pseudopotential*, $u(\mathbf{r})$, which simulates the effects of the true potential $v(\mathbf{r})$ on the electrons in the conduction band. In calculating the electrical resistivity, therefore, we replace the Fourier transform $v(q)$ of the bare or screened potential of an ion by the corresponding matrix element, $u(q)$, of the corresponding pseudopotential. Since the pseudopotential is a weak function, constructed so as to have no bound states, the Born approximation (10.10) for the transition rates between plane wave states is presumably valid.

Suppose, for example, that we solve the Schrödinger equation (without external boundary conditions) at energy $\mathscr{E}$ in the potential $v(\mathbf{r})$. Since this potential has a number of deep bound states, the wave function $\psi(\mathbf{r})$ must have several nodes within the *core radius* R_c (fig. 10.2). Suppose now we solve the same equation in a potential $u(\mathbf{r})$ that is identical with $v(\mathbf{r})$ for $r > R_c$, but which is otherwise much 'weaker'. The corresponding solution, $\phi(\mathbf{r})$, need have no such nodes; and by adjustment of $u(\mathbf{r})$ within the core region, we can make $\phi(\mathbf{r})$ and $\psi(\mathbf{r})$ join smoothly at R_c, and coincide thereafter. For an electron of this energy, from the outside the pseudopotential $u(\mathbf{r})$ will seem equivalent to the true potential $v(\mathbf{r})$. We see, moreover, that the *pseudo-wave function* $\phi(\mathbf{r})$ now looks very much like the simple plane waves (10.2) of a free electron. Thus, the *a priori* representation of the

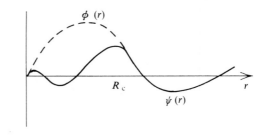

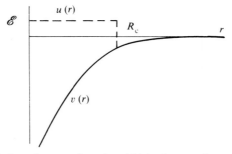

Fig. 10.2. Pseudo-wave function $\phi(\mathbf{r})$ in the pseudopotential $u(\mathbf{r})$ joins smoothly at R_c on to true wave function $\psi(\mathbf{r})$ in true atomic potential $v(\mathbf{r})$.

metal as an assembly of deep atomic potentials interacting very strongly with the conduction electrons is replaced schematically by the NFE model of an assembly of atomic pseudopotentials weakly perturbing the pseudo-plane waves of a gas of free electrons.

This artifice also meets the objection that the linear response approximation (10.28) is not valid in the rapidly varying potential within the core of an ion. We transform the bare ion potential $v_b(\mathbf{r})$ into a *bare ion pseudopotential* $u_b(\mathbf{r})$ *before* carrying out the dielectric response calculation. Since $u_b(\mathbf{r})$ can be made a smooth, weak function within R_c, its Fourier transform $u_b(\mathbf{q})$ must be quite small for large values of q, and the corresponding *screened pseudopotential* must have nice small matrix elements

$$u_s(q) = u_b(q)/\varepsilon(q), \tag{10.30}$$

which we put in place of $v(q)$ in (10.17). Since a bare ion looks like the valence charge at large distances, the bare ion pseudopotential $u_b(\mathbf{r})$ must be of the same form (10.20) for $r \gg R_c$ (fig. 10.3). We readily verify, from (10.23) and (10.30), that $u_s(q)$ looks like (10.25) for small q, so that in real space the screened pseudopotential $u_s(\mathbf{r})$ would behave like a screened

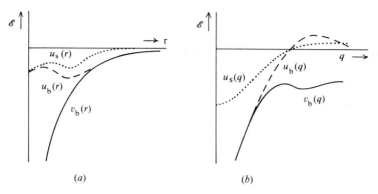

Fig. 10.3. Bare and screened pseudopotentials: (*a*) in real space; (*b*) in reciprocal space.

coulomb interaction (10.21) at large distances. Thus, the concept of plasma resistance remains unchanged in the pseudopotential representation.

Dielectric screening is actually produced by a redistribution of density in the electron gas. Approximate Hartree self-consistency is achieved by surrounding each bare positive ion by a 'cloud' of (negative) electron charge. Clouds on neighbouring ions may overlap and interpenetrate freely and when an ion is moved it carries its own cloud with it like a halo. The total negative charge in each cloud is exactly equal to the ionic charge $+Z|e|$, so that the metal always appears electrostatically neutral beyond one or two interatomic distances and there are no large residual electric fields that could produce radical redistribution of the conduction electron gas. General principles suggest that the radial distribution of electron density in each charge cloud must resemble the probability density of the valence electrons in the bound states of a free neutral *atom* of the same chemical species. Thus, if we were to write u_s in place of v_s in (10.29), we would see the system (fig. 10.4) as an assembly of quasi-independent *neutral pseudo-atoms* (Ziman 1964).

The screened pseudopotential formulation of the NFE model is a valuable 'zeroth order approximation' to the very difficult problem of calculating the electronic properties of metals. In one or another of its many variants it is often taken as a basis for quantitative theories of band structure, electronic transport, cohesion, etc. in both ordered and disordered systems (see e.g. Heine 1970; Cohen & Heine 1970; Heine & Weaire 1970). In the present context, where we are mainly concerned with the effects of structural disorder on these properties, it is very convenient to

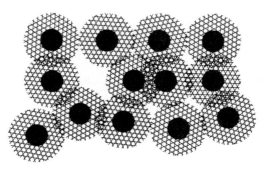

Fig. 10.4. Neutral pseudo-atoms.

appeal to the 'pseudopotential concept' and to assert that the potential to be used in the Schrödinger equation (10.1) can be written in the form (10.11), where each 'atomic potential' $v(\mathbf{r})$ can be treated as a small perturbation on a free electron system. The NFE formula (10.17) for the electrical resistivity and many more complicated formulae for other properties of liquid metals (see § 10.5) are based upon this assumption that a perturbation expansion in powers of $v(q)$ (e.g., the series of which the 'golden rule' (10.10) is the leading term) can always be made to converge.

For many simple metals this is not too far from the truth. But it is very important to keep in mind the fundamental limitations of the pseudopotential method (see e.g. Ziman 1971a), which render it almost useless as a practical computational technique for a wide variety of quite familiar metallic systems.

In the first place, *pseudopotentials are not uniquely defined.* Thus, in fig. 10.2 we may simulate the given potential $v(\mathbf{r})$ by *any* potential $u(\mathbf{r})$ whose pseudo-wave function $\phi(\mathbf{r})$ has the same logarithmic derivative at R_c as the true wave function $\psi(\mathbf{r})$. It is natural to choose a simple form of *model potential,* such as a constant value within R_c, for which the Schrödinger equation can easily be solved, but any other functional form is theoretically allowed. Although this arbitrariness would eventually be eliminated from the answers to real physical questions – for example, by summing a convergent perturbation series containing all the matrix elements of the pseudopotential to all orders – it is highly prejudicial to good mathematical order and discipline not to be dealing with invariant objects in the algebraic formulae.

It is well known, also, that *pseudopotentials are non-local operators.* This comes out very clearly in the derivation of *analytical pseudopotentials,* where use is made of the fact that an electron state function in the conduction band must be *orthogonal* to all the bound states of the ion cores. The pseudopotential then contains an operator that projects the bound state functions χ_b out of the wave function ψ, leaving a pseudo-wave function ϕ that can be expanded in simple plane waves. But such an operator cannot, in general, be represented by a local function $u(\mathbf{r})$. Similarly, in the model potential approach, the solution of the Schrödinger equation for a given value of $\mathscr{E}$ is not unique and the model potential whose pseudo-wave function is a good match to, say, an *s*-like wave function $\psi_0(\mathbf{r})$ need not be the same as the model potential that one would choose for a state $\psi_l(\mathbf{r})$ of higher angular momentum. In other words, the pseudopotential must contain operators that are sensitive to the rotational symmetry of the pseudo-wave function on which it is to act.

The fact that *pseudopotentials are energy-dependent* is obvious from the algebraic form of most expressions for analytical pseudopotentials. Similarly, in the scheme of fig. 10.2, there is no guarantee that the model potential $u(\mathbf{r})$ whose pseudo-wave function ϕ matches the true-wave function ψ at a given energy $\mathscr{E}$ must have the same property at any other energy. The best we can do is to contrive a form of $u(\mathbf{r})$ that is approximately invariant over the range of energies in which we are interested.

These deficiencies in the pseudopotential concept become intolerable in materials where the conduction electrons in a free band cannot be isolated from other electron states of the atom or ion. In the *transition metals,* for example, the atomic *d*-levels are not fully occupied and contribute electrons to the bands near the Fermi level. Such levels cannot be treated as core states to be eliminated by orthogonal projection from a pseudo-plane wave *s*-band, but must be included explicitly in the model. That is to say, the pseudopotential operator becomes so strongly dependent on energy and angular momentum that it cannot meet the fundamental conditions for the convergence of perturbation expansions for scattering, etc. in the NFE model.

Phenomenologically, such systems can be dealt with by postulating the existence of a tight-bound *d-band,* of the type discussed in §§ 8.1 and 9.1, that crosses and *hybridizes* with a nearly free s-band. Many of the characteristic physical properties of the transition metals and their alloys can be explained satisfactorily by this familiar *two-band model.* Indeed, a high proportion of the work on the *tight binding alloy* model, discussed at length in chapter 9, is intended to refer to systems of this kind, although the model is usually further simplified by reducing the five *d*-states of the atom to a single atomic state and ignoring any interactions with the *s*-band.

It must be emphasized, however, that this formulation of the electronic properties of transition metals, noble metals and quite a number of other metallic systems has not been put on a firm mathematical basis. Attempts to derive the band structure from first principles, using the LCAO representation (8.10) for the *d*-band, have not been successful quantitatively. For crystalline materials, it is possible to derive the two-band model from basic principles in the form of a semi-empirical *model Hamiltonian* whose matrix elements can be chosen to fit the band structure (see e.g. Ziman 1971*a*), but these coefficients cannot easily be calculated from the atomic potentials or state functions. There is no reason to suppose that the same coefficients would be appropriate for a disordered system, such as a liquid metal, where local symmetries and interatomic distances are not quite the same as in the regular crystal. The virtues of the two-band model are those of a very simple qualitative description, and cannot be raised to a higher level of quantitative precision.

10.3 Muffin-tin potentials

In the theory of electronic band structure, the fundamental limitations of the pseudopotential concept are avoided by representing the crystal as an array of *muffin-tin potentials* (fig. 10.5). Each ion is supposed to be at the centre of a sphere, within which the potential is spherically symmetrical. These spheres never overlap and the potential in the *interstitial regions* is supposed to be constant.

Physically speaking, this representation can be considered to be no more than an empirical approximation, constructed practically by a succession of compromises. The screening problem must first be dealt with. Given, say, the self-consistent Hartree–Fock electron-wave functions of the free atoms, one must build up a suitable one-electron potential function $\mathscr{V}(\mathbf{r})$ within the condensed material. To avoid the enormous effort and expense of going through subsequent cycles of self-consistency, it is desirable to use the best possible recipe from the very beginning and not to rely on, say, linear dielectric screening (§ 10.2) in the ion cores. All that need be said here is that the art of constructing such a potential 'from first principles' has been mastered in practice, although it is not well understood in all theoretical rigour.

This potential must then be dissected into *muffin-tin wells*. Naturally enough, the *muffin-tin radius* R_{MT} is chosen to be as large as possible without the spheres overlapping. For a regular crystal this is easy; the

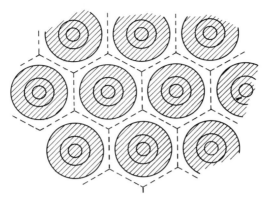

Fig. 10.5. Muffin-tin potentials.

inscribed sphere of each Wigner–Seitz cell (fig. 1.1(a)) will do. In a topologically disordered assembly, however, where the Voronoi polyhedra (fig. 2.42) are not identical, this raises difficulties. For systems such as metallic liquids, which can be represented quite well as random close packed assemblies of hard spheres (§§ 2.11, 6.7), the value of R_{MT} is well-defined, but in the limit of gas disorder (§ 2.15) the muffin-tin representation would not be a satisfactory approximation to the total potential (see § 13.4). In other words, in the absence of lattice translational symmetry (2.1), we rely upon the 'atomicity' (2.2) of the potential as a special property that is not to be found in the most general type of random medium (§ 3.1).

Finally, the potential within each sphere must be smoothed into a spherically symmetric potential $v_{MT}(\mathbf{r})$, and the *muffin-tin zero* $\mathscr{E}_{MTZ}$ defined by averaging $\mathscr{V}(\mathbf{r})$ over the interstitial region. These distortions and simplifications introduce errors which can be corrected computationally in the calculation of Bloch states (see e.g. Ziman 1971a) but which are almost always neglected in the much more complex algebra and geometry of a disordered system. Unfortunately, this approximation is unjustified in a wide variety of topologically disordered materials, such as tetrahedral glasses (§ 2.8), where the potential distribution in the interstitial regions may be far from locally uniform (fig. 10.6). It must also be remembered that long-range charge shifts associated with the deformation potential (10.26) are entirely ignored in the muffin-tin model, which is thus inconsistent with the continuum interpretation of plasma resistance in liquid metals.

Mathematically speaking, the great advantage of a muffin-tin model is that the whole space of the specimen is sharply divided into distinct regions,

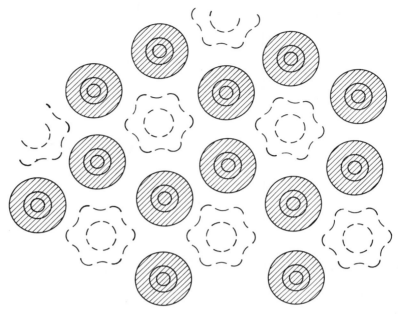

Fig. 10.6. Potential in interstitial regions of tetrahedrally coordinated material is not well represented by muffin-tin model.

in each of which the Schrödinger equation can be very easily solved. Within each muffin-tin we exploit the spherical symmetry of $v_{MT}(\mathbf{r})$ by using an *angular momentum representation*; at any chosen energy $\mathscr{E}$, we can write the wave function in the form

$$\psi(\mathbf{r}) = \Sigma_L A_L \mathscr{R}_l(r; \mathscr{E}) Y_L(\hat{\mathbf{r}}), \qquad (10.31)$$

where the coefficients A_L are arbitrary. This requires simply that we solve the *radial Schrödinger equation* for the *radial function* $\mathscr{R}_l(r; \mathscr{E})$ in the potential $v_{MT}(\mathbf{r})$ for each value of the total angular momentum l, and multiply by a *spherical harmonic* $Y_L(\hat{\mathbf{r}})$ with an appropriate value of magnetic quantum number m. In general this sum would run over all allowed pairs of $(l, m) \equiv L$; in practice only a few of the lowest values of $l = 0, 1, 2, \ldots$ need be considered. In the interstitial region, on the other hand, where the potential is flat, a representation in terms of plane waves, as in (10.2), is appropriate.

These features are fully exploited in the APW and KKR methods of band structure calculation (see e.g. Ziman 1971*a*) where functions of the two types are matched across the boundary of each muffin-tin sphere. Our

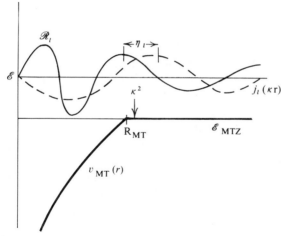

Fig. 10.7. Free-space solution $j_l(\kappa r)$ must be shifted in phase by amount η_l to match radial solution $\mathscr{R}_l(r)$ at boundary of muffin-tin well.

confidence in the reliability of muffin-tin potentials calculated from, say, self-consistent atomic orbitals rests almost entirely on the quantitative accuracy of 'first principles' band-structure computations using these methods.

Since each muffin-tin well is an isolated spherically symmetric potential, $v_{MT}(\mathbf{r})$, of finite range R_{MT}, immersed in a uniform medium of constant potential $\mathscr{E}_{MTZ}$, it can be characterized by its *scattering phase shifts* $\eta_l(\mathscr{E})$. In some respects, these are three-dimensional analogues of the phase shifts (8.30) produced by successive cells of a linear chain. Thus (fig. 10.7), the potential $v_{MT}(\mathbf{r})$ shifts the phase of a spherical wave, of angular momentum l and energy

$$\mathscr{E}=\mathscr{E}_{MTZ}=\kappa^2, \qquad (10.32)$$

by an amount $\eta_l(\mathscr{E})$. This means that the radial function $\mathscr{R}_l(r; \mathscr{E})$ can be extended into the interstitial region as a phase-shifted combination of spherical Bessel functions

$$\mathscr{R}_l(r; \mathscr{E})=j_l(\kappa r)-\tan\eta_l(\mathscr{E})\, n_l(\kappa r); \qquad r>R_{MT}. \qquad (10.33)$$

The phase shift is thus defined algebraically by matching logarithmic derivatives at the surface of the muffin-tin sphere, i.e.

$$\tan \eta_l(\mathscr{E})=\frac{\mathscr{R}_l(R_{MT}; \mathscr{E})\, j_l'(\kappa R_{MT})-\mathscr{R}_l'(R_{MT}; \mathscr{E})\, j_l(\kappa R_{MT})}{\mathscr{R}_l(R_{MT}; \mathscr{E})\, n_l'(\kappa R_{MT})-\mathscr{R}_l'(R_{MT}; \mathscr{E})\, n_l(\kappa R_{MT})}. \qquad (10.34)$$

In practice it is very easy to compute the solution of the radial Schrödinger equation in any given potential and, hence, to calculate the phase shifts.

The connection between the phase shift representation and the pseudo-potential concept can be understood by comparing figs. 10.2 and 10.6. The quantity $\eta_l(\mathscr{E})$ defined by (10.34) is arbitrary up to the addition of any integral multiple of π. In practice we should automatically choose a value within a standard range, such as $0 \leqslant \eta_l(\mathscr{E}) \leqslant \pi$. But this is equivalent to ignoring all the phase changes associated with the interior oscillations of $\mathscr{R}_l(r; \mathscr{E})$ – i.e. all the phase angle that would have been drawn into the muffin-tin sphere if we had supposed the potential $v_{MT}(\mathbf{r})$ to have grown continuously from zero over a period of time. In other words, we use 'reduced' phase shifts that might have been produced by the weak pseudo-potential $u(r)$ of fig. 10.2, for which the radial function would have no inner nodes.

The reduced phase shifts are genuine mathematical invariants of the muffin-tin potential. In the language of advanced quantum theory, the *t-matrix* of the muffin-tin well has diagonal elements

$$t_l = -\kappa^{-1} \sin \eta_l \exp i\eta_l \tag{10.35}$$

in the angular representation, which is an irreducible representation of the rotation group under which the potential is symmetrical. The transition rate for elastic scattering of plane waves by such an object is given *exactly* by the 'golden rule' (10.10), with *scattering amplitude*.

$$t(q) = -(4\pi N/\kappa)\Sigma_l(2l+1)\sin \eta_l \exp i\eta_l \, P_l(\cos \theta) \tag{10.36}$$

in place of a matrix element $v(q)$ of the potential. This expression is the sum, to all orders, of the perturbation series of which the Born approximation is merely the leading term.

For this reason, it is very tempting (Evans, Greenwood & Lloyd 1971) to use this exact and unambiguous formula for the atomic form factor in the NFE expressions for electrical resistivity, etc. The deficiencies of the simple muffin-tin model in the long-wave limit are more than compensated by replacing $v(q)$ in the integral (10.17) by the t-matrix (10.36), i.e.

$$\rho_L = \frac{3\pi}{\hbar} \frac{1}{j_F^2} \frac{1}{N} \int_0^1 |t(q)|^2 S(q) \, 4(q/2k_F)^3 \, \mathrm{d}(q/2k_F), \tag{10.37}$$

where, of course, $t(q)$ is evaluated for $\kappa^2 = \mathscr{E}_F$ and $q = 2k_F \sin \tfrac{1}{2}\theta$.

This plausible substitution extends the NFE formula to materials such as liquid transition metals where the pseudopotential concept is no longer valid. The essential point is that band-structure calculations based on muffin-tin potentials, such as the APW and KKR methods, work perfectly well for such metals without any mention of 'bound states' or '*s–d* hybridization'. The whole of the band structure – '*s*-bands' and '*d*-bands' alike – is dependent on the behaviour of the phase shifts. The characteristic feature of the transition metals – narrow *d*-bands crossing a broad NFE *s*-band – is derived entirely from a narrow *d-resonance* of the muffin-tin wells (see e.g. Ziman 1971*a*) located near the energy $\mathscr{E}_d$ of the *d*-states of the original atoms. Quantitative 'first principles' calculations confirm that the phase shift for $l=2$ passes rapidly through $\pi/2$, closely following the standard resonance formula

$$\eta_2(\mathscr{E}) \approx \tan^{-1} \frac{\frac{1}{2}\Gamma}{\mathscr{E}_d - \mathscr{E}}. \tag{10.38}$$

This *resonance representation* of the electronic structure of transition metals is physically and mathematically equivalent to the two-band model (§ 10.2). But it has the great advantage of conceptual unity and practical computability. It turns out, for example, that the width of the *d*-bands is governed mainly by the *resonance width* Γ in (10.38). This, in turn, is determined by the precise shape of the muffin-tin potential $v_{\mathrm{MT}}(\mathbf{r})$ and thus, *a priori*, by the self-consistent field of the atom/ion at this site. Values of $\mathscr{E}_d$ and of Γ for various metals can be calculated from first principles and checked by comparison of computed band structures with experimental Fermi surfaces.

The *t-matrix formula* (10.37) can thus be tested directly against experiment for a variety of liquid metals (see e.g. Busch & Guntherodt 1974). Various characteristic features of the formula, such as the large contribution to the thermoelectric power (10.19) from the strong energy-dependence (10.38) of the *t*-matrix (10.36), are open to quantitative confirmation (or otherwise). Although we have sworn not to discuss the results of such research, much of our interest in this sort of formula is occasioned by the fact that it does predict physically measurable quantities in terms of known parameters.

The resonance representation also throws new light on the theory of *transition metal alloys*, whether substitutionally disordered on a lattice or topologically disordered as in the liquid state. Despite its interest as a well-defined mathematical system, the TBA model (§ 9.1) is not a satisfactory description of the electronic structure of real transition metal alloys, where the complexities of '*s–d* hybridization' cannot be ignored. It is highly desirable, therefore (§ 10.7), to look again at such systems from the muffin-tin resonance point of view.

The choice between the pseudopotential and *t*-matrix formulations depends on the nature of the self-consistent potential $\mathscr{V}(\mathbf{r})$. Little progress can be made towards a theoretical connection between the electronic properties and atomic arrangement of a given material unless this potential is approximately 'atomic' in the sense of (2.2). To make use of diffraction information about the atomic arrangement (chapter 4) we are practically bound to assume that $\mathscr{V}(\mathbf{r})$ can be expressed as the sum of independent 'atomic' potentials, each spherically symmetrical about the nucleus of an atom. In general, however, an expression of the form (10.11) is only a rough approximation to the true physical description of the potential, and the 'atomic' potential functions $v(\mathbf{r} - \mathbf{R}_i)$ may not be uniquely defined. When, in fact, these potentials can be chosen so that they never overlap (fig. 10.5) we say that we have a muffin-tin potential, where there are decisive mathematical advantages in the canonical angular momentum/phase shift representation. In many cases, however, it would be physically misleading to exclude the effects of superposing contributions from neighbouring 'atomic' potentials at different distances or in different orientations (fig. 10.3). In such cases, the flexibility and lack of rigour of pseudopotential concepts, applied to a linear superposition formula such as (10.29), facilitate the development of interesting mathematical expressions which are by no means unconnected with reality.

It must be remembered, however, that the elementary NFE model for disordered metals, whether in its pseudopotential or t-matrix forms, is little better than a phenomenological fudge. Having indicated the main features of the potentials in which the electrons are supposed to move, we must search for more rigorous justifications of the formulae for their observable physical properties.

10.4 The electron spectrum

The NFE phenomenology is based upon the *ansatz* that the electron states are analogous to the plane wave eigenfunctions (10.2) of a highly degenerate gas of non-interacting free particles. This can be given a crude 'hand waving' plausibility in terms of pseudo-wave functions (§ 10.2); but we cannot ignore the effects on these functions of the atomic pseudopotentials or muffin-tin wells. Before embarking upon the calculation of more complicated physical properties such as the electrical resistivity we should determine the true *energy spectrum* $\mathscr{N}(\mathscr{E})$ of the electrons in a topologically disordered system such as a liquid metal.

For a crystalline material, the *geometry* of the system – i.e. the spatial arrangement of the atomic centres – is the dominant influence on this spectrum. Whatever the nature of the individual atomic centres, the electron waves are strongly affected by *coherent diffraction* from the lattice planes. This gives rise to splitting of the NFE energy levels whenever the wave vector **k** crosses a *zone boundary* and thus leads eventually to distortions of the *Fermi surface* and the characteristic singularities of the electronic *band structure*. As is well known, the simple NFE model breaks down completely in the neighbourhood of such singularities, and naive formulae such as (10.17) must be profoundly modified. Thus, for example, in many polyvalent systems the crystal structure may generate a low density of states or even an energy gap at the Fermi level, making the material a *semimetal* or *semiconductor*.

The question is whether there are analogous *geometrical effects* on the electron spectrum in a condensed system *without long-range order*. For example, when we melt a semimetal or semiconductor, do we expect the characteristic singularities of the band structure to persist as 'features' in the electron spectrum, such as *pseudo-gaps*? The answer to this question affects not only the calculation of the electronic transport properties but also many other physical phenomena, such as positron annihilation and the Knight shift, which depend directly on the spectral density $\mathcal{N}(\mathscr{E}_F)$ at the Fermi level.

Many characteristic features of the electronic band structure of solids arise, however, from *local* attributes of the electron potential. In a transition metal, for example, the narrow *d*-bands are associated with tight-bound *d*-levels of the individual ions (§ 10.2) or (equivalently) with strong *d*-resonances of the muffin-tin wells (§ 10.3) and are not produced solely by the long-range or short-range geometry of the atomic arrangements in the crystal. Such *non-structural* properties (Ziman 1967a) will be discussed in § 10.8; for the moment we concentrate on systems where each pseudo-atom or muffin-tin well is a weak scatterer, without actual or virtual bound states in the neighbourhood of the valence band. By restricting the discussion to such *simple metals* we can also ignore awkward questions about the convergence of perturbation expansions in powers of matrix elements of the pseudopotential.

In the conventional NFE theory of the band structure of *crystalline* metals, the (pseudo)potential $\mathscr{V}(\mathbf{r})$ is indeed treated as a perturbation on the plane-wave functions (10.2). In the notation of (10.9) and (10.12), the energy of a Bloch state of wave vector **k**, can be expanded as a *Rayleigh–Schrödinger series*:

$$\mathcal{E}(\mathbf{k}) = \mathcal{E}^0_{\mathbf{k}} + \langle \mathbf{k} | \mathscr{V} | \mathbf{k} \rangle + \Sigma_{\mathbf{q}} \frac{|\langle \mathbf{k} | \mathscr{V} | \mathbf{k}+\mathbf{q} \rangle|^2}{\mathcal{E}^0_{\mathbf{k}} - \mathcal{E}^0_{\mathbf{k}+\mathbf{q}}} + \cdots$$

$$= \mathcal{E}^0_{\mathbf{k}} + v(0) + \frac{1}{N} \Sigma_{\mathbf{q}} \frac{S(\mathbf{q})|v(\mathbf{q})|^2}{\mathcal{E}^0_{\mathbf{k}} - \mathcal{E}^0_{\mathbf{k}+\mathbf{q}}} + \cdots \qquad (10.39)$$

Dropping the average (pseudo)potential (10.9), we fix our attention on the second-order term, which represents the deviation of the *dispersion function* $\mathcal{E}(\mathbf{k})$ from the free-electron energy $\mathcal{E}^0_{\mathbf{k}}$.

For a perfect crystal, the structure factor $S(q)$, defined by (4.8), is zero unless $\mathbf{q}$ coincides with one of the reciprocal lattice vectors $\mathbf{g}$. The summation over all values of $\mathbf{g}$ gives a reasonable answer, except for regions of $\mathbf{k}$ space where one of the energy denominators can nearly vanish. Whenever we can find a particular reciprocal lattice vector $\mathbf{G}$ such that

$$\mathcal{E}^0_{\mathbf{k}} \approx \mathcal{E}^0_{\mathbf{k}+\mathbf{G}} \qquad (10.40)$$

– i.e. when $\mathbf{k}$ lies near a zone boundary – we must eliminate the degeneracy by diagonalizing a small matrix before trying to sum the perturbation series.

This complication does not arise in the *Brillouin–Wigner* perturbation expansion, where the 'perturbed' energy $\mathcal{E}(\mathbf{k})$ replaces the 'unperturbed' energy $\mathcal{E}^0_{\mathbf{k}}$ in each denominator. Near the zone boundary (10.40), the series is dominated by the term

$$\mathcal{E}(\mathbf{k}) - \mathcal{E}^0_{\mathbf{k}} \approx \frac{|v(\mathbf{G})|^2}{\mathcal{E}(\mathbf{k}) - \mathcal{E}^0_{\mathbf{k}+\mathbf{G}}}, \qquad (10.41)$$

which can be regarded as an equation to be solved for the variable $\mathcal{E}(\mathbf{k})$. It is easy to see that the free electron levels are split by $2|v(\mathbf{G})|$ at the zone boundary; in a one-dimensional system this would be seen as a jump discontinuity, corresponding to an energy gap of the type discussed in § 8.3.

In a *disordered* system the structure factor $S(\mathbf{q})$ is not restricted to the reciprocal lattice, but is a continuous function in momentum space. The Brillouin–Wigner analogue of (10.39) thus reads

$$\mathcal{E}(\mathbf{k}) \approx \mathcal{E}^0_{\mathbf{k}} + \frac{V}{8\pi^3 N} \int \frac{S(\mathbf{q})|v(\mathbf{q})|^2}{\mathcal{E}(\mathbf{k}) - \mathcal{E}^0_{\mathbf{k}+\mathbf{q}}} \, \mathrm{d}^3\mathbf{q}. \qquad (10.42)$$

Since $S(\mathbf{q})$, $v(\mathbf{q})$ and $\mathcal{E}^0_{\mathbf{k}}$ must all be spherically symmetrical functions, and the integral always converges, there is no great difficulty in treating this relationship as an integral equation which can be solved numerically (e.g. Shaw & Smith 1969) for $\mathcal{E}(\mathbf{k})$ as a function of the wave number $\mathbf{k}$.

It is tempting to regard the result of such an equation as a dispersion function for an electron of energy $\mathcal{E}$ propagating in the disordered metal. But this generalization of (10.41) is unsound in principle. We know, for example, that an actual eigenfunction of the Hamiltonian of a topologically disordered system must be regarded as a *Byzantine function* in real space – of such complexity as to defy compact mathematical description. Again, since there is no true translational symmetry, momentum is not a good quantum number and the Fourier transform of such an eigenfunction cannot have a perfectly sharp spectrum in reciprocal space. No precise meaning can therefore be given to a 'dispersion function' deduced from (10.42). The best that can be said is that $\mathscr{E}(\mathbf{k})$ applies to the eigenstates of a Hamiltonian which has already undergone ensemble averaging in the definition of its matrix elements (10.12) and is, therefore, much simpler and more homogeneous than the true Hamiltonian.

In these circumstances, the correct procedure (§ 9.2) is to calculate the average *Green function* $\langle G(\mathscr{E}) \rangle$ of the total electron Hamiltonian and to deduce the spectrum from the fundamental identity (9.10). This Green function is generated by the perturbation $\mathscr{V}$ acting on the propagator G_0 of the free-electron Hamiltonian $\mathscr{H}_0$ and is defined as in (9.19) and (9.33) by the identity

$$G = G_0 + G_0 \mathscr{V} G. \tag{10.43}$$

Because our system is statistically homogeneous and isotropic, this operator must be diagonal and spherically symmetrical in a momentum representation (9.12). Defining the *self-energy operator* Σ by

$$\langle G \rangle = \{G_0^{-1} - \Sigma\}^{-1}, \tag{10.44}$$

we find, as in (9.31), that the amplitude of $\langle G \rangle$ can be written

$$\mathscr{G}(\mathbf{k}; \mathscr{E}) = \{\mathscr{E} - \mathscr{E}_{\mathbf{k}}^0 - \Sigma(\mathbf{k}; \mathscr{E})\}^{-1}. \tag{10.45}$$

In other words, $\Sigma(\mathbf{k}; \mathscr{E})$ measures the effect of the perturbation on the eigenvalues $\mathscr{E}_{\mathbf{k}}^0$ of the free electron Hamiltonian.

Exactly as in (9.45), the self-energy can be expanded as a perturbation series in powers of $\mathscr{V}$ (§ 10.5). But excellent results can be obtained directly from the identity

$$\Sigma = \langle \mathscr{V} G \mathscr{V} \rangle \{1 + G_0 \langle \mathscr{V} G \mathscr{V} \rangle\}^{-1}, \tag{10.46}$$

which is an algebraic consequence of (10.43) and (10.44). Retaining only second-order terms in $\mathscr{V}$, and replacing the operator G by its ensemble average, we arrive at the *self-consistent internal propagator approximation* (Ballentine 1966, 1975)

$$\Sigma \approx \langle \mathscr{V} \langle G \rangle \mathscr{V} \rangle. \tag{10.47}$$

In the momentum representation (9.12), (10.8) this means that the self-energy function in (10.45) must satisfy

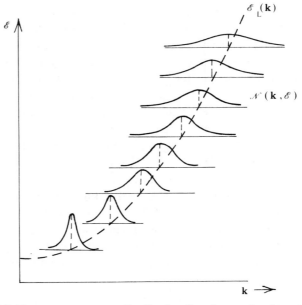

Fig. 10.8. Energy–momentum distribution function peaks along the curve $\mathscr{E}_L(\mathbf{k})$.

$$\Sigma(\mathbf{k};\,\mathscr{E}) \approx \frac{V}{8\pi^3 N} \int \frac{S(\mathbf{q})|v(\mathbf{q})|^2}{\mathscr{E} - \mathscr{E}^0_{\mathbf{k}+\mathbf{q}} - \Sigma(\mathbf{k}+\mathbf{q};\,\mathscr{E})} \, d^3\mathbf{q}, \qquad (10.48)$$

which is the correct form of (10.42).

In solving this integral equation we must remember (§ 9.3) that $\Sigma(\mathbf{k};\,\mathscr{E})$ is a complex quantity, whose real and imaginary parts are associated with shifts and broadening of the energy levels. To be precise, we construct the Green function (10.45) whose imaginary part

$$\mathscr{N}(\mathbf{k};\,\mathscr{E}) = \frac{1}{\pi}\mathrm{Im}.\ \mathscr{G}(\mathbf{k};\,\mathscr{E}+\mathrm{i}0) \qquad (10.49)$$

is the spectral density of states of energy $\mathscr{E}$ and apparent momentum $\mathbf{k}$. On the $(\mathscr{E},\,\mathbf{k})$ plane (fig. 10.8), this function looks like a ridge, as if a sharply defined dispersion function had been broadened in energy and/or momentum by incoherent scattering from the disordered assembly of atoms. Although the physical notion of a dispersion function for the electrons in a disordered metal might be represented by the trajectory $\mathscr{E}_L(\mathbf{k})$ of the peak of $\mathscr{N}(\mathbf{k};\,\mathscr{E})$, this curve has no special mathematical significance and is not, for example, the solution of (10.42) (Ballentine 1975).

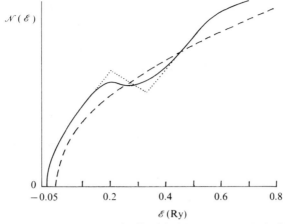

Fig. 10.9. Calculated spectrum for liquid Bi (———) compared with free electron parabola (– – – –) (Chan & Ballentine 1972). In a crystalline solid this spectrum would have van Hove singularities, e.g. (. . . . .).

The energy spectrum itself is obtained, as prescribed by (9.12), by integrating (10.49) over all values of $\mathbf{k}$. For a reasonably realistic model with a strong pseudopotential, the calculated spectrum (fig. 10.9) rises above the free electron curve and then drops below it in a manner reminiscent of the density of states for overlapping bands in a regular crystal. Going back to the integral equation (10.48), we find that the characteristic peak in the structure factor (fig. 4.1) at q_m (corresponding to constructive interference between waves diffracted from nearest neighbour atoms) gives rise to a kink in $\mathscr{E}_L(\mathbf{k})$ (fig. 10.10) in the region $\mathbf{k} \sim \frac{1}{2}q_m$. This could be interpreted as a smudged version of the discontinuity in $\mathscr{E}(\mathbf{k})$ at the corresponding zone boundary in the crystalline solid.

10.5 Many-atom scattering

The magnitude of the 'pseudo-gap anomaly' calculated for simple metals (Ballentine 1975) is disappointing. Even the strong pseudopotential of Bi, which nearly separates the bands in a crystalline model, produces only a modest dip in the spectral density of the liquid (fig. 10.9). In the absence of long-range order, the electron spectrum in a disordered metal almost reverts to the free electron spectrum and does not seem to be strongly influenced by the local geometry of atomic arrangements. The hypothesis that when a metal melts it retains a significant memory of its original

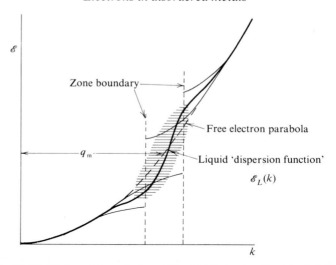

Fig. 10.10. Kink and broadening of liquid 'dispersion function' $\mathscr{E}_L(\mathbf{k})$ occurs in the region where there would be jump discontinuities of $\mathscr{E}(\mathbf{k})$ in the crystalline solid.

electronic band structure is not justified by the results of computations based on (10.48).

But this integral equation is no more than an approximation corresponding to the second-order term of a perturbation series such as (10.39) and really only takes account of the interference between waves scattered from pairs of atoms in the liquid. In (10.48), the statistical distribution of atomic centres appears only through the structure factor $S(\mathbf{q})$, which is the Fourier transform (4.9) of the pair correlation function $g(\mathbf{1}, \mathbf{2})$ of the atomic arrangement. When we take account of the *multiple scattering* of electrons by atoms we draw into the analysis the higher correlation functions, g_3, g_4, etc. as defined in (§ 2.6). The full effect of local geometry on the electrons cannot be assessed without reference to these terms in the perturbation series (Edwards 1962).

The representation of the self-energy (10.46) by an *Edwards series* of irreducible graphs is a straightforward exercise in diagrammatic analysis (see e.g. Ballentine 1975). Here again, as in §§ 5.10 and 6.4, the same formulae may be deduced algebraically by the method of *cumulants*. A direct adaptation (Ziman 1969*b*) of the derivation of the linked cluster theorem of quantum-field theory (Kubo 1962) tells us that we may write the self-energy in (10.44) as a series of the form

$$\Sigma = \sum_{n=1}^{\infty} \langle\langle \mathscr{V}(G_0\mathscr{V})^{n-1}\rangle\rangle_{\mathrm{cum}}, \tag{10.50}$$

where the usual averages over the statistical ensemble are combined into cumulant averages as in (5.167). Indeed, this is the basic theorem concerning the S-matrix of a quantum-mechanical Hamiltonian, from which the statistical mechanical analogues (5.166) and (6.138) can be deduced by the method of temperature Green functions.

In the momentum representation, each matrix element of the perturbing potential (10.11) can be written, as in (10.8), in the form

$$\langle \mathbf{k}|\mathscr{V}|\mathbf{k}+\mathbf{q}\rangle = v(\mathbf{q})\frac{1}{N}\sum_j \exp i\mathbf{q}\cdot\mathbf{R}_j$$

$$\equiv v(\mathbf{q})\sum_j X_j(\mathbf{q}). \tag{10.51}$$

In this representation, therefore, the nth term of (10.50), is the multiple integral

$$\Sigma_n(\mathbf{k};\,\mathscr{E}) = \int d^3\mathbf{q} \int d^3\mathbf{q}' \ldots \int d^3\mathbf{q}^{(n-1)}\, \delta(\mathbf{q}+\mathbf{q}'+\ldots \mathbf{q}^{(n-1)}) \times$$
$$\times v(\mathbf{q})G_0(\mathbf{k}+\mathbf{q})v(\mathbf{q}')G_0(\mathbf{k}+\mathbf{q}+\mathbf{q}') \ldots v(\mathbf{q}^{(n-1)}) \times$$
$$\times \langle \sum_j X_j(\mathbf{q}) \sum_{j'} X_{j'}(\mathbf{q}') \ldots \sum_{j^{(n-1)}} X_{j^{(n-1)}}(\mathbf{q}^{(n-1)})\rangle_{\mathrm{cum}}. \tag{10.52}$$

The ensemble cumulant average applies only to the last factor and can easily be written down in terms of the atomic position correlation functions. Consider, for example, the contribution to $\Sigma_3(\mathbf{k};\,\mathscr{E})$ from terms containing the product of factors corresponding to three distinct atoms. The integral will then acquire a factor

$$c_3(\mathbf{q},\mathbf{q}',\mathbf{q}'') \equiv N^3 \langle X_1(\mathbf{q})X_2(\mathbf{q}')X_3(\mathbf{q}'')\rangle_{\mathrm{cum}}$$
$$= \iiint g_3(1,2,3)\exp\{i(\mathbf{q}\cdot\mathbf{R}_1 + \mathbf{q}'\cdot\mathbf{R}_2 + \mathbf{q}''\cdot\mathbf{R}_3)\}d^3\mathbf{R}_1 d^3\mathbf{R}_2 d^3\mathbf{R}_3 \tag{10.53}$$

(in the notation of §2.7). Here the cumulant average is the same as the usual average, because we can drop all terms containing $\langle X_1(\mathbf{q})\rangle_{\mathrm{cum}}$. But the fourth order term will contain a more complicated factor

$$\langle\exp\{i\mathbf{q}\cdot\mathbf{R}_1 + \mathbf{q}'\cdot\mathbf{R}_2 + \mathbf{q}''\cdot\mathbf{R}_3 + \mathbf{q}'''\cdot\mathbf{R}_4\}\rangle_{\mathrm{cum}} =$$
$$= c_4(\mathbf{q},\mathbf{q}',\mathbf{q}'',\mathbf{q}''') - \{c_2(\mathbf{q},\mathbf{q}')\,c_2(\mathbf{q}'',\mathbf{q}'')\ldots \mathrm{etc}\}, \tag{10.54}$$

confirming the basic theorem on cumulants by vanishing if there are only pairwise correlations between the positions of the atoms. This definition of the nth term of the Edwards series thus singles out the genuine n-atom *clustering correlations* from the background of apparent correlations generated by mere probability products associated with independent clusters of lower order (cf. Schwartz 1973). This applies particularly at small wave numbers of any of the wave vectors $\mathbf{q}$, $\mathbf{q}'$ ... etc; these residual correlations must be vanishingly small for atoms that are a long distance apart in the liquid.

We know enough about the higher-order correlation functions for atomic positions in liquids (§§ 2.7, 2.12) to evaluate Σ_3 and Σ_4. A reasonable starting point, for example, would be the superposition approximation (2.27) which describes the behaviour of $g_3(1, 2, 3)$ fairly well except when all three atoms are close together (cf. (2.52)). Unfortunately, even this apparently simple expression leads to an integral (10.52) which does not factorize and which has not yet been evaluated numerically. It is not very difficult, of course, to write down an artificial formula for each correlation function, such as

$$g_n(1, 2 \ldots \mathbf{n}) = g_2(1, 2)g_2(2, 3) \ldots g_2(\mathbf{n}-1, \mathbf{n}), \qquad (10.55)$$

that allows this factorization to all orders. But this *chain approximation* is obviously not consistent with the superposition approximations (2.27) and (2.28) for g_3 and g_4, and does not satisfy the very important physical principle that the cluster correlation (10.54) should vanish over large separations (Ballentine 1975). The results obtained by using (10.55) are, therefore, unreliable. Without better evidence we can only assume that Σ_3 and Σ_4 do not make significant contributions to the spectral density of the electrons in metallic liquids and that unspectacular pseudo-gaps deduced from (10.48) are the full measure of the effects of local geometry on the band structure.

Nevertheless, there are circumstances in which the higher order terms of the Edwards series may be very important. Consider, for example, a *microcrystalline* specimen of a conventional semiconductor such as Si or Ge. For any one crystallite, considered on its own, the NFE Brillouin–Wigner perturbation formula (10.41) would describe correctly the band gaps produced by large values of the pseudopotential matrix element $|v(\mathbf{G})|$. But the pair distribution function (2.23) for a microcrystalline assembly is averaged over all orientations of the crystallites and must, therefore, be a series of concentric spherical shells, whose structure factor $S(\mathbf{q})$ can scarcely be distinguished from that of a typical liquid (§§ 2.6, 2.9, 4.1). Thus, if we were to calculate the electron spectrum pedantically from (10.48) we should find modest 'pseudo-gaps' where the real gaps should be. This paradox is resolved (Ziman 1969b) by taking the Edwards series as far as $\Sigma_4(\mathbf{k}; \mathscr{E})$ where the averaging over the relative orientations of *different* crystallites does not destroy the coherence of the diffraction effect due to long-range order within a crystallite. The fact that the 3- and 4-body cluster correlations in a microcrystalline assembly are *not* the same as those in a liquid and do *not* vanish at large distances thus has important consequences for the electronic properties (see § 11.5).

10.6 Scattering operators

In addition to the *many-atom* scattering terms, the cumulant expansion (10.52) for the self-energy includes cases where the coordinates of the *same* atom appear more than once in a product. Thus, for example, $\Sigma_3(\mathbf{k}; \mathscr{E})$ contains more terms than are described by the integral (10.53).

In the theory of substitutional disorder (§ 9.3), the effects of multiple scattering from the same site were dealt with by defining a *single site t-matrix* (9.25) to replace the disorder potential in the propagator series (9.24). By direct algebraic analogy we construct an operator satisfying the integral equation

$$t^i = v_i + v_i G^0 t^i, \tag{10.56}$$

which is used to eliminate the atomic potential v_i at the site $\mathbf{R}_i$ from the various perturbation expansions. The effect in the irreducible graphical series for $\Sigma(\mathbf{k}; \mathscr{E})$ is exactly the same as in the propagator series (9.26); by submitting t for v at each site we automatically take account of all *immediate* repetitions of the same site label and thus significantly reduce the number of apparently distinct diagrams to be evaluated in summing the series. As with (9.26), this device does not solve the problem of giving the correct statistical weight to terms in which the same site is revisited several times with intermediate excursions elsewhere (fig. 9.4), but it certainly helps to take account of many important terms in the series.

This procedure is obviously essential if the potential is too strong to allow the Born approximation (10.10) for the amplitude of single-site scattering. Quite apart from formal diagrammatic considerations it makes good physical sense (§ 10.3) to represent each atom by its correct scattering amplitude, including all multiple scattering effects. That is to say, the calculation of 'geometric' effects on the electron spectrum (§§ 10.4, 10.5) should be much improved by this substitution.

It is convenient for abstract algebraic manipulations to rewrite the Green function identities of § 9.2 in terms of *scattering operators* (see e.g. Gyorffy

1970; Schwarz & Ehrenreich 1971; Roth 1974). By analogy with (10.56) we define an *assembly T-matrix* satisfying

$$T = \mathcal{V} + \mathcal{V} G^0 T. \tag{10.57}$$

This is related to the assembly Green function (10.43) by the identity

$$G = G^0 + G^0 T G^0; \tag{10.58}$$

the spectral information (9.12) can thus be derived directly from the ensemble average $\langle T \rangle$. Indeed, this is approximately equal to the self energy (10.50); (9.26) is a special case of the identity

$$\Sigma = \langle T \rangle \{ \langle G \rangle \}^{-1} G^0 = \langle T \rangle \{ 1 + \langle T \rangle G^0 \}^{-1}. \tag{10.59}$$

Just as in § 9.3 we can expand (10.57) in powers of the perturbation and eliminate the individual atomic potentials in favour of single-site t-matrices (10.56). The assembly T-matrix can be written as a sum

$$T = \sum_{ij} \mathcal{T}^{ij}, \tag{10.60}$$

where each of the *scattering path operators* (Gyorffy 1972) satisfies the equation

$$\mathcal{T}^{ij} = t^i \delta_{ij} + \sum_{k \neq i} t^i G^0 \mathcal{T}^{kj}. \tag{10.61}$$

In other words, $\mathcal{T}^{ij}$ is a generalization of the usual single site t-matrix in that it measures the amplitude of the wave scattered from the atom at $\mathbf{R}_i$ when a wave is incident on the atom at $\mathbf{R}_j$.

It is easy to see that (10.61) generates a series expansion of the form (9.26) where the same index cannot immediately repeat itself. The crudest approximation (Stern 1973) is to take the leading term of this series, whose ensemble average is just the average atomic t-matrix $\langle t^i \rangle$. Interpreting (10.59) in a momentum representation we see that the self energy approximates to the sum of the traces of these matrices; the system behaves as a continuous medium with an 'effective potential'

$$\Sigma \approx N \bar{t}(0) \tag{10.62}$$

equal to the total forward scattering amplitude of the constituent atoms (Lax 1952).

This uniform correction to the energy zero of the spectrum is akin to the virtual crystal approximation (9.15) in the theory of substitutional disorder. To improve upon it, let us introduce the symbol

$$\mathcal{T}_i = \sum_j \mathcal{T}^{ij} \tag{10.63}$$

which stands for the total amplitude emitted from site i as a consequence of a wave incident on this and all other sites. We propose now to take an ensemble average of (10.61), which is summed to read

$$\mathcal{T}_i = t^i + t^i G^0 \sum_{j \neq i} \mathcal{T}_j. \tag{10.64}$$

But the interpretation of symbols like $\langle T \rangle$ requires some care. Strictly speaking, we are interested in *conditional averages*, such as

$$\langle \mathcal{T}_1 \rangle_1 \equiv \int \mathcal{T}_1 \, g(1, 2 \ldots N) \, d2 \ldots dN, \tag{10.65}$$

where (in the notation of § 2.7), the position of the central atom **1** is kept fixed whilst all other atoms may be found in all configurations with relative probability $g(1, 2 \ldots N)$. Carrying out this operation on (10.64), and using the normalization identities of the atomic distribution functions $g(1, 2)$, etc., we get

$$\langle \mathcal{T}_1 \rangle_1 = t' + t' G^0 (N-1) \int g(1, 2) \langle \mathcal{T}_2 \rangle_{21} \, d2. \tag{10.66}$$

This exact formula is not, unfortunately, a closed integral equation, since the symbol $\langle \mathcal{T}_2 \rangle_{21}$ is defined as the average of the operator (10.63) with both the central atom 2 and another atom 1 in fixed positions. If we were to seek an expression for this average, then we should encounter an integral containing $g(1, 2, 3)$ and $\langle \mathcal{T}_3 \rangle_{321}$ – and so on. Just as in the analytical theory of liquids (§ 2.12), this hierarchy of equations can only be topped by a *closure assumption* such as the superposition approximation (2.17) used in the BBGKY theory. Strictly speaking, the spectrum of electronic excitations must depend on the atomic distribution functions to all orders, as it does explicitly in the Edwards series (10.50).

But the simple and natural assumption (Lax 1952) that $\langle \mathcal{T}_2 \rangle_{21}$ is seldom different from $\langle \mathcal{T}_2 \rangle_2$ leads at once to the *quasi-crystalline approximation* (QCA):

$$\langle \mathcal{T}_1 \rangle_1 \equiv \mathcal{T}(1) = t^1 + t^1 G_0 N \int g(1, 2) \mathcal{T}(2) \, d2 \tag{10.67}$$

which could be solved for the average scattering operator and hence for the self energy (10.59) and the electron spectrum. This approximation is the equivalent, for a topologically disordered system, of the average t-matrix approximation (9.27) for the TBA model (Gyorffy 1970) and gives surprisingly good results for a one-dimensional 'liquid' (Peterson, Schwartz & Butler 1975). But the form of the integral equation (10.67) is only schematic and no solutions have been computed for any realistic model of a metallic liquid. It is not known, therefore, whether the results of solving it would be significantly better than would be obtained rather more directly from, say, the second-order perturbation expression (10.48) with t-matrices in place of Fourier components of the potential.

Naturally enough, considerable theoretical effort has been expended in the search for an improved theory in which the propagators and t-matrices are defined self-consistently, in the spirit of the coherent potential approximation (§ 9.4). But the problem is much more difficult than it is in the alloy model because the *medium propagator* $\tilde{G}$ and the effective site t-matrix $\tilde{t}$ cannot be related to one another so simply as in (9.46)–(9.48). In the *effective medium approximation* (EMA) of Roth (1974), these are defined so as to satisfy an 'averaged' version of (10.61), i.e.

$$\mathcal{T}(1, 2) = \tilde{t}(1)\{\delta(1, 2) + \int \tilde{G}(1, 3)\mathcal{T}(3, 2)\,\mathrm{d}3\}, \tag{10.68}$$

where the scattering path operators $\mathcal{T}(1, 2)$, etc. are themselves averages like (10.65). Climbing the hierarchy of identities like (10.66) to the next level, and making the superposition approximation (2.17) for the three-atom distribution function, one can write down two more self-consistency conditions from which the various unknown functions could, in principle, be deduced. Essentially the same level of approximation is achieved by Movaghar, Miller & Bennemann (1974) and Yonezawa & Watabe (1975). In diagrammatic language (Yonezawa, Roth & Watabe 1975) EMA, like CPA, takes into account all possible single-site diagrams and their corrections. But since nothing at all is known about the numerical solution of these equations, the intrinsic value of this development cannot be assessed. The application of this formalism to *tight-binding* models of topologically disordered systems (Movaghar & Miller 1975; Roth 1975) must also be considered somewhat academic, except in so far as it throws light on some of the problems associated with clusters and short-range order in the alloy problem (§ 9.5) and on substitutional systems with off-diagonal disorder (§ 9.8).

10.7 Partial-wave representations

For all its algebraic elegance, the t-matrix formalism of § 10.6 hides a paradox. Let us suppose that the assembly potential is of *muffin-tin* form (§ 10.3); each atomic potential $v_i(\mathbf{r})$ is spherically symmetrical within a sphere of radius R_{MT} that does not overlap its neighbours. According to (10.34) and (10.35) we could compute the t-matrix for such an object by solving the radial Schrödinger equations within the sphere and defining the scattering phase shifts $\eta_l(\mathcal{E})$ in terms of the logarithmic derivatives at $\mathcal{R}_l(\mathbf{r}; \mathcal{E})$ and R_{MT}. But consider any formula of § 10.6 in which the symbol t^i appears. In a momentum representation, we should need to evaluate expressions such as

$$\langle\mathbf{k}|tG^0t|\mathbf{k}\rangle \equiv \int \frac{|\langle\mathbf{k}|t|\mathbf{k}+\mathbf{q}\rangle|^2}{\mathcal{E} - \mathcal{E}^0_{\mathbf{k}+\mathbf{q}}}\,\mathrm{d}^3\mathbf{q}, \tag{10.69}$$

where $\langle \mathbf{k}|t|\mathbf{k}+\mathbf{q}\rangle$ means a matrix element of the atomic t-matrix t^i between free electron states of different energies $\mathscr{E}_{\mathbf{k}}^0$ and $\mathscr{E}_{\mathbf{k}+\mathbf{q}}^0$ (cf. (10.48)). But in the angular momentum representation (10.35), *the matrix elements of t are defined only for elastic scattering.* To use the formalism of § 10.6 we are apparently forced to construct the whole t-matrix 'off the energy shell', perhaps by solving the integral equation (10.56).

Yet this cannot be necessary. Given the phase shifts we can consider each atomic sphere as a 'black box', which prescribes the linear boundary conditions (10.34) on any wave function that reaches its surface from the outside and requires to be matched smoothly inwards. From this point of view, the solution of the Schrödinger equation in the potential $\mathscr{V}(\mathbf{r})$, at energy $\mathscr{E}$, reduces to finding a solution in the 'empty' interstitial region that satisfies these conditions on all its local boundaries. This is a perfectly well-posed mathematical problem; the only information we really need about each atomic sphere is its position $\mathbf{R}_i$ and its scattering phase shifts $\eta_l(\mathscr{E})$; knowledge of the t-matrices off the energy shell is thus superfluous.

This paradox is resolved by translating the various operator equations into *coordinate* language (Beeby & Edwards 1963; Beeby 1964; see e.g. Lloyd & Smith 1972). In this representation (10.56) stands for the *Lippmann–Schwinger equation*

$$t(\mathbf{r}, \mathbf{r}') = v(\mathbf{r})\delta(\mathbf{r} - \mathbf{r}') + \int v(\mathbf{r})\, G^0(\mathbf{r} - \mathbf{r}''; \mathscr{E})\, t(\mathbf{r}'', \mathbf{r}')\, d^3r'' \qquad (10.70)$$

The energy appears explicitly only as a parameter, $\mathscr{E}$, in the propagator G^0, and thus has a fixed value for each solution $t(\mathbf{r}, \mathbf{r}')$ of the equation. The t-matrix in the coordinate representation only exists on this energy shell, whose actual energy need not be shown in the subsequent analysis.

Since the atomic potential $v(\mathbf{r})$ has a finite range, the solution of (10.70) is identically zero if either coordinate lies outside the atomic sphere:

$$t(\mathbf{r}, \mathbf{r}') = 0 \quad \text{for } |\mathbf{r}| > R_{\text{MT}} \text{ or } |\mathbf{r}'| > R_{\text{MT}}. \qquad (10.71)$$

Moreover, the potential $v(\mathbf{r})$ is spherically symmetric, so that the t-matrix must be diagonal in an angular momentum representation. Algebraically, this means that we can expand $t(\mathbf{r}, \mathbf{r}')$ in spherical harmonics, $Y_L(\hat{\mathbf{r}})$, $Y_L(\hat{\mathbf{r}}')$, for the *directions* of $\mathbf{r}$ and $\mathbf{r}'$, in the form

$$t(\mathbf{r}, \mathbf{r}') = \sum_l Y_L(\hat{\mathbf{r}}) t_L(r, r') Y_L(\hat{\mathbf{r}}'). \qquad (10.72)$$

As in (10.31), the symbol L stands for the pair (l, m) of angular-momentum quantum numbers. We must now solve (10.72) for the function $t_L(r, r')$. But this is really no more than solving the radial Schrödinger equation for the radial functions (10.32). By various identities governing the t-matrix,

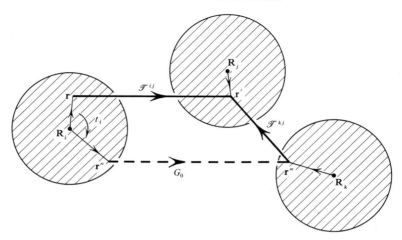

Fig. 10.11. Geometry of scattering path operators.

we connect (10.72) with the partial-wave formula (10.35) for the matrix elements in the *angular momentum representation*

$$t_L = \int_0^\infty \int_0^\infty j_L(\kappa r) t_L(r, r') j_L(\kappa r') r^2 \mathrm{d}r \, r'^2 \mathrm{d}r'. \tag{10.73}$$

In this identity, of course, $\kappa = \sqrt{\mathscr{E}}$ is the wave number of the spherical Bessel function solutions for an outgoing wave in free space. But there is no need to invert (10.73) to find $t_L(r, r')$ explicitly; from (10.72) and (10.73), formulae containing operator products involving t can be directly reduced in this *partial-wave representation* to formulae containing the scattering phase shifts $\eta_l(\mathscr{E})$ at the given value of the electron energy.

To illustrate this technique (Gyorffy 1972) consider the operator equation (10.61) defining the scattering path operators $\mathscr{T}^{ij}$. *In a coordinate representation*, measuring $\mathbf{r}$, etc. relative to the appropriate atomic centres $\mathbf{R}_i$ (fig. 10.11), etc. this reads

$$\mathscr{T}^{ij}(\mathbf{r}, \mathbf{r}') = t^i(\mathbf{r}, \mathbf{r}')\delta_{ij} +$$
$$+ \sum_{k \neq i} \int \mathrm{d}^3\mathbf{r}'' \int \mathrm{d}^3\mathbf{r}''' \, t^i(\mathbf{r}, \mathbf{r}'') G^0(\mathbf{R}_i + \mathbf{r}'' - \mathbf{R}_k - \mathbf{r}''') \mathscr{T}^{kj}(\mathbf{r}''', \mathbf{r}') \tag{10.74}$$

where, as in (10.70), the energy $\mathscr{E}$ would only appear as a fixed parameter in the propagator G^0.

From this definition it follows, as in (10.71), that the matrix elements vanish if either coordinate point lies outside its appropriate atomic sphere:

$$\mathscr{T}^{ij}(\mathbf{r}, \mathbf{r}') = 0 \quad \text{for } |\mathbf{r} - \mathbf{R}_i| > R_{\mathrm{MT}} \text{ or } |\mathbf{r} - \mathbf{R}_j| > R_{\mathrm{MT}}. \tag{10.75}$$

But because of the special axis $(\mathbf{R}_i - \mathbf{R}_j)$, this function cannot be spherically symmetric and its general expansion in spherical harmonics,

$$\mathcal{T}^{ij}(\mathbf{r}, \mathbf{r}') = \sum_{L,L'} Y_L(\hat{\mathbf{r}}) \mathcal{T}^{ij}_{LL'}(\mathbf{r}, \mathbf{r}') Y_{L'}(\hat{\mathbf{r}}'), \tag{10.76}$$

is not, like (10.72), diagonal. These functions, in their turn, are related to the matrix elements in the partial-wave representation by a generalization of (10.73), i.e.

$$\mathcal{T}^{ij}_{LL'}(\mathscr{E}) = \int_0^\infty \int_0^\infty j_L(\kappa r) \mathcal{T}^{ij}_{LL'}(r, r') j_{L'}(\kappa r') r^2 \mathrm{d}r \, r'^2 \, \mathrm{d}r'. \tag{10.77}$$

The symbol $\mathscr{E}$ appears in this equation simply to remind us that all the variables are defined on the same energy shell $\mathscr{E} = \kappa^2$.

The only other operator appearing in (10.74) is the free-space propagator G^0. This we know to be diagonal in the momentum representation (9.9). A Fourier transform to the coordinate representation, and systematic use of the standard identity

$$e^{i\mathbf{q}\cdot\mathbf{r}} = 4\pi \sum_l i^L j_L(qr) Y_L(\hat{\mathbf{q}}) Y_L(\hat{\mathbf{r}}), \tag{10.78}$$

permits the reduction to partial-wave form:

$$\begin{aligned}
G^0(\mathbf{R}+\mathbf{r}-\mathbf{R}'-\mathbf{r}'; \mathscr{E}) &= \frac{1}{(2\pi)^3} \int \frac{\exp\{i\mathbf{q}\,(\mathbf{R}+\mathbf{r}-\mathbf{R}'-\mathbf{r}')\}}{\mathscr{E} - q^2 + i\varepsilon} \, \mathrm{d}^3\mathbf{q} \\
&= \frac{2}{\pi} \sum_{L,L'} i^{L-L'} \int j_L(qr) \frac{\exp\{i\mathbf{q}\cdot(\mathbf{R}-\mathbf{R}')\}}{\kappa^2 - q^2 + i\varepsilon} j_{L'}(qr') \\
&\qquad\qquad Y_L(\hat{\mathbf{q}}) Y_{L'}(\hat{\mathbf{q}}) \, \mathrm{d}^3\mathbf{q} \, Y_L(\hat{\mathbf{r}}) Y_{L'}(\hat{\mathbf{r}}') \\
&= \sum_{L,L'} Y_L(\hat{\mathbf{r}}) j_L(\kappa r) G^0_{LL'}(\mathbf{R}-\mathbf{R}') j_{L'}(\kappa r') Y_{L'}(\hat{\mathbf{r}}').
\end{aligned} \tag{10.79}$$

The key point is that the integration over q (which has the important effect of singling out the wave number $q = \kappa$ at the poles of the integrand) is not valid unless we exclude the singularity at the origin of G^0 in the coordinate representation. But the atomic spheres centred on $\mathbf{r}_i$ and $\mathbf{R}_j$ do not overlap, so that $(\mathbf{R}_i + \mathbf{r} - \mathbf{R}_j - \mathbf{r}')$ never becomes zero for points where, by (10.75), the scattering operator $\mathcal{T}^{ij}(\mathbf{r}, \mathbf{r}')$ does not vanish. This is the step in the analysis where the 'muffin-tin' condition is found to be essential.

The actual form of the partial-wave matrix element of the propagator in (10.79) looks somewhat complicated, i.e.

$$G^0_{LL'}(\mathbf{R}_{ij}) = -8\pi i\kappa \sum_{L''} C^{L''}_{L'L} i^{L''} h^+_{L''}(\kappa R_{ij}) Y_{L''}(\hat{\mathbf{R}}_{ij}), \tag{10.80}$$

where the *Gaunt numbers* are defined as integrals of triple products of spherical harmonics –

$$C^{L''}_{L'L} = \int Y_L(\Omega) Y_{L''}(\Omega) Y_L(\Omega) \, \mathrm{d}\Omega \tag{10.81}$$

– and $h_l^+(x)$ is a spherical Hankel function. But these are standard functions which are tabulated or can be computed. The partial-wave representation of (10.61) or (10.74), i.e.

$$\mathcal{T}_{LL'}^{ij} = t_L^i \, \delta_{LL'} \, \delta_{ij} + \sum_{k \neq i} \sum_{L''} t_L^i G_{LL''}^0 (\mathbf{R}_i - \mathbf{R}_k) \mathcal{T}_{L''L'}^{kj}, \tag{10.82}$$

is thus perfectly explicit, with each matrix element to be evaluated at the same energy $\mathcal{E}$.

Any assembly of muffin-tin potentials may be dealt with in this way; any of the generalized operator identities of §§ 10.4–10.6 can be written as a matrix equation in the partial-wave representation. This applies, in particular, to the theory of substitutional disorder on a regular lattice, discussed at length in chapter 9. Physically speaking it is much more natural to think of a transition metal alloy as an assembly of atomic potentials with different d-resonances (§ 10.3) than as an LCAO/tight-binding system (§ 9.1). The algebra of the CPA method, for example § 9.4, can be generalized (Gyorffy 1972) to give a set of self-consistency conditions, analogous to (9.49), for a *coherent site t-matrix, $\tilde{t}$,* in the partial-wave representation. Indeed the algebraic similarity between the equations (10.82) for the scattering path operator and the simple TBA equations (9.1) for the excitation amplitude suggests that such a generalization must be possible in principle.

In practice, however, the implicit 9×9 matrix equations for up to $l=2$ are not so easily solved as the algebraic equation (9.60) for the simple '*s*-wave *t*-matrix' of the TBA model. As a first step (Gyorffy & Stocks 1974) consider the ATA formula (9.27), where we would write

$$\tilde{t} = c_A t^A + c_B t^B \tag{10.83}$$

for the effective site t-matrix in a mixture of A and B type atoms. With d-wave phase shifts η_2^A and η_2^B, respectively, the $L=2$ component of $\tilde{t}$ must be the sum of terms like (10.35), i.e.

$$\tilde{t}_2 = -\frac{1}{\kappa} \{ c_A \exp(2i\eta_2^A) + c_B \exp(2i\eta_2^B) - 1 \}. \tag{10.84}$$

But this diagonal element of the matrix $\tilde{t}$ is *not unitary;* it cannot, for example, be expressed in the form (10.35) with a real value of η. Physically, this corresponds to *absorptive scattering* by each site; the disorder of the system makes itself felt as a systematic tendency to transform a *coherent* excitation into *incoherent* waves as the electron interacts with the atoms. Unfortunately, this irreversible effect causes severe mathematical difficulties in the analysis.

10.8 The coherent-wave approximation

The problem of solving the Schrödinger equation for an electron in a disordered metal is a special case of the general problem of wave propagation through a disordered assembly of scattering objects. This problem arises in a variety of physical situations; we might be thinking of the *optical model* of nuclear physics (see e.g. Foldy & Walecka 1969), of light scattering by molecules or colloidal particles, or of macroscopic systems such as the propagation of a radar beam through a rain cloud or of a sonar beam through a mass of gas bubbles in a liquid (see e.g. Foldy 1945; Lax 1951; Waterman & Truell 1961; Burke & Twersky 1964). We are now in a position to connect this classical problem of applied mathematics with the quantum-mechanical problem of electron states in disordered systems.

The objective of the classical theory is to calculate the properties of the assembly, considered as a homogeneous medium, from knowledge of the elementary scatterers and their spatial distribution. The essential mathematical difficulty is to take full account of the multiple scattering (§§ 10.5, 10.6), which not only changes the effective 'dielectric constant' of the medium but also, through spatial disorder, eventually reduces any coherent excitation to zero over large distances.

The standard classical model is an assembly of *non-overlapping spheres,* through which is passing an excitation of standard frequency. In our language, this corresponds to the solution of the Schrödinger equation (10.1) in the field (10.11) of a muffin-tin potential (§ 10.3) at a standard energy $\mathscr{E} = \kappa^2$. In the macroscopic problem the 'wave function' can be directly observed; we start, therefore, from the integral form of the Schrödinger equation

$$\Psi(\mathbf{r}) = \int G^0(\mathbf{r} - \mathbf{r}') \sum_j v(\mathbf{r}' - \mathbf{R}_j) \Psi(\mathbf{r}') \, d^3\mathbf{r}'. \tag{10.85}$$

Making use of the muffin-tin property, we introduce a local coordinate $\mathbf{r}$, within each Voronoi polyhedron (§ 2.11), measured (as in (10.79)) relative to the nearest atomic site $\mathbf{R}_i$. The wave function in this cell is determined by the local potential and also by waves propagated from all other cells:

$$\Psi_i(\mathbf{r}) = \int G^0(\mathbf{r} - \mathbf{r}') v(\mathbf{r}') \Psi_i(\mathbf{r}') \, d^3\mathbf{r}' +$$
$$+ \sum_{j \neq i} \int G^0(\mathbf{R}_i - \mathbf{R}_j + \mathbf{r} - \mathbf{r}') v(\mathbf{r}') \Psi_j(\mathbf{r}') \, d^3\mathbf{r}'. \tag{10.86}$$

Since $v(\mathbf{r}') = 0$ for $r' \geqslant R_{MT}$, the terms in the sum on the right are all distinct and uncoupled.

Suppose, first, that the atomic sites $\mathbf{R}_i$ formed a regular lattice. The solution of the Schrödinger equation would then satisfy the Bloch theorem (1.2); there must be a real wave vector $\mathbf{k}$ such that

$$\Psi_i(\mathbf{r}) = \exp(i\mathbf{k} \cdot \mathbf{R}_i)\Psi_0(\mathbf{r}). \qquad (10.87)$$

Substituting into (10.86), we get an integral equation for the standard function $\Psi_0(\mathbf{r})$, i.e.

$$\Psi_0(\mathbf{r}) = \int G^0(\mathbf{r} - \mathbf{r}')v(\mathbf{r}')\Psi_0(\mathbf{r}')\, d^3\mathbf{r}' + \int G'_{\mathbf{k},\kappa}(\mathbf{r}, \mathbf{r}')v(\mathbf{r}')\Psi_0(\mathbf{r}')\, d^3\mathbf{r}' \qquad (10.88)$$

where the *incomplete Greenian,*

$$G'_{\mathbf{k},\kappa}(\mathbf{r}, \mathbf{r}') \equiv \sum_{j \neq i} G^0(\mathbf{R}_i - \mathbf{R}_j + \mathbf{r} - \mathbf{r}')\exp\{i\mathbf{k} \cdot (\mathbf{R}_j - \mathbf{R}_i)\}, \qquad (10.89)$$

is independent of the site label i. For a given value of $\kappa = \sqrt{\mathscr{E}}$, the integral equation (10.88) can be solved for $\psi_0(\mathbf{r})$ only for special values of $\mathbf{k}$: this is the basis of the *KKR method* for calculating the band structure of the crystalline solid (see e.g. Ziman 1971*a*). In practice, the wave function $\Psi_0(\mathbf{r})$ is expanded in spherical harmonics as in (10.31), with coefficients chosen to satisfy (10.88). This is equivalent to writing this integral equation as a matrix equation in the partial wave representation (§ 10.7); the atomic spheres appear through their scattering phase shifts, $\eta_l(\mathscr{E})$ whilst the so-called *structure constants* are simply the matrix elements of (10.89) in this representation and can be constructed by inspection from (10.78) to (10.81).

For a disordered system, however, general statements can only be made about statistical properties. Suppose, therefore, that we take an ensemble average of (10.86) over the distribution $g_N(1, 2 \ldots N)$ of atomic positions. Just as in (10.66), we can reduce the integrations to a relation of the form

$$\langle\Psi_1(\mathbf{r})\rangle_1 = \int G^0(\mathbf{r} - \mathbf{r}')v(\mathbf{r}')\langle\Psi_1(\mathbf{r}')\rangle_1 +$$

$$+ \int\int G^0(\mathbf{R}_1 - \mathbf{R}_2 + \mathbf{r} - \mathbf{r}')g_2(1, 2)\langle\Psi_2(\mathbf{r}')\rangle_{2,1}\, d^3\mathbf{R}_2\, v(\mathbf{r}')d^3\mathbf{r}' \qquad (10.90)$$

where the symbols are conditional averages, as defined in (10.65). But the corresponding equation for $\langle\Psi_2(\mathbf{r}')\rangle_{2,1}$ involves three-particle distributions: we face once more the problem of solving a hierarchy of equations for successively more complex functions of this type (cf. § 2.12).

But suppose we make the simplext closure approximation (Foldy 1945; Lax 1951), and write

$$\langle\Psi_2(\mathbf{r}')\rangle_{2,1} \approx \langle\Psi_2(\mathbf{r}')\rangle_2. \qquad (10.91)$$

Under these conditions, (10.90) is a homogeneous integral equation which is translationally invariant in the space of the atomic sites **1** and **2**. This equation must, therefore, have general solutions satisfying the analogue of the Bloch condition (10.87), with a *medium propagation vector* $\tilde{\mathbf{k}}$, i.e.

$$\langle \Psi_2(\mathbf{r}) \rangle_2 = \exp\{i\tilde{\mathbf{k}} \cdot (\mathbf{R}_2 - \mathbf{R}_1)\} \langle \Psi_1(\mathbf{r}) \rangle_1. \qquad (10.92)$$

Putting (10.91) and (10.92) into (10.90), we get an integral equation of precisely the form (10.88) with an incomplete Greenian, analogous to (10.89) (Phariseau & Ziman 1963; Ziman 1966),

$$G^0_{\tilde{\mathbf{k}},\kappa} = \int G^0(\mathbf{R}_1 - \mathbf{R}_2 + \mathbf{r} - \mathbf{r}') \exp\{i\tilde{\mathbf{k}} \cdot (\mathbf{R}_2 - \mathbf{R}_1)\} g_2(1, 2) \, d^3\mathbf{R}_2. \quad (10.93)$$

We thus study the propagation of a *coherent wave*, defined as in (10.92), and ignore the incoherently scattered radiation that averages to zero in the ensemble. The coherent part behaves as if it had the propagation vector $\tilde{\mathbf{k}}$, which is to be determined as a function of κ (i.e. of the energy), by looking for solutions of (10.88) and (10.93). The analogy with the KKR method for ordered systems is obvious.

In the absence of correlations between atomic positions (i.e. with $g_2(1, 2) \equiv 1$) the coherent wave equation has an exact solution which is easily constructed by reference to the general principles of scattering theory (Foldy 1945):

$$\tilde{k}^2 = \kappa^2 + N\bar{t}(0), \qquad (10.94)$$

where $\bar{t}(0)$ is the forward scattering amplitude (10.36) of a single sphere or muffin-tin well. But this is a *complex* quantity; the medium thus behaves as if it had a *complex dielectric constant*. The imaginary part of $\tilde{k}$ makes (10.92) into a decaying exponential function. This corresponds physically, of course, to the degradation of the coherent wave into incoherent radiation as it traverses the medium.

In the neighbourhood of a scattering resonance (10.38), where the phase shift goes through $\pi/2$, the real part of $\tilde{k}$ may deviate substantially from κ, whilst the inverse of the imaginary part – the *extinction distance* – may become very short. Yet this is a *non-structural effect* (§ 10.4); the propagation characteristics of the medium depend only on the scattering properties of the individual spheres and on their density in space, without reference to the geometry of their relative positions.

In the same spirit (Anderson & McMillan 1967), we may suppose that each atom of the system is at the centre of a Wigner–Seitz sphere of radius R_s embedded in a continuous medium of complex propagation constant $\tilde{k}$. Since the radius, R_{MT}, of each muffin-tin sphere must be somewhat less than R_s we may use (10.33) and the known values of $\eta_l(\mathscr{E})$ to calculate the logarithmic derivatives

$$\mathscr{L}_l(R_s; \mathscr{E}) = \mathscr{R}'_l(R_s; \mathscr{E}) / \mathscr{R}_l(R_s; \mathscr{E}) \qquad (10.95)$$

on the surface of this sphere. To match these derivatives to partial waves in the external medium, we use (10.34) to define modified 'phase shifts', $\tilde{\eta}_l$ on the surface at R_s. In other words, we use (10.95) for the logarithmic derivatives, and replace κ by the complex propagation constant $\bar{k}$, so that each $\tilde{\eta}_l$ will be a complex quantity (cf. (10.84)). To this phenomenological model we apply the CPA principle (9.49) and suppose that the atomic sphere should not produce any forward scattering in the embedding medium. This self-consistency condition,

$$\tilde{t}(\bar{\mathbf{k}}, \bar{\mathbf{k}}) \equiv -\frac{1}{\kappa} \sum_l (2l+1) \sin \tilde{\eta}_l \exp i\tilde{\eta}_l = 0, \qquad (10.96)$$

can be solved numerically for $\bar{k}$ as a function of $\mathscr{E}$. Unfortunately, the results of this computation (Olson 1975; Chang, Sher, Petzinger & Weisz 1975) have 'unphysical' features; as we saw in § 9.5, it is not easy to generalize the CPA method beyond s-wave scattering from a single site.

It is worth noting, however, that the solution of (10.96) for $\bar{k} = 0$ occurs at the energy $\mathscr{E}_0$ where the s-wave logarithmic derivative (10.95) vanishes. But this condition –

$$\mathscr{L}_0(R_s; \mathscr{E}_0) = 0 \qquad (10.97)$$

– is none other than the familiar Wigner–Seitz *ansatz* for the bottom of the free band in a crystalline metal. This simple 'non-structural' property of the electron distribution (Ziman 1967a) is evidently valid for liquid metals, where the Voronoi polyhedra can still be approximated by Wigner–Seitz spheres (§ 2.11).

To include 'geometrical' effects in this theory (cf. Lloyd & Berry 1967) we have to solve the equation (10.88) with an incomplete Greenian (10.93) containing a genuine pair correlation function $g_2(\mathbf{1}, \mathbf{2})$. Naturally we go to a partial wave representation (§ 10.7) in which the scattering potentials appear through their phase shifts and the analogues of the KKR structure constants contain functions like (10.80) integrated over interatomic distances. In the limit of small phase shifts, these formulae are consistent with calculations based upon a naive t-matrix theory (cf. (10.37)) for the extinction distance (Ziman 1966); but the fact that the 'coherent wave' (10.92) is *increasing* exponentially along the direction $-\bar{\mathbf{k}}$ leads to divergences and mathematical difficulties which have not been satisfactorily resolved (Morgan & Ziman 1967).

Indeed, the classical 'coherent-wave approximation' is mathematically equivalent to the 'quasi-crystalline approximation' of § 10.6 (Schwartz & Ehrenreich 1972a). It is obvious, for example, that the closure approximation (10.91) is just the same in principle as the approximation by which (10.66) was reduced to (10.67). The canonical language of scattering path operators (10.60), with precisely defined ensemble averages (10.65), is more satisfactory than any talk about the wave functions themselves. Since $\mathcal{T}^{ij}$ is simply the operator that transforms Ψ_i into Ψ_j, it already contains all the information we might need about the spectral properties of the disordered system.

Notice, for example, that the *forward scattering approximation* (10.94) in the coherent wave language is precisely equivalent to the simple formula (10.62); it is a matter of taste and mathematical convenience whether we interpret the imaginary part of a complex self-energy Σ as a measure of the *time* decay of a uniform spatial excitation or whether we call this the imaginary part of $\bar{k}^2 - \kappa^2$, giving rise to the *spatial* extinction of a steady coherent beam of radiation (Ballentine & Heine 1964). Again, the algebraic analysis of the coherent wave approximation (e.g. Ziman 1966) is no more than an attempt to solve the equations (10.82) for the scattering path operators in the partial-wave representation, using the quasi-crystalline approximation to close the hierarchy of ensemble averages.

10.9 Cluster scattering

The spectral density of any system can be written in the invariant form (9.7), i.e.

$$\mathcal{N}(\mathcal{E}) = -\frac{1}{\pi} \, \mathrm{Im}.[\mathrm{Tr}\{G^+(\mathcal{E})\}] \tag{10.98}$$

in the canonical operator formalism. The Green function (9.4) can be integrated from some arbitrary energy zero to give the *cumulated spectral density* (8.68):

$$\mathcal{D}(\mathcal{E}) \equiv \int^{\mathcal{E}} \mathcal{N}(\mathcal{E}) \, d\mathcal{E}$$

$$= -\frac{1}{\pi} \, \mathrm{Im}.[\mathrm{Tr}\{\int^{\mathcal{E}} (\mathcal{E} - \mathcal{H})^{-1} \, d\mathcal{E}\}]$$

$$= -\frac{1}{\pi} \, \mathrm{Im}.[\mathrm{Tr}\{\ln(\mathcal{E} - \mathcal{H})\}]. \tag{10.99}$$

In the present context it is natural to compare this spectrum with the corresponding function $\mathcal{D}^0(\mathcal{E})$ for a free electron system with Hamiltonian $\mathcal{H}^0$. The effect of the scattering potential $\mathcal{V}$ is to change the spectral density to

$$\mathcal{D}(\mathscr{E}) = \mathcal{D}^0(\mathscr{E}) - \frac{1}{\pi} \text{Im.}[\text{Tr}\{\ln(\mathscr{E} - \mathscr{H}) - \ln(\mathscr{E} - \mathscr{H}^0)\}]$$

$$= \mathcal{D}^0(\mathscr{E}) - \frac{1}{\pi} \text{Im.}[\text{Tr}\{\ln(1 - \mathscr{V}G^0)\}]. \qquad (10.100)$$

In principle this invariant expression can be evaluated in any matrix representation of the operators. But further manipulations with the T-matrix (10.57) yield

$$\mathcal{D}(\mathscr{E}) = \mathcal{D}^0(\mathscr{E}) - \frac{1}{\pi} \text{Im.}[\text{Tr}\{\ln(\mathscr{V}T^{-1})\}]$$

$$= \mathcal{D}^0(\mathscr{E}) + \frac{1}{\pi} \text{Im.}[\text{Tr}\{\ln T\}], \qquad (10.101)$$

since a real potential $\mathscr{V}(\mathbf{r})$ would not contribute to the imaginary part of the logarithm.

To see how this formula works consider scattering by a single atomic potential $v(\mathbf{r})$. In a partial wave representation this has a t-matrix with diagonal components (10.35). Dropping terms that refer simply to the choice of an origin of energy in (10.99) we get

$$\mathcal{D}(\mathscr{E}) = \mathcal{D}^0(\mathscr{E}) + \frac{1}{\pi} \text{Im.}[\text{Tr}\{\ln(\kappa^{-1} \sin\eta_l \, e^{i\eta_l} \delta_{ll'})\}]$$

$$= \mathcal{D}^0(\mathscr{E}) + \frac{1}{\pi} \text{Im.}\left[\sum_L \{\ln(\kappa^{-1} \sin\eta_l) + i\eta_l\}\right]$$

$$= \mathcal{D}^0(\mathscr{E}) + \frac{1}{\pi} \sum_l (2l+1)\eta_l(\mathscr{E}). \qquad (10.102)$$

This is immediately recognizable as the *Friedel sum rule* for the electron states drawn into the scattering potential by the phase shifts $\eta_l(\mathscr{E})$. Note, moreover, that this formula would also count *bound* states of $v(\mathbf{r})$ by recording the corresponding poles of the t-matrix in the negative-energy regime.

In the much more general case of multiple scattering by an assembly of non-overlapping spheres, this abstract formalism is surprisingly powerful. According to (10.60), the assembly T-matrix is a sum of scattering path operators, $\mathscr{T}^{ij}$, each of which is identically zero except for points lying in the cells about $\mathbf{r}_i$ and $\mathbf{R}_j$ respectively. In other words, $\mathscr{T}^{ij}$ is a matrix element of the total T-matrix in what we might call a *cell representation*.

The defining equations (10.61) for the scattering path operators were rewritten in (10.82) in a *partial-wave representation*. But the symbol $\mathscr{T}^{ij}_{LL'}$ defined in (10.77) is really a matrix element of the T-matrix between a state

$|i; L\rangle$ – of angular momentum L about $\mathbf{R}_i$ – and a state $|j; L'\rangle$ – of angular momentum L' about another site $\mathbf{R}_j$. In other words, (10.82) represents these operators in a *cell-wave representation,* where the labels run through all the cells of the assembly and all values of angular momentum.

In this language, the scattering spheres themselves are represented by the doubly diagonal site t-matrix

$$t = t_L^i \, \delta_{ij} \, \delta_{LL'}. \tag{10.103}$$

In (10.82) we also observe a sum involving the propagator $G^0(\mathbf{R}_i - \mathbf{R}_k)$ for all values of $k \neq i$. But this symbol, also, must be taken to be zero for any points except those in the ith and kth cells: in the cell representation it has exactly the same matrix elements as the *incomplete Greenian,*

$$G' = \sum_j \sum_{i \neq j} G^0(\mathbf{R}_i - \mathbf{R}_j), \tag{10.104}$$

whose Fourier transform appeared in (10.89) and (10.93).

With these conventions we see that (10.82) is simply the explicit form, in a cell-wave representation, of the matrix equation

$$T = t + tG'T, \tag{10.105}$$

which can be solved algebraically:

$$T = \{t^{-1} - G'\}^{-1}. \tag{10.106}$$

Whether or not this formula is meaningful in other representations, there is no objection to our putting this expression for T into the invariant formula (10.101) for the cumulated spectrum, to be evaluated in the cell-wave language where it is certainly valid. To be specific (Lloyd 1967), we could write (10.101) and (10.106) in the form

$$\mathscr{D}(\mathscr{E}) = \mathscr{D}^0(\mathscr{E}) - \frac{1}{\pi} \, \text{Im}.[\ln \, ||t_L^{-1}\delta_{LL'}\delta_{ij} - G_{LL'}'^{ij}||], \tag{10.107}$$

where the determinant $||T||$ appears by virtue of the matrix identity

$$\text{Tr}\{\ln T\} \equiv \ln ||T||. \tag{10.108}$$

In general, the *Lloyd determinant* is of infinite order, and cannot be evaluated exactly. But as an explicit representation of an invariant formula involving only 'on-energy shell' matrix elements of the t-matrix, it occupies a central position in the theory of scattering. In the derivation of (10.107) we assumed that we were dealing with a muffin-tin potential (§ 10.3), where each *cell potential* $v_i(\mathbf{r} - \mathbf{R}_i)$ is spherically symmetrical. Further scrutiny of the analysis (Ziesche 1974) reveals that the only necessary condition is that $\mathscr{V}(\mathbf{r})$ should have a unique cell representation, so that neighbouring cell potentials never overlap. In other words, we may divide our system into

Voronoi cells, separated only by infinitesimal 'interstitial' regions and allow $v_i(\mathbf{r}-\mathbf{R}_i)$ to occupy the whole of its cell, without restriction to a spherical muffin-tin well. In the formal theory this merely means that the cell t-matrices (10.103) are not necessarily diagonal in their partial-wave components, whilst the matrix elements of the incomplete Greenian (10.104) need more careful definition than was implied, for example, by (10.80). In general, however, the elements of the determinant can no longer be conveniently segregated into *structure constants, G'^{ij}*, which depend only on the geometrical arrangement of sites $\mathbf{R}_i$ and cells, and on *atomic parameters* such as t_L^{-1} which depend only on the potential in each cell.

In a regular lattice with translational symmetry the incomplete Greenian (10.104) must be diagonal in a Bloch representation. A Fourier transformation (1.42), on the cell vectors $\mathbf{R}_i$, $\mathbf{R}_j$ partially diagonalizes the Lloyd determinant. In this representation, the cumulated spectral density would be given by

$$\mathscr{D}(\mathscr{E})=\mathscr{D}^0(\mathscr{E})-\frac{1}{\pi}\int \mathrm{Im.}\{\ln||t_L^{-1}(\mathscr{E})\delta_{LL'}-G'_{\mathbf{k};\kappa;L,L'}||\}\mathrm{d}^3\mathbf{k} \quad (10.109)$$

since G' would not be diagonal in its partial-wave components, and would depend on the wave vector $\mathbf{k}$ and on the energy $\mathscr{E}=\kappa^2$. But the logarithm acquires an imaginary part $i\pi$ whenever the determinant passes through zero. Each root of

$$||t_L^{-1}(\mathscr{E})\delta_{LL'}-G'_{\mathbf{k};\kappa;LL'}||=0 \quad (10.110)$$

contributes a state to the cumulated spectrum. But (10.110) is precisely the condition for the existence of a Bloch function (10.87), of energy $\mathscr{E}$ and wave vector $\mathbf{k}$, satisfying the scattering equations (10.88) and (10.89) in the partial-wave representation. The Lloyd determinant thus automatically counts the Bloch states obtained by the *KKR method* of band structure calculation (Beeby 1964).

In the algebraic and numerical calculations associated with the Lloyd formula (10.107), the *transition matrix T* is often eliminated in favour of the corresponding *reaction matrix K*, defined by the operator formula

$$T^{-1}=K^{-1}+i\,\kappa. \quad (10.111)$$

This has certain technical advantages (Keller & Smith 1972) especially in the negative energy region below $\mathscr{E}_{\mathrm{MTZ}}$, but is not of any physical significance.

As a relatively compact canonical formula, valid in the strong scattering region, the Lloyd determinant may be used to resolve ambiguities in the interpretation of phenomenological models (e.g. Ducastelle 1975). Thus, for example, the forward scattering approximation (10.94) for the propagation constant of a dense disordered assembly is mathematically equivalent to the average t-matrix approximation (10.62). But this, in turn, is just what we get if we drop the Greenian entirely from (10.107); the spectrum of the whole system is just the aggregate of the Friedel sums (10.102) of its constituent atoms – a result that would be difficult to derive directly from the formula (10.94) for the complex propagation constant $\bar{\mathbf{k}}$ of a coherent wave. In this approximation an assembly of transition metal atoms would have a spectrum of d-bands, each of the same width Γ as the corresponding resonance (10.38) of its muffin-tin well. Appeal to this principle has proved useful in dealing with the energy dependence of the pseudopotential in self-energy formulae such as (10.48) (Ballentine 1975) and with the approximate solutions (10.82) of the CPA equations for a muffin-tin model of a binary alloy (Gyorffy & Stocks 1974).

But the real value of this formalism is that it provides a well-defined procedure for direct computation of *local environmental corrections* (cf. § 9.5) to the spectrum of a topologically disordered system such as a liquid metal or amorphous semiconductor (§ 11.5). All that we need do (Klima, McGill & Ziman 1970; McGill & Klima 1972) is to apply the theory systematically to a compact *cluster* of n atoms whose T-matrix can be calculated in a cell wave representation by numerical solution of momentum (10.82) up to some finite angular l_{max}. Then the density of states obtained by applying (10.101) to this matrix is precisely what we should get for an assembly of non-interacting clusters of the given type. In other words (fig. 10.12) we divide up the whole assembly into non-overlapping clusters and solve completely for all internal interactions and externally scattered waves produced by an external excitation. Because of the spherical symmetry of each atom, and the fact that the free-electron spectrum takes care of all partial waves with negligibly small phase shifts, this is a set of linear equations in $n(l_{max}+1)^2$ variables, with coefficients of the form (10.80). Although we neglect the interactions between clusters (i.e. along many nearest-neighbour links across cluster boundaries) we get a considerable amount of 'structure' in the spectrum that may properly be attributed to the local geometry of the system (fig. 10.13) (Keller 1971; Keller & Jones 1971; House & Smith 1973). Such calculations evidently take into account the relative positions of many atoms of the assembly, although not with the weights that would be assigned to the corresponding distribution functions in a systematic expansion such as the Edwards series (10.50). It would, of course, be desirable (Best & Lloyd 1975) to embed each cluster in a

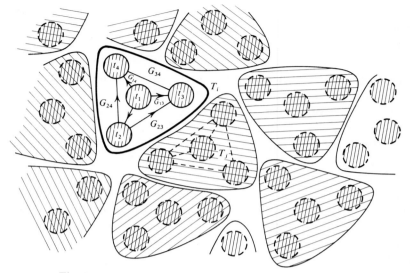

Fig. 10.12. Dissection of muffin-tin assembly into clusters.

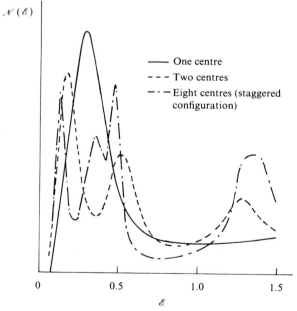

Fig. 10.13. Spectral density for clusters of 1, 2 and 8 'carbon' atoms in diamond-lattice configuration (McGill & Klima 1972).

'medium' with self-consistent properties in the spirit of the molecular CPA theories of tight-binding alloys (§ 9.5). But experience with these theories, and the complexities of a satisfactory *single site* CPA theory for liquids (§ 10.6), suggest the difficulties of this research programme.

10.10 Transport theory

From the spectral properties of electrons in metallic liquids (§§ 10.4–10.9) it was hoped that we might learn enough about the energy eigenstates to improve on the NFE formulae (10.17) and (10.37) for the electrical resistivity. But the derivation of these formulae relied upon the Boltzmann equation of elementary transport theory, which can certainly not be taken for granted when the electrons interact strongly with the disordered assembly of ions. The naive picture of electrons in NFE pseudo-wave functions being weakly scattered by neutral pseudo-atoms (§ 10.2) is reasonably plausible for systems such as liquid alkali metals, but the phenomenological formula (10.37) is not based upon theoretical first principles and little is known of the physical circumstances under which it might be supposed valid, nor of the corrections that should be applied when these conditions are not well satisfied.

In principle, any irreversible process may be described by a *master equation* of which the Boltzmann equation is a simplified version. It is not known, however, whether the electrons in liquid metals satisfy the mathematical conditions under which this simplification is a good approximation (see e.g. Chester 1963). Since the general theory of irreversible processes is itself a major field of mathematical physics we reluctantly draw back from entry into that happy realm.

Quite apart from abstract questions of existence and validity, a *practical* weakness of the Boltzmann method is the assumption that the total Hamiltonian of the system can be written in the form (10.1), i.e.

$$\mathscr{H} = \mathscr{H}^0 + \mathscr{V} \tag{10.112}$$

where $\mathscr{V}$ is a relatively weak, statistically incoherent perturbation on the freely propagating states of the unperturbed Hamiltonian $\mathscr{H}^0$. But these conditions are far from realistic for a system such as a liquid transition metal (§ 10.3), where the ionic potentials exert a very strong influence on the electron states (fig. 10.12). In such systems, however (§§ 10.6–10.9), the electron spectrum can often be derived by approximate calculations of the assembly Green function (10.43), or the corresponding T-matrix (10.57), without appeal to the weak scattering assumption and without trying to write down the eigenstates of $\mathcal{H}^0$ or $\mathcal{H}$. There is much interest, therefore, in invariant formulae for transport coefficients, valid in any matrix representation of the operators.

This formal invariance is a valuable characteristic of *linear-response formulae* for transport coefficients (see e.g. Kubo 1966). Thus, the *electrical conductivity tensor* at frequency ω can be written (Kubo 1956)

$$\sigma_{\mu\nu}(\omega) = \frac{1 - e^{-\beta\omega}}{2\omega} \int_{-\infty}^{\infty} \mathrm{Tr}\{\rho j_\nu(t) j_\mu(0)\} e^{-i\omega t} \, dt, \qquad (10.113)$$

where

$$\rho = \exp(-\beta\mathcal{H})\Big/\mathrm{Tr}\{\exp(-\beta\mathcal{H})\} \qquad (10.114)$$

is the *canonical density matrix* (cf. (5.2)) for the whole system and the Cartesian components of the *current operator*

$$j_\nu(t) \equiv e^{-i\mathcal{H}t} j_\nu(0) e^{i\mathcal{H}t} \qquad (10.115)$$

evolve with time in the interaction representation. This operator, which mediates the interaction between each electron charge and any external electric field, is defined precisely in terms of the momentum operator by

$$\mathbf{j} \equiv \left(\frac{e}{m}\right)\mathbf{p} \equiv \left(\frac{e}{m}\right)\frac{\hbar}{i}\,\boldsymbol{\nabla}. \qquad (10.116)$$

The reader is referred to the original papers or to standard treatises for the derivation of the Kubo formula (10.113). Physically speaking, this formula exemplifies the *fluctuation–dissipation theorem*. The linear response to an *external* force – the current induced by an oscillating electric field – is proportional to the time correlation of *internal* fluctuations of the system, calculated as if in thermodynamic equilibrium without the influence of such external forces. The Hamiltonian in (10.114) and (10.115) is thus the total Hamiltonian of the system in the absence of any imposed electromagnetic field.

For an assembly of independent electrons with Fermi–Dirac distribution $f(\mathscr{E}_\alpha)$ over the eigenstates $|\alpha\rangle$ of a one-electron Hamiltonian $\mathcal{H}$, the *static conductivity* can be deduced from (10.113) by going to the limit $\omega \to 0$ (Greenwood 1958):

$$\sigma_{\mu\nu} = \pi\hbar \sum_{\alpha,\alpha'} \{\langle\alpha|j_\mu|\alpha'\rangle\langle\alpha'|j_\nu|\alpha\rangle\delta(\mathscr{E}_\alpha - \mathscr{E}_{\alpha'})[-\partial f(\mathscr{E}_\alpha)/\partial\mathscr{E}_\alpha]\}. \quad (10.117)$$

Since $[-\partial f(\mathscr{E})/\partial\mathscr{E}]$ behaves like $\delta(\mathscr{E} - \mathscr{E}_F)$ at low temperatures relative to $\mathscr{E}_F/k$, the transport properties of metallic liquids derive, as usual from states near the Fermi level, $\mathscr{E}_F$.

Not having explicit formulae for the electron eigenstates in topologically disordered systems, we rewrite everything in terms of Green functions. The required sum over eigenstates can be generated from (9.6) by the operator

$$G^\times(\mathscr{E}) \equiv -\frac{1}{2\pi i}(G^+ - G^-) = \sum_\alpha |\alpha\rangle \delta(\mathscr{E} - \mathscr{E}_\alpha)\langle\alpha|, \qquad (10.118)$$

whose trace (9.7) is just the spectral density. To avoid having to calculate the matrix elements of the current (10.116), we naturally use a momentum representation for the eigenfunctions $|\mathbf{k}; \alpha\rangle$ and the corresponding Green functions. The double sum over eigenstates in (10.117) can be transformed into a double integral over momentum variables: by definition

$$\sigma_{\mu\nu} = \pi h \sum_{\alpha,\alpha'} \langle\alpha|j_\mu|\alpha'\rangle\langle\alpha'|j_\nu|\alpha\rangle\delta(\mathscr{E}_\alpha - \mathscr{E}_F)\,\delta(\mathscr{E}_{\alpha'} - \mathscr{E}_F)$$

$$= \pi h\left(\frac{eh}{m}\right)^2 \sum_{\alpha,\alpha'} \int\!\!\int \langle\alpha; \mathbf{k}|k_\mu|\mathbf{k}; \alpha'\rangle\langle\alpha'; \mathbf{k}'|k'_\nu|\mathbf{k}'; \alpha\rangle\,d^3k\,d^3k'$$

$$\delta(\mathscr{E}_\alpha - \mathscr{E}_F)\,\delta(\mathscr{E}_{\alpha'} - \mathscr{E}_F)$$

$$= \pi h\left(\frac{eh}{m}\right)^2 \int\!\!\int \{k_\mu\, G^\times(\mathscr{E}_F; \mathbf{k}, \mathbf{k}')\, k'_\nu\, G^\times(\mathscr{E}_F; \mathbf{k}', \mathbf{k})\}\,d^3k\,d^3k'. \qquad (10.119)$$

The integrand of (10.119) is no more than a product of momentum operators and assembly Green functions and is, therefore, equivalent ergodically to its ensemble average. Unfortunately, we cannot calculate the average of a product of *two* Green functions from knowledge of the average of a *single* Green function such as (9.13):

$$\langle G^\times G^\times\rangle \neq \langle G^\times\rangle\langle G^\times\rangle. \qquad (10.120)$$

This is the fundamental technical difficulty in calculating transport coefficients from linear response formulae such as (10.113), (10.117) or (10.119). The electrical conductivity depends upon correlations in the *two-particle Green functions* which are more subtle than can be determined from the part of the *single-particle propagator* that governs the energy spectrum. As we see from (10.117), the transport properties are sensitive to phase relations in the off-diagonal matrix elements of the current, which are easily lost by premature ensemble averaging.

To appreciate this important theoretical point, let us ignore these correlations and try to evaluate the conductivity as if (10.120) were an equality. In the momentum representation, an ensemble average Green function (9.13) is necessarily diagonal. From (9.31) or (10.45), we may write

$$\langle G^\times(\varepsilon_F; \mathbf{k}, \mathbf{k}')\rangle = \frac{1}{\pi}\frac{\Gamma}{(\varepsilon_F - \varepsilon_\mathbf{k})^2 + \Gamma^2}\,\delta_{\mathbf{k},\mathbf{k}'}, \qquad (10.121)$$

where Γ is the imaginary part of the self-energy $\Sigma(\mathbf{k}; \mathscr{E})$. The real part of Σ defines the renormalized energy $\mathscr{E}_\mathbf{k}$, about which (10.121) behaves approximately like a delta function. But (10.119) contains the square of this Green function; for an isotropic system, we calculate the *scalar conductivity*

$$
\begin{aligned}
\sigma &= \pi h \left(\frac{eh}{m}\right)^2 \int \tfrac{1}{3} k^2 \left\{\frac{\Gamma/\pi}{(\mathscr{E}_F - \mathscr{E}_\mathbf{k})^2 + \Gamma^2}\right\}^2 d^3\mathbf{k} \\
&= \tfrac{1}{3}\left(\frac{ehk_F}{m}\right)^2 \frac{h}{\pi} \int\!\!\int \left\{\frac{\Gamma}{(\mathscr{E}_F - \mathscr{E}_\mathbf{k})^2 + \Gamma^2}\right\}^2 \mathscr{N}(\mathscr{E}_\mathbf{k})\, d\mathscr{E}_\mathbf{k} \\
&= \tfrac{1}{3} j_F^2\, \mathscr{N}(\mathscr{E}_F) \frac{h}{\Gamma},
\end{aligned}
\tag{10.122}
$$

using (10.7). In other words, we arrive at the NFE formula (10.14) with a relaxation time

$$
\tau = h/\Gamma. \tag{10.123}
$$

Since Γ, the imaginary part of the self energy, is the decay constant for a coherent wave in the disordered medium (§ 10.8), this looks a plausible result. But suppose we calculate this self-energy from the forward-scattering approximation (10.62), which ought to be valid for, say, a 'gas' of independent scattering centres:

$$
\Gamma/h \equiv N \operatorname{Im}. t(0) = \int \mathscr{T}(\theta)\, d\Omega, \tag{10.124}
$$

since the imaginary part of the forward scattering amplitude is linked by the *optical theorem* to the total probability of scattering into any element of solid angle $d\Omega$. This looks very like (10.15) – but *lacks the factor* $(1 - \cos\theta)$. The correlations dropped when (10.120) is treated as an equality are those that arise from the *persistence of velocity* of the electrons after they are scattered.

The conductivity cannot, in fact, be derived correctly from a linear response formula such as (10.119) without considerable mathematical finesse (Edwards 1958; Langer 1960; Neal 1970). The key step is to derive by diagrammatic analysis, or write down as a special case of the *Bethe–Salpeter equation*, an integral equation for the two-particle Green function. Taking an ensemble average (10.120), we get the identity

$$
\langle G^\times(\mathbf{k}, \mathbf{k}')G^\times(\mathbf{k}', \mathbf{k})\rangle = \langle G^\times(\mathbf{k}, \mathbf{k}')\rangle\langle G^\times(\mathbf{k}', \mathbf{k})\rangle +
$$
$$
+ \int\!\int \langle G^\times(\mathbf{k}, \mathbf{k}')\rangle\langle G^\times(\mathbf{k}', \mathbf{k}'')\rangle w(\mathbf{k}'', \mathbf{k}''')\langle G^\times(\mathbf{k}''', \mathbf{k}')\rangle\langle G^\times(\mathbf{k}''', \mathbf{k})\rangle\, d^3\mathbf{k}''\, d^3\mathbf{k}'''.
\tag{10.125}
$$

But because $\langle G^\times \rangle$ is diagonal in the momentum representation (10.121), the integrations over $\mathbf{k}''$ and $\mathbf{k}'''$ are avoided. Multiplying by $\mathbf{k} \cdot \mathbf{k}'$ to reproduce the integrand of (10.119), we get

$$(\mathbf{k}\cdot\mathbf{k}')\langle G^\times(\mathbf{k},\mathbf{k}')G^\times(\mathbf{k}',\mathbf{k})\rangle = k^2\{\langle G^\times(\mathbf{k},\mathbf{k})\rangle\}^2\delta_{\mathbf{k},\mathbf{k}'} +$$
$$+\langle G^\times(\mathbf{k},\mathbf{k})\rangle\langle G^\times(\mathbf{k}',\mathbf{k}')\rangle(\mathbf{k}.\mathbf{k}')w(\mathbf{k},\mathbf{k}')\langle G^\times(\mathbf{k},\mathbf{k}')\rangle\langle G^\times(\mathbf{k}',\mathbf{k})\rangle.$$

$$(10.126)$$

But now, to evaluate the scalar conductivity we must carry out a double integral over $\mathbf{k}$ and $\mathbf{k}'$. The first term in the right of (10.126) gives (10.122); but in the second term we reproduce σ itself with an extra factor proportional to

$$\Gamma' = \int \frac{\mathbf{k}\cdot\mathbf{k}'}{k^2} w(\mathbf{k},\mathbf{k}')\, d^3\mathbf{k}'. \qquad (10.127)$$

Solving this equation, we get the NFE formula (10.14), with inverse relaxation time

$$\hbar/\tau = \Gamma - \Gamma'. \qquad (10.128)$$

Provided that we can identify the scattering kernel $w(\mathbf{k},\mathbf{k}')$ with the scattering probability $\mathscr{Q}(\theta)$, so that the factor $(\mathbf{k}\cdot\mathbf{k}')/k^2$ in (10.127) merely introduces a factor $\cos\theta$ into, say, the integrand (10.124), this is now precisely the result (10.15) deduced from the Boltzman equation. But the formal derivation of the Bethe–Salpeter equation (10.125) merely tells us that $w(\mathbf{k}'',\mathbf{k}''')$ is 'the sum of irreducible vertex parts', which is by no means obviously the same thing as the scattering process that governs the imaginary part of the self-energy $\Sigma(\mathbf{k};\mathscr{E})$.

This laborious calculation merely confirms that the linear response formula gives the same result as the Boltzmann method for the electrical conductivity of, say, a dilute metallic alloy. The proof does not guarantee the validity of the phenomenological formula (10.37) for the resistivity of a liquid transition metal, where the structure factor $S(q)$ must surely modulate the scattering t-matrix of the individual atoms. For almost all practical purposes, it seems safer to rely upon various investigations (Chester & Thellung 1959; Luttinger 1967) which have determined the conditions under which the linear response theory is equivalent to the Boltzmann method (e.g. Morgan 1969; Ashcroft & Schaich 1970) or to the solution of a more general master equation.

In the recent literature (see e.g. Huberman & Chester 1975) there is much discussion of an alternative formulation of the linear response theory in which *inverse transport coefficients* (e.g. the electrical resistivity) are expressed in terms of the *force–force correlations* seen by the electrons in the metal. Although this approach seems at first sight to give some nice simple formulae, it is now clear that the exact calculation of transport coefficients by this means is no less arduous and subtle than the derivation of the conductivity from the conventional Kubo formula (10.113).

The linear response formalism does, however, define unequivocally the symbol j_F in the NFE formulae (10.17) and (10.37). From (10.116), we see that the 'Fermi current' (10.7) is always proportional to the 'Fermi wave vector', k_F, even when the electron spectrum is very far from the free-electron ideal. In (10.122), this is selected by the peak of the quasi-delta function (10.121); k_F is the value of $|\mathbf{k}|$ for which $\mathscr{E}_k = \mathscr{E}_F$. For any system whose spectrum can be treated as a perturbed free-electron spectrum (§§ 10.4, 10.5), the standard game of counting $\mathbf{k}$-states in a Fermi sphere of radius k_F is still valid and j_F is given, to a very good approximation, by (10.7). It follows from (10.17) that the electrical conductivity of a liquid metal does not depend directly on the density of states $\mathscr{N}(\mathscr{E}_F)$ at the Fermi level, even when strong scattering may make this very different from the free electron value (Mott 1966; Faber 1966; Ziman 1967b; Faber 1972).

At first sight, this absence of any 'effective mass correction' in the NFE formula is surprising. The confusion arises (Ziman 1967b) by unjustified application of the standard formula for the current in terms of the *group velocity* for an electron in a crystal:

$$\mathbf{j_k} = e\mathbf{v_k} = (e/\hbar)\partial\mathscr{E}(\mathbf{k})/\partial\mathbf{k}. \qquad (10.129)$$

This theorem is true for a *Bloch function* ψ_k of wave vector $\mathbf{k}$ and energy $\mathscr{E}(\mathbf{k})$, because the momentum spectrum of ψ_k has discrete peaks at $\mathbf{k}$ and at all points $\mathbf{k} + \mathbf{g}$ reducible to $\mathbf{k}$ in an extended zone scheme (fig. 10.14(a)). The expectation value of the momentum (10.116) in such a state need not be the same as the reduced wave vector $\mathbf{k}$, and is given correctly by (10.129). For a disordered system, however, each eigenstate is a Byzantine function, for which momentum is not a good quantum number, and the dispersion function $\mathscr{E}_L(\mathbf{k})$ is no more than an ill-defined average over a statistical distribution of electronic excitations (fig. 10.8). The theorem (10.129) does not apply to this function and the current must be evaluated from the momentum spectrum (fig. 10.14(b)). If we are to believe (10.87), this spectrum is dominated by the real part of the wave vector $\mathbf{k}$ of the coherent part of the excitation, with phase-incoherent broadening by scattering from the disordered assembly.

In this context, it is worth recalling that the *Hall coefficient* for a simple NFE system of the type discussed in § 10.1 is given by

$$R = 1/nec, \qquad (10.130)$$

where n is the density of conduction electrons and c is the velocity of light.

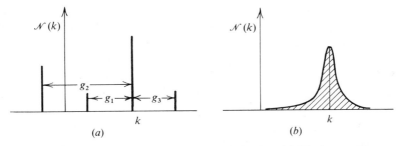

Fig. 10.14. Momentum spectrum of an excitation: (*a*) Bloch state in a crystal; (*b*) electronic state in a liquid.

This result does not depend on the relaxation time τ; for a crystalline specimen, elementary arguments show that it should be independent of any band-width parameter such as an 'effective mass'. But these arguments depend on the validity of the group-velocity theorem (10.129) and do not necessarily hold in, say, a liquid transition metal, where $\mathcal{N}(\mathcal{E}_F)$ may be far from its free electron value (Ziman 1967*b*). It is disappointing that strong scattering corrections to (10.130) have not yet been calculated from, say, the linear response formalism. Indeed the galvanomagnetic properties of liquid transition metals involve so many complicating factors, such as the magnetic moments of the constituent ions, that no theory of the phenomena may be regarded as well established.

Much of the experimental work on topologically disordered metals is concerned with the electrical properties of *liquid alloys* (see e.g. Faber 1972). In principle, the theory of the electronic spectrum and transport properties of such systems is merely a generalization of the theory for monatomic liquids developed in this chapter. Thus, for example, in the NFE formula (10.17) for the electrical resistivity we merely replace the square of the matrix element (10.12) by the corresponding function (4.38) already derived to describe the diffraction of X-rays or neutrons by a mixed liquid. The resulting expression, involving the pseudopotentials (or, presumably, t-matrices) of the constituents, and the various partial structure factors (4.36), looks very complicated, but can be somewhat simplified (cf. § 2.13) if the liquid can be regarded as a *random substitutional mixture* (Faber & Ziman 1965). Putting (4.40) into (10.17) or (10.37), for example, we see that the resistivity of the alloy can be written in the form

$$\rho_L = \bar{\rho}_L + \Delta\rho, \tag{10.131}$$

where $\bar{\rho}_L$ is the resistivity of a liquid with the mean pseudopotential $\bar{u}(q)$, whilst $\Delta\rho$ is proportional to the variance of the pseudopotentials as in (4.41) and is due to the substitutional disorder. Once more, it would be of value to put these formulae on more secure theoretical foundations.

In some cases the effect of alloying is much more drastic than would be described by such formulae, to the extent that the system may approach the condition of a *liquid semiconductor*. From all that we have learnt about 'geometrical' effects on the electron spectrum of metallic liquids (§ 10.5) it seems unlikely that this phenomenon is due to some very special structural rearrangement of the ions (cf. § 11.5) but is usually associated with electron transfer from the conduction band into localized states of one of the constituents to make compact negative ions. Although the role of *electron affinity* in disordered condensed materials is of considerable interest (see e.g. Greenwood & Ratti 1972) these phenomena lie somewhat outside the chosen scope of this work.

11

Excitations of a topologically disordered network

——

'Fine nets and stratagems to catch us in'
Herbert

11.1 Dynamics of liquids and glasses

The theory of the elastic vibrations of a topologically disordered atomic assembly, such as a liquid or a glass, presents a problem that does not arise in the theory of electron states in such systems (chapter 10). In brief, *there is no simple basis for a momentum representation of the collective excitations of the system.*

The model is elementary in principle. We are concerned, as in § 8.1, with small displacements $\mathbf{u}_i$ from the equilibrium 'sites' $\mathbf{R}_i$ of an assembly of atoms of mass M_i. The symbol $\mathbf{\Phi}_{ij}$ represents the tensor force acting in the ith atom as a result of a unit displacement of the jth atom and has the translational invariance (8.2). The eigenvalues of equations analogous to (8.3), i.e.

$$(\mathbf{\Phi}_{ii} - \omega^2 M_i \, \mathbf{1}) \cdot \mathbf{u}_i + \sum_{j \neq i} \mathbf{\Phi}_{ij} \cdot \mathbf{u}_j = 0, \tag{11.1}$$

give the frequencies ω of the classical modes of vibration, whose spectral and other physical properties we are seeking.

The lattice spectrum of a *crystal* is derived by means of the Bloch theorem (1.2), which ensures that a Fourier transform (1.42) reduces the equations of motion to a finite set. This representation in reciprocal space is also a valuable mathematical device for a *substitutionally* disordered system (§ 9.2) even though it does not automatically reduce the problem to finite terms. But as was made clear in chapter 2, a topologically disordered assembly of more than one dimension is not equivalent to a unique, regular lattice, so that there is no canonical basis for such a representation.

In the theory of electron states, we do not necessarily invoke the Bloch theorem, and need not assume the existence of an underlying regular crystal lattice: an electron wave function $\psi(\mathbf{r})$ is a continuous function throughout space, so that a momentum representation in terms of plane waves (10.2) with all values of **k** is always meaningful, if not necessarily mathematically fruitful. But in the dynamical equations (11.1) the atomic displacement variables u_i are only defined at the points R_i and would, therefore, be subject to complicated constraints if represented in a straightforward manner as Fourier integrals over a continuous spectrum of wave vectors. In the dynamical problem, moreover, we have no obvious 'unperturbed' state comparable to the NFE model (§ 10.1) of the electron theory of disordered metals: take away the atoms that define the disorder and nothing is left to vibrate.

Taking an even more austere mathematical viewpoint, what could we learn about the assembly itself by inspection of the coefficients Φ_{ij} in the equations (11.1)? We would soon recognize that the main non-vanishing coefficients must correspond to the elements of the *adjacency matrix* of a graph of 'sites' and 'bonds'. But it would not be immediately obvious that this network was equivalent to an actual three-dimensional assembly of atomic centres, with bonds joining neighbouring sites. The connectivity of the bonds would, in general, seem quite random by comparison with the cyclic ordering of the non-vanishing elements in the corresponding matrix of a regular lattice; it would not be easy to discern in the equations of motion of our model the valuable properties inherent in its geometrical structure. This is the fundamental difficulty with statistical geometry (§§ 2.10, 2.11). Equations defined on a *topologically disordered network* do not solve themselves 'automatically' by internal mathematical manipulations such as group theoretical transformations and representations; to find physically meaningful solutions we must rely heavily on insight derived from the actual assembly described by the equations.

For this reason, the results of numerical computations on various realistic models of amorphous semiconductors (Alben, Weaire, Smith & Brodsky 1975) and of silica glasses (see e.g. Bell 1972; Dean 1972; Böttger 1974) are mainly of interest for direct comparison with experiment. Thus, for example, a 'free end SiO_2' model (fig. 11.1) has a rich spectrum of modes that can be identified empirically with various types of atomic motion. Judging by the variations in the participation ratio (9.130) many of these modes are relatively localized whilst others are well spread through the system. But these calculations make little reference to the overall geometry of the model beyond the fact that Φ_{ij} becomes small or zero as the sites i and j get further apart and (Bell 1974) have not shown up any mysterious or 'pathological' behaviour comparable with the spectral singularities in alloy

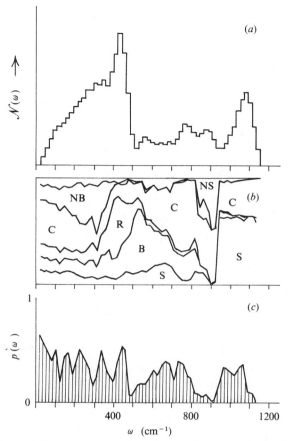

Fig. 11.1. (*a*) Frequency spectrum; (*b*) assignment to various modes; (*c*) participation ratio, for a free end SiO_2 model (Bell & Dean 1970).

models on regular lattices (figs. 8.10, 9.9). It seems, however (Bell, Bird & Dean 1974), that the spectra of two-dimensional glass models are more complex and more sensitive to the type of disorder than the corresponding three-dimensional systems.

11.2 The continuum limit

These mathematical complications show up when we study the equations of motion (11.1) in the long-wave limit. In this limit a glass is, like any other solid, a rigid material with normal, homogeneous, elastic properties and may be considered a macroscopic *elastic continuum* capable of supporting acoustic vibrations and waves. These modes should automatically appear as approximate solutions of the microscopic equations of motion (11.1) and should also show the physical effects of scattering from the underlying structural disorder. This mathematical programme, which should lead us to a theory of short-wave collective modes of glasses and liquids, turns out to be far from trivial.

To calculate the vibrations of an elastic medium of mass density ρ, we apply the classical theory of fields. The atomic displacement $\mathbf{u}_j$ at the site $\mathbf{R}_j$ becomes a continuous vector field, $\mathbf{u}(\mathbf{R})$, which measures the displacement of the medium in the neighbourhood of the point $\mathbf{R}$. The momentum $M_j\dot{\mathbf{u}}_j$ of the atom at this site is generalized into a local *momentum density*

$$\mathbf{p}(\mathbf{R}) = \rho\dot{\mathbf{u}}(\mathbf{R}). \tag{11.2}$$

Instead of writing down equations of motion analogous to (8.1) we set down the total Hamiltonian of our system schematically in the form

$$\mathcal{H} = \tfrac{1}{2}\int[\rho^{-1}\mathbf{p}(\mathbf{R})\cdot\mathbf{p}(\mathbf{R}) + \{\nabla\,\mathbf{u}(\mathbf{R})\}{:}\mathbf{\Phi}{:}\{\nabla\,\mathbf{u}(\mathbf{R})\}]\,d^3\mathbf{R}. \tag{11.3}$$

In this formula the first term is the kinetic energy, whilst the potential energy is quadratic in the components of the local *strain tensor*,

$$\{\nabla\,\mathbf{u}(\mathbf{R})\}_{\alpha\beta} = \partial u_\alpha/\partial R_\beta \quad (\alpha, \beta = x, y, z). \tag{11.4}$$

The components of the fourth rank cartesian tensor $\mathbf{\Phi}$ can easily be translated into the usual *elastic moduli* of the conventional symbolism.

The Hamiltonian is reduced to Bloch diagonal form by a Fourier integral transform. By analogy with (1.42) we introduce collective variables

$$\mathbf{U_q} = V^{\frac{1}{2}}\int\mathbf{u}(\mathbf{R})\,e^{i\mathbf{q}\cdot\mathbf{R}}\,d^3\mathbf{R}; \quad \mathbf{P_q} = V^{-\frac{1}{2}}\int\mathbf{p}(\mathbf{R})\,e^{-i\mathbf{q}\cdot\mathbf{R}}\,d^3\mathbf{R}, \tag{11.5}$$

and easily verify that the total Hamiltonian (11.3) is transformed to

$$\mathcal{H} = \tfrac{1}{2}\sum_{\mathbf{q}}\{\rho^{-1}\mathbf{P_q}\cdot\mathbf{P_q^*} + \mathbf{U_q}\cdot\mathbf{E(q)}\cdot\mathbf{U_q^*}\}, \tag{11.6}$$

where the elastic moduli are contained in the second rank cartesian tensor

$$\mathbf{E(q)} \equiv \mathbf{q}\cdot\mathbf{\Phi}\cdot\mathbf{q}. \tag{11.7}$$

This Hamiltonian is thus diagonal in the reciprocal space representation. For a given wave vector, the three eigenvalues of the matrix (11.7) yield the three mode frequencies; for an isotropic medium such as a glass, we can write this schematically

$$\omega_{\mathbf{q}}^{(\alpha)} = \{E^{(\alpha)}(\mathbf{q})/\rho\}^{\frac{1}{2}}$$
$$= \{\Phi^{(\alpha)}/\rho\}^{\frac{1}{2}}q, \tag{11.8}$$

showing that acoustic waves are propagated without dispersion in a continuum without microscopic structure.

This diagonalization technique is made feasible by the *orthogonality* of plane waves:

$$V^{-1} \int e^{i\mathbf{q}\cdot\mathbf{R}} e^{-i\mathbf{q}'\cdot\mathbf{R}} d^3\mathbf{R} = \delta_{\mathbf{q},\mathbf{q}'}. \tag{11.9}$$

This ensures that (11.6) does not contain products of terms with differing wave vectors and it also ensures that (11.5) defines variables that are *dynamically conjugate*. In the language of quantum field theory, the canonical commutation relations

$$[\mathbf{u}(\mathbf{R}), \mathbf{p}(\mathbf{R}')] = i\hbar\delta(\mathbf{R} - \mathbf{R}') \tag{11.10}$$

for each Cartesian component of the local displacement and momentum are immediately transformed into the corresponding relations

$$[\mathbf{U}_{\mathbf{q}}, \mathbf{P}_{\mathbf{q}}] = i\hbar\delta_{\mathbf{q}\mathbf{q}'} \tag{11.11}$$

for the new variables (11.5). In classical field theory the commutator symbols are, of course, Poisson brackets, with the same fundamental properties. These relations are essential if we are to reduce (11.6) to a sum of simple harmonic oscillator terms (1.45), with the frequencies (11.8).

Returning now to our atomistic model, we observe that the equations of motion (11.1) would come from a Hamiltonian

$$\mathcal{H} = \tfrac{1}{2}\sum_i M_i^{-1}\mathbf{p}_i \cdot \mathbf{p}_i + \tfrac{1}{2}\sum_{i,j}(\mathbf{u}_i - \mathbf{u}_j)\cdot \boldsymbol{\Phi}_{ij}\cdot(\mathbf{u}_i - \mathbf{u}_j), \tag{11.12}$$

which closely resembles (11.3). Can this Hamiltonian be reduced to the block diagonal form (11.6)? If the sites form a regular lattice this is trivial (§ 1.8). Each integral over $\mathbf{R}$ becomes a sum over lattice vectors $\mathbf{R}_i = \boldsymbol{l}_i$; we substitute

$$\mathbf{u}_i = N^{-\frac{1}{2}}\sum_{\mathbf{q}}\mathbf{U}_{\mathbf{q}}\exp\{i\mathbf{q}\cdot\boldsymbol{l}_i\} \tag{11.13}$$

into (11.12), and use the orthogonality condition

$$N^{-1}\sum_i \exp\{i\mathbf{q}\cdot\boldsymbol{l}_i\}\exp\{-i\mathbf{q}'\cdot\boldsymbol{l}_i\} = \delta_{\mathbf{q}\mathbf{q}'} \tag{11.14}$$

(which is valid if all wave vectors are reduced to the same Brillouin zone) to eliminate off-diagonal terms and to preserve the canonical commutation relations (11.11). In this representation, the normal mode frequencies (11.8) are deduced from the eigenvalues of

$$\mathbf{E}(\mathbf{q}) = 2\sum_{ij}\boldsymbol{\Phi}_{i,j}\{1 - \cos\mathbf{q}\cdot(\boldsymbol{l}_i - \boldsymbol{l}_j)\}. \tag{11.15}$$

This expression is well known to be the correct generalization of (11.7) from a continuum model in calculating the phonon spectrum of a crystal.

But when we try the same transformation for a *disordered* system we run into trouble. Suppose we have defined N *wave amplitudes* $\mathbf{U}_{\mathbf{q}}$ such that

$$\mathbf{u}_i = N^{-\frac{1}{2}}\sum_{\mathbf{q}}\mathbf{U}_{\mathbf{q}}\exp\{i\mathbf{q}\cdot\mathbf{R}_i\}, \tag{11.16}$$

exactly as in (11.13). When we substitute in, say, the potential energy term of (11.12) we fail to eliminate the off-diagonal terms corresponding to interactions between modes of different wave vectors. This failure may be traced to the fact that the *phase functions*

$$\psi_{\mathbf{q}}^{i} \equiv \exp\{i\mathbf{q}\cdot\mathbf{R}_{i}\} \tag{11.17}$$

do not form an orthogonal basis for the transformation: instead of (11.9) or (11.14) we get

$$N^{-1}\sum_{i}\psi_{\mathbf{q}}^{i}\psi_{\mathbf{q}'}^{i*} \equiv N^{-1}\sum_{i}\exp\{i(\mathbf{q}-\mathbf{q}')\cdot\mathbf{R}_{i}\}$$
$$= e^{i\xi}S(\mathbf{q}-\mathbf{q}'), \tag{11.18}$$

where ξ is an arbitrary phase variable that depends on the choice of origin and $S(\mathbf{q}-\mathbf{q}')$ is none other than the *structure factor* (4.8). This function certainly has a delta function singularity at $\mathbf{q}'=\mathbf{q}$, but, as we saw in §4.1, for topologically disordered systems it does not vanish automatically for all other values of $\mathbf{q}-\mathbf{q}'$.

The use of non-orthogonal plane-wave representations in the theory of liquids has a long history (Zubarev 1953; Green 1954; Tomonaga 1955; Percus & Yevick 1958, etc.) but is by no means thoroughly mastered. The fundamental difficulty is that a transformation to a non-orthogonal basis set plays havoc with the commutation relations (11.11). This can be avoided formally (Morgan 1969) by introducing a set of functions $\phi_{\mathbf{q}}^{i}$ that are *biorthonormal* to the phase functions (11.17) in the sense that they satisfy the conditions

$$N^{-1}\sum_{i}\psi_{\mathbf{q}}^{i}\phi_{\mathbf{q}'}^{i*}=\delta_{\mathbf{q}\mathbf{q}'}; \quad N^{-1}\sum_{\mathbf{q}}\psi_{\mathbf{q}}^{i}\phi_{\mathbf{q}}^{j*}=\delta_{ij}. \tag{11.19}$$

In terms of these new functions, the defining formulae

$$\mathbf{p}_{i}=N^{-\frac{1}{2}}\sum_{\mathbf{q}}\mathbf{P}_{\mathbf{q}}\phi_{\mathbf{q}}^{i*} \tag{11.20}$$

link the atomic momenta with collective momentum variables $\mathbf{P}_{\mathbf{q}}$ that satisfy the standard commutation relations (11.11) and are, therefore, canonically conjugate to the collective coordinates $\mathbf{U}_{\mathbf{q}}$.

Unfortunately, the inversion of the equations (11.19) to find the unknown functions $\phi_{\mathbf{q}}^{i}$ presents serious difficulties (Morgan 1969), so that we are not much nearer to an exact solution of the problem. Nevertheless, we can substitute from (11.16) and (11.20) into (11.12) and thus obtain an expression for the Hamiltonian in terms of the respectable mechanical coordinates or operators $\mathbf{P}_{\mathbf{q}}$, $\mathbf{U}_{\mathbf{q}}$. If approximations are to be made, it is better that these should be seen as the dropping of particular types of term from the Hamiltonian, rather than within the more mysterious sphere of influence of the commutation relations!

It turns out, in fact, that the diagonal terms in this representation of the Hamiltonian are the same as in (11.6) –

$$\mathbf{E}(\mathbf{q}) = 2 \sum_{ij} \mathbf{\Phi}_{ij} \{1 - \cos \mathbf{q} \cdot (\mathbf{R}_i - \mathbf{R}_j)\}, \tag{11.21}$$

– which is really just the same as (11.15). For an isotropic homogeneous medium, with radial distribution function $g(\mathbf{R})$ (§ 2.7), this means that the three mode frequencies can be calculated from the corresponding ensemble averages:

$$\{\omega_{\mathbf{q}}^{(\alpha)}\}^2 = \frac{2N}{M} \int g(\mathbf{R}) \, \Phi^{(\alpha)}(\mathbf{R}) \{1 - \cos(\mathbf{q} \, \mathbf{R})\} \, \mathrm{d}^3 \mathbf{R} \tag{11.22}$$

where $\Phi^{(\alpha)}(\mathbf{R})$ is an eigenvalue of the force-constant tensor $\mathbf{\Phi}_{ij}$ for sites where $\mathbf{R}_i - \mathbf{R}_j = \mathbf{R}$. In the long wave limit, this formula can be linked with the continuum formula (11.8) by means of identities such as (6.12) which relate the macroscopic elastic moduli to the microscopic interatomic forces. Within this approximation, therefore, our collective modes have the correct physical properties.

The *phonon dispersion formula* (11.22) can be arrived at by many other approximate methods. Thus, for example, (11.16) makes essentially the same assumption (Morgan 1969) as (10.87) in the *coherent-wave approximation* (§ 10.8) for *electron* states in disordered liquids. The long-wave limit is certainly correct; but the peak in $g(R)$ at typical nearest neighbour distances (fig. 2.27) should produce a maximum in $\omega_{\mathbf{q}}$ as the wavelength comes down to this length. In principle, this effect might be observed, in liquids or glasses, by *inelastic neutron diffraction* (§ 4.2). The experimental evidence for liquids, however (Copley & Lovesey 1975), will scarcely bear this interpretation, suggesting that the representation (11.16) in terms of collective modes breaks down completely when the wavelength becomes comparable with the scale of the microscopic arrangement of atoms in the system. We should then go over to a representation of the excitations in terms of more or less localized vibrations – not to mention the irreversible atomic movements that explain the *fluidity* of liquids.

Nevertheless, further information about the physical properties of the system can be obtained (Morgan 1969) by looking at more terms in the Hamiltonian (11.12) in the plane-wave representation (11.16). Thus, for example, the potential energy gives rise to terms like

$$Q_{\mathbf{q}\mathbf{q}'} \sim \Phi_{\mathbf{q}\mathbf{q}'} S(\mathbf{q} - \mathbf{q}') U_{\mathbf{q}} U_{\mathbf{q}'}^*, \tag{11.23}$$

which obviously contains the structure factor because of the non-orthogonality matrix element (11.18). Treating terms of this sort as perturbations on the diagonal part of the Hamiltonian (11.6), we can estimate the effects of interactions between the various plane-wave modes. Thus, from the point of view of a theory of *thermal conduction in glasses* (Morgan & Smith 1974) we can construct an approximate Boltzmann equation in which $Q_{qq'}$ gives rise to transitions between the 'phonon' modes of wave vectors q and q'. It is interesting to note that the rate of scattering would depend on the structure factor, just as in §§ 4.1 and 10.1. In other words, the collective excitations of the assembly are approximately plane waves, which suffer scattering from the microscopic disorder in much the same way as they would if they were a beam of external particles being diffracted from the specimen or if they were the conduction electrons of a metal. But this description is only valid for excitations of long wavelength, which are otherwise scarcely aware of the atomicity of the material.

At low temperatures, where only low frequency modes are excited, the elastic continuum model ought to be a good approximation. Nevertheless, the observed specific heat and thermal conductivity of various glassy materials do not behave as would be expected in the long wave limit (Zeller & Pohl 1971). To explain these discrepancies, it has been pointed out (Anderson, Halperin & Varna 1972; Phillips 1972) that a glassy structure (§§ 2.8–2.10) is not in true thermodynamic equilibrium and is not necessarily a perfectly rigid framework. It is likely to contain an appreciable concentration of nearly unstable local atomic configurations which can easily be thermally excited. At a *softon* (Economou, Ngai & Reinecke 1975) the potential energy of the system has a double minimum (fig. 11.2) as a function of some local coordinate such as the position of a particular atom. If the barrier to movement is not too high, the system may readily tunnel to another configuration of nearly the same energy. Such additional, low-energy, localized states could account phenomenologically for the linear specific heat of the glass at low temperatures and for the strong phonon scattering observed in the thermal conductivity and other physical properties.

This illustrates the difficulty of deducing the physical behaviour of a topologically disordered system from mathematical theories that emphasize statistical characteristics such as atomic correlation functions. Such a formulation cannot deal quantitatively with rare, but significant situations involving, say, a few dozen atoms at a time interacting through complicated interatomic forces. Perhaps this is the reason for the analytical

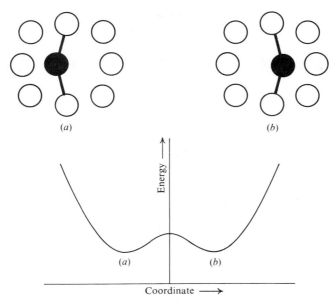

Fig. 11.2. Energy of a 'softon' has two shallow minima corresponding to two alternative equilibrium configurations.

obduracy of the theory of the *fluidity* of liquids, which may be described loosely as the production and motion of softons in a random, close packed assembly (§ 2.11) under the influence of shear stress.

11.3 Ideal tetrahedral coordination

In most glassy networks statistical variations in the local physical parameters, such as coordination number and bond length, generate so much 'noise' that the effects of purely topological disorder are masked. It is possible, however, both in principle and in practice, to produce a material that approximates closely to the ideal *tetrahedral glass model* (TGM), where each atom has exactly four neighbours, arranged with perfect tetrahedral symmetry at exactly the same distance (§§ 2.8–2.10). An ordered crystal having this local structure would be a diamond lattice or one of its many regular polytypes (Joannopoulos & Cohen 1973). From the point of view of any particular atom, the differences between these structures and a topologically disordered network would not be apparent until one reached third neighbours (fig. 2.28). The tetrahedral glass is thus almost perfectly

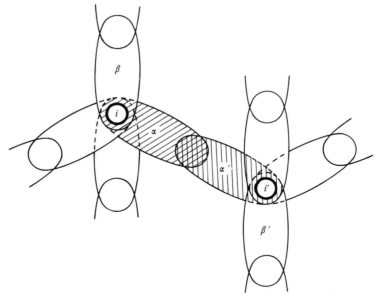

Fig. 11.3. Atomic orbitals in the tetrahedral glass model.

homogeneous on the atomic scale. If the interatomic force constants or other interaction coefficients do not reach beyond next-nearest neighbours, the spectrum of this model is governed entirely by the *connectivity* of the network, whether ordered or disordered.

Although relevant to lattice dynamics (Weaire & Alben 1972; Alben, Weaire, Smith & Brodsky 1975), the mathematical properties of the TGM are usually expressed in the language of electron states. In a tight-binding model Hamiltonian (§ 8.1), the matrix elements $V_{ij}^{(\alpha\beta)}$ would be restricted to interactions between the orbitals $|\alpha, i\rangle$, $|\beta, j\rangle$ on the same atom ($i=j$) or on neighbouring sites in the network. More specifically, we have in mind the familiar sp^3 hybridized orbitals of the carbon atom, which are directed outwards in the tetrahedral directions (fig. 11.3). In the TGM, any such orbital $|\alpha, i\rangle$ can be supposed to overlap strongly with just one other orbital $|\alpha', i'\rangle$ on a neighbouring site i', towards which it is directed as if to form a chemical bond. This, for example, defines the terms in a simple standard Hamiltonian (Weaire & Thorpe 1971)

$$\mathscr{H} = \sum_i \sum_{\alpha,\beta} V_1 |\alpha, i\rangle\langle i, \beta| + \sum_{i,i'} V_2 |\alpha, i\rangle\langle i', \alpha'|, \tag{11.24}$$

where all four orbitals on the same atom interact with matrix element V_1, and orbitals on the same 'bond' have interaction V_2.

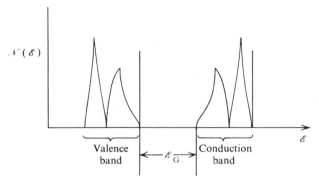

Fig. 11.4. Spectrum of diamond lattice in Weaire model.

On a regular diamond lattice, the spectrum of this Hamiltonian resembles the spectrum of the electron states in crystalline Si or Ge, and the two parameters V_1 and V_2 can be chosen so as to reproduce, say, the width of the 'valence band', and of the 'energy gap' separating it from the 'conduction band' (fig. 11.4). The spectrum of eigenstates of this Hamiltonian for a glassy network should correspond, therefore, to the energy band structure of the amorphous phases of these elements.

It turns out, indeed, *that the existence of a gap in the spectrum of the Weaire model Hamiltonian (11.24) does not depend on the long range order of the network* (Weaire 1971; Weaire & Thorpe 1971). The most direct proof of this theorem (Straley 1972) shows the significance of the regular co-ordination number, $z = 4$, in the tetrahedral glass.

We write (11.24) in the form

$$\mathcal{H} = zV_1\hat{A} + V_2\hat{T}. \tag{11.25}$$

The operator

$$\hat{A} \equiv \sum_i \left\{ z^{-1} \sum_{\alpha,\beta} |\alpha, i\rangle\langle i, \beta| \right\} \tag{11.26}$$

projects any function on to the subspace $\{A\}$ spanned by the local functions

$$|i\rangle \equiv \tfrac{1}{2}\sum_\alpha |\alpha, i\rangle\langle i, \alpha|. \tag{11.27}$$

$\hat{A}$ is, therefore, idempotent, with unit eigenvalues for eigenfunctions in $\{A\}$. The operator

$$\hat{T} \equiv \sum_{i,\ \alpha} |\alpha, i\rangle\langle i', \alpha'| \tag{11.28}$$

simply exchanges the two orbitals at the ends of each 'bond'. From this it

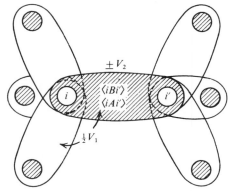

Fig. 11.5. Bond orbital representation in the tetrahedral glass model.

follows that $\hat{T}^2$ is the identity operator, and that the eigenvalues of $\hat{T}$ are ± 1.

Using these algebraic properties, we may verify the relation

$$(\mathscr{H}^2 - zV_1\mathscr{H})\hat{A} = zV_1V_2\hat{A}\hat{T}\hat{A} + V_2{}^2\hat{A}. \tag{11.29}$$

But $\hat{A}\hat{T}\hat{A}$ commutes with $\hat{A}$; these two operators have simultaneous eigenfunctions $|\varepsilon\rangle$ in the space $\{A\}$, satisfying

$$z\hat{A}\hat{T}\hat{A}|\varepsilon\rangle = \varepsilon|\varepsilon\rangle; \quad \hat{A}|\varepsilon\rangle = |\varepsilon\rangle. \tag{11.30}$$

Since $|\varepsilon\rangle$ is an eigenfunction of the right-hand side of (11.29), the corresponding eigenfunction of $\mathscr{H}$ must have energy $\mathscr{E}$ satisfying

$$\mathscr{E}^2 - zV_1\mathscr{E} = V_1V_2\varepsilon + V_2{}^2. \tag{11.31}$$

In other words, the real roots of

$$\mathscr{E} = \tfrac{1}{2}zV_1 \pm \{(\tfrac{1}{2}zV_1)^2 + V_1V_2\varepsilon + V_2{}^2\}^{\frac{1}{2}} \tag{11.32}$$

relate the spectrum of the two-band Hamiltonian (11.25) to the spectrum of a simpler *one-band Hamiltonian*

$$\hat{\mathscr{H}} \equiv z\hat{A}\hat{T}\hat{A} = \sum_{i,i'}|i\rangle\langle i'|. \tag{11.33}$$

This relationship does not take account of any eigenfunctions of $\mathscr{H}$ that happen to lie in the subspace orthogonal to the manifold $\{A\}$ spanned by the functions (11.27). But from the properties of $\hat{T}$, we immediately deduce that these can contribute to the eigenvalues of $\mathscr{H}$ at the two points

$$\mathscr{E} = +V_2, \quad \mathscr{E} = -V_2, \tag{11.34}$$

where they will show up in the spectrum as delta functions, just as in the case of the perfect crystal (fig. 11.5).

To complete the proof we note that the one-band Hamiltonian (11.33) is simply the *connectivity matrix* of the sites that are supposed to be joined by bonds and, therefore, contains z entries, of magnitude unity, in each row and each column, the remaining elements being zero. Using the *Perron–Frobenius theorem* (9.78), we deduce that the spectrum of this matrix must lie in the range

$$-z \leqslant \varepsilon \leqslant z, \tag{11.35}$$

which puts bounds on the two bands deduced from (11.32). It is easy to see that these bands are always separated by a gap of width $|zV_1 - 2V_2|$ which vanishes only for the special case where $V_2 = \frac{1}{2}zV_1$. For the tetrahedral glass, where $z = 4$, inspection of various cases confirms that the delta functions (11.34) do not lie in this gap for any values of V_1 and V_2. The theorem also holds for a diatomic tetrahedral network, where two types of atom occupy alternative sites along any chain of bonds, as in a compound semiconductor (§ 2.10). Alternative proofs of the theorem (Schwartz & Ehrenreich 1972b; Hulin 1972; Huang & Dy 1974) appear to be essentially equivalent to this elementary algebraic demonstration.

In more conventional chemical language (Heine 1971), one may rewrite the two-band Hamiltonian (11.25) in terms of *bond orbitals* $|ii'\rangle$ labelled according to the two sites they link. From the atomic orbitals $|\alpha, i\rangle$ we can construct *bonding* and *antibonding* combinations, respectively:

$$|iBi'\rangle \equiv 2^{-\frac{1}{2}}\{|\alpha, i\rangle + |\alpha', i'\rangle\}, \tag{11.36}$$

and

$$|iAi'\rangle \equiv 2^{-\frac{1}{2}}\{|\alpha, i\rangle - |\alpha', i'\rangle\}. \tag{11.37}$$

On a given bond, these two orbitals are separated in energy by their diagonal matrix elements, $+V_2$ and $-V_2$ respectively: V_2 is, of course, a negative quantity, so that the bonding combination lies lower. Each type is then broadened into a separate band by matrix elements of magnitude $\frac{1}{2}V_1$ between bond orbitals 'hinged' on the same site (fig. 11.5). Thus, the bonding orbitals (11.36) form the *valence band* of the material whilst the antibonding orbitals (11.37) combine to form a *conduction band*. The remaining matrix elements, such as those linking $|iBi'\rangle$ with $|iAi''\rangle$, do not close the gap between these two bands, hence confirming the theorem by another route (Heine 1971). The advantage of this representation of the TGM spectrum is that the parameters governing the widths, spacing and mutual interactions of the bond orbitals can be adjusted independently, to simulate more closely the valence and conduction bands of an actual semiconductor. For each of these bands, further terms can be added to the

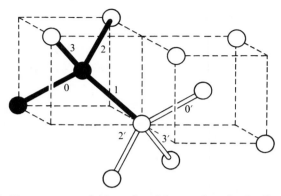

Fig. 11.6. Next-nearest-neighbour bond interactions in the diamond lattice are of two types: (*a*) between parallel bonds, 0 0′; (*b*) between non-parallel bonds, 0 2′.

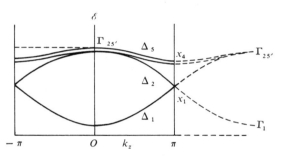

Fig. 11.7. Valence band of diamond lattice.

Hamiltonian to represent interactions between orbitals on more distant bonds. Thus (Ziman 1971*b*), a much more realistic structure for the valence band is obtained by including interactions between 'next-nearest' bonding orbitals (fig. 11.6), which broaden the 'δ-function' at the top of this band in the Weaire model into the familiar 'heavy hole' band, with its triple degeneracy at the centre of the zone (fig. 11.7). Unfortunately, the *conduction* bands of semiconductors are not represented so simply as this by empirical tight-binding formulae involving the *antibonding* orbitals.

The Weaire transformation (11.32) puts all the emphasis on the one-band Hamiltonian $\hat{\mathscr{H}}$, whose spectrum has been computed numerically for various disordered networks (Alben, Heimendahl, Galison & Long 1975). The visible effect of topological disorder (fig. 11.8) is to smudge the sharp peaks and internal zeros of the spectrum of the corresponding perfect

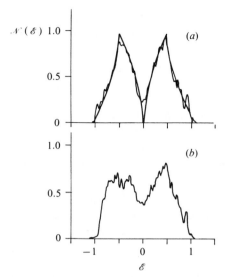

Fig. 11.8. Spectrum of (*a*) 525-unit diamond cluster (with exact lattice spectrum for comparison); (*b*) 500 unit computer-generated random tetrahedral network (Alben *et al.* 1975).

lattice. But the band edges remain well-defined. The upper end of the range (11.35) is always attained by the 'uniform' function

$$|\Psi(+)\rangle \equiv N^{-\frac{1}{2}}\sum_i |i\rangle, \tag{11.38}$$

which is obviously an eigenstate of (11.33). If the network has no odd-numbered circuits (§ 2.10), each site can be assigned odd or even parity p_i. The 'alternating' function

$$|\Psi(-)\rangle \equiv N^{-\frac{1}{2}}\sum_i (-)^{p_i}|i\rangle \tag{11.39}$$

is then also an eigenfunction, with eigenvalue $-z$. Under these conditions, therefore, the spectrum of the disordered system occupies the full range (11.35) and the band gap for the two-band Hamiltonian (11.25) is not affected by the loss of long-range order. These properties are not confined to the Weaire model. It is an amusing exercise in graph theory to show that the upper edge of the valence band (fig. 11.7) of a more general two-band Hamiltonian is similarly attained by 'alternating' combinations of bond orbitals and that the well-known triple degeneracy of the *p*-states at this point in the Brillouin zone can be reproduced without appeal to the Bloch theorem (Ziman 1971*b*).

Indeed, it seems clear that the *functional form of the spectrum of a TGM Hamiltonian near a band edge is almost independent of the long range connectivity of the network.* Consider, for example, 'wave-modulated' versions of (11.38) for various values of **q**, i.e.

$$|\Psi(\mathbf{q}, +)\rangle = N^{-\frac{1}{2}}\sum_{i}e^{i\mathbf{q}\cdot R_i}|i\rangle. \qquad (11.40)$$

For an ordered network these would, of course, be Bloch functions, and eigenfunctions of $\mathscr{H}$. For small values of q, these functions turn out to be very close to eigenstates of the disordered system on a 'local' basis and are very nearly mutually orthogonal (cf. § 11.2). Thus, the spectral density from the Bloch states near the centre of the Brillouin zone is approximately reproduced in the glassy network. A similar modulation of the alternating function (11.39) supplies the spectrum near the other band edge of the one-band Hamiltonian and the same argument can be used in the bond-orbital representation (Ziman 1971*b*). It is worth remarking that wave-modulated functions like (11.40) are extended functions, with nearly uniform amplitude throughout the material. If these are, indeed, good approximations to eigenstates of the Hamiltonian, then we may say that the electron states near the band edges of a tetrahedral glass model are *not localized.* Thus, the *band tails* and *mobility edges* found in the Anderson model (§ 9.9) are not to be expected in these materials.

For a more rigorous discussion of these principles, we may refer to the theory of spectral moments (§ 9.7). As shown formally by Domb, Maraduddin, Montroll & Weiss (1959) the analytical form of the spectral limits of a tight-binding Hamiltonian is governed by the asymptotic behaviour of the moments, μ_p, for large p. From (9.86) and (9.87) we learn that this depends on the number of closed paths of length p on the lattice. It is readily conjectured (Cyrot-Lackman 1972) that this number must depend on the dimensionality and coordination number of the network, but can scarcely be affected over long distances by lack of topological order. This conjecture is made plausible by appeal to the diffusion limit (§ 7.8) for long random walks (Thorpe, Weaire & Alben 1973; Lukes & Nix 1973; Carroll, Lukes, Nix & Ringwood 1974) but does not seem to have been proved rigorously.

11.4 Tree models

In the literature on the theory of amorphous semiconductors, there is some discussion of the electron spectrum of a Bethe lattice or *tree* model (§§ 5.4, 7.5, 9.10). Consider, for example, the one-band tight-binding Hamiltonian

(11.33) for a *regular* tree with coordination number z. The equations (9.32) for the Green functions $G_{ll'}$ in a site representation take the simple form

$$\varepsilon G_{l0} = G_{l-1,0} + (z-1)G_{l+1,0}, \tag{11.41}$$

going out to the lth site along any chain starting from a standard site at $l' = 0$. The homogeneity of the lattice allows us to solve these recurrence relations as a self-consistency condition for, say, the ratio,

$$\alpha = G_{l+1,0}/G_{l0}, \tag{11.42}$$

of successive Green functions. From (9.8) we get the spectral density (Thorpe & Weaire 1971)

$$\mathcal{N}(\varepsilon) = \frac{2}{\pi} \frac{\{4(z-1) - \varepsilon^2\}^{\frac{1}{2}}}{z^2 - \varepsilon^2}, \tag{11.43}$$

which is a simple function, but does not cover the whole range (11.35) of the spectrum of this Hamiltonian for a multiply-connected lattice.

Can anything be learnt about the effects of topological disorder from this model? From the point of view of graph theory, a tree is equivalent in many respects to a one-dimensional chain (§ 5.4). It is easy to see, for example, that the recurrence relations (11.41) may be written in the form of a *transfer matrix* operation (9.19), just as in the theory of the TBA model for a linear chain. In the light of the general theory of § 8.2, it is scarcely surprising that this formalism also applies (Dancz & Edwards 1973) to the *free electron network* model on the same lattice. In this model (see e.g. Montroll 1970) the electrons are supposed to move freely along the one-dimensional 'bonds' of the network, with continuity conditions at each node. But this is merely a generalization of the Kronig–Penney model in one dimension, with transfer matrix (8.24) or (8.27).

For this reason, it is impossible to simulate genuine topological disorder in such a model. A 'glassy tree' is like the 'linear liquid' or 'linear glass' models of §§ 2.2 and 8.2; the disorder introduced by, for example, varying the lengths of the bonds of a free electron network (Dancz, Edwards & March 1973), or by varying the diagonal site energies (Dancz & Edwards 1975), is essentially substitutional and cannot be equated with the irregular connectivity of an actual glassy material. It is, of course, interesting to note that the spectral density for any such model can be derived from the solution of an integral equation involving the probability distribution of, say, the logarithmic derivative of the wave function at each node along a path through the free electron network. But the fact that the equations of motion for all these models can be expressed in the transfer matrix formalism clearly indicates that this is none other than the Dyson–Schmidt condition (8.76) in unfamiliar garb (§ 9.11). The question of electron mobility in such a system is prejudiced by this 'one-dimensionality', since all the wave functions are probably localized by the mechanism of § 8.7.

In assessing the relevance of a tree model, one must also be wary of boundary effects (Runnels 1967). The number of 'surface' sites on any large specimen of such a material is comparable with the number of 'bulk' sites, so that the boundary conditions on the eigenstates play a significant part in determining the spectrum (Ziman 1973). Thus, for example (Nagle, Bonner & Thorpe 1972), the bulk spectrum (11.43) of extended states on an ordered tree must be supplemented with a discrete spectrum of surface states which is not negligible as $N \to \infty$. But these objections do not, perhaps, apply to the device of embedding a finite disordered cluster in an 'infinite' tree lattice, so as to avoid unphysical boundary conditions on the surface of the cluster (Joannopoulos & Yndurain 1974).

11.5 The band-gap paradox

The atomic arrangement of the tetrahedral glass model (§ 2.8) is closely realized in well-prepared specimens of amorphous Si and Ge. Having discussed some of the theoretical properties of models with this structure (§ 11.3) we naturally ask whether these properties are realized in practice. The fundamental question is whether such a material should be a semiconductor or metal. To pose this question dramatically: suppose that a material with this structure was not known on earth but was thought to exist on Mars; what theoretical prediction would be made to help an astronaut to find it?

The answer of theoretical *chemistry* is very easy. The specimen is a single saturated 'molecule', analogous to diamond. Each atom has exactly four neighbours to which it is attached by a typical covalent bond accommodating just two electrons. Since there are four electrons per atom, all the bonds are saturated. It is irrelevant whether or not the overall connectivity of these bonds has the translational symmetry of a crystal lattice. Energy is required to remove an electron from any bond and carry it to some distant point in the specimen. The material must, therefore, be an *insulator* or *semiconductor*.

On the other hand, the conventional wisdom of theoretical *physics* would regard the absence of long-range order as highly significant. Crystalline Si or Ge is a semiconductor because the valence electrons exactly fill the Bloch states in the lowest Brillouin zones of the reciprocal lattice. These states can be represented as nearly free-electron plane waves diffracted by the atomic pseudopotentials; the gaps in the energy spectrum are directly associated with the Bragg diffraction of these waves by the lattice planes of the crystal. In the absence of long range order, there are no lattice planes, no Bragg diffraction, no Brillouin zones – hence, presumably no band gaps. The material ought, surely, to be a *metal*.

The contradition between these two answers to a simple theoretical question is not to be dismissed lightly: it is a paradox that casts doubt upon many accepted ideas in the theory of condensed matter. Unfortunately, the various investigations recorded in this book do not fully resolve it.

Since the 'chemical' approach undoubtedly gives the answer that is consistent with experiment, it is tempting to adopt it without further discussion. This is the attitude implicit in the Weaire model (§ 11.3) where it is taken for granted that the electron states can be adequately represented as linear combinations of atomic orbitals. Since the valence bands in crystalline semiconductors can usually be represented quite satisfactorily by bonding orbitals (11.36), with relatively few interaction matrix elements, this is a tenable position. But the conduction bands in these materials cannot be represented empirically by correspondingly simple combinations of antibonding orbitals (11.37), so that there are too many parameters in the LCAO Hamiltonian (§ 8.1) to apply the Weaire–Thorpe theorem. Chemical theory does not, therefore, 'predict' that there should be a band gap in the disordered material, even though it is consistent, in a crude approximation, with this fact of nature.

In any case, although the electronic structure of a crystalline semiconductor can be simulated by the Bloch states of a tight-binding Hamiltonian, the numerous matrix elements in such a representation cannot easily be calculated from first principles. The 'physicist's method' of assigning to each atom an *empirical pseudopotential* or *form factor* (§ 10.2) is much closer to a practical computational scheme for a band-structure calculation starting from, say, realistic self-consistent atomic potentials. In principle, therefore, the spectrum of the amorphous material should be calculable directly from a model of nearly free electrons interacting with a 'glassy' assembly of such pseudo-atoms. As we saw in § 10.4, scattering by pairs of atoms has a disappointingly weak effect on the electron spectrum, and cannot produce band gaps. But to predict from this that the material ought to be metallic does not take account of higher-order terms in the Edwards series (§ 10.5), where strong correlations in the three-body or four-body distribution functions could have the desired effect. Indeed, it is precisely in

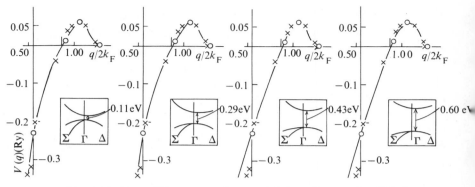

Fig. 11.9. Four different pseudopotential form factors that 'fit' the known values for germanium. In the insets are shown corresponding band structures for the wurtzite lattice near the Γ point (Aymerich & Smith 1973).

the angular variations of these distribution functions (§ 2.10), which are dominated locally by the tetrahedral bonding angle, that glassy assemblies differ from liquids or other types of disordered system. In so far as g_3 (**1, 2, 3**) and g_4 (**1, 2, 3, 4**) for an amorphous material do not differ from the corresponding functions for a *microcrystalline* assembly of the same substance, it might be expected to have a similar spectrum. The fact that the disordered material is a semiconductor thus has a plausible 'physical' explanation (Ziman 1969*b*).

It cannot be taken for granted, however, that local tetrahedral coordination is sufficient, in itself, to produce band gaps. At first sight this seems to follow from calculations of the electronic band structure of ordered *diamond polytypes* of Si and Ge. All the atoms in these hypothetical materials are perfectly tetrahedrally bonded, but the lattice has a large unit cell which may have no more symmetry than a small sample of a glassy assembly. Using the semi-empirical pseudopotential method, Joannopoulos & Cohen (1973) found band gaps in all these systems. But since the form factors could not be calculated exactly from first principles, they had to be obtained empirically from the pseudopotential matrix elements for the real crystalline elements in the diamond structure. These are defined for fewer wave numbers than are needed for the polytype calculations: it is not difficult to construct an alternative, but plausible, scheme of interpolation (fig. 11.9) which nearly closes the band gaps in the more complex structures (Aymerich & Smith 1973). Thus, the semiconducting properties are quite sensitive to the form of the atomic potentials, and in principle one could have a material that was semiconducting in microcrystalline form but metallic as a glass.

Yet another approach to the problem is to replace each atomic potential by a muffin-tin well (§ 10.3) with appropriate phase shifts. This is not a good model for even a typical crystalline semiconductor, because it does not include important effects from the interstitial regions (fig. 10.6); but it permits detailed computations of the local density of states for a quite large disordered 'cluster' of atoms (§ 10.9). A 'pseudo-gap' does begin to appear in this spectrum in the neighbourhood of the band gap of the corresponding crystalline materials (fig. 10.16) but there is no convincing demonstration that this would become an absolute energy gap in the limit of an infinite specimen. Conceptually speaking, this approach does take some account of the 'chemical' bonding mechanism through the p-wave scattering phase shifts, which have a broad resonance in the neighbourhood of the atomic p-levels, but the model is not sufficiently close to reality to give reliable 'predictions' of the properties of hypothetical materials.

From these remarks, we may console ourselves with the thought that the paradox is not a sharp one, since neither the 'chemical' nor 'physical' predictions are precise or reliable. Nevertheless, the theory of condensed matter is challenged by our present inability to reconcile these two points of view within a unified formal scheme.

12
Dilute and amorphous magnets

—

'Almost like a diluted madness'
Carlyle

12.1 The dilute Ising model

The statistical thermodynamics of order–disorder phenomena is best understood for what we might call *regular lattices of 'spins'* (chapter 5). All the sites of a regular lattice are assumed to be occupied by identical multi-state objects which interact uniformly with their neighbours. In other words, the mechanics of such a system is defined by a Hamiltonian (e.g. (5.1)) with the translational symmetry of a perfect infinite crystal.

In general, however, something is now known (chapters 8–11) of the excitation spectra of spatially disordered systems, where the underlying Hamiltonian lacks perfect lattice translational symmetry. It is of interest, therefore, to study the thermodynamic properties of *irregular spin assemblies,* whose sites are not equivalent in occupation, or in relation to their neighbours.

In discussing any such system, it is important to distinguish clearly between the 'disorder' in the spin states and the 'disorder' in the siting and interactions of these objects. In general, we assume that the latter is *quenched* into the system and cannot change. Thus, for example, in a *dilute ferromagnet,* 'magnetic' and 'non-magnetic' atoms are assigned at random to the sites of a lattice, without correlations, as if an alloy had been quenched from a perfectly disordered configuration at a high temperature to a (perhaps metastable) condition in which all the atoms are immobile. In other words, throughout the analysis, regardless of, say, the temperature assumed for the ensemble of spin states, the probability that a given site is occupied by a spin is independent of other occupations and remains constant. On the other hand, the probability distribution of the values of the spin components over the 'magnetic' sites is governed by statistical mechanics and varies with temperature T and magnetic field H; this

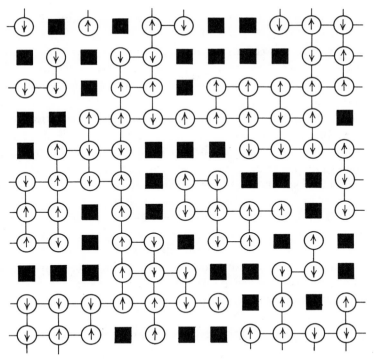

Fig. 12.1. Dilute Ising model.

disorder is always *annealed* to minimize, say, the free energy under the given constraints.

The theory of irregular spin assemblies is surely no less complicated than that of the ordered systems from which they derive. We have a choice from any of the basic models of spins and interactions (§ 5.1), with a variety of types of substitutional or topological disorder to be quenched into the Hamiltonian. It is not surprising, therefore, that attention has been concentrated on a few special cases, of which the *dilute Ising model* is obviously the most elementary (fig. 12.1). This may be defined by a Hamiltonian analogous to (1.18)

$$\mathcal{H}(p) = -\tfrac{1}{2}\sum_{l,l'} \xi_l \xi_{l'} J\sigma_l \sigma_{l'} - \bar{\mu}H \sum_l \xi_l \sigma_l \qquad (12.1)$$

where ξ_l is a random variable with values 0, 1 and average $\langle \xi_l \rangle = p$ describing the occupation of the lth site by a 'spin' whose state is defined, as usual, by $\sigma_l = \pm 1$.

This model (see e.g. Essam 1972) has quite a long history. It was early realized (Sato, Arrott & Kikuchi 1959; Elliott, Heap, Morgan & Rushbrooke 1960) that the critical temperature $T_c(p)$ for a phase transition must decrease monotonically as the magnetic concentration p is reduced and that $T_c(p)$ must in fact vanish at some non-zero value of p which would depend essentially on the topology of the embedding lattice. Indeed, it is quite obvious (Domb & Sykes 1961) that this must be simply the critical concentration p_c^S for *site percolation* (§ 9.10) in this lattice. To prove this formally (Kikuchi 1970) it is only necessary to remark that any isolated cluster containing a *finite* number of magnetic sites has its identical twin elsewhere in the (infinite) specimen. At zero temperature, in zero field, these two clusters must, by symmetry, be equally and oppositely magnetized. Thus, for $p < p_c^S$, when *all* such clusters are finite, the system has no net magnetization. On the other hand, for concentrations above the percolation threshold, there exists at least one infinite connected cluster of magnetic atoms. In fact, this cluster is unique, since there is only a vanishing probability that two infinite clusters could be so isolated from one another by non-magnetic atoms as not to have at least one connecting link. The magnetization of this infinite cluster is thus observable as the net magnetic moment of the system. Indeed, it is obvious that the average saturation spin polarization (5.5) per site at $T = 0$, $H = 0$ must be equal to the *percolation probability* function,

$$\langle \sigma \rangle = P(p), \tag{12.2}$$

as defined in § 9.10.

The corresponding, less physical, *quenched bond model,* in which a fraction $(1 - p)$ of the exchange interactions are artificially cut out, has analogous properties, with a critical concentration at the bond percolation threshold p_c^B for the lattice in question. Formal theorems (Griffiths & Lebowitz 1968; Griffiths 1972) confirm that this behaviour at $T = 0$ is a special case of a phase transition that must occur when the system is made to cross a critical curve $T_c(p)$ in the temperature–concentration plane (fig. 12.2)

But the detailed properties of this critical phenomenon are not yet well understood, and demand elaborate mathematical analysis. Thus, to locate the transition, it is necessary to apply the full machinery of graphical expansions (cf. § 5.10) extrapolating from high temperatures or high concentrations (Rapaport 1972). It seems fairly clear, however, that $T_c(p)$ rises with infinite slope from the percolation threshold p_c, (cf. fig. 9.21), but soon conforms quite closely to a straight line directed to the transition

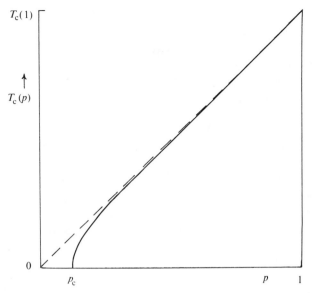

Fig. 12.2. Critical temperature of dilute Ising ferromagnet as a function of concentration.

point, $p = 1$, $T = T_c(1)$, for the 'perfect' system. In other words, to a good approximation

$$T_c(p) \approx pP(p)T_c(1) \tag{12.3}$$

as if our system consisted, as in (12.1), only of the spins in the 'percolating clusters', with mutual interactions uniformly reduced in proportion to the average magnetic dilution of the system.

Although scaling arguments (§ 5.12) suggest some properties of the critical exponents of the transition (Essam, Gwilym & Loveluck 1976), it is not absolutely certain that this is perfectly sharp (Harris 1974). Could it be that the quenched-in disorder rounds off the phase transition in the spin coordinates? The best argument against this is a formal transformation of the Hamiltonian (12.1) to a much more complicated expression with the translational invariance of the lattice (Grinstein & Luther 1976). To a good approximation, no peculiar behaviour shows up when the renormalization group technique is applied to this expression.

It can be shown, however (Griffiths 1969), that the magnetization of a dilute ferromagnet cannot be an analytic function of H at $H = 0$ at any temperature below $T_c(1)$. In other words, peculiar magnetic properties might be observed in the temperature range $T_c(p) < T < T_c(1)$, where normal paramagnetism would be expected. The essence of this argument is that there is always a *finite* probability $W_n(p)$ of finding a relatively compact interacting cluster of n magnetic atoms in the alloy, however big the value of n. Thus, for example, referring back to the derivation of the Lifshitz formula (9.83) for band tailing in a TBA model, which depends upon (9.82), we write

$$W_n(p) \sim e^{-An} \tag{12.4}$$

where A is a number like $\ln (1/p)$. As n increases, the magnetic properties of such a cluster must approximate more and more closely to those of an infinite regular specimen with critical temperature $T_c(1)$ – for example the susceptibility must be very nearly singular as the magnetic field passes through zero. Fortunately, this pathological behaviour cannot be of any practical significance (Wortis 1974; Harris 1975). To show any effect in a small field, of magnitude $|H|$, one would need a cluster of at least the size $n \sim \ell T_c/\bar{\mu}|H|$. From (12.4), we see that such clusters are extremely rare; mathematically speaking, the singularity in $M(H)$ has the form of a branch cut along the imaginary axis, with exponentially negligible weight $\exp\{-\text{const}/|H|\}$ for $|H| \to 0$.

12.2 Dilute Heisenberg magnets

By analogy with the dilute Ising model, we may consider a *dilute Heisenberg model,* with Hamiltonian

$$\mathcal{H}(p) = -\tfrac{1}{2}\sum_{l,l'}\xi_l\,\xi_{l'}\,J(\mathbf{R}_{ll'})\mathbf{S}_l\cdot\mathbf{S}_{l'} - \bar{\mu}\sum_l\xi_l\,\mathbf{S}_l\cdot\mathbf{H}, \tag{12.5}$$

as in (1.16) and (12.1). In this formula, the symbol $\mathbf{S}_l$ may be considered a classical vector or a quantum mechanical operator. But the thermodynamic properties of the classical Heisenberg model are known exactly only for the one-dimensional chain, whilst for the corresponding quantum-mechanical system a very elaborate analysis is required even to find the ground state in the antiferromagnetic case (§ 5.6). The properties of dilute Heisenberg models are thus only known very schematically.

There does not seem to be any objection to the general argument of Kikuchi (§ 12.1) that there can be no phase transition when the concentration of magnetic atoms falls below the site percolation threshold p_c^S. It is not perfectly obvious, however, that a magnetically polarized phase *must* be thermodynamically stable near $T = 0$ when p just exceeds p_c. It could be, for example (see e.g. Kirkpatrick 1973), that the topology of the infinite connected cluster has not sufficient 'dimensionality' to resist fluctuations away from the ordered Heisenberg ground state (cf. § 5.6). This theoretical possiblity cannot be eliminated (Kaneyoshi 1975) by appeal to any 'effective medium' (§§ 5.2, 9.4) or small cluster approximation (§§ 5.4, 10.9). Nor is it inconsistent with the form of $T_c(p)$ suggested by the extrapolation of the high-temperature expansions down to low values of p. It seems that *bond* dilution of the quantal Heisenberg ferromagnet drives $T_c(p)$ with almost perfect linearity towards zero at the bond percolation threshold p_c^B (Brown, Essam & Place 1975), whilst the corresponding line for the site-diluted model (Rushbrooke, Muse, Stephenson & Pirnie 1972) points clearly towards an intercept above p_c^S The effects of dilution on the critical properties of cooperative assemblies thus depend on the nature of the model and there is no obvious universal scaling principle related to the onset of percolation.

The stability of the ordered phase against low-energy excitations can also be studied. The long-range order of a regular Heisenberg ferromagnet below T_c is reduced by the excitation of *magnons* or *spin waves* (§ 8.1). These are usually represented in terms of spin deviation operators (8.5), i.e.

$$S^{\pm} \equiv S_l^{(x)} \pm i \, S_l^{(y)}, \tag{12.6}$$

which measure small deviations of the spin on the lth site away from the magnetization direction z. For weak magnon excitations, the Hamiltonian (1.16) can be represented by the quadratic terms in these variables. For our diluted Hamiltonian (12.5) we suppose that the labels i, i' apply only to pairs of 'magnetic' sites of the lattice, linked by a nearest-neighbour exchange parameter J. From the approximate Hamiltonian

$$\mathcal{H}(p) \approx -\sum_{i,i'} J\{S^2 + S(S_i^+ S_{i'}^- + S_i^- S_{i'}^+ - S_i^- S_i^+ - S_{i'}^- S_{i'}^+)\}, \tag{12.7}$$

we may deduce linearized equations of motion for these operators in the form (8.6), i.e.

$$i\hbar \frac{d \, S_i^-(t)}{dt} = 2SJ \sum_{i'} \{S_i^-(t) - S_{i'}^-(t)\}, \tag{12.8}$$

which are to be solved for the magnon modes.

From the point of view of graph theory, the magnetic sites form a topologically disordered network, represented by a *connectivity matrix* (11.33). In the notation of § 11.3, the solutions of the equations of motion (12.8) for a given frequency ω_α must be an eigenfunction of the *differencing matrix,*

$$\Delta \equiv \sum_{i,i'} \{|i\rangle\langle i| - |i\rangle\langle i'|\}, \tag{12.9}$$

whose components measure the differences of any excitation between connected sites. In other words, $\hbar\omega_\alpha = 2SJ\lambda_\alpha$, where

$$\lambda_\alpha|i; \alpha\rangle = \sum_j \Delta_{ij}|j; \alpha\rangle. \tag{12.10}$$

By definition, the differencing matrix is a direct sum of separate matrices for disconnected clusters: since we are interested in the spin waves that reduce the long range magnetic polarization of the specimen we need only consider the eigenstates (12.10) defined on the infinite cluster that exists above the percolation threshold. In general, these will not be eigenfunctions of the crystal momentum, but we can always evaluate a macroscopic 'momentum vector'

$$\mathbf{q}_\alpha = \langle\alpha; \mathbf{R}|(\hbar/i)\,\nabla\,|\mathbf{R}; \alpha\rangle \tag{12.11}$$

where $\mathbf{R}$ is the actual position of the ith site in space. In the continuum limit, where the lattice constant, a tends to zero, the differencing matrix tends to the Laplacian operator $-a^2\nabla^2$. For small values of $\mathbf{q}_\alpha$, the eigenvalues of this matrix must be quadratic in the 'momentum', i.e.

$$\lambda_\alpha \approx a^2 q_\alpha^2 D(p) \tag{12.12}$$

where $D(p)$ is a dimensionless quantity that depends only on the structure of the underlying crystal lattice and on the magnetic site concentration p. In the case of a regular ferromagnet where $D(1)=z$, (12.12) reduces to the magnon dispersion formula (1.47).

Now, on the other hand, consider *maze conduction* (§ 9.11) on this same topologically disordered network. To each 'magnetic' site we assign a capacitance C, and to each connecting link a conductance σ_0. From Kirchhoff's laws, the equations of motion for a time-dependent potential $V_i(t)$ at the ith node is

$$\frac{\mathrm{d}V_i(t)}{\mathrm{d}t} = -(\sigma_0/c)\sum_{i'}\{V_i(t) - V_{i'}(t)\}. \tag{12.13}$$

Apart from a factor i on the left, these equations are of the same form as the spin-deviation equation (12.8). The solution of the maze-conduction problem can thus be related to the properties of the eigenfunctions $|\alpha\rangle$ of the differencing matrix (12.9).

The 'macroscopic' static conductivity $\sigma(p)$ of this network must be the zero-frequency, long-wave limit of the dissipative part of the response function to an external field. The classical analogue of the fluctuation–dissipation theorem leads directly (Brenig, Wölfle & Dohler 1971) to an expression equivalent to the Kubo–Greenwood formula (10.117):

$$\sigma(p) = \lim_{\omega > 0} \left\{ \lim_{q \to 0} \left[\left(\frac{\omega^2 C^2}{q^2} \right) \sum_\alpha |\langle \mathbf{q} | \alpha \rangle|^2 \frac{\lambda_\alpha}{\lambda_\alpha^2 + \omega^2 C^2} \right] \right\}. \quad (12.14)$$

In the long wave limit, the Fourier transform $\langle \mathbf{q}|\alpha \rangle$ of an eigenfunction of Δ tends to that of a plane wave of momentum defined by (12.11). But this wave only exists on the sites of the infinite connected cluster and therefore introduces, as normalization factor, the percolation probability function defined in § 9.10. Using (12.12) and taking the limits in (12.11) we get a remarkable relationship (Kirkpatrick 1973)

$$\sigma(p) = \sigma_0 \, a^2 D(p) P(p) \quad (12.15)$$

between the *spin wave stiffness parameter* $D(p)$ of the dilute ferromagnet and the *bulk conductance* $\sigma(p)$ of the corresponding network.

As we saw in § 9.11, the 'percolative mobility',

$$\mu(p) = \sigma(p)/P(p), \quad (12.16)$$

is not constant but decreases to zero as we approach p_c (see Butcher 1974a). From (12.15) we deduce that the magnetic ordering of the dilute Heisenberg ferromagnet becomes unstable to long-wave magnon excitations precisely at the percolation threshold p_c, in support of the well-founded conjecture that this is, indeed, the magnetic transition point of the system, where $T_c \to 0$ (Last 1972).

Away from the percolation threshold, where the properties of the system are no longer dominated by the connectivity of very large clusters, the spin-wave spectrum can be determined approximately by the general methods of chapter 9. Since magnon excitations in ferromagnets and antiferromagnets are mathematically similar to phonon and electron states (§ 8.1), the spectral theory of the TBA model leading to the CPA method (§ 9.4) can be applied with appropriate modifications and elaborations (Jones & Edwards 1971; Buyers, Pepper & Elliott 1972; Elliott & Pepper 1973; Harris, Leath, Nickel & Elliott 1974; Holcomb 1974, 1976). Improvements to this approximation to take into account the effects of the local environment (Nickel 1974; Theumann 1974; Roth 1976; Salzberg, Gonçalves da Silva & Falicov 1976) lead into the same mathematical area as the corresponding discussion for lattice waves and electron states in substitutional alloys (§§ 9.5–9.7).

12.3 Amorphous ferromagnets and spin glasses

In itself, *topological* disorder can scarcely have any effect on a magnetic phase transition. Suppose, for example, that we had a *tetrahedral glass*

(§§ 2.8, 11.3), with each site occupied by a magnetic ion interacting ferro-magnetically with its four neighbours. There is little doubt that such a material would behave in almost all respects as a typical regular ferromagnet. Comparison with the theory of electron states in such systems (§ 11.3) suggests that the only effects of the lack of long range crystalline order would be a smearing of the peaks and troughs in the spin wave spectrum, leaving the transition temperature T_c practically unchanged.

In practice, *amorphous ferromagnets* are *amorphous metals* or *glassy alloys* (§ 2.13) in which magnetic and non-magnetic ions are randomly packed, as in liquids. Quite apart from any effects of magnetic dilution (§ 12.1), the local disorder varies the spacing between pairs of magnetic ions, hence varying the exchange integrals J_{ij} between the corresponding spins. But provided these interactions are predominantly ferromagnetic, there is no difficulty in applying the mean field approximation (§ 5.2) at each site and taking averages over various local configurations of sites to estimate the magnetic transition temperature of the system (Gubanov 1960; Handrich 1969; Kaneyoshi 1973*a*, *b*). The delicate question whether the 'off-diagonal disorder' (§ 9.8) in the spin Hamiltonian can affect the sharpness of the transition has not been fully settled, but an application of the renormalization group technique (§ 5.12) to a rather general model suggests that a small amount of randomness in the interaction parameters does not change the character of the phase transition (Lubensky 1975).

In extreme cases, however, the interaction parameter J_{ij} may be such a sensitive function of distance that the interaction between the spins at $\mathbf{R}_i$ and $\mathbf{R}_j$ may have no predominant sign. This is found, for example, in dilute random alloys (1–10 per cent) of certain transition metals (e.g. Mn) in certain noble metals (e.g. Cu). The long-range, oscillatory 'RKKY' inter-action between the spins acts over a substantial volume of the gas-like distribution of magnetic ions; any given spin is thus subject to a large number of interactions, adding up to an effective field

$$\mathbf{H}_i^{eff} = \sum_{j \neq i} J_{ij}\, \mathbf{S}_j, \qquad (12.17)$$

which is random in magnitude and direction.

We may suppose, nevertheless (Marshall 1960), that this assembly has a magnetic ground state in which each spin vector has a definite direction $\mathbf{S}_i^0$ which is parallel to the effective field $\mathbf{H}_i^0$ produced by the other spins in *their* preferred orientations. But because the interaction parameters J_{ij} are seem-ingly random, with zero mean value, this state does not correspond to a ferromagnet, where all the spins are polarized in the same direction, nor to

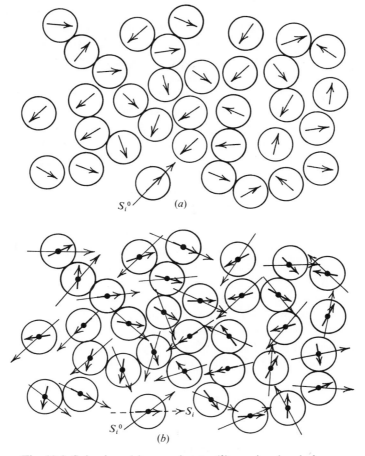

Fig. 12.3. Spin glass: (*a*) ground state; (*b*) quasi-ordered phase.

an antiferromagnet, where they point in opposite directions in equal numbers. In other words, at $T = 0$, a *spin glass* is *mictomagnetic* or *sperimagnetic* (Coey & Readman 1973), with no net magnetization nor any regular pattern of local ordering of the spins (fig. 12.3).

In this supposition that such a ground state is unique we ignore, of course, the rotational symmetry of each interaction term $J_{ij}\,\mathbf{S}_i\cdot\mathbf{S}_j$, which would allow the whole system of spins $\mathbf{S}_i^0$ to rotate in unison. Presumably this 'symmetry' of the sperimagnetic phase is 'broken' by dipolar and other interactions of the kind responsible for the anisotropy fields in ferromagnets (Harris, Plischke & Zuckermann 1973). We also neglect the problems associated with defining the quantum ground state of the Heisenberg Hamiltonian when this is not of maximum total spin (§ 5.6). The fundamental question is whether this is the lowest state of a *thermodynamically* stable phase, analogous to an ordered magnetic phase, to which there is a sharp transition at a well-defined temperature T_c.

This is an exceedingly difficult problem of statistical mechanics to which there is not yet a definite answer. But if there really is a phase transition, it is easy to estimate the critical temperature (Edwards & Anderson 1975). The essential point is that each spin $\mathbf{S}_i$ may still 'know' its *own* 'preferred orientation' $\mathbf{S}_i^0$, regardless of the fact that there is no correlation between these orientations for spins on *different* atoms. Thus, for the thermodynamic ensemble of spin configurations in equilibrium at temperature T, we can define a 'local' order parameter (§ 5.2).

$$\mathscr{S}_i \equiv \langle \mathbf{S}_i \cdot \mathbf{S}_i^0 \rangle / |\mathbf{S}_i^0|^2 \tag{12.18}$$

which is not necessarily zero. This is deemed to be an 'ordered' phase, to the extent that the thermodynamic average of the component of each spin along its preferred direction does not vanish, as it would in a normal paramagnetic phase (fig. 12.3 (*b*)).

The complete partition function for this ensemble cannot be calculated exactly. But the effective *mean field* (12.17) that polarizes $\mathbf{S}_i$ can be estimated approximately (cf. § 5.2) by giving each of the other spins $\mathbf{S}_j$ its average magnitude $\mathscr{S}_j \mathbf{S}_j^0$ (Sherrington 1975; Southern 1975; Kaneyoshi 1976):

$$\langle \mathbf{H}_i \rangle \approx \sum_{j \neq i} J_{ij} \, \mathscr{S}_j \mathbf{S}_j^0. \tag{12.19}$$

From the general formula (5.4) for the thermodynamic average of a single spin $\mathbf{S}_i$ in this field at the temperature $kT = 1/\beta$, we can compute

$$\mathscr{S}_i = \mathscr{F}\left\{ \beta \sum_{j \neq i} J_{ij} (\mathbf{S}_i^0 \cdot \mathbf{S}_j^0) \mathscr{S}_j \right\} \tag{12.20}$$

where $\mathscr{F}(\beta\mathscr{E})$ is a function that depends on whether we regard the symbol $\mathbf{S}_i$ as an Ising spin as in (5.5), a classical vector as in (5.72), or a true quantum-spin operator.

The complete set of equations (12.20) are to be solved for a non-zero set of values of the order parameters $\mathscr{S}_i$. To decide whether this is possible, even for small values of $\mathscr{S}_j$, let us recall that $\mathscr{F}(\beta X)$ is always linear for small values of X: in this neighbourhood, we write (12.20) in the form

$$\mathscr{S}_i \approx \alpha\beta \sum_{j \neq i} J_{ij} \mathbf{S}_i^0 \mathbf{S}_j^0 \mathscr{S}_j \tag{12.21}$$

where α is a number, of the order of unity, depending on the nature of the spins.

Now suppose we try to solve these equations by iteration:

$$\mathscr{S}_i \approx (\alpha\beta)^2 \sum_{j\neq i} \sum_{k\neq j} (J_{ij}\,\mathbf{S}_i^0 \cdot \mathbf{S}_j^0)(J_{ik}\,\mathbf{S}_j^0 \cdot \mathbf{S}_k^0)\mathscr{S}_k + \ldots \tag{12.22}$$

In setting up our model, we assumed that the exchange parameters J_{ij}, J_{jk} between pairs of spins are essentially independent random variables, with mean zero. Let us take averages, such as

$$\mathscr{S} = N^{-1} \sum_i \mathscr{S}_i \equiv \{\mathscr{S}_i\}_{\text{specimen}}, \tag{12.23}$$

over all spin sites of the specimen. In any such averaging of (12.22), the only terms that survive on the right are those for which $k = i$. Thus, the over-all order parameter (12.23) must satisfy a self-consistency condition

$$\mathscr{S} \approx (\alpha/kT)^2 \left\{ \sum_{j\neq i} (J_{ij}\,\mathbf{S}_i^0 \cdot \mathbf{S}_j^0)^2 \right\}_{\text{specimen}} \mathscr{S}. \tag{12.24}$$

By analogy with (5.5), we see that this locates the phase transition at the critical temperature

$$kT_c \approx \alpha \left\{ \sum_{j\neq i} (J_{ij}\,\mathbf{S}_i^0 \cdot \mathbf{S}_j^0)^2 \right\}_{\text{specimen}}^{\frac{1}{2}}. \tag{12.25}$$

This result is obviously of great interest for comparison with experiment. But as with other thermodynamic formulae deduced within the mean field approximation (cf. § 5.11), further calculations concerning the shape of the susceptibility curve near the critical temperature and the behaviour of the magnetic specific heat at low temperatures cannot be taken very seriously. It is evident that several steps in the above derivation are very sensitive to fluctuations of the various thermodynamic and structural variables about their mean values, which might drastically modify the physical behaviour of the system.

13

Electrons in 'gases'

13.1 Gas-like disorder

An electron in condensed matter can usually be supposed to see a one-electron potential which is approximately the sum of contributions from distinct 'atomic' centres. Schematically, we write (once more!)

$$\mathscr{V}(\mathbf{r}) = \sum_i v(\mathbf{r} - \mathbf{R}_i), \tag{13.1}$$

where, for simplicity, we assume that each centre $\mathbf{R}_i$ carries the same spherically symmetrical potential $v(\mathbf{r})$. Since the total potential $\mathscr{V}(\mathbf{r})$ is bounded, $v(\mathbf{r})$ may be assumed to have a finite range r_p, beyond which it makes no appreciable contribution to (13.1).

In systems such as substitutional alloys (chapter 9), liquid metals (chapter 10) and glassy semiconductors (chapter 11), where the atomic arrangements are relatively close packed, we have usually been able to exploit the 'atomicity' of the total potential (§ 2.1). Even when our assembly is topologically disordered, the *weak cellular approximation* (2.2) may still apply to $\mathscr{V}(\mathbf{r})$ over most of the volume of the material. This is obvious if r_p is not much larger than the geometrical 'hard sphere' radius σ, of an atom; each 'cell' of the material then contains just the potential $v(\mathbf{r})$ (cf. § 10.3). But this approximation remains valid for atomic potentials of longer range, provided that the volume density of packing is relatively homogeneous locally, as in typical models of liquid disorder (§ 2.11).

But these assumptions break down as the assembly approaches the ideal of perfect *gas-like disorder* (§ 2.15). In this model, the atomic centres $\{\mathbf{R}_i\}$ are distributed at random, with average atomic volume

$$V/N = \frac{4}{3} \pi r_s^3 \tag{13.2}$$

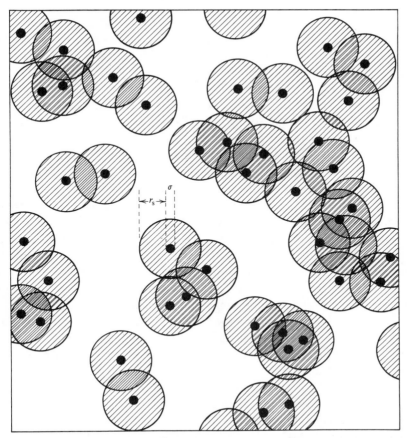

Fig. 13.1. Gas of atoms of radius σ (black circles) showing overlaps of 'atomic spheres' of radius r_s.

without constraint on the distance of nearest approach. Physically speaking, we can never shrink our atoms into geometrical points; but when the *packing fraction* (§ 2.11)

$$\eta = (\sigma/r_s)^3 \tag{13.3}$$

is small this is a fair approximation. Under these circumstances, the Voronoi cells (fig. 2.42) fluctuate wildly in volume about the mean value (13.2) and the cellular approximation (2.2) is no longer valid for $\mathscr{V}(\mathbf{r})$ (fig. 13.1).

Everything now depends on the nature of the 'atomic' potential $v(\mathbf{r})$. If this has a central singularity capable of accommodating an electron in an atomic state $\psi_a(\mathbf{r})$, the electronic properties of the system are dominated by the spatial distribution of these singularities in $\mathscr{V}(\mathbf{r})$. We may assume that a perturbation expansion in the *locator* representation (§ 9.3) is valid and apply the tight-binding phenomenology (§ 9.1). This is essentially the approach taken in §§ 13.2 and 13.3.

On the other hand, if $v(\mathbf{r})$ is a weak *pseudopotential* (§ 10.2) without bound states, we naturally apply the NFE method used for simple liquid metals (§§ 10.4, 10.4); in other words, a *propagator* representation is appropriate (§ 9.3). But even when $v(\mathbf{r})$ has no bound states, the validity of the NFE perturbation expansion depends not so much on the magnitude of $v(\mathbf{r})$ as on the magnitude of the *total* potential $\mathscr{V}(\mathbf{r})$ relative to the electron energy $\mathscr{E}$. This, in its turn, depends on the number density N/V of atomic centres in the gas. Thus, when the typical interatomic distance $2r_s$ becomes smaller than the range r_p of each atomic potential, we may be sure that (13.1) usually contains contributions from many different centres and an expansion such as (10.50) in powers of $N\langle v^2 \rangle$ may not be convergent.

In such circumstances, it is preferable to ignore the underlying atomicity and to define $\mathscr{V}(\mathbf{r})$ as a continuous random function (chapter 3) with a characteristic spectral distribution (3.9). A considerable literature on the theory of *wave propagation in random media* (e.g. Howe 1973; Uscinski 1974) consists essentially of a reformulation of the propagator representation of electron states in liquid metals (chapter 10) in terms of the diffraction of electromagnetic or acoustic waves in a disordered continuum (§ 4.4). But these formalisms do not go further along the perturbation series than we have already reached in §§ 10.4–10.8, and the gas-like disorder underlying the random potential is so featureless that we may expect little of interest in the spectral or transport properties of such a system. We need not consider this case further.

To assess the convergence of such a perturbation expansion, we may estimate the *relaxation time* τ for scattering by the fluctuations of $\mathscr{V}(\mathbf{r})$, just as in § 10.1. The expansion should be valid if the *mean free path* of an electron (or other excitation) is considerably longer than the *correlation length* of the random field (§ 3.3). This is equivalent to saying that the uncertainty in electron wave number is much smaller than the spectral range of the perturbing potential, so that a Fourier representation of the Schrödinger equation is a good starting point.

But when this condition is not well satisfied, the fluctuations of the potential cannot be treated as perturbations on simple plane-wave states. At some points, for example, $\mathscr{V}(\mathbf{r})$ may become larger than $\mathscr{E}$, so that a *classical* electron would be completely excluded from certain regions of the specimen. Despite the fact that a *quantum mechanical* electron could tunnel through such regions, this raises interesting and novel mathematical issues, which we enter in § 13.4.

13.2 The metal–insulator transition

The first point to note about the tight-binding method is that it also must fail at high densities of the atomic gas. The scale length in this case is not the range of each atomic potential, but the characteristic radius, r_a of each atomic wave function ψ_a. For an impurity level in a semiconductor, for example, this would be a few times the Bohr radius a_H of the hydrogen-like states (§ 2.15). When r_s is comparable with r_a, we have many 'di-atomic', 'tri-atomic' and more complex clusters of singularities of $\mathscr{V}(\mathbf{r})$, whose bound states would merge into 'molecular' levels, and would not be counted as distinct (fig. 13.2).

To discuss the statistical distribution of such clusters we must suppose that the radius r_a defines the sphere of influence occupied by each atomic state and that two such states are quite distinct unless their spheres touch or overlap. An electron put on to any atomic centre is thus assumed to have access only to the region occupied by a connected cluster of mutually overlapping spheres. This assumption makes no allowance for tunnelling through the potential barriers between neighbouring atomic states (§ 13.3); but since the wave functions ψ_a fall off exponentially with distance this is not entirely unrealistic as a zeroth-order approximation.

It is obvious that this model defines a characteristic *percolation* problem (Holcomb & Rehr 1969). This system will behave macroscopically like an insulator until the gas density N is sufficiently high to allow the appearance of an infinite cluster of overlapping spheres through which, one supposes, the electron would move freely. In other words, this percolation threshold would be seen as the critical density for a transition to metallic conduction, analogous to the Anderson transition (§ 9.9) and the maze conduction transition (§ 9.11) on a regular lattice.

Unfortunately, the mathematical results obtained for site and bond percolation on lattices (§ 9.10) cannot be generalized by rigorous arguments to include the present case, although quite good estimates of the critical density can be obtained by putting the atomic centres on a regular lattice of somewhat finer grain than the distance $2r_a$ within which they are assumed to interact (Holcomb & Rehr 1969). Indeed (Shante & Kirkpatrick

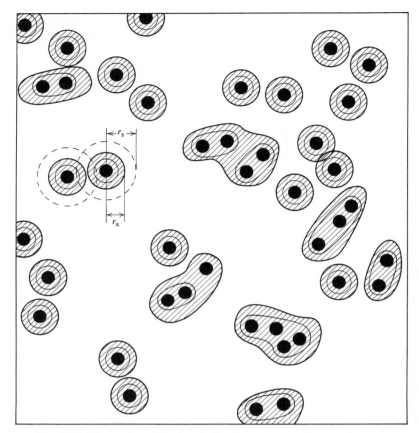

Fig. 13.2. Formation of 'molecular' orbitals around clusters of atoms in a random gas.

1971), a good estimate of this quantity had already been obtained by Domb & Dalton (1966), using graphical methods (§ 5.10) to determine the transition temperature of a dilute ferromagnet (§ 12.1) with identical interactions between all lattice sites in a relatively large neighbourhood.

From Monte Carlo computations on gas-like assemblies containing several thousand spheres (e.g. Pike & Seager 1974) it seems that the critical radius for percolation in a gas-like assembly is given by

$$r_c \approx 0.70 \, r_s. \tag{13.4}$$

In other words, for a given density of centres, the radius of interaction, r_a for each sphere must exceed r_c. This means that the volume of each 'sphere

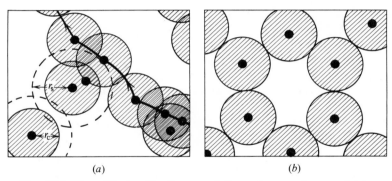

Fig. 13.3. The critical radius for percolation r_c in a random gas (*a*) is approximately equal to the packing radius in a regular diamond lattice (*b*).

of influence' of each atomic state is not much less than the average volume (13.2) available to each atom in the assembly:

$$\eta_c \equiv (r_c/r_s)^3 \approx 0.35. \tag{13.5}$$

It is interesting to note that this is close to the packing fraction of a regular diamond lattice; if we were to order the gas into this very open crystalline form the spheres of radius r_c would just touch! (fig. 13.3).

The model is not sufficiently well defined to permit any inferences concerning the nature of the high-density metallic state. In real systems the actual physical phenomena, such as the behaviour of the conductivity as one passes through the transition, must be governed by much more complicated mechanisms such as electron–electron correlation, which lie outside our present discussion (see e.g. Mott & Twose 1961; Mott 1974). It is apparent, however, that the tight-binding method would be quite inappropriate for any theory of electron states in a gas-like assembly at any concentration approaching the regime where $r_a \sim r_c$. This is a severe restriction on the theoretical discussion of systems, such as metallic vapours, where these conditions could not be taken for granted.

13.3 Hopping Conduction

The crude assumption (§ 13.2) that each atomic bound state ψ_a has a sharply defined sphere of interaction is not well justified. Generally speaking, an energy eigenfunction in a spherically symmetrical potential decays exponentially at large distances:

$$\psi_a(\mathbf{r}) \sim \exp(-r/a), \tag{13.6}$$

where a is a characteristic 'Bohr radius' for the state in question (§ 2.15). States centred at $\mathbf{R}_i$ and $\mathbf{R}_j$ must interact through an *overlap integral* (§ 8.1), such as

$$\int \psi_a^*(\mathbf{r}-\mathbf{R}_i)\psi_a(\mathbf{r}-\mathbf{R}_j)\, d^3r \sim \exp\{-|\mathbf{R}_i-\mathbf{R}_j|/a\}, \tag{13.7}$$

which becomes very small when the centres are far apart, but which never actually vanishes.

At densities well below the metal–insulator transition, the effects of such interactions on the electron *spectrum* can be ignored. But in a material that is supposed to be an insulator it is easy to detect very small currents due to electrons *tunnelling* from site to site through this overlap of wave functions. The actual mechanism of electron *hopping* from impurity to impurity in a semiconductor is quite complicated (Holstein 1959; Miller & Abrahams 1960) but the effect of spatial disorder can be treated as a problem in its own right. Thus, as a first approximation, the probability of a transition from centre $\mathbf{R}_i$ to centre $\mathbf{R}_j$ can be supposed to be proportional to the square of the overlap integral (13.7), i.e.

$$\sigma_{ij}=\exp\{-2|\mathbf{R}_i-\mathbf{R}_j|/a\}, \tag{13.8}$$

without regard to other factors.

In this model the gas is represented by a random distribution of nodal points, linked by 'conductances' varying exponentially with internodal distance. What is the macroscopic conductivity of such a network? Elementary scaling arguments suggest that this should behave like

$$\bar{\sigma}\sim\exp(-4\alpha r_s/a), \tag{13.9}$$

as a function of the gas density (13.2); but the constant α cannot be deduced by elementary phenomenological arguments such as an appeal to 'paths of least resistance' (Miller & Abrahams 1960). This is obviously a percolation problem, dominated by the possibility of finding a cat's cradle of critical paths of *infinite* length (Ambegaokar, Halperin & Langer 1971). But because each conductance (13.8) is a continuous function of internodal distance, we are not faced with the non-analyticities that arise at the percolation threshold in the standard models of maze conduction (§ 9.11).

For a phenomenological solution to this problem, we refer back to the theory of percolation in a gas of overlapping spheres (§ 13.2). As pointed out by Ambegaokar, Cochran & Kurkijärvi (1973), the over-all conductivity (13.9) would be dominated by the percolation paths that would just be opened up at this gas density (fig. 13.3(*a*)): in other words, the characteristic interaction distance $2\alpha r_s$ would be the distance $2r_c$ between two critical

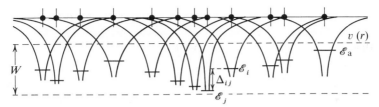

Fig. 13.4. Overlap of impurity potentials varies bound-state energies, as in the Anderson model.

spheres that could just touch and allow the path to continue. From (13.4) we thus deduce that we should write

$$\alpha \approx 0.70 \qquad (13.10)$$

in (13.9): this conjecture is very closely confirmed numerically by Monte Carlo computations (Seager & Pike 1974). The non-exponential factor omitted from (13.9) can also be estimated by reference to such computations (Kurkijärvi 1974).

In setting up this model, however, we neglected an important characteristic of the overall potential $\mathscr{V}(\mathbf{r})$. Although each atomic bound state $\psi_a(\mathbf{r} - \mathbf{R}_i)$ is essentially an eigenfunction of a single 'atomic' potential $v(\mathbf{r})$, the corresponding energy eigenvalue $\mathscr{E}_i$ is very sensitive to the 'tails' of potentials overlapping from other centres (fig. 13.4). The gas-like disorder modulates the bound-state energies, spreading them into a random distribution with a characteristic width W. In the locator representation this is the site-diagonal disorder characteristic of the Anderson model (§ 9.9) which is sometimes regarded as a model of a semiconductor impurity band.

In the present case, however, the overlap integrals (13.7) are so small compared with W that we are well within the regime of Anderson localization. But as remarked in § 9.11, the variations of $\mathscr{E}_i$ from site to site inhibit hopping transitions. For the overlap (13.7) between wave functions to have its effect we must find a means of neutralizing the energy difference

$$\Delta_{ij} = \mathscr{E}_i - \mathscr{E}_j. \qquad (13.11)$$

In practice, this is taken care of by the absorption or emission of a phonon. But the intrinsic rate of *phonon-assisted hopping* (Miller & Abrahams 1960; Emin 1975) depends on the temperature through a thermal factor,

$$f_{ij} \sim \exp\{-|\Delta_{ij}|/kT\}, \qquad (13.12)$$

which may have a very significant effect at low temperatures.

To estimate this effect, let us suppose, first of all, that the electron attempts to follow the percolation path assumed in calculating (13.9). For each hop we introduce a factor such as (13.12); the product of all such factors would be equivalent to reducing every transfer probability (13.8) by a factor

$$\langle f \rangle \sim \exp\{-\langle |\Delta| \rangle / \& T\}. \tag{13.13}$$

Thus, for example, if we assume that $\mathscr{E}_i$ is distributed uniformly over the range (9.108) with width W we easily calculate the expectation value of $|\Delta_{ij}|$ and replace (13.9) by

$$\langle \sigma \rangle \sim \exp\{-4\alpha r_s/a - \tfrac{1}{3} W/\& T\}. \tag{13.14}$$

But this answer is not correct because it does not allow for the possibility that the electron might choose a path avoiding 'bad' thermal factors. Suppose, indeed (Ambegaokar, Halperin & Langer 1971), that unfavourable hops are excluded – for example, that the electron only visits the fraction ξ of the atomic sites, where $\mathscr{E}_i$ lies within a band of width ξW chosen from the original distribution (fig. 13.5). These sites are, on the average, further apart than those of the whole gas, by a factor $\xi^{-\frac{1}{3}}$, so that the conductivity would behave like a modified version of (13.14), i.e.

$$\sigma(\xi) \sim \exp\{-4\alpha\xi^{-\frac{1}{3}}r_s/a - \tfrac{1}{3}\xi W/\& T\}. \tag{13.15}$$

Under the influence of an external field, the electron will presumably choose to make the best hop available to it in the neighbourhood (Mott 1968). This is equivalent to choosing ξ so as to maximize (13.15), giving for our best estimate of the overall conductivity of the system

$$\bar{\sigma} \sim \exp\left\{-\frac{4}{3}\left(\frac{4\alpha r_s}{a}\right)^{\frac{3}{4}}\left(\frac{W}{\& T}\right)^{\frac{1}{4}}\right\}, \tag{13.16}$$

with α as in (13.10).

This formula is not to be taken as numerically reliable, because of many arbitrary features in the model and various mathematical approximations in the derivation. But the characteristic 'exponential $T^{-\frac{1}{4}}$' behaviour of *variable-range hopping* (Mott 1968) has been verified with computer models (Seager & Pike 1974) and is capable of comparison with experiment (e.g. Mott 1974).

Thermally activated hopping also makes its contributions to the AC transport properties of the material (Butcher 1972). For these properties, theories based upon *continuous time random walks* (Scher & Lax 1973; Moore 1974; Butcher 1974*b*) give satisfactory answers, regardless of the characteristics of the disorder. But in the DC limit, these methods fail (Butcher 1974*c*). As we saw in connection with (13.14), the actual frozen-in disorder is not then sampled over all possible steps; the only important steps are those lying on infinite percolation paths, which have the highest

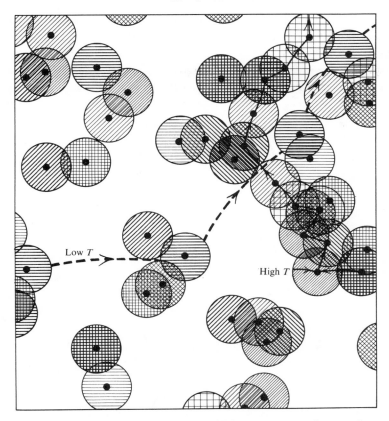

Fig. 13.5. Variable range hopping. At high temperatures the percolation path can pass through any impurity level. At low temperatures an electron can hop only to an impurity level of the same energy (as indicated by the depth of shading).

overall conductivity available in the circumstances. There is a subtle mathematical difference between a *diffusion process,* where the transition probabilities are re-randomized after each step, regardless of the position of the particle, and a *percolation process,* where the branching probabilities or restrictions are associated with the actual node reached by the particle at any given moment, and do not otherwise vary with time.

13.4 Semi-classical electrons in a random potential

In the case of a *dense* gas of *weak* scatterers, each atomic potential has sufficient range, r_p, to include many atomic spheres of radius r_s, but is not strong enough to bind an electron. The total potential (13.1) is now a superposition of many overlapping contributions and, therefore, behaves like a *Gaussian random field* (§ 3.3). Let us assume, for formal simplicity, that each atomic potential has zero mean as in (3.17); then according to (3.16), we may treat $\mathscr{V}(\mathbf{R})$ as a continuous random function, with probability distribution

$$P(\mathscr{V}) = \frac{1}{(2\pi)^{\frac{1}{2}}\mathscr{W}} \exp\left\{\frac{-\mathscr{V}^2}{2\mathscr{W}^2}\right\} \tag{13.17}$$

about the same energy zero,

$$\langle \mathscr{V}(\mathbf{R}) \rangle = 0. \tag{13.18}$$

From the atomic potentials $v(\mathbf{r})$ we can deduce the width $\mathscr{W}$ of the disorder: from (3.18) this is just

$$\mathscr{W}^2 = \langle |v|^2 \rangle \equiv N \int |v(\mathbf{R})|^2 \, d^3\mathbf{R}. \tag{13.19}$$

The same function also governs the auto-correlation function (3.19), i.e.

$$\Gamma(\mathbf{R}) = \langle v^*(\mathbf{R}' + \mathbf{R}) \, v(\mathbf{R}') \rangle / \langle |v|^2 \rangle, \tag{13.20}$$

from which we may deduce the two-point distribution function (3.14) and further properties of $\mathscr{V}(\mathbf{R})$.

In practice, however, the theory of electron states and electron transport in this sort of potential is so idealized and schematic that we are seldom interested in relating the properties of $\mathscr{V}(\mathbf{R})$ to such genuine physical objects as atoms with well-defined (pseudo)potentials $v(\mathbf{r})$. The essence of (13.20) is that $\Gamma(\mathbf{R})$ satisfies (3.5), and is characterized by a *correlation length*

$$L \sim r_p \sim 1/q_c, \tag{13.21}$$

which is comparable with the 'range' of each atomic potential. This would also be the minimum wavelength corresponding to the *spectral range* $0 < q \lesssim q_c$ of wave numbers in the Fourier representation (3.6) of the potential $\mathscr{V}$.

Let us now consider an electron of energy $\mathscr{E}$ propagating in the field $\mathscr{V}(\mathbf{R})$. If $\mathscr{E} \gg \mathscr{W}$, the electron always overrides the fluctuations of potential and the Schrödinger equation can be solved, to a good approximation, by treating $\mathscr{V}$ as a perturbation on propagating free-electron states (§§ 10.1, 10.4). But this approach is not valid at lower energies, where the potential $\mathscr{V}(\mathbf{R})$ may actually exceed $\mathscr{E}$ over some regions of the field. Considered as a *classical* particle, the electron is moving at 'height' $\mathscr{E}$ over a 'landscape' $\mathscr{V}(\mathbf{R})$, and cannot penetrate the regions above this contour (fig. 13.6). The

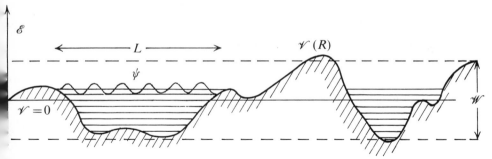

Fig. 13.6. Semi-classical electron in a random potential cannot penetrate barriers.

topography of the random function $\mathcal{V}(\mathbf{R})$ (§ 3.4) thus plays a significant role in the solutions of the Schrödinger equation, which must conform to the classical trajectories in the correspondence-principle limit (Ziman 1968).

If many de Broglie wavelengths of the electron could be fitted into each segment of the classical trajectories, the *semi-classical approximations* to the solutions of the Schrödinger equation would be valid (e.g. Berry & Mount 1972). Making reasonable assumptions about the functional form of $\Gamma(\mathbf{R})$, we can show (§ 3.4), that the characteristic spatial dimension of the 'topographic features' of $\mathcal{V}(\mathbf{R})$ is the correlation length L. In a typical 'valley', of 'energy depth' $\mathcal{W}$, the semi-classical approximation is valid if L exceeds the typical de Broglie wavelength $\hbar/\sqrt{(2m\mathcal{W})}$ – i.e. when

$$\mathcal{W}L^2 \gg 1, \tag{13.22}$$

in atomic units.

Under these conditions the electron energy spectrum is given to a good approximation (Kane 1963; Lifshitz 1968) by the *Thomas–Fermi formula*: in a region where the electron potential energy is $\mathcal{V}$, the *cumulated spectral density* (8.68) is given by

$$\mathcal{D}(\mathcal{E}; \mathcal{V}) = \frac{1}{3\pi^2} [2(\mathcal{E} - \mathcal{V})]^{\frac{3}{2}}. \tag{13.23}$$

But $\mathcal{V}$ has the probability distribution $P(\mathcal{V})$: the total cumulated spectrum for the whole system must be of the form

$$\mathcal{D}(\mathcal{E}) = \frac{1}{3\pi^2} \int_{-\infty}^{\mathcal{E}} [2(\mathcal{E} - \mathcal{V})]^{\frac{3}{2}} P(\mathcal{V}) \, d\mathcal{V}. \tag{13.24}$$

The density of electron states in a random potential $\mathcal{V}(\mathbf{R})$ with the Gaussian distribution (13.17) should approximate to

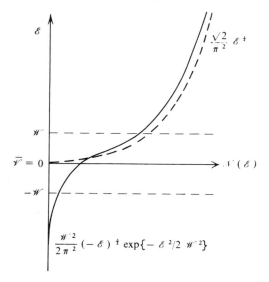

Fig. 13.7. Thomas–Fermi spectral density in a Gaussian random potential.

$$\mathcal{N}_{TF}(\mathscr{E}) = \pi^{-\frac{1}{2}} \mathscr{W}^{-1} \int_{-\infty}^{\mathscr{E}} (\mathscr{E} - \mathscr{V})^{\frac{1}{2}} \exp\{-\mathscr{V}^2/2\mathscr{W}^2\}\, d\mathscr{V}, \quad (13.25)$$

in the semi-classical limit.

The behaviour of this integral (which can be expressed analytically as a parabolic cylinder function) is shown in fig. 13.7: At high energies ($\mathscr{E} \gg \mathscr{W}$) it tends to the standard free electron formula

$$\mathcal{N}_{TF}(\mathscr{E}) \approx \frac{\sqrt{2}}{\pi^2} \mathscr{E}^{\frac{1}{2}}. \quad (13.26)$$

But for energies below the conventional zero (13.18), the spectral density does not vanish but tends, for $\mathscr{E} \ll -\mathscr{W}$, to

$$\mathcal{N}_{TF}(\mathscr{E}) \approx \frac{\mathscr{W}^2}{2\pi^2} (-\mathscr{E})^{-\frac{3}{2}} \exp\{-\mathscr{E}^2/2\mathscr{W}^2\}. \quad (13.27)$$

In other words, the spectrum acquires a modified Gaussian *tail* (§ 9.6) associated with electron states in deep fluctuations of the potential.

In this approximation, the electronic transport properties of the system are dominated by the fact that a classical electron cannot penetrate any region where $\mathscr{V}(\mathbf{R})$ exceeds $\mathscr{E}$. It is intuitively obvious, however, that the topology of the 'allowed' regions must change as we go from low to high energies (Ziman 1968). Imagine water poured over the 'landscape' (fig. 13.8). For small values of $\mathscr{E}$, only the deeper minima are occupied, forming

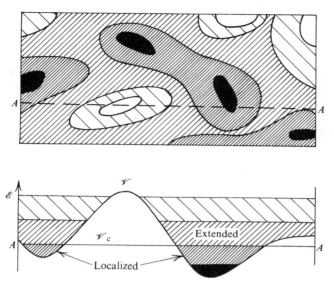

Fig. 13.8. Energy contours of a random potential (*a*) and a typical section (*b*) showing regions of localized states.

isolated 'ponds' or 'lakes'. At this energy, therefore, classical or semi-classical electrons would all be *localized*. But as the water level rises, these lakes begin to coalesce and eventually form a connected body permeating the whole system. Above the critical contour $\mathscr{E} = \mathscr{V}_c$, the specimen would contain *extended* electron states, and be electrically conducting along the allowed channels.

The problem of determining the threshold contour $\mathscr{V}_c$ for *percolation on a continuum* has an exact solution in two dimensions (Zallen & Scher 1971). From the symmetry of $\mathscr{V}(\mathbf{R})$ for positive and negative excursions from the zero mean (13.18), it follows that the topology of 'allowed' regions at energy $\mathscr{E}$ is the same as the topology of 'forbidden' regions at the energy $-\mathscr{E}$. But by the argument leading to the exact solution $p_c^S = \frac{1}{2}$ for site percolation on a planar triangular net (§ 9.10), the two types of region cannot be percolative simultaneously. Thus, we must have $\mathscr{V}_c = 0$, corresponding to a situation where the allowed region occupies exactly half the available volume.

This argument cannot be extended to three-dimensional fields, but it is plausible (Scher & Zallen 1970) that percolation occurs in the continuum model when the 'allowed' regions occupy the same *critical volume fraction*

η_c as for regular lattices of spheres. As we noted in table 9.1, this quantity is nearly constant for several different lattices: the conjecture

$$\eta_c \approx 0.15 \tag{13.28}$$

for three-dimensional random fields is consistent with Monte Carlo computations (Skal, Shklovskii & Efros 1973). By integration of (13.17) up to a level encompassing this proportion of the total volume, we find the approximate rule of thumb,

$$\mathscr{V}_c \approx -\mathscr{W}, \tag{13.29}$$

for the position of the percolation threshold in a Gaussian random potential of variance $\mathscr{W}$. This might be regarded as an estimate of the *mobility edge* (§ 9.9) for electron transport in such a system.

It is worth remarking, however, that (13.28) is not a good estimate of the 'allowed' volume fraction for percolation in an apparently similar system – a gas of overlapping spheres (§ 13.2). Even after making allowances for the regions contained within two or more spheres (Shante & Kirkpatrick 1971), the percolation threshold (13.5) would correspond to a critical volume fraction almost twice the value (13.28) in this system. This exemplifies the general principle (Beran 1968) that the macroscopic properties of a *composite material,* containing two or more phases with very different physical properties, cannot be deduced from the relative volume fractions of the constituents, but is also extremely sensitive to the geometry and topology of the boundary surfaces between the phases. Unfortunately, the parameters of typical boundary surfaces in a 'random mixture' are not related in any simple way to the lowest-order, point-distribution functions for the constituents (§ 3.2), so that the mathematical problem of calculating the over-all properties of the system is very ill-posed. The Gaussian random field has the virtue of being a standard model for such a system with topographical properties that can be explored analytically in some detail (§ 3.4).

For these reasons the theory of, say, electrical conduction in composite materials is in a very rudimentary state and has little to teach us concerning the electronic transport properties of microscopically disordered systems. On the contrary, the mathematical effort that has been devoted to such simplified models as the tight-binding alloy (chapter 9) has proved fruitful (Stroud 1975) in drawing attention to the power of the *effective medium* approach, which is at the heart of the CPA method (§ 9.4) and its modifications. At energies substantially above the percolation threshold (13.29) this should give the same excellent results as it does for maze conduction

(§ 9.11) and should also apply to the theory of composite materials (cf. Bruggemann 1935; Landauer 1952) where its capabilities have been for too long ignored.

13.5 Spectral tails in a choppy random potential

Let us suppose now that we make the random potential more 'choppy' by reducing the wave-length, L, of characteristic fluctuations. If the variance $\mathscr{W}^2$ of the Gaussian distribution (13.17) is kept fixed we can no longer satisfy the condition (13.22) for the semi-classical spectrum (13.25). Genuine wave-mechanical effects then reduce the spectral density in the negative energy tails (13.27).

Suppose, for example, that we have a deep fluctuation of the field down to a minimum at $\mathscr{V}_m$. This creates a local potential well, of diameter L; the total number of electron states up to energy $\mathscr{E}$ in such a well can be estimated from the Thomas–Fermi density (13.23) integrated over this volume, i.e.

$$n(\mathscr{E};\ \mathscr{V}_m) \sim (\mathscr{E} - \mathscr{V}_m)^{\frac{3}{2}} L^3. \tag{13.30}$$

If there is to be even a single bound quantum state in the well, this number must exceed unity, i.e. we must have

$$0 > \mathscr{E} \gtrsim \mathscr{V}_m + 1/L^2. \tag{13.31}$$

In other words, the well must be deep enough to accommodate the quantum-mechanical, *zero-point energy*

$$\mathscr{E}_0 \sim \hbar^2/2mL^2 \tag{13.32}$$

needed to localize an electron within this volume (fig. 13.9). This may be taken into account in the Thomas–Fermi spectral density (13.23), by shifting the energy zero: phenomenologically we write

$$\mathscr{D}(\mathscr{E};\ \mathscr{V}_m) \approx \frac{1}{3\pi^2}\{2(\mathscr{E} - \mathscr{V}_m - \chi/L^2)\}^{\frac{3}{2}} \tag{13.33}$$

where χ is a factor of order unity in atomic units and the function only exists for positive values of $\mathscr{E} - \mathscr{V}_m - \chi/L^2$. From (3.38) we confirm that $\mathscr{V}_m$ has a Gaussian distribution just like $\mathscr{V}(\mathbf{R})$, so that the semi-classical formula (13.25) is replaced by a spectrum of the form

$$\mathscr{N}(\mathscr{E}) \sim \mathscr{W}^{-1} \int_{-\infty}^{\mathscr{E} - \chi/L^2} (\mathscr{E} - \mathscr{V}_m - \chi/L^2)^{\frac{1}{2}} \exp\{-\mathscr{V}_m^2/2\mathscr{W}^2\}\, d\mathscr{V}_m \tag{13.34}$$

For negative values of $\mathscr{E}$ this behaves like

$$\mathscr{N}(\mathscr{E}) \sim (|\mathscr{E}| + \chi/L^2)^{-\frac{3}{2}} \exp\{-(|\mathscr{E}| + \chi/L)^2/2\mathscr{W}^2\}, \tag{13.35}$$

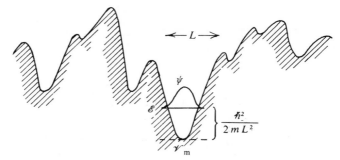

Fig. 13.9. Bound state in a potential minimum.

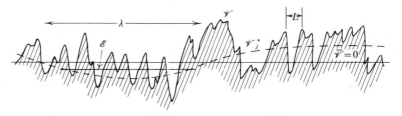

Fig. 13.10. Bound states in a minimum of the 'smoothed' potential $\mathscr{V}^*$.

which rapidly becomes much smaller than (13.27) as we approach the *white noise limit* where the correlation length tends to zero. Thus, as

$$L \to 0, \qquad\qquad (13.36)$$

we expect quantum effects to destroy the tail of states below the spatial average (13.18) of the random potential $\mathscr{V}(\mathbf{R})$.

But this argument does not take account of interactions between neighbouring minima of $\mathscr{V}(\mathbf{R})$. As the valleys of the potential landscape become narrower so do the ridges between them (fig. 13.10). It may still be possible to construct a local-wave function in the negative-energy region by allowing it to spread over many such features. This basic idea (Halperin & Lax 1966, 1967; Lifshitz 1968; Halperin 1973) is merely an extension of the Lifshitz method (9.83) for band tails in substitutional alloys and has a simple phenomenological formulation.

Suppose, schematically, we divide the whole volume of the specimen into relatively large 'cells', each of size λ, much greater than the correlation length L. The average potential $\mathscr{V}^*$ in each cell must also be a Gaussian random function, distributed as in (13.17) whose variance,

$$\mathscr{W}^{*2} \sim (L/\lambda)^3 \, \mathscr{W}^2, \qquad\qquad (13.37)$$

is reduced in inverse proportion to the smoothing volume λ^3. Any cell with a sufficiently deep value of $\mathscr{V}^*$ may bind a state ψ^* whose zero point energy need not be greater than $\hbar^2/2m\lambda^2$. We may thus replace L by λ in the Thomas–Fermi formula (13.33), and arrive at a spectral density of the form (13.34), i.e.

$$\mathscr{N}(\mathscr{E}; \lambda) \sim \mathscr{W}^{*-1} \int_{-\infty}^{\mathscr{E}-\chi/\lambda^2} (\mathscr{E} - \mathscr{V}^* - \chi/\lambda^2)^{\frac{1}{2}} \exp\{-\mathscr{V}^{*2}/2\mathscr{W}^{*2}\} \, d\mathscr{V}^*.$$

(13.38)

Although the subdivision into cells seems quite arbitrary what we are really doing is guessing that Ψ^* is a 'smooth' function whose optimum radius, λ, may be estimated by appeal to the usual variational principle for the expectation value of the energy. In the present circumstances the function to be maximized (Lloyd & Best 1975) is neither the spectral density, $\mathscr{N}(\mathscr{E})$, nor the integrated spectral density, $\mathscr{D}(\mathscr{E})$, but the *pressure functional*,

$$\mathscr{P}(\mathscr{E}; \lambda) = \int^{\mathscr{E}} \mathscr{D}(\mathscr{E}; \lambda) \, d\mathscr{E}.$$

(13.39)

In the negative energy tail of (13.38), however, where the analogue of (13.35) would be a satisfactory approximation, differention of $\mathscr{N}(\mathscr{E}; \lambda)$ with respect to λ yields the asymptotic form

$$\mathscr{N}(\mathscr{E}) \sim \exp\{-8(\chi/3)^{\frac{3}{2}}|\mathscr{E}|^{\frac{1}{2}}/\mathscr{W}^2 L^3\}.$$

(13.40)

The spectrum of these states thus falls off much less rapidly than (13.35) as $\mathscr{E} \to -\infty$.

This tail does not disappear in the *white noise limit* (13.36) where the system is supposed to approach an infinitely *dense* assembly of infinitesimally *weak* scatters, each of which contributes a delta function singularity to the potential. But as we go to this limit, the auto-correlation function $\Gamma(\mathbf{R})$ tends to a delta function, i.e.

$$\langle \mathscr{V}(\mathbf{R}) \mathscr{V}(\mathbf{R}') \rangle = \tfrac{1}{2}\xi\delta(\mathbf{R} - \mathbf{R}'),$$

(13.41)

whose finite residue eventually appears in the denominator of the exponent in (13.30). In other words, for a d-dimensional system, the variance of the Gaussian (13.17) is allowed to increase in such a way that

$$\mathscr{W}^2 L^d \to \xi \quad \text{as } L \to 0.$$

(13.42)

In this limit, the *one-dimensional* problem can, in fact, be solved exactly (Frisch & Lloyd 1960). The method looks complicated, but is essentially a sophisticated generalization of the *Dyson–Schmidt method* (§ 8.5) which is always available as a rigorous technique for calculating the spectrum of any one-dimensional disordered system. This is because the arrival of the electron at each successive delta-function potential can be regarded as a Markov process. Thus, the 'phase variable',

$$z(x) = \psi'(x)/\psi(x), \tag{13.43}$$

(cf. (8.28)), can be treated as a stochastic variable whose probability distribution is stationary and must satisfy a self-consistency condition analogous to (8.76). In the actual calculations, this integral equation is found to be equivalent to a partial differential equation analogous to the Fokker–Planck equation (Halperin 1965) with a solution in the negative energy region of the form

$$\mathcal{N}(\mathcal{E}) \sim \frac{8|\mathcal{E}|}{\pi\xi} \exp\left\{ -\frac{4}{3} \frac{|2\mathcal{E}|^{\frac{3}{2}}}{\xi} \right\}. \tag{13.44}$$

Since this is the functional form that would have been obtained for a one-dimensional model by the method leading to (13.40), we confirm the validity of the Lifshitz *ansatz* even in this extreme case. The same type of expression can also be deduced in quite a different way by approximately decoupling the average Green functions in a conventional coordinate representation (Dallacasa 1975).

There is, thus, little doubt of the validity of (13.40) in the general three-dimensional case as the limiting form of the spectral tail. The parameter χ and the pre-factors of the exponential can be estimated by more detailed consideration (Halperin & Lax 1966, 1967; Zittartz & Langer 1966; Friedberg & Luttinger 1975) of the optimum form of ψ^* in relation to the minima of a smoothed version of $\mathcal{V}(\mathbf{R})$ in which it sits. It must be noted, however, that (13.40) would not hold for values of $\mathcal{E}$ that would make $\lambda < L$.

For energies above the percolation threshold (13.29) of the random potential $\mathcal{V}(\mathbf{R})$, a many-branched extended state permeating the 'allowed' regions might be a better guess than a compact localized state such as ψ^*. Doubts concerning the mutual consistency of the phenomenological assumptions underlying (13.40) (Hernandez & Ziman 1973; Eggarter & Cohen 1974; Jones & McCubbin 1975) remain unresolved. In all the applications of the *path integral formalism* (§ 7.9) to these problems (Edwards 1970; Bezak 1970; Abram & Edwards 1972; Freed 1972*b*; Saitoh & Edwards 1974; Samathiyakanit 1974), a similar symmetry-breaking *localization ansatz* has to be made to draw the states in the tail out of the continuum of extended states in the positive energy region or else the equations must be crudely approximated by postulating a self-consistent field acting on the electron as it propagates through the material. Since this formalism is of considerable mathematical complexity, and has not yet given a rigorous answer to the somewhat academic question of the transport properties of electrons in this range of energies in this grossly simplified and contrived model, I may perhaps be forgiven if I do not pursue the matter further, and take the opportunity of a natural break to announce that this is, as far as I am concerned, THE END.

References

—

[*The number of the section in which the reference is cited is given in square brackets.*]

Abou-Chacra, R., Anderson, P.W. & Thouless, D.J. (1973) *J. Phys. C* **6**, 1734–52. [9.9]
Abou-Chacra, R. & Thouless, D.J. (1974) *J. Phys. C* **7**, 65–75. [9.9]
Abram, R.A. & Edwards, S.F. (1972). *J. Phys. C* **5**, 1183–95, 1196–206. [13.5]
Abrikosov, A.A., Gorkov, L.P. & Dzyaloshinski, I.E. (1963) *Methods of Quantum Field Theory in Statistical Physics.* New Jersey: Prentice Hall. [5.10]
Adams, D.J. & Matheson, A.J. (1972) *J. Chem. Phys.* **56**, 1489–94. [2.11]
Agacy, R.L. & Borland, R.E. (1964) *Proc. Phys. Soc.* **84**, 1017–26. [8.4] [8.5]
Akhiezer, N.I. (1965) *The Classical Moment Problem.* London: Oliver & Boyd. [9.4]
Alben, R. & Boutron, P. (1975) *Science,* **187**, 430–2. [2.8]
Alben, R., von Heimendahl, L., Galison, P. & Long, M. (1975) *J. Phys. C* **8**, L468–72. [11.3]
Alben, R., Weaire, D., Smith, J.E. Jr & Brodsky, M.H. (1975) *Phys. Rev. B* **11**, 2271–96. [11.1] [11.3]
Alder, B.J. (1955) *J. Chem. Phys.* **23**, 263–71. [2.13]
Alder, B.J. (1964a) *Phys. Rev. Lett.* **12**, 317–19. [2.11]
Alder, B.J. (1964b) *J. Chem. Phys.* **40**, 2724. [2.13]
Alder, B.J. & Hoover, W.G. (1968) in *Physics of Simple Fluids* (eds. Temperley, Rowlinson & Rushbrooke), pp. 81–113. Amsterdam: North Holland. [6.5] [6.6]
Alder, B.J., Hoover, W.G. & Young, D.A. (1968) *J. Chem. Phys.* **49**, 3688–96. [2.11] [6.8]
Alder, B.J. & Wainwright, T.E. (1957) *J. Chem. Phys.* **27**, 1208. [6.6] [6.7]
Alder, B.J. & Wainwright, T.E. (1959) *J. Chem. Phys.* **31**, 459. [6.6]
Alder, B.J. & Wainwright, T.E. (1960) *J. Chem. Phys.* **33**, 1439. [6.6] [6.7]
Alder, B.J. & Wainwright, T.E. (1962) *Phys. Rev.* **127**, 359–61. [6.7]
Ambegaokar, V., Cochran, S. & Kurkijärvi, J. (1973) *Phys. Rev. B* **2**, 3682–8. [13.3]
Ambegaokar, V., Halperin, B.I. & Langer, J.S. (1971) *Phys. Rev. B* **4**, 2612. [13.3]
Anderson, P.W. (1950) *Phys. Rev.* **80**, 922. [5.4]
Anderson, P.W. (1958) *Phys. Rev.* **109**, 1492–505. [9.9]
Anderson, P.W. (1970) *Comments on Solid State Phys.* **2**, 193–8. [9.9]
Anderson, P.W. (1972) *Proc. Nat. Acad. Sci.* **69**, 1097–9. [9.9]
Anderson, P.W., Halperin, B.I. & Varma, C.M. (1972) *Phil. Mag.* **25**, 1. [11.2]
Anderson, P.W. & McMillan, W.L. (1967) in *Theory of Magnetism in Transition Metals* (ed. Marshall). New York: Academic Press. [10.8]

Ashcroft, N.W. & Langreth, D.C. (1967a) *Phys. Rev.* **156**, 685–91. [2.13]
Ashcroft, N.W. & Langreth, D.C. (1967b) *Phys. Rev.* **159**, 500–10. [4.5]
Ashcroft, N.W. & Lekner, J. (1966) *Phys. Rev.* **145**, 83. [2.12]
Ashcroft, N.W. & Schaich, W. (1970) *Phys. Rev. B* **1**, 1370–9. [10.10]
Athreya, K., Subramanian, R.R. & Kumar, N. (1973) *Curr. Sci.* **41**, 867–8. [9.9]
Aymerich, F. & Smith, P.V. (1973) *J. Phys. C* **6**, L41–5. [11.5]
Ballentine, L.E. (1966) *Can. J. Phys.* **44**, 2533–52. [10.4]
Ballentine, L.E. (1975) in *Non-Simple Liquids* (eds Prigogine & Rice), **31**, 263–327. New York: Wiley. [10.4][10.5][10.9]
Ballentine, L.E. & Heine, V. (1964) *Phil. Mag.* **9**, 617–22. [10.8]
Barber, M.N. & Baxter, R.J. (1973) *J. Phys. C* **6**, 2913–21. [5.8]
Barker, J.A. & Henderson, D. (1967) *J. Chem. Phys.* **47**, 2856, 4714–21. [6.4]
Barker, J.A. & Henderson, D. (1972) *Ann. Rev. Phys. Chem.* **23**, 439–84. [6.4]
Bartenev, G.M. (1970) *The Structure and Mechanical Properties of Inorganic Glasses.* Groningen: Wolters-Noordhat. [2.8][2.9]
Batchelor, G.K. (1953) *The Theory of Homogeneous Turbulence.* Cambridge: University Press. [3.1]
Baxter, R.J. (1972) *Ann. Phys. (USA)* **70**, 193–228, 323–37. [5.8]
Beckmann, P. & Spizzichino, A. (1963) *The Scattering of Waves from Rough Surfaces.* Oxford: Pergamon Press. [3.1][3.4]
Beeby, J.L. (1964) *Proc. Roy. Soc. A* **279**, 82–96. [10.7][10.9]
Beeby, J.L. (1973) *J. Phys. C* **6**, L283–7. [9.9]
Beeby, J.L. & Edwards, S.F. (1963) *Proc. Roy. Soc. A* **274**, 395–411. [10.7]
Beeman, W.W., Kaesberg, P., Anderegg, J.W. & Webb, M.B. (1957) *Handb. Phys.* **XXXII**, 320–442. [4.4]
Bell, R.J. (1972) *Rep. Prog. Phys.* **35**, 1315–409. [11.1]
Bell, R.J. (1974) *J. Phys. C* **7**, L265–9. [11.1]
Bell, R.J., Bird, N.F. & Dean, P. (1974) *J. Phys. C* **7**, 2457–66. [11.1]
Bell, R.J. & Dean, P. (1966) *Nature,* **212**, 1354–6. [2.8]
Bell, R.J. & Dean, P. (1970) *Disc. Farad. Soc.* **50**, 55–61. [9.9][11.1]
Beran, M.J. (1968) *Statistical Continuum Theories.* New York: Interscience. [3.2][3.3][13.4]
Bergmann, A. & Halperin, V. (1974) *J. Phys. C* **7**, 289–98. [9.8]
Berk, N.F., Shazeer, D.J. & Tahir-Kheli, R.A. (1973) *Phys. Rev. B* **8**, 2496–504. [9.5]
Berk, N.F. & Tahir-Kheli, R.A. (1973) *Physica,* **67**, 501–20. [9.5]
Berlin, T.H. & Kac, M. (1952) *Phys. Rev.* **86**, 821. [5.9]
Bernal, J.D. (1959) *Nature,* **183**, 141. [2.11]
Bernal, J.D. (1960) *Nature,* **185**, 68. [2.11]
Bernal, J.D. (1960) *Scientific American,* **203**, 124. [2.11]
Bernal, J.D. (1964) *Proc. Roy. Soc. A* **280**, 299. [2.11]
Bernal, J.D. & King, S.V. (1968) in *Physics of Simple Fluids* (eds Temperley, Rowlinson & Rushbrooke), pp. 230–52. Amsterdam: North Holland. [2.11]
Berry, G.C. & Casassa, E.F. (1970) *Macromolecular Reviews,* **4d**, 1–66. [7.4][7.7]
Berry, M.V. (1973) *Phil. Trans. A* **273**, 611–58. [3.1][3.3]
Berry, M.V. & Doyle, P.A. (1973) *J. Phys. C* **6**, L6–9. [4.7]
Berry, M.V. & Mount, K.E. (1972) *Rep. Prog. Phys.* **35**, 315–98. [3.4][13.4]

Bertaut, E.F. (1963) in *Magnetism* (eds Rado & Suhl), Vol. III, pp. 149–209. London: Academic Press. [1.6]

Best, P.R. & Lloyd, P. (1975) *J. Phys. C* **8**, 2219–34. [10.9]

Bethe, H.A. (1931) *Z. Phys.* **71**, 205. [5.6]

Bethe, H.A. (1935) *Proc. Roy. Soc. A* **216**, 45. [5.4]

Bezak, V. (1970) *Proc. Roy. Soc. A* **315**, 339–54. [13.5]

Bhatia, A.B., Hargrove, W.H. & March, N.H. (1973) *J. Phys. C* **6**, 621–30. [4.6]

Bhatia, A.B. & Krishnan, K.S. (1948) *Proc. Roy. Soc. A* **194**, 185. [10.1]

Bhatia, A.B. & Thornton, D.E. (1970) *Phys. Rev. B* **2**, 3004–12. [4.6]

Bishop, A.R. (1973) *J. Phys. C* **6**, 2089–93. [9.10]

Bishop, A.R. (1975) *J. Phys. C* **8**, 3317–27. [9.7]

Bishop, A.R. & Mookerjee, A. (1974) *J. Phys. C* **7**, 2165–79. [9.7]

Blair, D.G. (1967) *Proc. Phys. Soc.* **91**, 736–8. [8.6]

Blair, D.G. (1972) *Phys. Rev. B* **5**, 2097–102. [8.7]

Blinc, R. (1959) *J. Phys. Chem. Solids*, **13**, 204. [1.4]

Blinc, R. & Ribaric, M. (1963) *Phys. Rev.* **130**, 1816. [1.4]

Blinc, R. (1968) *Theory of Condensed Matter* (eds Bassani, Cagliotti & Ziman), pp. 395–442. Vienna: IAEA. [1.4]

Blinc, R. & Svetina, S. (1966) *Phys. Rev.* **147**, 423–9, 430–8. [1.5]

Bloch, F. (1930) *Z. Phys.* **61**, 206. [2.4]

Block, R. & Schommers, W. (1975) *J. Phys. C* **8**, 1997–2002. [2.11]

Bondi, A. (1968) *Physical Properties of Molecular Crystals, Liquids and Glasses.* New York: Wiley. [2.14]

Bonner, J.C. & Fisher, M.E. (1964) *Phys. Rev.* **135**, 640–58. [5.6]

Borland, R.E. (1961) *Proc. Phys. Soc.* **77**, 705–11; **78**, 926–31. [2.2] [8.2] [8.4]

Borland, R.E. (1963) *Proc. Roy. Soc. A* **274**, 529–45. [8.7]

Born, M. & Wolf, E. (1970) *Principles of Optics*, 4th edn. Oxford: Pergamon Press. [4.7]

Böttger, H. (1974) *Phys. Status Solidi, B* **62**, 9. [11.1]

Bowers, R.G. & McKerrell, A. (1973) *J. Phys. C* **6**, 2721–32. [5.10] [7.8]

Brenig, W., Wölfle, P. & Döhler, G. (1971) *Z. Phys.* **246**, 1. [12.2]

Brillouin, L. (1962) *Science and Information Theory*, 2nd edn. New York: Academic Press. [6.8]

Broadbent, S.R. & Hammersley, J.M. (1957) *Proc. Camb. Phil. Soc.* **53**, 629–41. [9.10]

Brouers, F. (1971) *J. Phys. C* **4**, 773–82. [9.4] [9.9]

Brouers, F. & Ducastelle, F. (1975) *J. Phys. F* **5**, 45. [9.5]

Brouers, F., Cyrot, M. & Cyrot-Lackman, F. (1973) *Phys. Rev. B* **7**, 4370–3. [9.5]

Brouers, F., Ducastelle, F. & van der Rest, J. (1973) *J. Phys. F* **3**, 1704–15. [9.8]

Brout, R. (1965) *Phase Transition.* New York: Benjamin. [5.1] [5.2] [5.10] [6.2] [6.5]

Brown, E., Essam, J.W. & Place, C.M. (1975) *J. Phys. C* **8**, 321–35. [12.2]

Brown, G.H., Doane, J.W. & Neff, V.D. (1970) *Critical Reviews in Solid State Sciences*, **1**, 303–79. [2.14]

Bruggeman, D.A.G. (1935) *Ann. Phys.* **24**, 636. [13.4]

Brush, S.G., Sahlin, H.L. & Teller, E. (1960) *J. Chem. Phys.* **45**, 2102. [2.12]

Buff, F.P. & Brout, R. (1955) *J. Chem. Phys.* **23**, 458. [2.12]

Buhner, H.F. & Steeb, S. (1970) *Z. Naturforsch.* **25A**, 1862–7. [4.5]

Burke, J.E. & Twersky, V. (1964) *J. Res. Nat. Bur. Stand.* **68D**, 500–40. [10.8]

Burley, D.M. (1972) in Domb & Green (1972) Vol. 2, pp. 329–74. [5.4]

Busch, G. & Güntherodt, H.-J. (1974) *Solid State Physics*, **29**, 235–313. [10.3]

Butcher, P.N. (1972) *J. Phys. C* **5**, 1817–29. [13.3]

Butcher, P.N. (1974*a*) *J. Phys. C* **7**, 3533–40. [12.2]

Butcher, P.N. (1974*b*) *J. Phys. C* **7**, 879–92. [13.3]

Butcher, P.N. (1974*c*) *J. Phys. C* **7**, 2645–54. [13.3]

Butler, W.H. (1973) *Phys. Rev. B* **8**, 4499–510. [9.5]

Buyers, W.J.L., Pepper, D.E. & Elliott, R.I. (1973) *J. Phys. C* **6**, 1933–52. [12.2]

Capek, V. (1971) *Phys. Stat. Solid, (b)* **43**, 61–72. [9.5]

Capek, V. (1972) *Phys. Stat. Solid, (b)* **52**, 399–406. [9.5]

Cargill, G.S. III (1975) *Solid State Phys.* **30**, 227–320. [2.13]

Carnahan, N.F. & Starling, K.E. (1969) *J. Chem. Phys.* **51**, 635. [6.3]

Carroll, C.E. (1975) *Phys. Rev. B* **12**, 4142–5. [9.6]

Carroll, C.E., Lukes, T., Nix, B. & Ringwood, G.A. (1974) *J. Phys. A* **7**, 1958–64. [11.3]

Chan, T. & Ballentine, L.E. (1972) *Can. J. Phys.* **50**, 813–20. [10.4]

Chandrasekhar, S. (1976) *Rep. Prog. Phys.* **39**, 613–92. [2.14]

Chang, K.S., Sher, A., Petzinger, K.G. & Weisz, G. (1975) *Phys. Rev. B* **12**, 5506–13. [10.8]

Chaudhari, P., Graczyk, J.F., Henderson, D. & Steinhard, P. (1975) *Phil. Mag.* **31**, 727–32. [2.11]

Chen, Y.D. & Steele, W.A. (1969) *J. Chem. Phys.* **50**, 1428. [2.14]

Chester, G.V. (1963) *Proc. Phys. Soc.* **81**, 938–48. [10.10]

Chock, D.P., Resibois, P., Dewel, G. & Dagonnier, R. (1971) *Physica*, **53**, 364–92. [1.4]

Clapp, P.C. (1964) *Phys. Letters,* **13**, 305–6. [5.4]

Coey, J.M.D. & Readman, P.W. (1973) *Nature*, **246**, 476–8. [12.3]

Cohen, M.L. & Heine, V. (1970) *Solid State Phys.* **24**, 37–249. [10.2]

Cole, G.H.A. (1967) *An Introduction to the Statistical Theory of Classical Simple Dense Fluids.* Oxford: Pergamon Press. [2.12]

Collins, R. (1972) in Domb & Green (1972) Vol. 2, pp. 271–303. [2.11] [6.8]

Connell, G.A.N. (1975) *Solid State Comm.* **16**, 109–12. [2.11]

Connell, G.A.N. & Temkin, R.J. (1974) *Phys. Rev. B* **9**, 5323–6. [2.10]

Cooper, A.R. & Aubourg, P.F. (1972) in *Amorphous Materials* (eds Douglas & Ellis), pp. 310–12. London: Wiley, Interscience. [2.2] [2.6] [2.14]

Copley, J.R.P. & Lovesey, S.W. (1975) *Rep. Prog. Phys.* **38**, 461–563. [11.2]

Cotterill, R.M.J. & Pedersen, L.B. (1972) *Solid State Comm.* **10**, 439–41. [2.5] [2.11]

Cowley, J.M. (1950) *Phys. Rev.* **77**, 669. [5.4]

Cowley, J.M. (1965) *Phys. Rev.* **138**, A1384–9. [5.4]

Croxton, C.A. (1974) *Liquid State Physics.* Cambridge: University Press. [2.12]

Cubiotti, G., Donato, E. & Jacobs, R.L. (1975) *J. Phys. F* **5**, 2068–78. [9.8]

Cyrot-Lackman, F. (1972) *J. Phys. C* **5**, 300–5. [9.6] [9.7] [11.3]

Dallacasa, V. (1975) *J. Phys. C* **8**, 114–20. [13.5]

Dancz, J. & Edwards, S.F. (1973) *J. Phys. C* **6**, 3413–29. [11.4]

Dancz, J. & Edwards, S.F. (1975) *J. Phys. C* **8**, 2532–48. [11.4]

Dancz, J., Edwards, S.F. & March, N.H. (1973) *J. Phys. C* **6**, 873–9. [11.4]

Date, M., Yamazaki, H., Motokawa, M. & Tazawa, S. (1970) *Prog. Theor. Phys. Supp.* **46**, 194–209. [2.3]

Dean, P. (1960) *Proc. Roy. Soc. A* **254**, 507. [8.4]

Dean, P. (1972) *Rev. Mod. Phys.* **44**, 127. [11.1]

Dean, P. & Bacon, M.D. (1963) *Proc. Phys. Soc.* **81**, 642. [8.7]

Debye, P. & Bueche, F. (1952) *J. Chem. Phys.* **20**, 1337. [7.4]

De Gennes, P.G. (1963*a*) *Solid State Comm.* **1**, 132. [1.4]

De Gennes, P.G. (1963*b*) in *Magnetism* (eds Rado & Suhl), Vol. III, pp. 115–47. [1.8]

De Gennes, P.G. (1969) *Rep. Prog. Phys.* **32**, 187–205. [7.9]

De Gennes, P.G. (1972) *Phys. Letters,* **38A**, 339–40. [7.7]

De Gennes, P.G. (1974) *The Physics of Liquid Crystals.* Oxford: Clarendon Press. [2.14] [5.11]

De Jongh, L.J. & Miedema, A.R. (1974) *Adv. in Phys.* **23**, 1–260. [2.3] [5.1]

Des Cloiseaux, J. & Pearson, J.J. (1962) *Phys. Rev.* **128**, 2131–5. [5.6]

Dobson, G.R. & Gordon, M. (1964) *J. Chem. Phys.* **41**, 2389. [7.5]

Dobson, G.R. & Gordon, M. (1965) *J. Chem. Phys.* **43**, 705. [7.5] [7.6]

Dodds, J.A. (1975) *Nature,* **256**, 187–9. [2.13]

Doman, B.G.S. & Ter Haar, D. (1962) *Phys. Letters,* **2**, 15–16. [5.4]

Domb, C. (1954) *Proc. Camb. Phil. Soc.* **50**, 586–91.

Domb, C (1960) *Adv. in Phys.* **9**, 149–361. [1.6] [5.2] [5.4] [5.7] [5.10]

Domb, C. (1969) *Adv. Chem. Phys.* **15**, 229. [5.10] [7.7] [7.8]

Domb, C. (1970*a*) *Adv. in Phys.* **19**, 339–70. [5.10]

Domb, C. (1970*b*) *J. Phys. C.* **3**, 255–83. [5.10]

Domb, C., Barrett, A.J. & Lax, M. (1973) *J. Phys. A* **6**, L82–7. [7.7] [7.8]

Domb, C., & Dalton, N.N. (1966) *Proc. Phys. Soc.* **89**, 859. [13.2]

Domb, C. & Green, M.S. (eds.) *Phase Transitions and Critical Phenomena* (1972) Vol. 1, (1972) Vol. 2, (1974) Vol. 3, (1976) Vol. 5(*a*). London: Academic Press. [5.1] [5.10]

Domb, C. & Joyce, G.S. (1972) *J. Phys. C* **5**, 956–76. [7.7] [7.8]

Domb, C., Maradudin, A.R., Montroll, E.W. & Weiss, G.H. (1959) *Phys. Rev.* **115**, 24–36. [8.4] [11.3]

Domb, C. & Sykes, M.F. (1961) *Phys. Rev.* **122**, 77–8. [12.1]

Doob, J.L. (1973) *Stochastic Processes,* p. 108. New York: Wiley. [9.9]

Ducastelle, F. (1971) *J. Phys. C* **4**, L75–7. [9.4]

Ducastelle, F. (1972) *J. Phys. F* **2**, 468–86. [9.5]

Ducastelle, F. (1974*a*) *J. Phys. C* **7**, 1795–816. [9.4] [9.5]

Ducastelle, F. (1974*b*) *J. Physique,* **35**, 983–8. [9.6]

Ducastelle, F. (1975) *J. Phys. C* **8**, 3297–316. [10.9]

Dunn, A.G., Essam, J.W. & Ritchie, D.S. (1975) *J. Phys. C* **8**, 4219–35. [9.10]

Dunn, A.G., Essam, J.W. & Loveluck, J.M. (1975) *J. Phys. C* **8**, 743–50. [9.10]

Duwez, P. (1967) *Trans. Am. Soc. Metals,* **60**, 606. [2.13]

Duwez, P., Willens, R.H. & Klement, W. (1960) *J. App. Phys.* **31**, 1136. [2.13]

Dyson, F.J. (1953) *Phys. Rev.* **92**, 1331. [8.5]

Dyson, F.J. (1969) *Comm. Math. Phys.* **12**, 91, 212. [5.5]

Economou, E.N. & Cohen, M.H. (1971) *Phys. Rev. B* **4**, 396–403. [9.7]

Economou, E.N. & Cohen, M.H. (1972) *Phys. Rev. B* **5**, 2931–48. [9.9]

Economou, E.N., Ngai, K.L. & Reinecke, T.L. (1975) *Proceedings of the NATO Advanced Studies Institute on Linear and Nonlinear Electronic Transport in Solids.* Antwerp. [11.2]

Edwards, F.G., Enderby, J.E., Howe, R.A. & Page, D.I. (1975) *J. Phys. C* **8**, 3483–90. [2.13] [4.5]

Edwards, J.T. & Thouless, D.J. (1971) *J. Phys. C* **4**, 453–7. [9.9]

Edwards, J.T. & Thouless, D.J. (1972) *J. Phys. C* **5**, 807–20. [9.9]

Edwards, S.F. (1958) *Phil. Mag.* **3**, 1020. [10.10]

Edwards, S.F. (1962) *Proc. Roy. Soc. A* **267**, 518–40. [10.5]

Edwards, S.F. (1965) *Proc. Phys. Soc.* **85**, 613. [7.9]

Edwards, S.F. (1967) *Proc. Phys. Soc.* **91**, 513–9. [7.10]

Edwards, S.F. (1968) *J. Phys. A* **1**, 15–27. [7.10]

Edwards, S.F. (1969) *J. Phys. C* **2**, 1–10. [7.6]

Edwards, S.F. (1970) *J. Phys. C* **3**, L30–31. [13.5]

Edwards, S.F. & Anderson, P.W. (1975) *J. Phys. F* **5**, 965–74. [12.3]

Edwards, S.F. & Freed, K.F. (1969) *J. Phys. A* **2**, 145. [7.10]

Edwards, S.F. & Freed, K.F. (1970) *J. Phys. C* **3**, 739–49, 750–9, 760–8. [7.10]

Egelstaff, P.A. (1967) *An Introduction to the Liquid State*. Oxford: Pergamon Press. [2.12]

Egelstaff, P.A., Page, D.I. & Heard, C.R.T. (1971) *J. Phys. C* **4**, 1453–65. [2.12]

Egelstaff, P.A., Page, D.I. & Heard, C.R.T. (1969) *Phys. Letters*, **30A**, 376. [2.12]

Egelstaff, P.A. & Ring, J.W. (1968) in *Physics of Simple Liquids* (eds Temperley, Rowlinson & Rushbrooke), pp. 253–97. Amsterdam: North Holland. [6.2]

Eggarter, T.P. (1973) *Phys. Rev. B* **7**, 1727–30. [9.7]

Eggarter, T.P. & Cohen, M.H. (1974) *J. Phys. C* **7**, L103–6. [13.5]

Elliott, R.J., Heap, B.R., Morgan, D.J. & Rushbrooke, G.S. (1960) *Phys. Rev. Lett.* **5**, 366. [12.1]

Elliott, R.J., Krumhansl, J.A. & Leath, P.L. (1974) *Rev. Mod. Phys.* **46**, 465–543. [9.3] [9.4]

Elliott, R.J. & Marshall, W. (1958) *Rev. Mod. Phys.* **30**, 75–89. [1.7] [4.6]

Elliott, R.J. & Pepper, D.E. (1973) *Phys. Rev. B* **8**, 2374–8. [12.2]

Emin, D. (1975) *Adv. in Phys.* **24**, 305–48. [13.3]

Enderby, J.E. (1968) in *Physics of Simple Fluids* (eds Temperley, Rowlinson & Rushbrooke), pp. 611–44. Amsterdam: North Holland. [4.1]

Enderby, J.E., North, D.M. & Egelstaff, P.A. (1966) *Phil. Mag.* **14**, 961. [4.5]

Epstein, N. & Young, M.J. (1962) *Nature*, **196**, 885–6. [2.13]

Essam, J.W. (1972) in Domb & Green (1972) Vol. 2, pp. 197–270. [9.10] [12.1]

Essam, J.W., Gwilym, K.M. & Loveluck, J.M. (1976) *J. Phys. C* **9**, 365–78. [12.1]

Essam, J.W., Place, C.M. & Sondheimer, E.H. (1974) *J. Phys. C* **7**, L258–60. [9.11]

Evans, R., Greenwood, D.A. & Lloyd, P. (1971) *Phys. Letters*, **35A**, 57–8. [10.3]

Even, U. & Jortner, J. (1974) *Phil. Mag.* **30**, 325–34. [2.15]

Eyring, H. & Jhon, M.S. (1969) *Significant Liquid Structures*. New York: Wiley. [2.9]

Faber, T.E. (1966) *Adv. in Phys.* **15**, 547–81. [10.10]

Faber, T.E. (1972) *An Introduction to the Theory of Liquid Metals*. Cambridge: University Press. [6.8] [10.1] [10.10]

Faber, T.E. & Ziman, J.M. (1965) *Phil. Mag.* **11**, 153. [10.10]

Feller, W. (1967) *An Introduction to Probability Theory and its Applications. I and II*, 3rd edn, p. 360. New York: Wiley. [2.2] [7.4] [7.8]

Finney, J.L. (1970) *Proc. Roy. Soc. A* **319**, 479–93, 495–507. [2.11] [4.4]

Finney, J.L. (1975) *J. de Phys.* **36**, C2–1. [2.11]

Fisher, I.Z. & Kopeliovich, B.L. (1960) *Sov. Phys. Dokl.* **5**, 761–3. [2.6] [2.12]

Fisher, M.E. (1964) *Proc. IBM Symposium on Combinatorial Problems*. [6.8] [9.10]

Fisher, M.E. (1967) *Rep. Prog. Phys.* **30**, 615–730. [1.5] [5.1] [5.11] [5.12] [6.2]

Fisher, M.E. (1972) *Essays in Phys.* **4**, 43–89. [5.1]

Fisher, M.E. (1974) *Rev. Mod. Phys.* **46**, 597–616. [5.1] [5.12]

Fisher, M.E. & Camp, W.J. (1972) *Phys. Rev. B* **5**, 3730–7. [9.7]

Fisher, M.E. & Essam, J.W. (1961) *J. Math. Phys.* **2**, 609. [7.5] [9.10]

Fixman, M. (1966) *J. Chem. Phys.* **45**, 785. [7.7]

Fletcher, N.H. (1970) *The Chemical Physics of Ice.* Cambridge: University Press. [1.4] [2.8]

Flory, P.J. (1941) *J. Chem. Phys.* **9**, 660. [7.1]

Flory, P.J. (1942) *J. Chem. Phys.* **10**, 51. [7.1]

Flory, P.J. (1949) *J. Chem. Phys.* **17**, 303. [7.7]

Flory, P.J. (1953) *Principles of Polymer Chemistry.* Ithaca: Cornell University Press. [7.4] [7.5]

Flory, P.J. (1969) *Statistical Mechanics of Chain Molecules.* New York: Interscience. [7.1] [7.2] [7.3] [7.4]

Flory, P.J. (1970) *Disc. Farad. Soc.* **49**, 7–29. [7.1] [7.2] [7.5]

Flory, P.J. & Fisk, S. (1966) *J. Chem. Phys.* **44**, 2243. [7.7]

Foldy, L.L. (1945) *Phys. Rev.* **67**, 107–19. [10.8]

Foldy, L.C. & Walecka, J.D. (1969) *Ann. of Phys. N.Y.* **54**, 447–504. [10.8]

Fortuin, C.M. (1972) *Physica,* **58**, 393–418. [9.10]

Fortuin, C.M. & Kasteleyn, P.W. (1971) *Physica,* **57**, 536. [9.10]

Fosdick, L.D. & James, H.M. (1953) *Phys. Rev.* **91**, 1131. [5.4]

Fournet, G. (1957) *Handb. Phys.* **XXXII**, 238–319. [4.4]

Fowler, R.H. & Guggenheim, E.A. (1949) *Statistical Thermodynamics.* Cambridge: University Press. [6.2]

Frank, F.C. (1958) *Disc. Farad. Soc.,* **25**, 19. [2.14]

Freed, K.F. (1971) *J. Phys. C* **4**, L331–5. [7.9]

Freed, K.F. (1972*a*) *Adv. in Chem. Phys.* **22**, 1–128. [7.9]

Freed, K.F. (1972*b*) *Phys. Rev. B* **5**, 4802–26. [13.5]

Friedberg, R. & Luttinger, J.M. (1975) *Phys. Rev. B* **12**, 4460–75. [13.5]

Frisch, H. (1965) *Trans. Rheology Soc., Pt. 9,* **1**, 293. [3.3]

Frisch, H.L. & Hammersley, J.M. (1963) *J. Soc. Indust. Appl. Math.* **11**, 894–918. [9.10]

Frisch, H.L., Hammersley, J.M. & Welsh, D.J.A. (1962) *Phys. Rev.* **126**, 949–51. [9.10]

Frisch, H.L. & Lloyd, S.P. (1960) *Phys. Rev.* **120**, 1175. [13.5]

Fritchie, C.J. (1966) *Acta. Cryst.* **20**, 892. [2.3]

Fujita, T. & Hori, J.I. (1972) *J. Phys. C* **5**, 1059–66. [9.9]

Fujita, T. & Hori, J.I. (1973) *J. Phys. C* **6**, 51–6. [9.9]

Furstenberg, H. (1963) *Trans. Amer. Math. Soc.* **108**, 377. [8.7]

Gaspard, J.P. & Cyrot-Lackmann, F. (1973) *J. Phys. C* **6**, 3077–96. [9.7]

Gehlen, P.C. & Cohen, J.B. (1965) *Phys. Rev.* **139**, A844–55. [1.7]

Gerstenkorn, H. (1952) *Ann. Phys.* **10**, 49. [10.1] [10.2]

Gilbert, E.N. (1962) *Annals of Math. Statistics,* **33**, 958–72. [3.3]

Gillan, M., Larsen, B., Tosi, M.P. & March, N.H. (1976) *J. Phys. C* **9**, 889–908. [2.13]

Good, I.J. (1963) *Proc. Roy. Soc. A* **272**, 54. [7.5]

Gordon, M., Love, A. & Pugh, D. (1969) *J. Chem. Phys.* **49**, 4680. [7.6]

Gordon, M. (1962) *Proc. Roy. Soc. A* **268**, 240. [7.5]

Gordon, M. & Malcolm, G.N. (1966) *Proc. Roy. Soc. A* **295**, 29–54. [7.5]

Green, H.S. (1952) *The Molecular Theory of Fluids.* Amsterdam: North Holland. [4.4]

References   499

Green, M.S. (1954) *J. Chem. Phys.* **22**, 398–413. [11.2]
Greenholz, M. & Kidron, A. (1970) *Acta Cryst. A* **26**, 311–4. [1.7]
Greenwood, D.A. (1958) *Proc. Phys. Soc.* **71**, 585. [10.10]
Greenwood, D.A. & Ratti, V.R. (1972) *J. Phys. F* **2**, 289–96. [10.10]
Griffiths, R.B. (1969) *Phys. Rev. Letters*, **23**, 17. [12.1]
Griffiths, R.B. (1971) in *Statistical Mechanics and Field Theory* (eds de Witt & Stora), pp. 241–79. New York: Gordon & Breach. [2.4] [5.5]
Griffiths, R.B. (1972) in Domb & Green (1972) Vol. 1, pp. 7–109. [2.4] [12.1]
Griffiths, R.B. & Lebowitz, J.L. (1968) *J. Math. Phys.* **9**, 1284. [12.1]
Grigorivici, R. & Mamaila, R. (1969) *J. Non-Cryst. Solids*, **1**, 371. [2.8] [2.10]
Grinstein, G. & Luther, A. (1976) *Phys. Rev. B* **13**, 1329–43. [12.1]
Groeneveld, J. (1962) *Phys. Letters*, **3**, 50. [6.5]
Gubanov, A.I. (1960) *Sov. Phys. Solid State* **2**, 468. [12.3]
Gubanov, A.I. (1965) *Quantum Electron Theory of Amorphous Conductors.* New York: Consultants Bureau. [2.6]
Gubernatis, J.E. & Taylor, P.L. (1971) *J. Phys. C* **4**, L94–6. [8.5]
Gubernatis, J.E. & Taylor, P.L. (1973) *J. Phys. C* **6**, 1889–94. [8.5]
Guggenheim, E.A. (1935) *Proc. Roy. Soc. A* **148**, 304. [5.3]
Guggenheim, E.A. & McGlashan, M.C. (1951) *Proc. Roy. Soc. A* **206**, 335. [5.4]
Gunton, J.D. & Buckingham, M.J. (1968) *Phys. Rev.* **166**, 152. [5.9]
Gürsey, F. (1950) *Proc. Camb. Phil. Soc.* **46**, 182. [6.1]
Guttman, L. (1956) *Solid State Phys.* **3**, 145–223. [1.5] [1.6] [1.7]
Gyorffy, B.L. (1970) *Phys. Rev.* **1**, 3290–9. [10.6]
Gyorffy, B.L. (1972) *Phys. Rev. B* **5**, 2382–4. [10.6] [10.7]
Gyorffy, B.L. & Stocks, G.M. (1974) *J. de Phys.* **35**, C4–75–80. [10.7] [10.9]
Halder, N.C. & Wagner, C.N.J. (1967) *J. Chem. Phys.* **47**, 4385–91. [4.5]
Halperin, B.I. (1965) *Phys. Rev.* **139 A**, 104–17. [13.5]
Halperin, B.I. (1973) *Physica Fennica*, **8**, 215–51. [9.9] [13.5]
Halperin, B.I. & Lax, M. (1966) *Phys. Rev.* **148**, 722–40. [13.5]
Halperin, B.I. & Lax, M. (1967) *Phys. Rev.* **153**, 802–14. [13.5]
Hammersley, J.M. (1957) *Proc. Camb. Phil. Soc.* **53**, 642. [7.8]
Hammersley, J.M. (1961) *J. Math. Phys.* **2**, 728–33. [9.10]
Handrich, K. (1969) *Phys. Stat. Solid.* **32**, K55. [12.3]
Hansen, J.P. (1970) *Phys. Rev. A* **2**, 221. [6.7]
Hansen, J.P. & Verlet, L. (1969) *Phys. Rev.* **184**, 151–61. [6.7]
Harary, F. (ed.) (1967) *Graph Theory and Theoretical Physics.* London: Academic Press. [5.4]
Harris, A.B. (1973) *Phys. Rev. B* **8**, 3661–4. [9.6]
Harris, A.B. (1974) *J. Phys. C* **7**, 1671–92. [12.1]
Harris, A.B. (1975) *Phys. Rev. B* **12**, 203–7. [12.1]
Harris, A.B., Leath, P.L., Nickel, B.G. & Elliott, R.J. (1974) *J. Phys. C* **7**, 1693–718. [12.2]
Harris, R., Plischke, M. & Zuckermann, M.J. (1973) *Phys. Rev. Lett.* **31**, 160–2. [12.3]
Harris, T.E. (1960) *Proc. Camb. Phil. Soc.* **56**, 13–20. [9.10]
Harris, T.E. (1963) *The Theory of Branching Processes.* Berlin: Springer. [7.5]
Hasegawa, H. & Nakamura, M. (1969) *J. Phys. Soc. Japan*, **26**, 1362. [9.4]
Haydock, R., Heine, V. & Kelly, M.J. (1972) *J. Phys. C* **5**, 2845–58. [9.7]

Haydock, R., Heine, V. & Kelly, M.J. (1975) *J. Phys. C* **8**, 2591–605. [9.7]

Haydock, R. & Mookerjhee, A. (1974) *J. Phys. C* **7**, 3001–12. [9.9]

Heine, V. (1970) *Solid State Phys.* **24**, 1–36. [10.2]

Heine, V. (1971) *J. Phys. C* **4**, L221–3. [11.3]

Heine, V. & Weaire, D. (1970) *Solid State Phys.* **24**, 250–463. [10.2]

Henderson, D.J., Barker, J.A. & Watts, R.O. (1970) *IBM J. Res. Dev.* **14**, 668–76. [6.3]

Henderson, D. & Herman, F. (1972) *J. Non-Cryst. Solids*, **8–10**, 359–63. [2.8]

Herbert, D.C. & Jones, R. (1971) *J. Phys. C* **4**, 1145–61. [8.7]

Hernandez, J.P. & Ziman, J.M. (1973) *J. Phys. C* **6**, L251–3. [13.5]

Herzfeld, K.F. & Goeppert-Mayer, M. (1934) *J. Chem. Phys.* **2**, 38–45. [6.1]

Hijman, S.J. & de Boer, J. (1955–6) *Physica*, **21**, 471, 485, 499; **22**, 408. [5.4]

Hill, T.L. (1956) *Statistical Mechanics.* New York: McGraw Hill. [4.4] [5.10] [6.1] [6.5]

Hirota, T. & Ishii, K. (1971) *Prog. Theor. Phys.* **45**, 1713. [8.7]

Hirota, T. (1973) *Prog. Theor. Phys.* **50**, 1240–7. [8.7]

Holcomb, D.F. & Rehr, J.J. (1969) *Phys. Rev.* **183**, 773–6. [13.2]

Holcomb, W.K. (1974) *J. Phys. C* **7**, 4299–313. [12.2]

Holcomb, W.K. (1976) *J. Phys. C* **9**, 1771–8. [12.2]

Holstein, T. (1959) *Ann. Phys. N.Y.* **8**, 325, 343. [13.3]

Hoover, W.G. & Alder, B.J. (1967) *J. Chem. Phys.* **46**, 686. [6.7]

Hoover, W.G. & Ree, F.H. (1967) *J. Chem. Phys.* **47**, 4873–8. [6.7]

Hoover, W.G. & Ree, F.H. (1968) *J. Chem. Phys.* **49**, 3609. [6.7] [6.8]

Hori, J. (1968) *Spectral Properties of Disordered Chains and Lattices.* Oxford: Pergamon Press. [8.2] [8.3] [8.4]

Hosemann, R. & Bagchi, S.N. (1902) *Direct Analysis of Diffraction by Matter.* Amsterdam: North Holland. [2.9] [2.10]

House, D. & Smith, P.V. (1973) *J. Phys. F* **3**, 753–8. [10.9]

Howe, M.S. (1973) *Phil. Trans. A* **274**, 523–49. [13.1]

Huang, I.L. & Dy, K.S. (1974) *Phys. Rev. B* **9**, 5316–17. [11.3]

Huang, K. (1963) *Statistical Mechanics.* New York: Wiley. [5.7]

Hubbard, J. (1963) *Proc. Roy. Soc. A* **276**, 238–57. [9.4]

Hubbard, J. (1964) *Proc. Roy. Soc. A* **277**, 237–59, **281**, 401–19. [9.4]

Hubbard, J. (1971) *J. Chem. Phys.* **55**, 1382. [6.6]

Huberman, M. & Chester, G.V. (1975) *Adv. in Phys.* **24**, 489–514. [10.10]

Huggins, M.L. (1941) *J. Chem. Phys.* **9**, 440. [7.1]

Huggins, M.L. (1958) *Physical Chemistry of High Polymers.* New York: Wiley. [7.4] and fig. 7.3.

Hulin, M. (1972) *Phys. Stat. Solid. B* **52**, 119. [11.3]

Hulthén, L. (1938) *Arkiv Mat. Astron. Fyz.* **26A**, No. 11. [5.6]

Isherwood, S.P. & Orton, B.R. (1969) *J. App. Cryst.* **2**, 219–23. [4.5]

Ishii, K. (1973) *Supp. Prog. Theor. Phys.* **53**, 77–138. [8.7]

Isihara, A. (1950) *J. Phys. Soc. Japan,* **5**, 201. [7.4]

Isihara, A. (1968) *J. Phys. A* **1**, 539–48. [6.4]

Isihara, A. & Hayashida, T. (1951) *J. Phys. Soc. Japan,* **6**, 40. [2.14]

Jacobs, R.L. (1973) *J. Phys. F* **3**, 933–43. [9.7]

Jacobs, R.L. (1974) *J. Phys. F* **4**, 1351–8. [9.7]

James, H.M. & Guth, E. (1943) *J. Chem. Phys.* **11**, 455–81. [7.6]

James, H.M. & Guth, E. (1947) *J. Chem. Phys.* **15**, 669.   [7.6]

Joannopoulos, J.D. & Cohen, M.L. (1973) *Phys. Rev. B* **7**, 2644–56.   [11.3][11.5]

Joannopoulos, J.D. & Yndurain, F. (1974) *Phys. Rev. B* **10**, 5164–74.   [11.4]

Johnson, M.D. & March, N.H. (1973) *Phys. Letters*, **3**, 313.   [2.12]

Jona, F. & Shirane, G. (1962) *Ferroelectric Crystals.* Oxford: Pergamon Press.   [1.4]

Jones, H. (1973) *Rep. Prog. Phys.* **36**, 1425–97.   [2.13]

Jones, P.L., Khor, K.E., Roberts, A.P. & Smith, P.V. (1970) *J. Phys. C* **3**, 1211–20.   [8.6]

Jones, R. & McCubbin, W.L. (1975) *J. Phys. C* **8**, L321–3.   [13.5]

Jones, R.C. & Edwards, S.F. (1971) *J. Phys. C* **4**, L193–7.   [12.2]

Joyce, G.S. (1972) in Domb & Green (1972) Vol. 2, pp. 375–442.   [5.8]

Kac, M. (1968) in *Brandeis Lectures 1966* (eds Chrétien, Gross & Desser). New York: Gordon & Breach.   [5.5]

Kac, M. & Ward, J.C. (1952) *Phys. Rev.* **88**, 1332.   [5.7][5.10]

Kadanoff, L.P. (1966) *Physics,* **2**, 263.   [5.12]

Kadanoff, L.P. (1976) in Domb & Green (1976) Vol. 5a, pp. 1–34.   [5.12]

Kadanoff, L.P. *et al.* (1967) *Rev. Mod. Phys.* **39**, 395–431.   [5.11][5.12]

Kane, E.O. (1963) *Phys. Rev.* **131**, 78,   [13,4]

Kaneyoshi, T. (1973a) *J. Phys. C* **6**, L19–23.   [12.3]

Kaneyoshi, T. (1973b) *J. Phys. C* **6**, 3130–8.   [12.3]

Kaneyoshi, T. (1975) *J. Phys. C* **8**, 3415–26.   [12.2]

Kaneyoshi, T. (1976) *J. Phys. C* **9**, L289–92.   [12.3]

Känzig, W. (1957) *Solid State Phys.* **4**, 1–197.   [1.4]

Kaplow, R., Strong, S.L. & Auerbach, B.L. (1965) *Phys. Rev.* **138A**, 1336–45.   [2.9]

Kasteleyn, P.W. (1967) in *Graph Theory and Theoretical Physics* (ed. Harary), pp. 43–110. London: Academic Press.   [5.10][7.8]

Katz, I. & Chandler, D. (1972) *Phys. Rev. Letters,* **29**, 247–9.   [6.3]

Kawasaki, K. (1967) *Prog. Theor. Phys.* **38**, 1082.   [1.7][1.8]

Kawasaki, K. & Mori. H. (1962) *Prog. Theor. Phys.* **28**, 690.   [1.8]

Keating, D.T. (1963) *J. App. Phys.* **34**, 923.   [4.5]

Keller, J. (1971) *J. Phys. C* **4**, 3143–54.   [10.9]

Keller, J. & Jones, R. (1971) *J. Phys. F* **1**, L33–6.   [10.9]

Keller, J. & Smith, P.V. (1972) *J. Phys. C* **5**, 1109–22.   [10.9]

Khor, K.E. (1971) *J. Phys. C* **4**, 150–2.   [8.6]

Khor, K.E. & Smith, P.V. (1971a, b) *J. Phys. C* **4**, 2029–40, 2041–51.   [9.9]

Kihara, T. (1953) *Rev. Mod. Phys.* **25**, 831.   [2.14]

Kikuchi, M. (1970) *J. Phys. Soc. Japan,* **29**, 246–301.   [9.9]

Kikuchi, R. (1951) *Phys. Rev.* **81**, 988.   [5.4]

Kikuchi, R. (1970) *J. Chem. Phys.* **53**, 2713–8.   [12.1]

Kikuchi, R. & Brush, S.G. (1967) *J. Chem. Phys.* **47**, 195.   [5.4]

Kirkpatrick, S. (1971) *Phys. Rev. Letters,* **27**, 1722–5.   [9.11]

Kirkpatrick, S. (1973) *Rev. Mod. Phys.* **45**, 524.   [9.11][12.2]

Kirkpatrick, S. & Eggarter, T.P. (1972) *Phys. Rev. B* **6**, 3598–609.   [9.4][9.9][9.10]

Kirkwood, J.G. (1950) *J. Chem. Phys.* **18**, 380.   [6.8]

Kirkwood, J.G., Maun, E.K. & Alder, B.J. (1950) *J. Chem. Phys.* **18**, 1040–7.   [2.12]

Klima, J., McGill, T.C. & Ziman, J.M. (1970) *Disc. Farad. Soc.* **50**, 20–6.   [10.9]

Kotze, I.A. & Kuhlmann-Wilsdorf, D. (1971) *Phil. Mag.* **23**, 1133.   [2.11]

Kosterlitz, J.M. & Thouless, D.J. (1972) *J. Phys.* C **5**, L124–6.   [2.5]

Kosterlitz, J.M. & Thouless, D.J. (1973) *J. Phys.* C **6**, 1181–203.   [2.5]

Kovalenko, N.T. & Fisher, I.Z. (1973) *Soviet Phys. Uspekhi*, **15**, 592–607.   [6.3]

Kramers, H.A. & Wannier, G.H. (1941) *Phys. Rev.* **60**, 252, 263.   [5.4] [5.10] [9.10]

Krogmann, K. (1969) *Angew. Chem. Int. Ed. Engl.* **8**, 35.   [2.3]

Krumhansl, J.A. & Wang, S.S. (1972) *J. Chem. Phys.* **56**, 2034–41, 2179–80.   [2.11]

Krupicka, S. & Sternberk, J. (eds) (1968) *Elements of Theoretical Magnetism*. London: Iliffe Books.   [1.6] [1.8]

Kubo, R. (1956) *Can. J. Phys.* **34**, 1274–7.   [10.10]

Kubo, R. (1962) *J. Phys. Soc. Japan,* **17**, 1100–20.   [1.5] [5.10] [6.5] [10.5]

Kubo, R. (1966) *Rep. Prog. Phys.* **29**, 255–84.   [10.10]

Kuhn, W. (1934) *Kolloid Z.* **68**, 2.   [7.4]

Kuhn, W. (1936) *Kolloid Z.* **76**, 258.   [7.6]

Kumar, A.P. & Baskaran, G. (1973) *J. Phys.* C **6**, L399–401.   [9.9]

Kumar, N. & Subramanian, R.R. (1974) *J. Phys.* C **7**, 1817–21.   [9.9]

Kumaravadivel, R., Evans, R. & Greenwood, D.A. (1974) *J. Phys.* F **4**, 1839–48.   [2.12]

Kurata, M., Kikuchi, R. & Watari, T. (1953) *J. Chem. Phys.* **21**, 434–48.   [5.4]

Kurkijärvi, J. (1974) *Phys. Rev.* B **9**, 770–4.   [13.3]

Kuse, D. & Zeller, H.R. (1971) *Phys. Rev. Letters,* **27**, 1060–3.   [2.3]

Kyselka, A. (1974) *J. Phys.* A **7**, 315–17.   [7.9]

Landau, L.D. & Lifshitz, E.M. (1958) *Statistical Physics*. London: Pergamon Press.   [5.11] [6.2]

Landauer, R. (1952) *J. App. Phys.* **23**, 779.   [13.4]

Langer, J.S. (1960) *Phys. Rev.* **120**, 714–25.   [10.10]

Last, B.J. (1972) *J. Phys.* C **5**, 2805–12.   [12.2]

Last, B.J. & Thouless, D.J. (1971) *Phys. Rev. Lett.* **27**, 1719–21.   [9.11]

Last, B.J. & Thouless, D.J. (1974) *J. Phys.* C **7**, 715–31.   [9.9]

Lax, M. (1951) *Rev. Mod. Phys.* **23**, 287–310.   [10.8]

Lax, M. (1952) *Phys. Rev.* **85**, 621–9.   [10.6]

Leadbetter, A.J. & Wright, A.C. (1972) *J. Non-Cryst. Solids,* **7**, 23–36, 37–52, 141–55, 156–67.   [2.9]

Leath, P.L. (1968) *Phys. Rev.* **171**, 725–7.   [9.4]

Leath, P.L. (1970) *Phys. Rev.* B **2**, 3078–87.   [9.4]

Leath, P.L. (1973) *J. Phys.* C **6**, 1559–71.   [9.4] [9.5]

Leath, P.L. (1974) *J. de Phys.* **35**, C4–99–101.   [9.5]

Leblanc, O. (1967) *Physics and Chemistry of the Organic Solid State*, Vol. III, p. 133. New York: Interscience, Wiley.   [2.3]

Lebowitz, J.L. (1964) *Phys. Rev.* **133**, A895.   [2.13]

Lebowitz, J.L. & Rowlinson, J.S. (1964) *J. Chem. Phys.* **41**, 133.   [2.13] [6.3]

Lehmann, G. (1973) *J. Phys.* C **6**, 1881–8.   [9.9]

Lennard-Jones, J.E. & Devonshire, A.F. (1937) *Proc. Roy. Soc.* A **163**, 53.   [6.8]

Levesque, D. (1966) *Physica,* **32**, 1985.   [2.12]

Licciardello, D.C. & Economou, E.N. (1975) *Phys. Rev.* B **11**, 3697–717.   [9.9]

Lieb, E.H. (1963) *J. Math. Phys.* **4**, 671–8.   [6.5]

Lieb, E.H. (1967) *Phys. Rev.* **162**, 162–72.   [5.8]

Lieb, E.H. (1971) in *Statistical Mechanics and Quantum Field Theory* (eds de Witt & Stora), pp. 283–385. New York: Gordon & Breach. [5.8]

Lieb, E.H. & Wu, E.Y. (1972) in Domb & Green (1972) Vol. 2, pp. 331–490. [5.8]

Lieb, E.H. & Mattis, D.C. (1966) *Mathematical Physics in One Dimension*. New York: Academic Press. [2.3]

Lifshitz, I.M. (1964) *Adv. in Phys.* **13**, 483–536. [8.4] [9.6]

Lifshitz, I.M. (1968) *Soviet Physics JETP,* **26**, 462–9. [13.4] [13.5]

Lloyd, P. (1967) *Proc. Phys. Soc.* **90**, 297–316, 317–32. [10.9]

Lloyd, P. (1969) *J. Phys. C* **2**, 1717–25. [9.9]

Lloyd, P. & Berry, M.V. (1967) *Proc. Phys. Soc.* **91**, 678–88. [10.8]

Lloyd, P. & Best, P.R. (1975) *J. Phys. C* **8**, 3752–66. [13.5]

Lloyd, P. & Smith, P.V. (1972) *Adv. in Phys.* **21**, 69–142. [10.7]

Longuet-Higgins, H.C. (1951) *Proc. Roy. Soc. A* **205**, 247. [2.13]

Longuet-Higgins, M.S. (1957) *Phil. Trans. A* **249**, 321. [3.1] [3.4]

Longuet-Higgins, M.S. (1960) *J. Opt. Soc. Am.* **50**, 838–44, 845–50, 851–6. [3.1] [3.4]

Lowry, G.G. (ed.) (1970) *Markov Chains and Monte Carlo Calculations in Polymer Science*. New York: Dekker. [7.3]

Luban, M. (1976) in Domb & Green (1976) Vol. 5(*a*), pp. 35–86. [5.11]

Lubensky, T.C. (1975) *Phys. Rev. B* **11**, 3573–80. [12.3]

Lukes, T. & Nix, B. (1973) *J. Phys. A* **6**, 1534–42, 1888–95. [11.3]

Lumley, J.L. (1970) *Stochastic Tools in Turbulence*. New York: Academic Press. [3.1]

Luttinger, J.M. (1951) *Philips Res. Reps,* **6**, 303. [8.4]

Luttinger, J.M. (1967) in *Mathematical Methods in Solid State Superfluid Theory* (eds Clark & Derrick), pp. 157–93. Edinburgh: Oliver & Boyd. [10.10]

McAlister, S.P. & Turner, R. (1972) *J. Phys. F* **2**, L51–4. [4.6]

McDonald, I.R. & Singer, K. (1970) *Quart. Rev.* **24**, 238. [6.6]

McFarlane, S.C. (1975) *J. Phys. C* **8**, 2819–36. [4.7]

McGill, T.C. & Klima, J. (1972) *Phys. Rev. B* **5**, 1517–29. [10.9]

McIntyre, D. & Sengers, J.V. (1968) in *Physics of Simple Fluids* (eds Temperley, Rowlinson & Rushbrooke), pp. 447–506. Amsterdam: North Holland. [4.1] [4.4]

Ma, S. (1973) *Rev. Mod. Phys.* **45**, 581. [5.12]

Mader, S., Widmer, H., d'Heurle, F.M. & Nowick, A.S. (1963) *App. Phys. Letters,* **3**, 201. [2.13]

Maier, W. & Saupe, A. (1959) *Z. Naturforsch.* **14A**, 882. [2.14]

Maier, W. & Saupe, A. (1960) *Z. Naturforsch.* **15A**, 287. [2.14]

Makinson, R.E.B. & Roberts, A.P. (1962) *Proc. Phys. Soc.* **79**, 222–3. [8.4] [8.7]

Mangelsdorf, P.C. & Washington, E.L. (1960) *Nature,* **187**, 930. [2.13]

March, N.H. (1968) *Liquid Metals*. Oxford: Pergamon Press. [2.12]

Marshall, W. (1960) *Phys. Rev.* **118**, 1519–23. [12.3]

Marshall, W. & Low, G.G. (1968) *Theory of Condensed Matter*, pp. 501–38. Vienna: IAEA. [4.6]

Martin, D.H. (1967) *Magnetism in Solids*. London: Iliffe Books. [1.6] [1.8]

Martin, J.W. & Ziman, J.M. (1970) *J. Phys. C* **3**, L75–7. [2.5]

Matheson, A.J. (1974) *J. Phys. C* **7**, 2569–76. [2.11]

Matsubara, T. & Kaneyoshi, T. (1966) *Prog. Theor. Phys.* **36**, 695–711. [9.3]

Matsubara, T. & Toyozawa, Y. (1961) *Prog. Theor. Phys.* **26**, 739–56.   [9.3]
Matsubara, T. & Yonezawa, F. (1967) *Prog. Theor. Phys.* **37**, 1346–7.   [9.4]
Matsuda, H. (1964) *Prog. Theor. Phys.* **31**, 161.   [8.4]
Matsuda, H. & Ishii, K. (1970) *Prog. Theor. Phys. Supp.* **45**, 56.   [8.7]
Mayer, J.E. (1962) *J. Phys. Chem.* **66**, 591.   [2.12]
Mayer, J.E. & Mayer, M.G. (1940) *Statistical Mechanics.* New York: Wiley.   [6.5]
Mazur. J. (1970) in Lowry (1970) chap. 6, pp. 153–85.   [7.7]
Meijering, J.L. (1953) *Phillips Res. Reps*, **8**, 270.   [2.11]
Meijering, J.L. (1957) *Phillips Res. Reps*, **12**, 333–50.   [5.8]
Melby, L.R. (1965) *Can. J. Chem.* **43**, 1448.   [2.3]
Mermin, N.D. (1968) *Phys. Rev.* **176**, 250–4.   [2.4]
Mermin, N.D. & Wagner, H. (1966) *Phys. Rev. Letters*, **17**, 1133–6.   [2.4] [5.9]
Metropolis, N.A., Rosenbluth, A.W., Rosenbluth, M.N., Teller, A.H. & Teller, E. (1953) *J. Chem. Phys.* **21**, 1087.   [6.6]
Meyer, H., Weinhaus, F., Maraviglia, B. & Mills, R.L. (1972) *Phys. Rev. B* **6**, 1112–21.   [1.3]
Miller, A. & Abrahams, E. (1960) *Phys. Rev.* **120**, 745.   [13.3]
Miwa, H. (1974) *Prog. Theor. Phys.* **52**, 1–17.   [9.5]
Montroll, E.W. (1970) *J. Math. Phys.* **11**, 635–48.   [11.4]
Montroll, E.W., Potts, R.B. & Ward, J.C. (1963) *J. Math. Phys.* **4**, 308.   [5.7]
Montroll, E.W. & Weiss, G.H. (1965) *J. Math. Phys.* **6**, 167.   [7.8]
Mookerjee, A. (1973) *J. Phys. C* **6**, 1340–9.   [9.7]
Mookerjee, A. (1974) *J. Phys. C* **7**, 4069–72.   [9.9]
Mookerjee, A. (1975) *J. Phys. C* **8**, 1524–34, 2688–94, 2943–52.   [9.7]
Moore, E.J. (1974) *J. Phys. C* **7**, 339–51.   [13.3]
Moorjani, K., Tomoyasu, T., Sokoloski, M.M. & Bose, S.M. (1974) *J. Phys. C* **7**, 1098–116.   [9.8]
Morgan, G.J. (1969) *J. Phys. C* **2**, 1446–53, 1454–64.   [10.10] [11.2]
Morgan, G.J. & Smith, D. (1974) *J. Phys. C* **7**, 665–80.   [11.2]
Morgan, G.J. & Ziman, J.M. (1967) *Proc. Phys. Soc.* **91**, 689–700.   [10.8]
Moss, S.C. (1965) in *Local Atomic Arrangements Studied by X-ray Diffraction* (eds Cohen & Hilliard), pp. 95–122. New York: Gordon & Breach.   [1.7]
Moss, S.C. & Clapp, P.C. (1968) *Phys. Rev.* **171**, 764–77.   [1.7]
Mott, N.F. (1966) *Phil. Mag.* **13**, 989–1014.   [10.10]
Mott, N.F. (1967) *Adv. in Phys.* **16**, 49–144.   [9.9]
Mott, N.F. (1968) *Rev. Mod. Phys.* **40**, 677–83.   [13.3]
Mott, N.F. (1974) *Metal-Insulator Transitions.* London: Taylor & Francis.   [9.9] [13.2]
Mott, N.F. & Davis, E.A. (1971) *Electronic Processes in Non-Crystalline Materials.* Oxford: Clarendon Press.   [2.8] [2.15] [9.9]
Mott, N.F. & Twose, W.D. (1961) *Advan. Phys.* **10**, 107–63.   [2.15] [3.3] [8.7] [13.2]
Movaghar, B. & Miller, D. (1975) *J. Phys. F* **5**, 261–77.   [10.6]
Movaghar, B., Miller, D.E. & Bennemann, K.H. (1974) *J. Phys. F* **4**, 687–702.   [10.6]
Müller-Hartmann, E. (1973) *Solid State Comm.* **12**, 1269–70.   [9.4]
Münster, A. (1962) *Handb. Phys*, **XIII**, 1–396.   [1.5] [1.6] [1.7]
Münster, A. (1965) *Fluctuation Phenomena in Solids* (ed. Burgess). New York: Academic Press.   [1.5] [1.6] [1.7] [4.5] [4.6]
Muto, T. & Takagi, Y. (1955) *Solid State Phys.* **1**, 193–282.   [1.5]

References 505

Nabarro, F.R.N. (1967) *Theory of Crystal Dislocations*. Oxford: University Press. [2.5]

Nagle, J.F. (1966) *J. Math. Phys.* **7**, 1484, 1492. [1.4][5.8]

Nagle, J.F. (1974) *Proc. Roy. Soc. A* **337**, 569–89. [7.8]

Nagle, J.F., Bonner, J.C. & Thorpe, M.F. (1972) *Phys. Rev. B* **5**, 2233–41. [11.4]

Neal, T. (1970) *Phys. of Fluids* **13**, 249. [10.10]

Newell, G.F. & Montroll, E.W. (1953) *Rev. Mod. Phys.* **25**, 353–89. [5.6][5.10]

Nickel, B.G. (1974) *J. Phys. C* **7**, 1719–34. [12.2]

Nickel, B.G. & Butler, W.H. (1973) *Phys. Rev. Letters* **30**, 373–7. [9.5]

Nicoll, J.F., Chang, T.S., Hankey, A. & Stanley, H.E. (1975) *Phys. Rev. B* **11**, 1176–87. [6.2]

Niemeijer, T. & van Leeuwen, J.M.J. (1973) *Phys. Rev. Letters*, **31**, 1411–14. [5.12]

Niemeijer, T. & van Leeuwen, J.M.J. (1974) *Physica*, **71**, 17–40. [5.12]

North, D.M. & Wagner, C.N.J. (1970) *Phys. & Chem. of Liquids* **2**, 87–113. [4.5]

Olson, J.J. (1975) *Phys. Rev. B* **12**, 2908–16. [10.8]

Onodera, Y. & Toyozawa, Y. (1968) *J. Phys. Soc. Japan*, **24**, 341–55. [9.4]

Onsager, L. (1944) *Phys. Rev.* **65**, 117. [5.7]

Onsager, L. & Kaufman, B. (1949) *Phys. Rev.* **76**, 1232, 1244. [5.7]

Ornstein, L.S. & Zernike, F. (1914) *Proc. Acad. Sci., Amsterdam*, **17**, 793. [2.12][4.4]

Page, D.I. & Mika, K. (1971) *J. Phys. C* **4**, 3034. [4.5]

Paul, W., Connell, G.A.N. & Temkin, R.J. (1973) *Adv. in Phys.* **22**, 531–580, 581–642, 643–665. [2.10]

Pearson, F.J. & Rushbrooke, G.S. (1957) *Proc. Roy. Soc. Edin. A* **64**, 305. [2.13]

Peierls, R.E. (1935) *Ann. Inst. Henri Poincaré*, **5**, 177–222. [2.4]

Peierls, R. (1936) *Proc. Camb. Phil. Soc.* **32**, 477. [2.4][5.4][5.5]

Penrose, O. (1963) *J. Math. Phys.* **4**, 1312–20, 1488. [6.5]

Percus, J.K. (1962) *Phys. Rev. Letters*, **8**, 462. [2.12]

Percus, J.K. & Yevick, G.J. (1958) *Phys. Rev.* **110**, 1. [2.12][11.2]

Peterson, H.K., Schwartz, L.M. & Butler, W.H. (1975) *Phys. Rev. B* **11**, [10.6]

Phariseau, P. & Ziman, J.M. (1963) *Phil. Mag.* **8**, 1487–501. [10.8]

Phillips, W.A. (1972) *J. Low Temp. Phys.* **7**, 351. [11.2]

Pike, G.E., Camp, W.J., Seager, C.H. & McVay, G.L. (1974) *Phys. Rev. B* **10**, 4909–17. [9.10]

Pike, G.E. & Seager, C.H. (1974) *Phys. Rev. B* **10**, 1421–34. [9.10][13.2]

Pings, C.J. (1968) in *Physics of Simple Fluids* (eds Temperley, Robinson & pp. 387–445. Amsterdam: North Holland. [2.8][4.1]

Polk, D.E. (1971) *J. Non-Cryst. Solids*, **5**, 365–76. [2.8][2.10]

Polk, D.E. & Boudreaux, D.S. (1973) *Phys. Rev. Letters*, **31**, 92–5. [2.10]

Rahman, A. (1964a) *Phys. Rev. Letters*, **12**, 575. [2.11]

Rahman, A. (1964b) *Phys. Rev.* **136**, A405. [6.6]

Rahman, A. (1966) *J. Chem. Phys.* **45**, 2585. [2.11][6.6]

Rahman, A. (1968) *Neutron Inelastic Scattering*, pp. 561–72. Vienna: IAEA. [2.11]

Rahman, A. & Stillinger, F.H. (1971) *J. Chem. Phys.* **55**, 3336–59. [2.8][6.6]

Rapaport, D.C. (1972) *J. Phys. C* **5**, 1830–58. [12.1]

Raveche, H.J. & Mountain, R.D. (1970) *J. Chem. Phys.* **53**, 3101. [7.12]

Raveche, H.J. & Mountain, R.D. (1972) *J. Chem. Phys.* **57**, 3987–92. [2.12][2.13]

Ree, F.H. (1971) in *Physical Chemistry: an Advanced Treatise* (eds Eyring, Henderson & Jost), Vols 8A–B, chapter 3. [6.6]

Ree, F.H. & Hoover, W.G. (1964) *J. Chem. Phys.* **40**, 939. [6.5]

Reiss, H. (1967) *J. Chem. Phys.* **47**, 186. [7.9]

Rice, O.K. (1944) *J. Chem. Phys.* **12**, 1–18. [2.11]

Rice, S.O. (1944) *Bell System Tech. J.* **23**, 282. [3.1][3.2][3.3][3.4]

Rice, S.O. (1945) *Bell System Tech. J.* **24**, 41. [3.1][3.2][3.3][3.4]

Rigby, M. (1970) *J. Chem. Phys.* **53**, 1021–3. [2.14]

Roberts, A.P. (1971) *J. Phys. F* **1**, 404–15. [8.4]

Roberts, A.P., Jones, P.L., Khor, K.E. & Smith, P.V. (1969) *J. Phys. C* **2**, 1502–11. [8.6]

Roberts, A.P., Jones, P.L. & Smith, P.V. (1968) *J. Phys. C* **1**, 549–51. [8.6]

Roberts, A.P. & Makinson, R.E.B. (1962) *Proc. Phys. Soc.* **79**, 630–51. [8.7]

Ross, M. & Alder, B.J. (1966) *Phys. Rev. Letters*, **16**, 1077. [6.7]

Ross, R.G. & Greenwood, D.A. (1969) *Prog. in Materials Science*, **4**, 173–242. [2.15]

Roth, L.M. (1974) *Phys. Rev. B* **9**, 2476–84. [10.6]

Roth. L.M. (1975) *Phys. Rev. B* **11**, 3769–79. [10.6]

Roth, L.M. (1976) *J. Phys. C* **9**, L159–62. [12.2]

Rowlinson, J.S. (1967) *Disc. Farad. Soc.* **43**, 243. [2.12]

Rowlinson, J.S. (1969) *Liquids and Liquid Mixtures*, 2nd edn. London: Butterworth. [2.12] [2.13]

Rowlinson, J.S. (1970) *Disc. Farad. Soc.* **49**, 30–42. [2.11][2.13]

Rudd, W.G., Salsburg, Z.W., Yu, A.P. & Stillinger, F.H. (1968) *J. Chem. Phys.* **49**, 4857. [6.8]

Rudee, M.L. & Howie, A. (1972) *Phil. Mag.* **25**, 1001–7. [4.7]

Ruelle, D. (1969) *Statistical Mechanics: Rigorous Results*. New York: Benjamin. [5.5]

Runnels, L.K. (1967) *J. Math. Phys.* **8**, 2081. [5.4][11.4]

Rushbrooke, G.S. (1968) in *Physics of Simple Fluids* (eds Temperley, Rowlinson & Rushbrooke), pp. 25–58. Amsterdam: North Holland. [2.12][2.13]

Rushbrooke, G.S., Muse, R.A., Stephenson, R.L. & Pirnie, K. (1972) *J. Phys. C* **5**, 3371–86. [12.2]

Rys, F. (1963) *Helv. Phys. Acta*, **36**, 537. [1.4]

Saitoh, M. (1970) *Phys. Letters*, **33A**, 44–5. [9.9]

Saitoh, M. & Edwards, S.F. (1974) *J. Phys. C* **7**, 3937–40. [13.5]

Salzberg, J.B., Goncalves da Silva, C.E.T. & Falicov, L.M. *Phys. Rev. B* **14**, 1314–22. [12.2]

Samathiyakanit, V. (1974) *J. Phys. C* **7**, 2849–76. [13.5]

Sato, H., Arrott, A. & Kikuchi, R. (1959) *J. Phys. Chem. Solids*, **10**, 19. [12.1]

Saxon, D.S. & Hutner, R.A. (1949) *Philips Res. Reports*, **4**, 81. [8.4]

Scher, H. & Lax, M. (1973) *Phys. Rev. B* **7**, 4491–502, 4502–20. [13.3]

Scher, H. & Zallen, R. (1970) *J. Chem. Phys.* **53**, 3759. [9.10][13.4]

Schmidt, H. (1957) *Phys. Rev.* **105**, 425–41. [8.4][8.5]

Schofield, P. (1966) *Proc. Phys. Soc.* **88**, 149–70. [2.12]

Schonhammer, K. & Brenig, W. (1973) *Phys. Letters*, **42A**, 447. [9.9]

Schultz, T.D., Mattis, D.C. & Lieb, E.H. (1964) *Rev. Mod. Phys.* **36**, 856–71. [5.7]

Schwartz, L.H. (1965) in *Local Atomic Arrangements Studied by X-ray Diffraction* (eds Cohen & Hilliard), pp. 123–58. New York: Gordon & Breach. [1.7]

Schwartz, L.M. (1973) *Phys. Rev. B* **7**, 4425–35. [10.5]

Schwartz, L.M. & Ehrenreich, H. (1971) *Ann. Phys. (N.Y.)* **64**, 100. [10.6]

Schwartz, L.M. & Ehrenreich, H. (1972a) *Phys. Rev. B* **6**, 2923–30. [10.8]

Schwartz, L.M. & Ehrenreich, H. (1972b) *Phys. Rev. B* **6**, 4088–90. [11.3]

Schwartz, L.M., Krakauer, H. & Fukuyama, H. (1973) *Phys. Rev. Letters*, **30**, 746–9. [9.8]

Scott, G.D. & Mader, D.L. (1964) *Nature*, **201**, 382–3. [2.11]

Seager, C.H. & Pike, G.E. (1974) *Phys. Rev. B* **10**, 1435–46. [13.3]

Sen, P.N. & Yndurain, F. (1976) *Phys. Rev. B* **13**, 4387–95. [9.5]

Seward, III, T.P. & Uhlmann, D.R. (1972) in *Amorphous Materials* (eds Douglas & Ellis), pp. 327–36. London: Wiley. [4.4]

Shante, V.K.S. & Kirkpatrick, S. (1971) *Adv. in Phys.* **20**, 325–58. [9.10] [13.2] [13.4]

Shaw, R.W. & Smith, H.V. (1969) *Phys. Rev.* **178**, 985–97. [10.4]

Sherrington, D. (1975) *J. Phys. C* **8**, L208–12. [12.3]

Shiba, H. (1971) *Prog. Theor. Phys.* **46**, 77. [9.8]

Shohat, J.A. & Tamarkin, J.D. (1943) *The Problem of Moments*. Providence, RI: Am. Math. Soc. [9.4] [9.7]

Silbert, M., Umar, I.H., Watabe, M. & Young, W.H. (1975) *J. Phys. F* **5**, 1262–76. [6.4]

Skal, A.S., Shklovskii, B.I. & Efros, A.L. (1973) *JETP Lett.* **17**, 377–9. [13.4]

Smalley, I.J. (1962) *Nature*, **194**, 1271. [2.2]

Smart, J.S. (1966) *Effective Field Theories of Magnetism*. Philadelphia and London: Saunders. [1.6]

Smith, W.R., Henderson, D. & Barker, J.A. (1970) *J. Chem. Phys.* **53**, 508–15. [6.4]

Solomon, H. (1954–55) *Fifth Berkeley Symposium on Mathematical Statistics and Probability*, Vol. 3, pp. 119–34. Univ. Cal. Press. [2.2]

Southern, B. (1975) *J. Phys. C* **8**, L213–15. [12.3]

Soven, P. (1967) *Phys. Rev.* **156**, 809–13. [9.4]

Stanley, H.E. (1968) *Phys. Rev.* **176**, 718–22. [5.9]

Stanley, H.E. (1971) *Introduction to Phase Transitions and Critical Phenomena*. Oxford: Clarendon Press. [1.5] [4.4] [5.1] [5.9] [5.12] [6.2]

Steiner, M., Villain, J. & Windsor, C.G. (1976) *Adv. in Phys.* **25**, 87–209. [2.3]

Stell, G. (1969) *Phys. Rev.* **184**, 135. [5.9]

Stern, E.A. (1973) *Phys. Rev. B* **7**, 1303–11. [10.6]

Stillinger, F.H. & Rahman, A. (1972) *J. Chem. Phys.* **57**, 1281. [2.8] [6.6]

Stillinger, F.H. & Rahman, A. (1974) *J. Chem. Phys.* **60**, 1545–57. [2.8] [6.6]

Stinchcombe, R.B. (1973) *J. Phys. C* **6**, L1–5. [9.11]

Stinchcombe, R.B. (1974) *J. Phys. C* **7**, 179–203. [9.11]

Stinchcombe, R.B. & Watson, R.P. (1976) *J. Phys. C* **9**, 3221–48. [9.11]

Stinson, T.W. & Lister, J.D. (1973) *Phys. Rev. Letters*, **30**, 688–92. [4.4]

Straley, J.P. (1972) *Phys. Rev. B* **6**, 4086–8. [11.3]

Strieb, B., Callen, H.B. & Horwitz, B. (1963) *Phys. Rev.* **130**, 1798. [5.4]

Stroud, D. (1975) *Phys. Rev. B* **12**, 3368–73. [13.4]

Stroud, D. & Ashcroft, N.W. (1972) *Phys. Rev. B* **5**, 371–82. [6.3] [6.7]

Sutherland, B. (1970) *J. Math. Phys.* **11**, 3183. [1.4]

Suzuki, M. & Fisher, M.E. (1971) *J. Math. Phys.* **12**, 235–46. [1.4] [1.5]

Sweet, J.R. & Steele, W.A. (1967) *J. Chem. Phys.* **47**, 3022–9. [2.14]

Sykes, M.F. & Essam, J.W. (1964) *Phys. Rev.* **133**, A310–15. [9.10]

Symons, M.C.R. (1972) *Nature*, **239**, 257–9. [2.8]

Syozi, I. (1972) in Domb & Green (1972) Vol. 1, pp. 269–329. [5.7]

Takahashi, H. (1941) *Proc. Phys. Math. Soc. Japan*, **23**, 1069. [1.4]

Takahashi, H. (1942) *Proc. Phys. Math. Soc. Japan*, **24**, 60. [6.1]

Tanaka, M. & Fukui, Y. (1975) *Prog. Theor. Phys.* **53**, 1547–65. [2.11]

Tani, K. & Tanaka, H. (1967) *Phys. Letters*, **26A**, 68. [1.8]

Taylor, D.W. (1967) *Phys. Rev.* **156**, 1017–29. [9.4]

Temperley, H.N.V. (1961) *Proc. Phys. Soc.* **77**, 630. [5.4]

Temperley, H.N.V. (1972) in Domb & Green (1972) Vol. 1, pp. 227–67. [5.7] [5.10]

Temperley, H.N.V. & Lieb, E.H. (1971) *Proc. Roy. Soc. A* **322**, 251–80. [9.10]

Theumann, A. (1974) *J. Phys. C* **7**, 2328–46. [12.2]

Thiele, E. (1963) *J. Chem. Phys.* **39**, 474. [2.12]

Thompson, C.J. (1968) *J. Math. Phys.* **9**, 1059. [5.9]

Thompson, C.J. (1972) in Domb & Green (1972) Vol. 1, pp. 177–226. [5.5] [5.6]

Thorpe, M.F. & Weaire, D. (1971) *Phys. Rev. B* **4**, 3518–26. [11.4]

Thorpe, M.F., Weaire, D. & Alben, R. (1973) *Phys. Rev. B* **7**, 3777–87. [11.3]

Thouless, D.J. (1969) *Phys. Rev.* **187**, 732. [5.5]

Thouless, D.J. (1970) *J. Phys. C* **3**, 1559–66. [9.9]

Thouless, D.J. (1971) *J. Phys. C* **4**, L92–4. [9.9]

Thouless, D.J. (1972) *J. Phys. C* **5**, 77–81. [8.7]

Thouless, D.J. (1973) *J. Phys. C* **6**, L49–52. [8.7]

Thouless, D.J. (1974) *Phys. Reports 13C*, 93–142. [9.9] [9.11]

Tomonaga, S. (1955) *Prog. Theor. Phys.* **13**, 467–81, 482–95. [11.2]

Tong, B.Y. (1968) *Phys. Rev.* **175**, 710–22. [8.4]

Tonks, L. (1936) *Phys. Rev.* **50**, 955, [6.1]

Treloar, L.R.G. (1943) *Trans. Farad. Soc.* **39**, 36, 241. [7.6]

Treloar, L.R.G. (1958) *The Physics of Rubber Elasticity*, 2nd edn. Oxford: Clarendon Press. [7.6]

Tsukada, M. (1972) *J. Phys. Soc. Japan*, **32**, 1475–85. [9.5]

Uhlenbeck, G.E. (1960) *Physica (Supp.)* **26**, 17–29. [5.10] [6.5]

Uhlenbeck, G.E. & Ford, G.W. (1962) *Studies in Statistical Mechanics*, Vol. 1, pp. 123–211. Providence, RI: Am. Math. Soc. [5.10] [6.5]

Uscinski, B.J. (1974) *Proc. Roy. Soc. A* **336**, 379–92. [13.1]

van Hove, L. (1950) *Physica*, **16**, 137. [6.1]

van Hove, L. (1954) *Phys. Rev.* **95**, 1374–84. [1.8]

Velicky, B., Kirkpatrick, S. & Ehrenreich. H. (1968) *Phys. Rev.* **175**, 747–66. [9.4]

Verlet, L. (1967) *Phys. Rev.* **159**, 98. [6.6]

Verlet, L. (1968) *Phys. Rev.* **165**, 201. [2.11] [6.6]

Vicentini-Missoni, M. (1972) in Domb & Green (1972), Vol. 2, pp. 39–78. [6.2]

Vineyard, G.H. (1958) *Liquid Metals and Solidification*, p. 1. Cleveland, Ohio: Am. Soc. for Metals. [4.5]

Visscher, W.M. & Bolsterli, M. (1972) *Nature*, **239**, 504–7. [2.11] [2.13]

Volkenstein, M.V. (1963) *Configurational Statistics of Polymeric Chains*. New York: Interscience. [7.1] [7.2] [7.3] [7.5]

von Heimburg, J. & Thomas, H. (1974) *J. Phys. C* **7**, 3433–43. [5.4]

Vyssotsky, V.A., Gordon, S.B., Frisch, H.L. & Hammersley, J.M. (1961) *Phys. Rev.* **123**, 1566. [9.10]

Wagner, C.N.J., Halder, N.C. & North, D.M. (1967) *Phys. Letters,* **25A,** 663–4. [4.5]

Walatka, V.V., Labes, M.M. & Perlstein, J.H. (1973) *Phys. Rev. Letters,* **31,** 1139. [2.3]

Wall, F.T. (1942) *J. Chem. Phys.* **10,** 485–8. [7.6]

Wall, H.S. (1948) *Analytic Theory of Continued Fractions.* New York: Chelsea Publishing Co. [9.7]

Walter, J. & Eyring, H. (1941) *J. Chem. Phys.* **9,** 393–7. [2.9]

Wannier, G.H. (1959) *Solid State Theory.* Cambridge: University Press. [2.4]

Warren, B.E. (1972) in *Amorphous Materials* (eds. Douglas & Ellis), pp. 263–7. London: Wiley. [2.8]

Watabe, M. & Young, W.H. (1974) *J. Phys. F* **4,** L29–31. [6.4]

Waterman, P.C. & Truell, R. (1961) *J. Math. Phys.* **2,** 512–37. [10.8]

Watson, B.P. & Leath, P.L. (1974) *Phys. Rev. B* **9,** 4893–7. [9.11]

Weaire, D. (1971) *Phys. Rev. Lett.* **26,** 1541–3. [11.3]

Weaire, D. & Alben, R. (1972) *Phys. Rev. Lett.* **29,** 1505. [11.2]

Weaire, D. & Thorpe, M.F. (1971) *Phys. Rev. B* **4,** 2508–19. [11.3]

Weiss, P.R. (1948) *Phys. Rev.* **74,** 1493. [5.4]

Wertheim, M.S. (1963) *Phys. Rev. Lett.* **10,** 321. [2.12]

Wertheim, M.S. (1964) *J. Math. Phys.* **5,** 643. [2.12]

Widom, B. (1965) *J. Chem. Phys.* **43,** 3892, 3898. [5.12]

Widom, B. (1967) *Science,* **157,** 375–82. [6.2]

Wilson, K.G. (1971*a*) *Phys. Rev. B* **4,** 3174–83. [5.12]

Wilson, K.G. (1971*b*) *Phys. Rev. B* **4,** 3184–205. [5.12]

Wilson, K.G. (1975) *Rev. Mod. Phys.* **47,** 773–840. [5.12]

Wilson, K.G. & Kogut, J. (1974) *Physics Reports C* **12,** 75–200. [5.12]

Wilson, R.J. (1972) *Introduction to Graph Theory.* Edinburgh: Oliver & Boyd. [5.4]

Wood, W.W. (1968) in *Physics of Simple Fluids* (eds Temperley, Rowlinson & Rushbrooke), pp. 115–230. Amsterdam: North-Holland. [2.11] [6.6] [6.7] [6.8]

Wood, W.W. & Jacobson, J.D. (1957) *J. Chem. Phys.* **27,** 1207. [6.7]

Woodcock, L.V. (1972) *Proc. Roy. Soc. A* **328,** 83–95. [2.13]

Wortis, M. (1974) *Phys. Rev. B* **10,** 4665–71. [12.1]

Wu, T.T. (1966) *Phys. Rev.* **149,** 380–401. [1.7]

Yaglom. A.M. (1962) *An Introduction to the Theory of Stationary Random Functions.* Englewood Cliffs, NJ: Prentice-Hall. [3.1]

Yamakawa, H. (1972) *Pure & Appl. Chem. (G.B.).* **31,** 179–99. [7.7]

Yang, C.N. (1952) *Phys. Rev.* **85,** 809. [5.7]

Yang, C.N. & Lee, T.D. (1952) *Phys. Rev.* **87,** 404, 410. [6.2] [6.5]

Yang, C.N. & Yang, C.P. (1968*a*) *Phys. Rev.* **150,** 321–7, 327–39. [5.6]

Yang, C.N. & Yang, C.P. (1968*b*) *Phys. Rev.* **151,** 258–64. [5.6]

Yndurain, F. & Yndurain, F.J. (1975) *J. Phys. C* **8,** 434–44. [9.7]

Yonezawa, F. (1968) *Prog. Theor. Phys.* **40,** 734–57. [9.4]

Yonezawa, F. (1973) *Phys. Rev. B* **7,** 5170. [9.4]

Yonezawa, F. & Matsubara, T. (1966) *Prog. Theor. Phys.* **35,** 357–79, 759–76. [9.3]

Yonezawa, F. & Morigaki, K. (1973) *Supp. Prog. Theor. Phys.* **53,** 1–76. [9.3] [9.4]

Yonezawa, F., Roth, L.M. & Watabe, M. (1975) *J. Phys. F* **5,** 435–42. [10.6]

Yonezawa, F. & Watabe, M. (1975) *Phys. Rev. B* **11,** 4746–52, 4753–62. [10.6]

Young, A.P. & Stinchcombe, R.B. (1975) *J. Phys. C* **8,** L535–40. [9.10]

Zachariasen, W.H. (1932) *J. Am. Chem. Soc.* **54**, 3841. [2.8]

Zallen, R. & Scher, H. (1971) *Phys. Rev. B* **4**, 4471–8. [13.4]

Zeller, R.C. & Pohl, R.O. (1971) *Phys. Rev. B* **4**, 2029–41. [11.2]

Ziesche, P. (1974) *J. Phys. C* **7**, 1085–97. [10.9]

Ziman, J.M. (1960) *Electrons and Phonons.* Oxford: University Press. [2.5]

Ziman, J.M. (1961) *Phil. Mag.* **6**, 1013. [10.1] [10.2]

Ziman, J.M. (1964) *Adv. in Phys.* **13**, 89–138. [10.2]

Ziman, J.M. (1966) *Proc. Phys. Soc.* **88**, 387–405. [10.8]

Ziman, J.M. (1967a) *Proc. Phys. Soc.* **91**, 701–23. [10.4] [10.8]

Ziman, J.M. (1967b) *Adv. in Phys.* **16**, 551–80. [10.10]

Ziman, J.M. (1968) *J. Phys. C* **1**, 1532–8. [9.10] [9.11] [13.4]

Ziman, J.M. (1969a) *J. Phys. C* **2**, 1230–47. [9.9] [9.10]

Ziman, J.M. (1969b) *J. Phys. C* **2**, 1704–16. [10.5] [11.5]

Ziman, J.M. (1971a) *Solid State Physics*, **26**, 1–101. [10.2] [10.3] [10.8]

Ziman, J.M. (1971b) *J. Phys. C* **4**, 3129–42. [11.3]

Ziman, J.M. (1973) *J. Phys. C* **6**, L361–2. [11.4]

Zimm, B.H. (1946) *J. Chem. Phys.* **14**, 164–79. [7.7]

Zittartz, J. (1974) *Solid State Comm.* **14**, 51–3. [9.5]

Zittartz, J. & Langer, J.S. (1966) *Phys. Rev.* **148**, 741–7. [13.5]

Zubarev, D.N. (1953) *Zh. Eksp. Teor. Fiz.* **25**, 548–59. [11.2]

Zwanzig, R.W. (1954) *J. Chem. Phys.* **22**, 1420–6. [6.4]

Index

adjacency matrix 440
ADP 15–16
aggregate, close-packed 100
alkali halides 101–2
alkali metal 431
allowed band 299, 312, 318–19
alloy
  binary chain 287–9, 301–7, 310, 315, 319, 338, 344, 346, 348
  binary random 327, 330, 335–40, 346, 348, 354, 370, 376, 379
  diffraction by 134
  dilute 5, 468
  glassy 100–1, 134, 235, 468
  Ising model 6, 9, 19–20, 28
  Kronig–Penney 292–3
  liquid 96–102, 133–6, 139–40, 209, 248, 437–8
  multi-component 338
  order–disorder in 2, 5, 22, 27, 134
  tight-binding 289, 387, 397
  transition metal 397, 403, 420, 468
  see also substitutionally disordered lattice excitations
amorphon 75
Anderson localization 358, 361–8, 370, 379, 384–6, 479
Anderson model 327, 335, 338, 348, 359, 361–9, 384–6, 454, 479
angular momentum representation 400, 402, 417–18
annealed disorder 461
anti-Curie point 154
antiferroelectrics 15, 16, 20, 139, 146, 181
antiferromagnetism 7, 20, 22–3, 25, 27, 48, 143, 146
  ground state 22–4, 105, 166–71, 180, 288, 464

in one dimension 43, 166–71
antiferromagnon 169
antiphase domains 24–5
APW method 400, 402
argon 78–9, 86, 92, 95, 227, 232
ATA see average t-matrix approximation
atomicity 36–7, 40, 56, 108, 122, 399, 404, 446, 472, 474
Au–Sn alloys 135
auto-correlation function 111, 112, 113, 117, 119–20, 482, 489
average t-matrix approximation 328–31, 369, 415, 420, 429
azimuthal angle 75–7

band edge 109, 119, 313–14, 347, 350, 363–4, 383, 453–4
band gap see gap, spectral
band-gap paradox 456–9
band structure, electronic 248, 288, 290, 324, 353, 387, 395, 405, 422, 428, 449, 451, 457
  see also spectral density, electronic
band width 290, 336, 346–7, 353, 359, 368, 429
BBGKY method 89–90, 94, 97, 149, 244, 281, 415
beat length 313–14, 347
Bernal model 78–87, 101, 237, 243
Bethe ansatz 168, 180–1, 264, 309
Bethe–Hulthén method 166–71
Bethe lattice 152–3, 195, 264, 454
Bethe method 151–5, 379
Bethe order parameter 17
Bethe–Peierls approximation 151–2, 343, 345, 356
Bethe–Salpeter equation 434–5
Bi 409

biorthonormal functions 444–5
Blinc model 16, 21
Bloch function 45, 53, 298, 313, 317, 319,
    323–4, 329, 358, 436, 454
Bloch–Mermin–Wagner theorem 49
Bloch representation 4, 32, 169, 175, 180,
    428, 442
Bloch state 289, 336, 363, 399, 405, 428, 456
Bloch wall 31, 50–1, 105
block Hamiltonian 201–2, 203
block spin variables 201–2, 204, 207
Boltzmann equation 389, 431, 435
bond, chemical 2, 248, 259–60, 448, 455–6
    dangling 67, 75
    hydrogen 11, 13, 255
    saturated 456
    tetrahedral 64, 71, 448, 458
    trihedral 66
bond angle 66, 69, 72–3, 75–6, 253, 257, 458
bond energy 15, 254, 257
bond network disorder 64–77, 100, 439–59
bond orbital 45, 450–4
bond percolation 371–90, 462, 465
Borland model 39–40, 303
Born approximation 123, 131, 393, 402, 413
Born series 123, 402, 405, 413
Bose–Einstein condensation 185
bound states, electron 393, 402, 405, 426,
    474–7, 479, 487–8
boundary conditions
    cyclic 168–9, 173–4, 295–6, 366, 456
    linear chain 296, 308, 317
    molecular dynamics 228, 231
    polymer coil 282
    radial function 417, 424
BRA *see* alloy, binary random
Bragg orientation 141
Bragg plane 4, 457
Bragg—Williams method 145
Bragg–Williams order parameter 24
branching number 373
branching processes 259–65
Bravais lattice 2, 273
Brillouin–Wigner series 406, 412
Brillouin zone 4, 33, 183, 273, 290, 353, 389,
    405, 443, 453, 456
Byzantine function 407, 436

Campbell's theorem 114, 115
car-parking problem 40
cascade process 261
catastrophe theory 205
Cayley transform 299–301, 304, 319
Cayley tree 152
cell, atomic 2, 84, 86, 93, 109, 130, 243, 294,
    472

cell-wave representation 426–9
cellular approximation 37, 472–3
cellular disorder 1–35, 137, 139, 322
    see also substitutional disorder
Central Limit theorem 113, 114, 258, 278,
    313
chain approximation 412
chain, disordered, excitations of 291–30
    boundary conditions 296, 308, 317
    cluster approximation 344
    coherent potential approximation 338
    continued fraction method 355–6
    Dyson–Schmidt method 309–12, 322, 356
    electron propagation 292–4
    local density approximation 312–15
    localization 315–20, 379, 456
    mean free path 320
    percolation 379
    phase-angle representation 295–301
    special frequencies 304–7
    spectral density 307–20, 338, 344
    spectral gaps 302–7, 406
    tight binding alloy 297, 300, 302, 455
    transfer matrix 291–5, 455
    transmission constant 317, 319
chain, linear
    antiferromagnetic 166–71
    dimensionality 152, 309, 312
    disordered 39–42, 63, 209–12, 245, 255
    ferromagnetic 165–6
    Heisenberg 165–71, 181
    Ising 161–5, 192
    metal 45
    ordered 39, 41, 164, 298–303, 313, 316,
        319, 353
    stability 47, 49, 164, 166
    systems 43–7, 144
chain clusters 160–1
cluster expansion 89, 91–2, 225–6, 237,
    244–5, 410
cluster integral 103, 225
cluster method 151–61, 176, 341–5, 352, 456,
    465
cluster scattering 429–30
coherent field approximation 145
coherent potential approximation 143, 330,
    332–45, 364, 416, 420, 467
    analytical properties 334–5, 341, 343
    band splitting 336–7
    cluster generalization 341–5, 354, 356–7
    consistency condition 333–4, 341–2, 344,
        383, 424
    graphical expansion 333
    liquid 416
    local environment correction 341–5, 429,
        467

coherent potential approximation (*cont.*)
  locator expansion 333–4
  mean field method 332, 341
  molecular 342–5, 431
  non-local 343, 357
  off-diagonal disorder 357
  one-dimension 338
  single-site 333–4, 341–3, 354, 356
  spectral density 335–40, 369
coherent-wave approximation 420–5, 429, 436, 445
collineations 80
colloids 132
commutation relations 443–5
compact representations 154, 161, 187, 195, 330, 332, 344
composite materials 486–7
compressibility 95, 130, 132, 213, 219–20, 282, 392
computer simulation 226–32, 245–6, 306, 337–9, 365–6, 384, 440–1, 452–3
  see also Monte Carlo method; molecular dynamics method
condensation 211–12, 216, 218, 226, 249
conduction band 449, 451–3, 457
conductivity 432–46, 466–7, 477, 488
configuration, macromolecular 254–7
configuration integral 88, 245
configuration space 230–1, 239
configurational entropy 220, 237, 241, 243
confusion, interval of 234, 239
conjugate variables 443–4
connective constant 196, 275, 363, 373, 379
connectivity
  bond network 71, 74–6, 87, 440, 448
  branching chain 264, 373
  cluster 384
  graph 152, 190
  lattice 152, 158–9, 196, 309, 322, 353, 373, 375, 454, 455
connectivity matrix 451, 465
continued fraction 333, 351–4, 365
continuity, range of 111, 119, 120
continuum disorder 4, 37–8, 56, 104, 108–21, 130, 138–9, 207, 252, 474, 482–92
  elasticity 441–7
  liquid metal 392, 441–7
  macromolecular chain 252, 277–85
  magnetic 466
  one-dimensional 294–6
  percolation 485–7, 490
  random field 294–5, 474, 482–92
  topography 118–21, 483
coordinate representation 417–19
coordination number 20, 68, 72, 84, 101, 148,
152, 165, 196, 243, 252, 261, 274, 377, 447, 449, 454
coordination shell 68–9, 80, 84, 90, 101–2, 125, 151, 214
copolymerization 259–60, 263–4
copper benzoate 43, 44
copper dipyridine dichloride 43, 44
correlation
  atomic motion 228, 245
  long range 7, 201
  many atom 411–12
  orientational 103
  short range 7, 17–22, 146–7, 163
  three spin 152–4, 281
correlation function
  angular 74, 77, 81–2
  atomic 17, 26–8, 31, 58–64, 244
  concentration 138–9
  density 132–9
  direct 91–3, 98, 126, 131, 186, 194
  fluctuation 150, 199–200, 203, 285
  four-body 61–4, 71, 77, 82, 89, 411–13, 457–8
  higher order 31, 110, 222, 281, 283, 410–13
  indirect 194–5
  N-body 88
  pair 24, 31, 59–60, 71, 86–9, 93, 125, 130, 137, 213, 391, 410, 412
  partial 97, 101, 134–9
  random function 110, 114–16, 118–21, 446
  spin 21, 33–4, 48, 150, 163, 165, 176, 185–6, 194, 196, 201, 281
  total 68, 75, 91, 93, 98, 111, 125
  triplet 31, 61–4, 71, 76, 80, 83, 89, 94, 213, 411–13, 415, 422, 457–8
  van Hove 128, 139
correlation length 26, 111–12, 115, 132, 150, 177, 186, 199–205, 474, 482–3, 488–90
corresponding states, principle of 96, 97, 217
Coulomb liquid 93
Cowley order parameter 17, 21
CPA *see* coherent potential approximation
critical concentration *see* percolation, threshold for
critical curve 200
critical phenomena 26–7, 32, 143–4, 196
  alloy 138
  cluster method 159–61, 341
  dilute Ising model 462
  ferroelectric 181
  ferromagnetic 27, 32, 154, 196, 216
  first order 200, 237
  graphical expansion 187, 195, 197, 376
  helix–coil 255
  Landau theory 197–200, 272
  liquid crystal 104

critical phenomena (*cont.*)
 liquid–vapour 140, 200, 211, 213, 215–20
 magnetism 140
 mean field approximation 104, 145, 147, 150, 272
 melting 227, 232, 239, 276–7
 mixing 247
 one dimension 163–6, 170–1
 Ornstein–Zernike 131–2, 140, 150, 186, 196, 198–200
 renormalization 203–8
 scaling 200–3
 second order 198–9
 spherical model 184–7
 spin glass 469–71
 thermodynamics 197–200
 two-dimensional Ising model 176–7, 181
critical pressure 217
critical temperature 27, 32, 197
 amorphous magnet 468
 dilute magnet 462–7, 476
 ferroelectric model 15, 187
 graphical expansion 194, 196, 378
 Ising model 49, 159, 194, 196, 378
 Kikuchi method 158–9, 161, 176, 184, 196
 liquid crystal 104
 liquid–vapour 216–26
 mean field approximation 145, 150, 154, 176
 mixing 247
 Osager solution 176, 194
 quasi-chemical approximation 147, 152, 158, 176, 196, 379
 spherical ferromagnet 184, 186
 spin glass 470–1
crystal 1–4, 83, 249
crystallite 56–7, 63, 70, 72, 412
Cu–Mn alloys 468
Cu–Sn alloys 135–6
Cu$_3$Au 22, 28, 30
CuCl 135
cumulants 18, 189–90, 195, 221–4, 330, 333, 382, 410–11, 413
Curie temperature 27, 35
Curie–Weiss theory 145
current operator 432, 436–7

d-band 46, 388, 397–8, 403, 405, 420, 429
deformation potential 109
deltahedra, canonical 82
density, critical 217
density fluctuations 130, 138–9, 217–18, 392
density of states *see* spectral density, electronic
density operator 128
deuterium, solid 11

diagonal disorder 287–91, 321, 356, 359, 368, 455, 479
diamond 456
 polytypes of 447, 458–9
 see also lattice, diamond
dielectric function 392, 394–5, 398, 421, 423
differencing matrix 465–6
diffraction 123–7, 446, 474
 coherent 127, 134, 137, 389, 405, 413
 crystal 4, 124, 126, 128, 389, 405, 413, 457
 imaging 140
 incoherent 127, 134, 137
 inelastic 127–9, 131, 140, 445
 mixture 133–7
 neutron 68, 124, 127–9, 134–7, 139, 389, 391, 445
 observation of disorder 27, 94–5, 122–41
 optical 130–1, 138, 141
 substitutional disorder 137–40
 X-ray 67–8, 71, 124, 129–37, 389, 391
diffusion 128, 228, 277–82, 359, 380, 454, 481
dimensionality
 space 47–51, 73, 144, 164–5, 177, 184, 186, 196, 203–4, 208, 212, 236, 245, 255, 271–5, 291, 322, 347, 355, 377, 454, 490
 spin 186, 196, 203–4, 208
 tree 152
dimer graph 188–90
dimer problem 192
director vector 104–6
Dirichlet regions 84
disclination 105
dislocation disorder 51–6, 57, 59, 78, 141, 236, 239–40
disorder 1
dispersion function
 electron energy 406–8, 410, 436
 magnon 34, 466–7
 phonon 443–5
distribution function 58
 radial 60–1, 67–71, 76, 79–81, 89–90, 94, 101, 112, 125–6, 213–14, 222, 445
 see also correlation function
domain, ordered 24, 29, 31, 49–51, 105
domain boundary energy 31, 49–51
droplets 205–6
Dyson equation 262, 325, 328, 330, 345, 353, 357, 368
Dyson–Schmidt method 309–12, 316, 322, 356, 365, 380–1, 455–6, 490

eclipsed configuration 75–7
Edwards series 410–12, 415, 429, 457
effective medium approximation 332, 381, 383, 416, 431, 465, 486–7
eight-vertex model 16, 143, 181

elasticity
  macroscopic 442–7
  rubber 266–8
electron
  amorphous semiconductor 429–30
  angular momentum representation 400, 404, 417
  average t-matrix 415
  band tails 486–91
  bound states 393, 395, 397, 487–8
  cell waves 427–8
  cluster scattering 428–31
  coherent potential 332–45
  coherent wave 421–5
  composite material 486–7
  conduction 289, 397
  coordinate representation 417–20
  crystalline metal 368, 388–9, 395, 398, 400–1, 404–6, 409–10, 412, 422, 428, 436
  current and momentum 436–7, 440
  d-band 397, 403
  disordered chain 288–9, 291
  disordered metal 387–443
  Edwards series 410–12
  effective medium 416
  effective potential 414, 431
  forward scattering 423, 429, 434
  geometric effects 404–5, 413, 424, 429–30, 438
  Green function 407–8
  impure semiconductor 107, 289, 475–81
  LCAO representation 288–9, 321, 387, 398, 420, 457
  liquid alloys 437–8
  liquid metal *see* metal, liquid
  localization 358–86, 485, 490–1
  microcrystalline semiconductors 412–13
  muffin–tin potentials 398–400, 416
  nearly free 388–91
  partial wave 416–20, 426–8
  percolation 370–86, 485–7
  perturbation expansion 405–6, 410–11
  phase shift 401–2
  pseudopotential 393–8, 404
  quasi-crystalline approximation 415–16
  random potential 482–92
  resonance band 403, 429
  scattering operator 413–16
  semiclassical 483–7
  t-matrix 403–4, 416–18
  tight-binding alloy 288–90, 397
  transport 387, 390–1, 431–8
electron density 129, 395, 399
electron–electron interaction 2, 5, 47, 93, 107, 240, 391–3, 477

electron microscope 72, 141, 277
EMA, *see* effective medium approximation
ensemble
  canonical 88, 228, 231
  grand canonical 94
  micro-canonical 228
  thermostatic 88, 142
entanglements 259, 283–5
entropy
  cluster expansion for 244–5
  communal 241–5
  configurational 220, 237, 241, 243, 250, 270
  crystal 237–41
  dilution 270–1
  disorientational 250, 253, 265–6
  excess, of liquid 220, 240–6
  ice 11, 13–14, 180–1
  informational 243
  macromolecular solution 250–2
  melting 241–5
  mixing 247, 250–1
  rubber 266–7
equation of state
  compressibility 219–20
  hard disc 235, 245
  hard sphere 218–20, 233–5, 237, 245
  Lennard-Jones 220
  linear fluid 211–12
  Percus–Yevick 218–20
  rubber 266
  van der Waals 215–17, 219, 222, 235
  virial series 225
ergodic hypothesis 18, 110, 112, 124, 228, 232, 308, 438
exchange interaction 18, 20, 23, 34, 139, 161
  anisotropic 177, 288
  effective 148
  long range 164
  random 468, 471
excitation ratio 295–7, 299
excitations
  chain disorder 291–320
  elastic continuum 441–7
  electronic 288–9, 321, 356, 448–56
  exciton 289–90
  lattice dynamic 286–7, 290, 303, 321, 326, 336–40, 356, 439–47
  localized 315–20, 339–40
  magnetic 287–8, 290, 321
  network disorder 439–59
  ordered lattice 323–4, 353
  substitutionally disordered lattice 321–86
  tight binding alloy 289–90, 321–2
excitons 289–90
expanded representation 161, 330, 345

exponents
  critical 26, 32, 132, 150, 161, 177, 181, 196, 199, 203, 208, 218, 272, 275, 463
  percolation 378, 380, 383
extended state 317–18, 358–9, 361, 366, 370, 379, 454, 456, 490

'F-model' 15, 180, 181
Fermi current 388–9, 436–7
Fermi level 388, 390, 405, 432, 436
Fermi radius 388, 390, 436
Fermi surface 388, 403, 405
ferrimagnetism 7, 24
ferroelectrics 15, 16, 20–1, 139, 143–4, 146, 178, 180–1, 276–7, 373
ferromagnetism 7, 9, 18, 20, 25, 143, 287
  amorphous 105, 467–71
  Curie–Weiss theory 145
  dilute 460–7, 476
  ground state 22, 24, 34, 105, 165–6, 169, 465, 468
  phase transition 27, 32, 154, 196, 216
  spherical model 182–7
  stability of 48, 166, 465–7
Feynman path integral 281, 491
fixed point 205
Flory formula 271–2, 275, 282
Flory–Huggins formula 250, 270, 276
fluctuation-dissipation theorem 432, 466
fluctuations, critical 9, 27, 32, 131–2, 150, 198–200, 205, 217, 341
fluid phases 213, 215, 218
fluidity 66, 78, 122, 282, 445, 447
form factor, atomic 124, 129, 402, 405
forward scattering approximation 423, 429, 434
free band 387, 397
free rotation model 253
freezing 83, 93, 101, 233–5, 240, 247
Friedel sum rule 308, 426, 429
fugacity 94, 225–6
functional integral 279–81, 283–4
functionality 261–2, 264

galvanomagnetic properties 437
gap, spectral 109, 298–307, 310, 315–18, 345–7, 388–9, 406, 449, 451, 453, 456–9
gas
  dilute 89
  imperfect 223
  one-dimensional 39, 209–12, 242, 245
gas disorder 37, 84–7, 106–7, 117, 289, 399, 472–81
  Anderson model 479
  clustering 475
  diffraction by 124

  electrons in 472–91
  hopping conduction 478–81
  percolation 475–7, 479–81
  potential in 472–4, 479, 482
  random potential model 482–92
gauche configuration 254, 276–7
Gaussian distribution 113, 258–9, 267–8, 271, 274, 277, 313
Gaussian linear liquid 39, 309
Gaussian model 208
Gaussian random function 113–21, 132, 141, 168, 482–91
Ge
  amorphous 66–7, 68, 76, 141, 456–9
  crystalline 412–13, 449, 456–7
gel fraction 263, 372
gelation 260–5, 372
Gibbs–Bogoliubov inequality 222
glass
  alloy 100–1, 235
  borate 66
  chalcogenide 67
  diatomic 76, 451
  diffraction 125, 130, 133, 141
  disorder 57, 70–4, 130, 440
  dynamics 128, 439–48
  elasticity 109
  electron states 448–54
  global analysis 205
  grain boundary 56, 57, 72
  local order 61, 141, 446
  macromolecular 249
  one-dimensional 39–40, 291, 302, 455
  silica 66, 440–1
  silicate 64, 68
  specific heat 446
  tetrahedral 67, 86–7, 399–400, 447–54
  thermal conduction 446
  tree model 455–6
  trihedral 66, 73–5
  two-dimensional 65, 73, 441
graph
  adjacency matrix of 440
  bubble 195
  closed 191–3, 355, 454
  connected 189–90, 195, 223–4, 349
  connectivity matrix of 451, 465
  correlation 195
  differencing matrix of 465–6
  dimer 188
  equivalent 188
  irreducible 224–5, 351, 413
  labelled 190
  ladder 195, 223, 273
  moment 349
  multi-bond 195

graph (*cont.*)
  polygon 195
  propagator 326
  ring 195
  self-avoiding 363
graph theory 73, 152, 187, 190, 195, 197, 225, 373, 453, 455, 465
graphical expansion 89, 196–7, 199, 208, 262, 345
  Born series 402, 405, 413
  Brillouin–Wigner 405–12
  coherent potential 333
  convergence of 360–2, 364–5
  Edwards series 410–11
  electron in metal 396, 405, 410
  excluded volume 272
  fluid thermodynamics 223–6
  high-temperature 188–9, 223–6, 375, 465
  Ising model 178, 187–97, 203
  locator 330–1, 333, 353, 360
  low temperature 192–7, 375
  moment 348
  percolation 376
  propagator 326–8, 330, 333, 413
  random walk 275
  Rayleigh–Schrödinger 405–6
  renormalized 363–5, 367
  scattering operator 414
Green function 154, 320, 322–4, 335–6, 359–60, 407–9, 414, 421–2, 425–7, 433, 455
  assembly 414, 431, 433
  average 324, 331–3, 336, 341–3, 351, 354, 407–8, 433, 490
  lattice 273
  local 353–4
  one electron 407–9, 433
  retarded 368
  site-diagonal 359, 364–5, 368–9
  temperature 190, 411
  total 331, 359, 362
  two-particle 433–4
Greenian, incomplete 422–3, 427–9
group theory 1, 2, 33, 174, 204–8, 440
Gubanov model 57
Guinier approximation 133
Gürsey model 212

Hadamard–Gerschgorin theorem 345
Hall coefficient 437
hard disc model 83–5, 225, 227, 235–7, 240, 242, 245
hard dumbbell model 103
hard sphere atoms 2, 68
hard sphere crystal 231, 237
hard sphere liquid 77–83, 90, 92–3, 107, 127, 130, 210–11, 218–20, 222, 225–7, 229–30, 232–6, 240–5, 391, 399
hard sphere mixture 97–9, 135, 246
hard square gas 143
harmonic lattice 47
Heisenberg model 6, 18, 48–9, 103, 182, 191
  classical 22, 27, 143, 145, 165, 186, 203, 209–10, 267
  dilute 464–7
  one-dimensional 165–71, 180, 209–10, 267, 464
  quantal 23–4, 143, 145, 154, 165–71
  spin–glass 468–71
Heisenberg spin 7, 16, 170
helical order 7–8, 23, 146
helix–coil transition 255
herglotz function 334–5, 341, 343, 351
hierarchy of distribution functions 64, 89, 90, 149, 154, 281, 415, 422
Hilbert transform 350–1
hole theory of liquids 6
homogeneity
  scaling 202–3
  spatial 4, 21, 31, 58–9, 64, 66, 68, 72, 85, 110–11, 124, 132, 260, 324, 355, 445, 448, 472
hopping 367, 478–81
hot solid disorder 42, 68, 70, 85, 212, 243
hydrogen, solid 11
hyper-netted chain approximation 92, 96

ice
  square 178–81
  vitreous 66
ice condition 11–12, 14, 16, 20–1, 178–9, 276
ice disorder 11–17, 20, 139, 143, 178–81
ice structure 11, 64
imaging 141
impurity centres 107, 116, 475, 478
impurity mode 306, 318
InSb 76
insulator 456, 478
interactor 330–1, 334, 357
interatomic forces 78, 88–90, 92–4, 209–14, 216
  attractive 213, 216, 221, 226, 232
  average force potential 89, 93
  effective 109, 414
  Lennard-Jones 83, 92–3, 96–7, 220, 222, 227, 232, 240
  linear chain 286
  liquid crystal 104
  macromolecule 247, 252, 265–6, 276, 280
  metallic vapours 107
  molten salts 101
  repulsive core 214, 222, 226, 232, 245

interatomic forces (*cont*.)
　rare gas 78
　soft 220, 222, 227, 241
　square well 211, 227
　triplet 231
interference function 124
internal field approximation 145
interstitial regions 2, 37, 40, 66, 98, 100, 129,
　398–400, 417, 428, 459
Ising model 6, 19, 21, 28, 33–4, 117, 143, 269,
　272, 327, 373, 378
　alloy 6, 139, 246
　Bethe method for 151–2
　critical fluctuations of 205–7
　dilute 460–4
　graphical expansions for 178, 187–97, 223,
　375
　ice disorder 16
　Kikuchi method for 159, 176
　magnetic 9, 16, 22, 216, 218
　mean field approximation for 145–6, 176–7
　one-dimensional 48–9, 148, 161–5, 170,
　192
　Onsager solution for 171–8, 181, 192, 199,
　203, 207, 270
　quasi-chemical approximation for 147–8,
　176–7
　random phase approximation for 149, 281
　three-dimensional 159, 161, 173, 182, 184,
　187, 194, 203, 208, 322, 358
　two-dimensional 49–51, 159, 171–8, 181,
　185, 187, 192, 199, 203, 207, 275, 281
Ising spin 6, 9, 16, 18, 24, 104, 145, 187
isotherm, fluid 212, 216–17, 219–20
isotopic disorder 286–7, 290, 300, 303, 321,
　326
isotropy 60, 64, 105, 111, 112, 125, 434, 442,
　445

K–Hg alloys 139–40
Kadanoff construction 203, 207
KDP 14–16, 143, 180–1
Kikuchi method 155–61, 352
KKR method 400, 402, 422–4, 428
Knight shift 405
knots 284–5
Kronig–Penney model 40–1, 292–5,
　298–300, 303, 306, 312, 314, 322, 455
Kubo–Greenwood formula 432, 436, 466–7

ladder cluster 159–61
ladder graph 195, 223, 273
Landau theory 198–200, 218, 272
Langevin function 145, 267–8
lattice 1–3, 36, 38, 143, 152, 178

binary alloy 287, 290
block 204, 207–8
body centred cubic 22, 85, 196, 203, 374
compound 301
continuum limit 441–7
covering 374–5
diamond 11, 14, 67–8, 72, 75, 87, 374, 378,
　430, 447, 449, 452, 458–9, 477
dislocated crystal 53
dual 193, 375
face centred cubic 2, 8, 9, 22, 87, 196, 203,
　231, 374, 378
gas 6, 19, 21, 215, 218
glass 439–47
layer systems 43
linear chain 40, 286, 290
liquids 439–47
vibrations 47
LCAO representation 288–9, 321, 387, 398,
　420, 457
LDA *see* local density approximation
Lennard-Jones–Devonshire model 244–5
Lennard-Jones potential 83, 92–3, 96–7, 220,
　222, 227, 232, 240
levelled exponentials 224
Lie algebra 173
Lieb method 180
Lifshitz formula 348, 464, 489–90
Lindemann criterion 240
linear-response formula 432–7
Lippmann–Schwinger equation 417
liquids
　analytical theory 81, 87–96, 221–6
　BBGKY method 89–90, 149, 244, 281
　compressibility 95, 130
　diffraction 124, 130, 135
　dislocation model 53, 55, 78, 236
　dynamics 128, 142, 228, 392, 439–47
　hole theory 6, 78
　hyper-netted chain 92
　ionic 101
　Kronig–Penney 292–3
　liquid crystals 103–6, 132, 200
　local order 61
　metallic *see* metal, liquid
　mixtures 96–102, 134–6, 209, 246
　molecular dynamics 227–9
　Monte Carlo method 228–32
　non-spherical molecules 102–6
　one-dimensional 39–42, 209–12, 291, 302–3,
　307–8, 314–15, 348, 415, 455
　perturbation method 221–3
　random close packed 77–87
　semiconducting 388, 438
　structure factor 126, 214, 218, 391
　thermodynamics 88, 142, 200

thermodynamics (*cont.*)
  topological disorder 2, 57, 213
  transition metal 403
  two-dimensional 83, 236
  van der Waals 213–18
  virial expansion 223–6
  water 66, 228
LJ '6:12' *see* Lennard-Jones potential
Lloyd determinant 427–9
Lloyd model 368–9
local density approximation 312–15, 347–8, 367
localization
  Anderson 358–70, 384–6
  criterion for 361, 363–6
  electron 109, 119, 358–70, 379, 384–6, 438, 485–7, 490–1
  excitation 315–20, 339–40, 350, 440, 445, 454, 456
  percolation 370, 379, 384–6, 485–7
  range of 318, 320
localization constant 318, 320
locator, medium 334, 354–5
locator expansion 330–4, 336, 341, 343, 351, 357, 360, 362–3, 365, 474, 479
Lorentzian spectrum 116, 132, 368
LRO *see* order, long range

macromolecular chains 18, 249
  branching 259–65
  conformation 254, 276
  continuous 277–85
  end-to-end distance 256–8, 265–8, 270–2
  entangled 282–5
  entropy 265, 268, 270
  equivalent 258–9, 268–9, 278
  ferroelectric model 276–7
  finite 278
  flexible 277
  freely jointed 253, 256–7, 259, 267, 272
  knotted 284–5
  Markovian 270, 278, 280–1
  n-mer 262–4
  network 264–5
  radius of gyration 256–8, 264, 271, 275, 282
  random flight 253, 257, 278
  rotational isomeric 255, 257, 276
  self-avoiding 269, 281–2
  skeletal 253
macromolecular disorder 246–85
magnetic chain 43, 47
magnetic disorder 6–11, 18–22, 35–6, 129, 139, 143–4, 321
magnetism

amorphous 467–71
dilute 460–7
magnetization, spontaneous 162, 176, 181, 185–6, 197, 199, 216, 218, 378, 462–4
magnon 9, 16, 34, 48, 56, 113, 140, 150, 167–70, 287–90, 465–7
many body theory 391
Markov process 255, 269–70, 276, 278–9, 319, 490
mass difference parameter 300–1, 304–6, 336
master equation 431, 436
Matsuda–Ishii theorem 319
Mayer cluster expansion 190, 223–6
Mayer function 91, 93
maze conduction 380–6, 466–7, 478, 487
MCPA *see* coherent potential approximation, molecular
mean field approximation 104, 143–8, 151–5, 165, 176–7, 198, 203, 208, 215–16, 246–8, 250, 272, 332, 356, 383, 468–71
mean free path 56, 319, 474
medium propagation vector 423
melting 88, 99, 209, 212, 232–45
  computer simulation 227, 229, 232–40
  dislocation theory of 53, 239
  entropy of 241–5
  latent heat of 241
  Lennard-Jones–Devonshire theory 244–5
  Lindemann criterion for 240
  polymer 276
  in two dimensions 235–6, 242
  volume change 241
Mermin theorem 47, 197
metal, amorphous 100–1, 468
metal, liquid, 37, 40, 81, 86, 93, 115, 117, 129, 222, 240, 472
  alloys 437–8
  band structure 404–9, 412
  electron theory 387–438
  muffin–tin potentials 398–404
  nearly free electrons 388–93, 424
  NFE formula 390–2, 404, 434–6
  non-structural properties 405, 423–4
  pseudopotentials 391–8, 404
  resonances 403, 405
  t-matrix 403–4
  transport theory 432–8
metal, one-dimensional 45–7
metal–insulator transition 7, 116, 475–8
metallic vapour 106
metastable states 142, 232, 235, 446, 470
MFA *see* mean field approximation
microcrystalline disorder 53–4, 56–64, 68, 72, 83, 85, 122, 126, 141, 227, 236, 412–13, 458
mictomagnetic 469

mixing, thermodynamics of 247–8, 251–2
Mn 468
mobility
　carrier 367–8, 456
　percolation 380, 383, 467
mobility edge 320, 358–9, 361–3, 366–8, 370, 386, 454, 486
model Hamiltonian 288, 398
model potential 396–7
molecular dynamics method 66, 79, 83, 92, 98–9, 102, 219, 227–9, 231, 232, 237
molecule, non-spherical 102–6, 248
moment expansion 187, 189, 221, 347–9
moment
　magnetic 7, 18, 33, 37, 144, 177
　spectral 348–51, 454
momentum, crystal 4, 407, 436–7, 466
momentum operator 436, 442, 466
momentum representation 4, 33, 324, 329, 331, 388, 406, 411, 439, 442–4
monomer 259, 261
Monte Carlo method 27–9, 39, 73, 79, 83, 90, 92, 101, 228–32, 237, 272, 309, 313–14, 318–19, 376, 378, 476, 479
Mott transition 7, 116, 475–8
muffin-tin potential 398–401, 403–5, 416–21, 427–9, 459

Na 240
Na–K alloys 139–40
NaCl 135
nearest neighbour interaction 19–20, 34, 49, 139, 144, 182, 202, 290, 357
nearly free electrons 321–2, 387–91, 395, 405, 440, 457, 474
Néel temperature 27
nematic order 103–4, 106
network, disordered 64, 67–72, 122, 264, 399–400, 447–54
　band-gap paradox 456–9
　continuum limit 441–7, 454
　excitations of 439–59
　free electron 455
　localization 454
　non-orthogonal representation 444
　tetrahedral glass model 67–72, 447–54
　tree model 454–6
　Weaire model 448–54, 465–6
next-nearest neighbour interactions 22, 27
NFE *see* nearly free electrons
NFE formula 390–2, 396, 402, 404, 431, 434–7
(NMP) (TCNQ) 46
noble metals 398, 468
noise, theory of 109, 114
non-central forces 88

non-crystalline solids 64, 72
non-Gaussian field 116
non-orthogonal representation 444, 446
nucleation 83, 235–6
nuclei, atomic 36, 108

O-spheres 39–40, 210
off-diagonal disorder 287–91, 321–2, 326, 354, 356–8, 416, 468
Onsager solution 171–8, 192, 194, 196, 199, 203, 270
opalescence, critical 132, 217
optical model 421
orbital
　antibonding 451–2, 457
　atomic 288–9, 321, 448
　bond 45, 450–4
　bonding 451–2, 457
order
　crystalline 1, 47–8, 244
　fluctuations of 198–200
　intermediate range 26
　local 27, 51, 59, 61, 199
　long range 4, 22–5, 31–2, 48–51, 59, 85–6, 104, 138, 148, 163, 170, 176–7, 185, 211, 413, 449
　orientational 103–6
　propagation of 26, 149, 154, 195, 198
　range of 25–7, 29, 32, 35, 53, 59, 68, 70, 129, 139, 147, 150, 196
　short range 17, 26, 32, 104, 138, 147–50, 163, 195, 248, 327
　spontaneous 163, 195, 248, 327, 154, 166, 176–7, 181
order–disorder transition 11, 15
　see also critical phenomena
order parameter 29
　continuum 111
　liquid crystal 104
　local 470–1
　long range 24–5, 27, 32, 34, 132, 145–6, 152, 216
　off-diagonal 21
　short range 17, 25, 27, 32, 137, 144, 149, 154, 185
orientational disorder 11, 53, 56, 60
Ornstein–Zernike relation 91–2, 98, 149, 154
Ornstein–Zernike spectrum 131–2, 139–40, 150, 177, 186, 196, 198–200
overlap integrals 288–90, 321, 346, 356–7, 359, 448, 478–9

packing constraints 38
packing density 39, 77–8, 85, 98, 102, 115, 472
packing fraction 85

packing fraction (*cont.*)
  body centred lattice 85
  close packed lattice 85, 93, 237
  diamond lattice 477
  liquid 90, 92–3, 218
  melting 236, 238, 243, 245
  partial 98
  percolation 377, 486
  Poisson pattern 117
  random close packed 85, 237, 245
Padé approximants 351
paracrystalline model 71, 74, 78
paramagnetic disorder 7, 9, 267, 464, 470
partial wave representation 402, 418–20, 422, 426–7, 429
participation ratio 366, 440–1
Pauling model 11–13, 181
Peierls–Mermin theorem 47–8
percolation 29, 121, 367, 370–86, 462–7
  bond 371–80, 402, 465
  continuum 485–7, 490
  diffusion and 481
  dilute ferromagnet 462–7
  exponents for 378, 383
  gas disorder 475–9, 486
  gelation and 261–4
  hopping conduction 478–81
  Ising model and 373, 378–9, 462–4
  localization and 370, 379, 384–6
  maze conduction 379–86, 466, 478
  mobility 380, 383, 467
  probability function 371–3, 377–8, 462, 467
  self-avoiding walk 373–4
  site 370–80, 462, 464
  threshold for 261–4, 337, 373–9, 383, 462, 464–6, 478, 482, 486–7, 490
  tree 264, 373, 374, 378, 380–3
  two-dimensional 376, 380, 482
Percus–Yevick (PY) method 91–4, 96, 98–9, 107, 126, 131, 135, 218–20, 222, 226, 240, 245, 391
Perron–Frobenius theorem 347, 451
perturbation method 221–3
perturbation theory 57, 314, 325, 363, 431, 474
  see also graphical expansion
Pfaffians 192
phase-angle representation 295–301, 307–9, 313, 316
phase function 444
phase separation 22, 24, 216, 220, 247
phase shift
  scattering 401–3, 417, 420–9, 459
  transfer matrix 296, 298, 301, 308–10, 401
phase transition *see* critical phenomena

phonons 16, 42, 47, 56, 108, 113, 128, 240, 244, 286–7, 290, 303, 336, 338–40, 443, 446, 480–1
plasma physics 106
plasma resistance 392, 395, 399
Poisson pattern 117–18
Polk model 76–7
Polya's theorem 275
polygonization 53
polymers 246–85
  chemistry 249, 255, 258, 266
  condensed 249, 259, 265, 276, 282
  cross-linked 249, 264–8, 282, 284–5
  crystallization 249
  electronic structure 45
  melting 276–7
  solutions 102–3, 248–52, 269
  unlinked 249
  see also macromolecular chains
poly(sulphur nitride) 45
positron annihilation 405
potential, electronic
  Anderson model 479
  atomic 288, 293–4
  average 389, 399, 406
  deformation 392, 399
  gas disorder 472–4
  glass disorder 399–400
  impurity centre 475, 479
  metal 388, 391, 413
  model 396–7
  muffin-tin 398–400
  neutral pseudo-atom 395–6
  overlapping 40–1, 479, 482
  pseudopotential 393–7, 402, 406, 409, 474
  random field 294, 322, 482
  screened Coulomb 392–3, 395
  superposition of 40–1, 109, 114–16, 124, 129, 133, 288, 389, 392–3, 395–6, 399, 404, 482
powder pattern 126
power law states 367
power spectrum 112, 118
premelting phenomena 239–40
pressure 94–5, 99, 211, 213
  melting 238, 245
  partial vapour 247–8, 252
pressure functional 489
propagator 325, 407, 433
  medium 332–4, 336, 341, 357, 416
  virtual crystal 325, 329, 332–3, 368–9
propagator expansion 325, 327–8, 333, 341, 343, 474
propagation in random medium 109, 119, 292, 406, 421–5, 429, 474
pseudo-assembly 156–61

pseudo-atoms 395–6, 405, 431, 457
pseudo-gap 318–20, 405, 409, 412, 459
pseudopotential 115, 127, 393–7, 404, 406, 409, 429, 457–8

QCA *see* quasi-chemical approximation *or* quasi-crystalline approximation
quantal gas 21, 143
quartz 64–5
quasi-chemical approximation 147–8, 150, 152, 154, 155, 176–7, 196, 379
quasi-crystalline approximation 416, 424–5
quasi-crystalline disorder 57, 71
quenched disorder 460, 462

radial function 400–2, 417
radial Schrödinger equation 400, 402, 416–17
radius of gyration 256, 258, 269
random close packing 56, 78–87, 93, 98, 100–1, 130, 220, 227, 232, 237, 245, 282, 399, 447, 468
random coil 255–9, 269–70, 282
random flight 253
random medium 109–10, 114, 118–20, 474, 482–7
random phase approximation 143, 149, 196, 281, 308
random series convergence 361, 364–5
Raoult's Law 247–8, 251–2
rare-earth metals 22–3, 388
rare gas 78, 248
Rayleigh scattering 133
Rayleigh–Schrödinger series 405–6
Rb 95
RCP *see* random close packing
RDF *see* radial function
reaction matrix 428
reciprocal lattice 388, 406
reciprocal space 4, 33, 48, 111–12, 124, 208, 324, 329, 442
recursion method 352–4, 365
reduced temperature 202, 212, 217
reference fluid 221–3
relaxation time 390, 434–5, 474
renormalization group 204–8, 272, 378, 384, 463, 468
resistivity 389, 436
   NFE formula 390, 393, 431, 436–8
   t-matrix formula 402–3
resonance band 321, 356, 403, 405, 429
resonance width 403, 429
RPA *see* random phase approximation
rubber 259–60, 264–8, 271, 282, 284–5

s-band 397, 403

s-wave 420
salts, molten 96, 101–2, 136
Saxon–Hutner theorem 302–3, 307, 315, 345–6
SAXS 130
scaling 200–8, 218, 313, 378, 463, 465
scattering 123, 421
   absorptive 420
   classical 421
   cluster 429–31
   coherent wave 422–4
   critical 140
   disorder 329, 336, 389, 408, 413, 421, 446, 474
   elastic 123, 417
   electron, by atoms 393, 401–2
   forward 423, 429, 434
   Lloyd determinant for 427–9
   multiple 123, 410, 413, 426
   neutron 127
   optical 130–1, 138, 259, 272, 392, 421
   partial waves 402, 418, 426–7
   Rayleigh 133
   resonance 403, 420, 423
   small angle 129–33, 138, 255–6, 421
   sphere assembly 421, 426–7
scattering amplitude 123, 402, 414
scattering length 127, 134–5
scattering operator 413–18
scattering path operator 414, 418–20, 425–6
scattering phase shifts 401–2, 416, 418, 420, 426
scattering *t*-matrix 402, 413–15, 417
scattering vector 123, 389
Schrödinger equation 280–1, 284
   radial 400, 402
screening, electron 391–6, 398
segment, monomer 253–8, 265, 277, 279
self-energy operator 407, 410–11, 413–14, 429, 434
semi-classical approximation 483–7
semiconductors
   amorphous 67, 76, 109, 119, 141, 388, 440, 449–59, 472
   band structure of 405, 412, 449
   compound 76, 451, 458
   covalent bonded 457–9
   glassy *see* amorphous *subentry*
   impurities in 107, 116, 119, 289, 475–7
   liquid 388, 438
semi-invariants 189
semimetal 405
Shiba condition 356–7
Si
   amorphous 66–7, 68, 76, 100, 141, 456–9
   crystalline 412, 449, 456–7

significant structure theory 70–1, 78
silica 64–6, 440–1
simple metals 405
simplicial graph 82, 84
single-occupancy model 237–8, 241–3
site percolation 370–80, 462, 464
site representation 323, 325, 359
Slater model 15, 16
smectic order 106
softon 446–7
sol fraction 261–4
solution 96
  conformal 96, 98, 134, 139
  ideal 247–9
  macromolecular 102, 248–52, 270
  regular 246–9
sp^3 hybridization 448, 453
special frequency 304–7, 310–12, 315, 319, 338–40, 346, 348, 370, 379, 440–1
specific heat 32, 144, 146–7, 162, 194, 198–9, 203, 217, 446
spectral bounds 346–7, 451, 454
spectral density, electronic
  amorphous semiconductors 429–30, 447–59
  Anderson model 366, 369
  band tail 348, 484, 486–91
  chain disorder 307–20, 338, 441
  cluster scattering 425–31, 459
  coherent potential approximation 335–40
  coherent wave 425
  conductivity and 436
  continued fractions 351–4
  cumulated 308, 314, 350, 425–8, 483, 489
  diamond lattice 449
  disordered metal 404–9
  Dyson–Schmidt method 309–12, 490
  Edwards series 410–12
  free band 387, 409, 436, 484
  glass dynamics 441
  Green function formalism 323–4, 329, 332, 335, 407–8, 411, 425, 433
  Lloyd determinant 427–8
  Lloyd model 369
  local density approximation 312–15, 347, 367
  moment expansion 348–51
  normal modes 287, 439, 443
  normalized 323
  one dimensional white noise 490
  Padé approximants 351
  pressure functional 489
  random potential 483–5, 488–90
  special frequencies 310–12, 315, 338–40, 358, 366, 370, 379, 440–1
  sum rule 335

tree model 455
Weaire model 450–3
spectral density of disorder 112, 115–16, 124, 128, 138, 149, 170
spectral disorder 32–5, 42, 113, 115, 168, 182, 205, 482, 489–90
spectral variable 289, 296, 321, 323, 327, 331
sperimagnetic 469
spherical model 143, 182–7, 203, 273
spin deviation 9, 23, 29, 34–5, 104, 168–9, 174, 287, 465–6
spin-diffusion model 40
spin disorder 10, 23, 144, 182, 332
  diluted 460–7
spin glass 469–71
spin wave *see* magnon
spinor representation 174
spiral order 23
split bands 336–40, 346, 370, 379
stability
  dilute ferromagnet 465–7
  mechanical 239–40
  spin glass 470–1
  spontaneous order 47–51, 164–6
  thermodynamic 226, 240
staggered configuration 75–7, 109, 254
state ratio 299
statistical geometry 72–7, 87, 93, 440
statistical mechanics 142–3, 190
statistical topography 109, 118–21
steepest descents, method of 183–4
step surface 116–18, 132
Stosszahlansatz 113
structure constants 422, 428
structure factor 124–33, 223, 240, 389, 391–2, 406, 409–10, 444–6
  partial 98, 99, 101, 134–40, 246, 437
substitutional disorder 5–6, 37, 40, 42, 108, 111, 134, 137, 142–208, 218, 272, 277, 286, 291, 321, 437, 460–7
substitutionally disordered lattice excitations 321–86
  Anderson localization 358–70, 379
  average *t*-matrix approximation 328–31
  cluster methods 341–5
  coherent potential approx 332–45
  continued fraction method 351–4
  Green function formalism 322–4
  Lloyd model 368–9
  locator expansion 330–1
  maze conduction 379–86
  mobility 367–8
  moment expansion 348–51
  off-diagonal disorder 356–7
  percolation theory 370–9
  propagator expansion 325–8

substitutionally disordered (*cont.*)
  spectral bounds 345–7
  spectral singularities 338–40, 346
  tight binding alloy 321–2
  virtual crystal approximation 325–6
supercooling 239
supercritical fluid 87, 106
superheating 239
superposition approximation 61,76, 81–2, 89, 95–7, 149, 222, 244, 281, 412, 415–16
  higher order 63–4, 77, 90, 97, 244, 285
surface
  liquid 83
  random 109, 120–1
  rough 109, 118
  step 116–17
surface states 458
surface tension 31
susceptibility, magnetic 32, 144, 146, 162–3, 194, 203, 464
symmetry, broken 24, 35

t-matrix 402–3
  assembly 414, 426, 431
  average 328–31
  cluster 342–4
  coherent site 420
  single-site 328–9, 332–3, 413–14, 417–18, 426
tail, band 310, 313, 318, 336, 338, 347–8, 350, 361, 367, 370, 454, 484, 488–90
TBA *see* tight-binding model, alloy
TBM *see* tight-binding model
Te 100
tetracyanoplatinates 45
tetragonal symmetry 14, 15
tetrahedral coordination 11, 14, 16, 64–7, 76, 109, 178, 253, 399–400, 447
tetrahedral glass model 68, 72–7, 86–7, 447–54, 468
TGM *see* tetrahedral glass model
thermal conduction 446
thermodynamic disorder 142
thermoelectric power 390, 403
Thomas–Fermi formula 483–4, 487–9
three-body potentials 88
tie-line 236, 237
tight-binding model 40, 288–90
  alloy 297, 320–2, 324, 340, 345–7, 356–9, 384, 387, 397, 403, 420, 474
  glass 448, 452, 454, 457
  liquid 416
TMMC 43
topography of random field 119–21
topological disorder 2, 36–107, 209–45, 246, 249, 347, 377, 439–59, 465

one-dimensional 40–1, 209, 291, 379
trans configuration 254, 276–7
transfer function 141
transfer matrix
  chain excitation 295–330, 322, 455
  electron chain 291–5
  Heisenberg model 165, 185
  ice model 178, 180
  Ising model 162–4, 171–4, 176, 181, 322
  macromolecular chain 255, 269–70
transformation tensor 256–7
transition matrix 428
transition metals 7, 388, 397–8, 402–3, 405, 420, 429, 431, 435, 437, 468
translational 1–3, 36
transport properties 7, 53, 56, 78, 81, 101, 109, 315, 320, 329, 368, 380–6, 389–90, 395, 402–3, 405, 431–8
transport theory 389, 431–8
tree graph 152–3, 158, 373
tree lattice 309, 353, 454–6
tree percolation 263–4, 372, 380–3
tridymite, high 64
trihedral bonding 66, 73–5
tunnelling 318, 320, 386, 446–7, 475, 478
turbulence, homogeneous 109
turning point, classical 119

Uhlenbeck theorem 190, 225
universality hypothesis 203, 218
Ursell–Mayer expansion 223–5

vacancy 6
valence band 449, 451–3, 456–9
van der Waals approximation 213–19, 221–2, 235, 272
van Hove correlation function 128
van Hove singularity 313, 350, 353, 409
vapour phase 213, 215
VCA *see* virtual crystal approximation
virial 213, 218
virial coefficient 103, 190, 225–6
virial series 219, 223–6, 245, 272
virtual crystal approximation 312, 325–6, 329, 335–6, 414
viscosity 256, 272
vitreous solids *see* glass
volume
  atomic 86, 109, 215, 472–3, 477
  excluded 215, 252, 269–72, 275–6, 281, 373
  free 244–5
  melting change of 241
  mixing change of 98–100
  percolation 377, 486
Voronoi cell 2–3, 84–5, 117, 399, 421, 424, 428, 473

vulcanization 259–60, 264, 267, 282

walk, random 102, 454
  continuous 277–82, 480
  lattice 272–7
  Markovian 278
  one-dimensional 272–3
  returning to origin 274–5
  self-avoiding 102, 195–6, 230–1, 252, 272,
    275–7
  unrestricted 272–6
walk, self-avoiding 152, 363, 373
  see also under walk, random

Wannier representation 289
water, structure of 66, 102, 228
wave vector 4, 33, 111, 123, 290
Weaire model 448–54, 457
Weaire–Thorpe theorem 449, 457
white noise limit 488–90
Wiener–Kintchine theorem 112, 115, 119, 128,
  132
Wigner–Seitz cell 2–3, 399, 423–4
wurtzite structure 72, 458

zone, ordered 29
zone boundary 405–6, 409–10